Notation Index

Numerical Methods

THE PRINDLE, WEBER & SCHMIDT SERIES IN MATHEMATICS

Swokowski and Cole, *Fundamentals of Trigonometry, Eighth Edition*

Swokowski and Cole, *Algebra and Trigonometry with Analytic Geometry, Eighth Edition*

Swokowski, *Precalculus: Functions and Graphs, Sixth Edition*

Swokowski, *Calculus, Fifth Edition*

Swokowski, *Calculus, Fifth Edition, Late Trigonometry Version*

Swokowski, *Calculus of a Single Variable*

Tan, *Applied Finite Mathematics, Third Edition*

Tan, *Calculus for the Managerial, Life, and Social Sciences, Second Edition*

Tan, *Applied Calculus, Second Edition*

Tan, *College Mathematics, Second Edition*

Trim, *Applied Partial Differential Equations*

Venit and Bishop, *Elementary Linear Algebra, Alternate Second Edition*

Venit and Bishop, *Elementary Linear Algebra, Third Edition*

Wiggins, *Problem Solver for Finite Mathematics and Calculus*

Willard, *Calculus and Its Applications, Second Edition*

Wood and Capell, *Arithmetic*

Wood and Capell, *Intermediate Algebra*

Wood, Capell, and Hall, *Developmental Mathematics, Fourth Edition*

Zill, *Calculus, Third Edition*

Zill, *A First Course in Differential Equations, Fifth Edition*

Zill and Cullen, *Elementary Differential Equations with Boundary-Value Problems, Third Edition*

Zill and Cullen, *Advanced Engineering Mathematics*

THE PRINDLE, WEBER & SCHMIDT SERIES IN ADVANCED MATHEMATICS

Brabenec, *Introduction to Real Analysis*

Ehrlich, *Fundamental Concepts of Abstract Algebra*

Eves, *Foundations and Fundamental Concepts of Mathematics, Third Edition*

Keisler, *Elementary Calculus: An Infinitesimal Approach, Second Edition*

Kirkwood, *An Introduction to Real Analysis*

Patty, *Foundations of Topology*

Ruckle, *Modern Analysis: Measure Theory and Functional Analysis with Applications*

Sieradski, *An Introduction to Topology and Homotopy*

Numerical Methods

J. Douglas Faires
Youngstown State University

Richard L. Burden
Youngstown State University

PWS-KENT PUBLISHING COMPANY

BOSTON

PWS-KENT
Publishing Company

20 Park Plaza
Boston, Massachusetts 02116

PWS-KENT Publishing Company is a division of Wadsworth, Inc.

Library of Congress Cataloging-in-Publication Data
Faires, J. Douglas.
    Numerical methods / J. Douglas Faires, Richard L. Burden.
       p.   cm.
    Includes bibliographical references and index.
    ISBN 0-534-93136-7
    1. Numerical analysis.   I. Burden, Richard L.   II. Title.
QA297.F35   1993
519.4—dc20   92-37392
                           CIP

Printed in the United States of America
93 94 95 96 97—10 9 8 7 6 5 4 3 2 1

Sponsoring Editor Steve Quigley
Developmental Editor Kelle Karshick
Assistant Editor Marnie Pommett
Production and Interior Design Pamela Rockwell
Cover Design Julia Gecha
Composition Publication Services
Manufacturing Coordinator Lisa Flanagan
Cover Printer John Pow Company
Text Printer/Binder R. R. Donnelley & Sons, Crawfordsville

Cover Art James Frederick Housel

 This book is printed on acid-free, recycled paper.

Contents

Preface

Numerical appproximation techniques are taught to undergraduates in a variety of ways. The traditional numerical analysis course emphasizes both the approximation methods and the mathematical analysis that produces them. A numerical methods course is generally more concerned with the choice and application of techniques to solve problems in engineering and the physical sciences than with the derivation of the methods.

The leading numerical analysis books are mature; they have been through a number of editions, which ensures a more accurate and consistent presentation of material, and they have a wealth of proven examples and exercises. They are also written for a full-year coverage of the subject, so they have methods that can be used for reference even when there is not sufficient time for discussing these methods in the course.

The books used in the numerical methods courses differ widely in both intent and content. Often a book written for numerical analysis is adapted for a numerical methods course by deleting the more theoretical topics and derivations. The weakness of this approach is that students often have difficulty determining which material is important in the course and which is tangential.

The second type of book used for a numerical methods course is one that is specifically written for a service course. These books follow the established line of service-oriented mathematics books, similar to those written for business, life sciences, and technical calculus, and the statistics books designed for students in economics, psychology, and business. However, the engineering and science students for whom the numerical methods course is designed have a much stronger mathematical background than their counterparts in other disciplines. They are quite capable of mastering the material in a numerical analysis course, but they have neither the time for nor, often, the interest in the theoretical aspects of such a course. What they need is a sophisticated introduction to the approximation techniques that are used to solve the problems that arise in science and engineering. They also need to know why the methods work, what type of error to expect, and when an application might lead to difficulties. Finally, they need information, with recommendations, regarding the availability of high-quality software for numerical approximation routines. In such a course the mathematical analysis is reduced for lack of time, not because of the mathematical abilities of the students.

The emphasis in this numerical methods book is on the intelligent application of approximation techniques to the type of problems that commonly occur in engineering and the physical sciences. The book is designed for a one-semester course, but it contains at least 50% more material than needed, so instructors have flexibility in topic coverage and students have a reference for future work. The techniques covered are essentially the same as those included in our book designed for a numerical analysis course (see Burden and Faires, *Numerical Analysis*, 5th ed., PWS-KENT Publishers, 1993). However, the emphasis in the two books is quite different.

In *Numerical Analysis*, a book with about 700 text pages, each technique is given a mathematical justification before the implementation of the method is discussed. If some portion of the justification is beyond the mathematical level of the book, then it is referenced, but the book is, for the most part, mathematically self-contained. In this numerical methods book, each technique is motivated and described from an implementation standpoint. The aim of the motivation is to convince the student that the method is reasonable both mathematically and computationally. A full mathematical justification is included only if it is concise and adds to the understanding of the method. The structure of *Numerical Methods* closely parallels that of our numerical analysis book, the primary reference for material not included here.

Software is included with and is an integral part of *Numerical Methods*. The software is written both in FORTRAN and in Pascal and is distributed with the book. The programs permit students to generate all the results that are included in the examples and to modify the programs to generate solutions to problems of their choice. The intent of the software is to provide students with programs that will solve most of the problems they are likely to encounter in their studies. Occasionally, however, exercises in the text contain problems for which the programs do not give satisfactory solutions. These illustrate the difficulties that can arise in the application of approximation techniques and show the need for the flexibility provided by the standard general-purpose software packages that are available for scientific computation. Information about the standard general-purpose software packages is discussed in the text. Included are those packages distributed by the International Mathematical and Statistical Library (IMSL), those produced by the National Algorithms Group (NAG), the specialized techniques in EISPACK and LINPACK, and the routines in MATLAB.

The following chart shows the chapter dependencies in the book. We have tried to keep the prerequisite material to a minimum to allow greater flexibility.

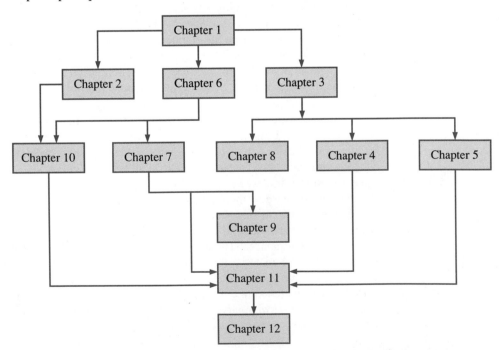

Acknowledgments

The authors of a new book rely heavily on the advice of those who are chosen to review the manuscript. We were most fortunate in constructing this book to have a collection of reviewers who understood the philosophy of the project and made many helpful suggestions for its improvement. This list includes

Neil Berger	University of Illinois in Chicago
Richard T. Bumby	Rutgers University
Curtis Cooper	Central Missouri State University
Michael Dellomo	The Mitre Corporation
L. Vincent Edmonson	Central Missouri State University
James F. Epperson	University of Alabama at Huntsville
Alejandro Garcia	University of California at Los Angeles
Leslie Glaser	University of Utah
Richard Goodrick	The Evergreen State University
Gregory Griffin	Louisiana State University
Nathaniel Grossman	University of California at Los Angeles
David R. Hill	Temple University
Joseph Hudson	General Motors Engineering and Management Institute
Leonard J. Lipkin	University of North Florida
Fritz Keinert	Iowa State University
Morris Newman	University of California at Santa Barbara
Jeffrey B. Puinom	Rensselaer Polytechnic Institute
Phil Schmidt	University of Akron
Gary Stanek	Youngstown State University
John G. Stevens	Montclair State College
Joan Wyzkoski Weiss	Fairfield University

This was our first experience producing a complete manuscript using TEX, the technical document system developed by Donald Knuth. The results were rewarding, but the labor was intensive. We are most grateful to our student assistants for making the work go smoothly. They were Genevieve Bundy, Sharyn Campbell, Beth Eggens, Melanie George, and Kim Graft.

Finally, we would like to thank Carl Leets and the rest of the staff of the Media Center at Youngstown State University for answering, at short notice, the multitude of production questions we constantly posed.

J. Douglas Faires
Richard L. Burden

Numerical Methods

Mathematical Preliminaries and Error Analysis

1.1 Introduction

This book examines problems that can be solved by methods of approximation, techniques we call *numerical methods*. To begin, we need to consider some of the mathematical and computational topics that describe difficulties that arise when approximating a solution to a problem.

This chapter begins by listing the concepts from calculus used in the book. Since nearly all the problems whose solutions can be approximated involve continuous functions, calculus is the principal tool for deriving numerical methods and verifying that they solve the problems. The definitions and results included in this section should be used to refresh your knowledge of calculus. The section also provides a handy reference when these concepts are needed later in the book.

There are two things to consider when applying a numerical technique to solve a problem. The first and most obvious is to obtain the approximation. The equally important second objective is to determine a safety factor for the approximation: some assurance, or at least a sense, of the accuracy of the approximation. The next two sections of this chapter deal with a standard difficulty that occurs when applying techniques to approximate the solution to a problem: Where and why is computational error produced and how can it be controlled?

The final section describes various types and sources of mathematical software for implementing numerical methods.

1.2 Review of Calculus

Fundamental to the study of calculus are the concepts of limit and continuity of a function. Suppose f is a function defined on a set X of real numbers. Then f has the **limit** L at x_0 in X, written $\lim_{x \to x_0} f(x) = L$, if, given any real number ϵ, there exists a real number $\delta > 0$ such that $|f(x) - L| < \epsilon$ whenever $0 < |x - x_0| < \delta$. (See Figure 1.1.) This definition ensures that values of the function will be close to L whenever x is sufficiently close to x_0.

Figure 1.1

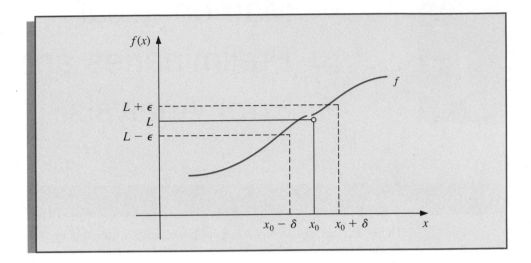

The function f is **continuous** at x_0 if $\lim_{x \to x_0} f(x) = f(x_0)$ and is **continuous on the set** X if it is continuous at each number in X. We will use $C(X)$ to denote the set of all functions that are continuous on X. When X is an interval of the real line, the parentheses in this notation are omitted. For example, the set of all functions continuous on the closed interval $[a, b]$ is denoted $C[a, b]$.

The limit of a sequence of real or complex numbers can be defined in a similar manner. An infinite sequence $\{x_n\}_{n=1}^{\infty}$ **converges** to a number x (called the limit) if, for any $\epsilon > 0$, there exists a positive integer $N(\epsilon)$ such that $n > N(\epsilon)$ implies $|x_n - x| < \epsilon$. The notation $\lim_{n \to \infty} x_n = x$, or $x_n \to x$ as $n \to \infty$, means that the sequence $\{x_n\}_{n=1}^{\infty}$ converges to x.

Continuity and Sequence Convergence

If f is a function defined on a set X of real numbers and $x_0 \in X$, then the following are equivalent:

(a) f is continuous at x_0.

(b) If $\{x_n\}_{n=1}^{\infty}$ is any sequence in X converging to x_0, then

$$\lim_{n \to \infty} f(x_n) = f(x_0).$$

The functions we consider when discussing numerical methods are continuous, since this is a minimal requirement for predictable behavior. Functions that are not continuous can skip over points of interest, which is not a satisfactory trait when attempting to approximate a solution to a problem. More sophisticated assumptions about a function generally lead to better approximation results. For example, a function with a smooth graph would normally behave more predictably than one with numerous jagged features. Smoothness relies on the concept of the derivative.

If f is a function defined in an open interval containing x_0, then f is **differentiable** at x_0 when

$$f'(x_0) = \lim_{x \to x_0} \frac{f(x) - f(x_0)}{x - x_0}$$

exists. The number $f'(x_0)$ is called the **derivative** of f at x_0. The derivative of f at x_0 is the slope of the tangent line to the graph of f at $(x_0, f(x_0))$, as shown in Figure 1.2.

Figure 1.2

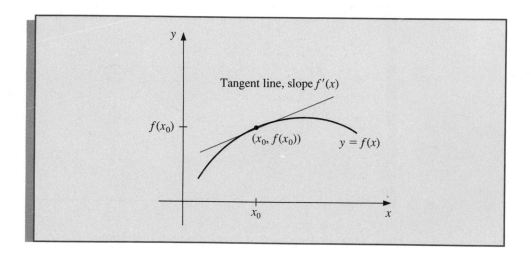

A function that has a derivative at each number in a set X is **differentiable** on X. Differentiability is a stronger condition on a function than continuity in the following sense:

Differentiability Implies Continuity

If the function f is differentiable at x_0, then f is continuous at x_0.

The set of all functions that have n continuous derivatives on X is denoted $C^n(X)$, and the set of functions that have derivatives of all orders on X is denoted $C^\infty(X)$. Polynomial, rational, trigonometric, exponential, and logarithmic functions are in $C^\infty(X)$, where X consists of all numbers at which the functions are defined.

The next results are of fundamental importance in deriving methods for error estimation. The proofs of most of these theorems can be found in any standard calculus text.

Mean Value Theorem

If $f \in C[a, b]$ and f is differentiable on (a, b), then a number c in (a, b) exists such that

$$f'(c) = \frac{f(b) - f(a)}{b - a}.$$

(See Figure 1.3.)

Figure 1.3

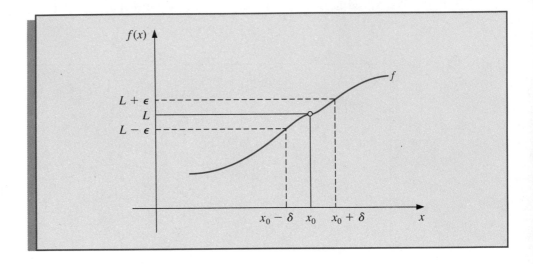

The following result is frequently used to determine bounds for error formulas:

Extreme Value Theorem

If $f \in C[a, b]$, then c_1 and c_2 in $[a, b]$ exist with $f(c_1) \leq f(x) \leq f(c_2)$ for all x in $[a, b]$. If, in addition, f is differentiable on (a, b), then the numbers c_1 and c_2 occur either at endpoints of $[a, b]$ or where f' is zero.

The other basic concept of calculus that is used is the integral. The **Riemann integral** of the function f on the interval $[a, b]$ is the following limit, provided it exists:

$$\int_a^b f(x)\, dx = \lim_{\max \Delta x_i \to 0} \sum_{i=1}^{n} f(z_i)\, \Delta x_i,$$

where the numbers $x_0, x_1, \ldots, x_n$ satisfy $a = x_0 \leq x_1 \leq \cdots \leq x_n = b$ and where, for each $i = 1, 2, \ldots, n$, $\Delta x_i = x_i - x_{i-1}$ and z_i is arbitrarily chosen in the interval $[x_{i-1}, x_i]$.

A function f continuous on an interval $[a, b]$ is Riemann integrable on the interval. This permits us to choose, for computational convenience, the points x_i to be equally spaced in $[a, b]$ and, for each $i = 1, 2, \ldots, n$, to choose $z_i = x_i$. In this case

$$\int_a^b f(x)\, dx = \lim_{n \to \infty} \frac{b-a}{n} \sum_{i=1}^{n} f(x_i),$$

where the numbers shown in Figure 1.4 as x_i are in this case chosen with $x_i = a + i(b-a)/n$.

Figure 1.4

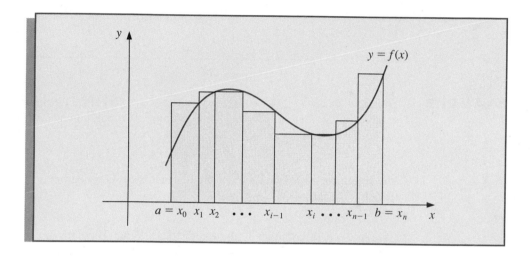

Two more basic results are needed in our study of numerical methods. The first is a generalization of the usual Mean Value Theorem for Integrals.

Mean Value Theorem for Integrals

If $f \in C[a, b]$, g is integrable on $[a, b]$, and $g(x)$ does not change sign on $[a, b]$, then there exists a number c in (a, b) for which

$$\int_a^b f(x)g(x)\, dx = f(c) \int_a^b g(x)\, dx.$$

The next result is the Intermediate Value Theorem. Although its statement is not difficult, the proof is beyond the scope of the usual calculus course.

Intermediate Value Theorem

If $f \in C[a, b]$ and K is any number between $f(a)$ and $f(b)$, then there exists a number c in (a, b) for which $f(c) = K$. (See Figure 1.5.)

Figure 1.5

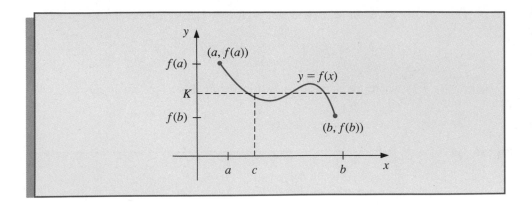

EXAMPLE 1   To show that $x^5 - 2x^3 + 3x^2 - 1 = 0$ has a solution in the interval $[0, 1]$, consider $f(x) = x^5 - 2x^3 + 3x^2 - 1$. The function f is a polynomial and hence is continuous on $[0, 1]$. Since

$$f(0) = -1 < 0 < 1 = f(1),$$

the Intermediate Value Theorem implies that there is a number x, with $0 < x < 1$, for which $x^5 - 2x^3 + 3x^2 - 1 = 0$.   ■ ■ ■

As seen in Example 1, the Intermediate Value Theorem is used to help determine when solutions to certain problems exist. It does not, however, give an efficient means for finding these solutions. This topic is considered in Chapter 2.

The final theorem in this review from calculus describes the development of the Taylor polynomials. The importance of the Taylor polynomials to the study of numerical analysis cannot be overemphasized. The following result will be used repeatedly:

Taylor's Theorem

Suppose $f \in C^n[a, b]$ and $f^{(n+1)}$ exists on $[a, b]$. Let x_0 be a number in $[a, b]$. For every x in $[a, b]$, there exists a number $\xi(x)$ between x_0 and x with

$$f(x) = P_n(x) + R_n(x),$$

where

$$P_n(x) = f(x_0) + f'(x_0)(x - x_0) + \frac{f''(x_0)}{2!}(x - x_0)^2 + \cdots + \frac{f^{(n)}(x_0)}{n!}(x - x_0)^n$$

$$= \sum_{k=0}^{n} \frac{f^{(k)}(x_0)}{k!}(x - x_0)^k$$

and

$$R_n(x) = \frac{f^{(n+1)}(\xi(x))}{(n+1)!}(x - x_0)^{n+1}.$$

Here, $P_n(x)$ is called the **nth Taylor polynomial** for f about x_0, and $R_n(x)$ is called the **remainder term** (or **truncation error**) associated with $P_n(x)$. The infinite series obtained by taking the limit of $P_n(x)$ as $n \to \infty$ is called the **Taylor series** for f about x_0. In the case $x_0 = 0$, the Taylor polynomial is often called a **Maclaurin polynomial** and the Taylor series is called a **Maclaurin series**.

The term *truncation error* refers to the error involved in using a truncated (that is, finite) summation to approximate the sum of an infinite series. This terminology will be reintroduced in subsequent chapters.

EXAMPLE 2 Determine (a) the second and (b) the third Taylor polynomials for $f(x) = \cos x$ about $x_0 = 0$, and use these polynomials to approximate $\cos(0.01)$. (c) Use the third Taylor polynomial and its remainder term to approximate $\int_0^{0.1} \cos x \, dx$. Since $f \in C^\infty(\mathbb{R})$, where $\mathbb{R}$ denotes the set of all real numbers, Taylor's Theorem can be applied for any $n > 0$.

(a) For $n = 2$ and $x_0 = 0$, Taylor's Theorem gives

$$\cos x = 1 - \frac{x^2}{2} + \frac{x^3}{6} \sin \xi(x),$$

where $\xi(x)$ is a number between 0 and x. With $x = 0.01$, the Taylor polynomial and remainder term is

$$\cos 0.01 = 1 - \frac{(0.01)^2}{2} + \frac{(0.01)^3}{6} \sin \xi(x) = 0.99995 + (0.1\bar{6}) \times 10^{-6} \sin \xi(x),$$

where $0 < \xi(x) < 0.01$. (The bar over the six in $0.1\bar{6}$ is used to indicate that this digit repeats indefinitely.) Since $|\sin \xi(x)| < 1$, we have

$$|\cos 0.01 - 0.99995| \le 0.1\bar{6} \times 10^{-6},$$

so the approximation 0.99995 matches at least the first five digits of $\cos 0.01$. Using standard tables we find that $\cos 0.01 = 0.99995000042$, which gives agreement through the first nine digits.

(b) Since $f'''(0) = 0$, the third Taylor polynomial and remainder term about $x_0 = 0$ is

$$\cos x = 1 - \frac{x^2}{2} + \frac{x^4}{24} \cos \hat{\xi}(x),$$

where $0 < \hat{\xi}(x) < 0.01$. The approximating polynomial remains the same, and the approximation is still 0.99995, but we now have a much better accuracy assurance, since

$$\left| \frac{x^4}{24} \cos \hat{\xi}(x) \right| \le \frac{(0.01)^4}{24}(1) \approx 4.2 \times 10^{-10}.$$

(c) Since

$$\cos x = 1 - \tfrac{1}{2}x^2 + \tfrac{1}{24}x^4 \cos \hat{\xi}(x),$$

$$\int_0^{0.1} \cos x \; dx = \int_0^{0.1} \left(1 - \frac{1}{2}x^2\right) dx + \frac{1}{24} \int_0^{0.1} x^4 \cos \hat{\xi}(x) \; dx$$

$$= \left[x - \frac{1}{6}x^3\right]_0^{0.1} + \frac{1}{24} \int_0^{0.1} x^4 \cos \hat{\xi}(x) \; dx$$

$$= 0.1 - \frac{1}{6}(0.1)^3 + \frac{1}{24} \int_0^{0.1} x^4 \cos \hat{\xi}(x) \; dx.$$

Therefore,

$$\int_0^{0.1} \cos x \; dx \approx 0.1 - \frac{1}{6}(0.1)^3 = 0.09983\overline{3}.$$

A bound for the error in this approximation can be determined from the integral of the Taylor remainder term:

$$\frac{1}{24} \left| \int_0^{0.1} x^4 \cos \hat{\xi}(x) \; dx \right| \le \frac{1}{24} \int_0^{0.1} x^4 \left| \cos \hat{\xi}(x) \right| \; dx$$

$$\le \frac{1}{24} \int_0^{0.1} x^4 \cdot 1 \; dx = 8.\overline{3} \times 10^{-8}.$$

Since the true value of this integral is

$$\int_0^{0.1} \cos x \; dx = \left[\sin x \right]_0^{0.1} = \sin 0.1 \approx 0.099833417,$$

the error for this approximation is within the bound.   ■ ■ ■

Figure 1.6

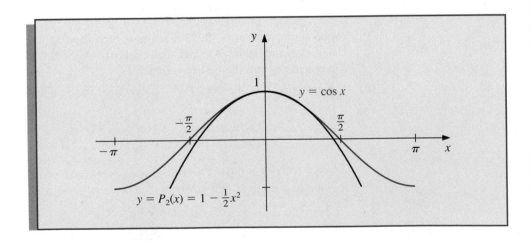

Figure 1.6 compares the Taylor polynomial and the cosine function. Parts (a) and (b) of the example show how two techniques can produce the same approximation but have

differing accuracy assurances. Remember that determining approximations is only part of our objective. The equally important other part is to determine the accuracy or error bounds for the approximations.

EXERCISE SET 1.2

1. Show that the following equations have at least one solution in the given intervals:
   (a) $x \cos x - 2x^2 + 3x - 1 = 0$,   [0.2, 0.3] and [1.2, 1.3]
   (b) $(x - 2)^2 - \ln x = 0$,   [1, 2] and [e, 4]
   (c) $2x \cos(2x) - (x - 2)^2 = 0$,   [2, 3] and [3, 4]
   (d) $x - (\ln x)^x = 0$,   [4, 5]

2. Find intervals containing solutions to the following equations:
   (a) $x - 3^{-x} = 0$ (b) $4x^2 - e^x = 0$
   (c) $x^3 - 2x^2 - 4x + 3 = 0$ (d) $x^3 + 4.001x^2 + 4.002x + 1.101 = 0$

3. Find $\max_{a \le x \le b} |f(x)|$ for the following functions and intervals:
   (a) $f(x) = (2 - e^x + 2x)/3$,   [0, 1]
   (b) $f(x) = (4x - 3)/(x^2 - 2x)$,   [0.5, 1]
   (c) $f(x) = 2x \cos(2x) - (x - 2)^2$,   [2, 4]
   (d) $f(x) = 1 + e^{-\cos(x-1)}$,   [1, 2]
   (e) $f(x) = 160 \cos(2x) - 64x \sin(2x)$,   [0, 0.4]
   (f) $f(x) = (\ln x)^x$,   [4, 5]

4. Let $f(x) = 2x \cos(2x) - (x - 2)^2$ and $x_0 = 0$.
   (a) Find the third Taylor polynomial, $P_3(x)$, and use it to approximate $f(0.4)$.
   (b) Use the error formula in Taylor's Theorem to find an upper bound for the error $|f(0.4) - P_3(0.4)|$. Compute the actual error.
   (c) Find the fourth Taylor polynomial, $P_4(x)$, and use it to approximate $f(0.4)$.
   (d) Use the error formula in Taylor's Theorem to find an upper bound for the error $|f(0.4) - P_4(0.4)|$. Compute the actual error.

5. Let $f(x) = x \cos x - 2x^2 + 3x - 1$ and $x_0 = 0$.
   (a) Find the third Taylor polynomial, $P_3(x)$, and use it to approximate $f(0.25)$.
   (b) Use the error formula in Taylor's Theorem to find an upper bound for the error $|f(0.25) - P_3(0.25)|$. Compute the actual error.
   (c) Find the fourth Taylor polynomial, $P_4(x)$, and use it to approximate $f(0.25)$.
   (d) Use the error formula in Taylor's Theorem to find an upper bound for the error $|f(0.25) - P_4(0.25)|$. Compute the actual error.

6. Find the fifth Taylor polynomial, $P_5(x)$, for $f(x) = \ln x$ expanded about $x_0 = 1$. Find a bound for the maximum error in using $P_5(x)$ to approximate $\ln x$ on [1, 1.3].

7. Find the fourth Taylor polynomial, $P_4(x)$, for $f(x) = e^{2x} \cos x$ expanded about $x_0 = 0$. Find a bound for the maximum error in using $P_4(x)$ to approximate $e^{2x} \cos x$ on [−0.5, 0.5].

8. Find the lowest-degree Taylor polynomial, $P_n(x)$, expanded about $x_0 = \pi/3$ that will approximate $\sin 58°$ to within 10^{-6}.

9. Find the lowest-degree Taylor polynomial, $P_n(x)$, expanded about $x_0 = \pi/6$ that will approximate $\cos 32°$ to within 10^{-6}.

10. Find for an arbitrary positive integer n, the nth Maclaurin polynomial, $P_n(x)$, for $f(x) = \arctan x$.

11. Let $f(x) = 5x^3 - 2x + 7$.
    (a) Find the Taylor polynomials of degree 1, 2, 3, 4 for $f(x)$ expanded about $x_0 = 0$.
    (b) Find the Taylor polynomials of degree 1, 2, 3, 4 for $f(x)$ expanded about $x_0 = 1$.

12. The polynomial $P_2(x) = 1 - \frac{1}{2}x^2$ is to be used to approximate $f(x) = \cos x$ in $\left[-\frac{1}{2}, \frac{1}{2}\right]$. Find a bound for the maximum error.

13. The polynomial $P_4(x) = x^2 + 2x^3 + 2x^4$ is to be used to approximate $f(x) = x^2 e^{2x}$ on $\left[0, \frac{1}{2}\right]$. Find a bound for the maximum error.

1.3 Computer Arithmetic

The arithmetic performed by a calculator or computer is different from the arithmetic that we use in our algebra and calculus courses. From past experience we expect that we always have as true statements such things as $2 + 2 = 4$, $4 \cdot 4 = 16$, and $(\sqrt{3})^2 = 3$. In standard computational arithmetic we have the first two but not always the third. To understand why this is true we must explore the world of finite-digit arithmetic.

In our traditional mathematical world we permit numbers with an infinite number of digits. The arithmetic we use in this world defines $\sqrt{3}$ as the unique positive number that when multiplied by itself produces the integer 3. In the computational world, however, each representable number has only a fixed, finite number of digits. This means, for example, that only rational numbers can be represented exactly, and not even all those can be. Since $\sqrt{3}$ is not rational, it is given an approximate representation within the machine, a representation whose square will not be precisely 3, although it will likely be sufficiently close to 3 to be acceptable in most situations. In most cases, then, this machine representation and arithmetic is satisfactory and passes without notice or concern, but we must be aware that problems may arise because of this discrepancy.

Round-off error is the error that occurs when a calculator or computer is used to perform real-number calculations. Such error arises because the arithmetic performed in a machine involves numbers with only a finite number of digits, with the result that calculations are performed with approximate representations of the actual numbers. In a typical computer, only a relatively small subset of the real number system is used for the representation of all the real numbers. This subset contains only rational numbers, both positive and negative, and stores the fractional part, called the *mantissa*, together with an exponential part, called the *characteristic*. For example, a single-precision floating-point number used in IBM mainframe machines consists of a 1-binary-digit (*bit*) sign indicator, a 7-bit exponent with a base of 16, and a 24-bit mantissa. Since 24 binary digits correspond to between 6 and 7 decimal digits, we can assume that this number has at least 6 decimal digits of precision for the floating-point number system. The exponent of 7 binary digits gives a range of 0 to 127. However, using only positive integers for the characteristic does not permit an adequate representation of numbers with small magnitude. To ensure that numbers with small magnitude are equally representable, 64 is subtracted from the characteristic, so that the range of the exponential part is actually from -64 to $+63$.

Consider, for example, the machine number

$$0 \quad 1000010 \quad 1011001100000100000000000.$$

The leftmost bit is zero, which indicates that the number is positive. The next seven bits, 1000010, are equivalent to the decimal number

$$1 \cdot 2^6 + 0 \cdot 2^5 + 0 \cdot 2^4 + 0 \cdot 2^3 + 0 \cdot 2^2 + 1 \cdot 2^1 + 0 \cdot 2^0 = 66$$

and are used to describe the characteristic, 16^{66-64}. The final 24 bits indicate that the mantissa is

$$1 \cdot \left(\tfrac{1}{2}\right)^1 + 1 \cdot \left(\tfrac{1}{2}\right)^3 + 1 \cdot \left(\tfrac{1}{2}\right)^4 + 1 \cdot \left(\tfrac{1}{2}\right)^7 + 1 \cdot \left(\tfrac{1}{2}\right)^8 + 1 \cdot \left(\tfrac{1}{2}\right)^{14}.$$

As a consequence, this machine number precisely represents the decimal number

$$+ \left[\left(\tfrac{1}{2}\right)^1 + \left(\tfrac{1}{2}\right)^3 + \left(\tfrac{1}{2}\right)^4 + \left(\tfrac{1}{2}\right)^7 + \left(\tfrac{1}{2}\right)^8 + \left(\tfrac{1}{2}\right)^{14}\right] 16^{66-64} = 179.015625.$$

However, the next smallest machine number is

$$0 \qquad 1000010 \qquad 10110011000000111111111 = 179.0156097412109175,$$

and the next largest machine number is

$$0 \qquad 1000010 \qquad 10110011000001000000001 = 179.0156402587890625.$$

This means that our original machine number represents not only 179.015625 but also half of the real numbers that are between this number and its two nearest machine number neighbors. To be precise, it is used to represent any real number in the interval

$$[179.01561737060546875, 179.01563262939453125).$$

To ensure uniqueness of representation and obtain all the available precision of the system, a normalization is imposed by requiring that at least one of the four leftmost bits of the mantissa of a machine number is a 1. Hence 15×2^{28} numbers of the form

$$\pm 0.d_1 d_2 \ldots d_{24} \times 16^{e_1 e_2 \ldots e_7}$$

are used by the system to represent all real numbers. With this representation, the number of binary machine numbers used to represent $[16^n, 16^{n+1}]$ is constant independent of n within the limit of the machine—that is, for $-64 \leq n \leq 63$. This requirement also implies that the smallest normalized, positive machine number that can be represented is

$$0 \qquad 0000000 \qquad 000100000000000000000000 = 16^{-65} \approx 10^{-78},$$

and the largest is

$$0 \qquad 1111111 \qquad 111111111111111111111111 = 16^{63} \approx 10^{76}.$$

Numbers occurring in calculations that have a magnitude of less than 16^{-65} result in *underflow* and are generally set to zero, whereas numbers greater than 16^{63} result in *overflow* and typically cause the computations to halt.

The arithmetic used on microcomputers differs somewhat from that used on mainframe computers. In 1985, the IEEE (Institute for Electrical and Electronic Engineers) published a report called *Binary Floating Point Arithmetic Standard 754–1985*. Formats were specified for single, double, and extended precisions, and these standards are generally followed by all microcomputer manufacturers who use floating-point hardware. For example, the numerical coprocessor for IBM compatible microcomputers implements a 64-bit representation for a real number, called a *long real*. The first bit is a sign indicator, denoted s. This is followed by an 11-bit exponent c and a 52-bit mantissa f. The base for the exponent is 2 and, to obtain numbers with both large and small magnitude, the actual exponent is $c - 1023$. In addition, a normalization is imposed that requires that the units digit be 1, and this digit is not stored as part of the 52-bit mantissa. Using this system gives a floating-point number of the form

$$(-1)^s * 2^{c-1023} * (1 + f),$$

which provides between 15 and 16 decimal digits of precision and a range of approximately 10^{-308} to 10^{308}. This is the form that compilers use with the numeric coprocessor; it is referred to as *double precision*.

The use of binary digits tends to complicate the computational problems that occur when a finite collection of machine numbers is used to represent all the real numbers. To examine these problems, we now assume, for simplicity, that machine numbers are represented in the normalized decimal form

$$\pm 0.d_1 d_2 \ldots d_k \times 10^n, \quad 1 \le d_1 \le 9, \quad 0 \le d_i \le 9$$

for each $i = 2, \ldots, k$. Numbers of this form are called *k-digit decimal machine numbers*.

Any positive real number, within numerical range of the machine, can be normalized to achieve the form

$$y = 0.d_1 d_2 \ldots d_k d_{k+1} d_{k+2} \ldots \times 10^n.$$

The **floating-point form** of y, denoted by $fl(y)$, is obtained by terminating the mantissa of y at k decimal digits. There are two ways of performing the termination. One method is to simply chop off the digits $d_{k+1} d_{k+2} \ldots$ to obtain

$$fl(y) = 0.d_1 d_2 \ldots d_k \times 10^n.$$

This is quite accurately called *chopping* the number. The other method is to add $5 \times 10^{n-(k+1)}$ to y and then chop the result to obtain $fl(y)$. This is called *rounding* the number. When rounding we add 1 to d_k to obtain $fl(y)$ whenever $d_{k+1} \ge 5$; that is, we round up. When $d_{k+1} < 5$, we simply chop off all but the first k digits, so we round down.

EXAMPLE 1   The irrational number π has the infinite decimal expansion $\pi = 3.14159265\ldots$. Written in normalized decimal form, we have

$$\pi = 0.314159265 \ldots \times 10^1.$$

The five-digit floating-point form of π using chopping is

$$fl(\pi) = 0.31415 \times 10^1 = 3.1415.$$

Since the sixth digit of the decimal expansion of π is a 9, the five-digit floating-point form of π using rounding is

$$fl(\pi) = (0.31415 + 0.00001) \times 10^1 = 0.31416 \times 10^1 = 3.1416.$$

■ ■ ■

The error that results from replacing a number with its floating-point form is called *round-off error* (regardless of whether the rounding or chopping method is used). There are two common methods for measuring approximation errors.

The approximation p^* to p has **absolute error** $|p - p^*|$ and **relative error** $|p - p*|/|p|$, provided that $p \neq 0$.

EXAMPLE 2

(a) If $p = 0.3000 \times 10^1$ and $p^* = 0.3100 \times 10^1$, the absolute error is 0.1 and the relative error is $0.333\overline{3} \times 10^{-1}$.

(b) If $p = 0.3000 \times 10^{-3}$ and $p^* = 0.3100 \times 10^{-3}$, the absolute error is 0.1×10^{-4}, but the relative error is again $0.333\overline{3} \times 10^{-1}$.

(c) If $p = 0.3000 \times 10^4$ and $p^* = 0.3100 \times 10^4$, the absolute error is 0.1×10^3, but the relative error is still $0.333\overline{3} \times 10^{-1}$.

This example shows that the same relative error can occur for widely varying absolute errors. As a measure of accuracy, the absolute error can be misleading and the relative error can be more meaningful, since the relative error takes into consideration the size of the true value. ■ ■ ■

The arithmetic operations of addition, subtraction, multiplication, and division performed by a computer on floating-point numbers also introduce error. These arithmetic operations involve manipulating binary digits by various shifting and logical operations, but the actual mechanics of the arithmetic is not pertinent to our discussion. To illustrate the problems that can occur, we will simulate this *finite-digit arithmetic* by first performing, at each stage in a calculation, the appropriate operation using exact arithmetic on the floating-point representations of the numbers. We then convert the result to the appropriate finite-digit (that is, decimal machine number) representation. The error produced in this way is the round-off error. The most common round-off error producing arithmetic operation involves the subtraction of nearly equal numbers.

EXAMPLE 3 Suppose we use four-digit decimal chopping arithmetic to simulate the problem of performing the computer operation $\pi - \frac{22}{7}$. The floating-point representations of these numbers are

$$fl(\pi) = 0.3141 \times 10^1 \quad \text{and} \quad fl\left(\tfrac{22}{7}\right) = 0.3142 \times 10^1.$$

Performing the exact arithmetic on the floating-point numbers gives

$$fl(\pi) - fl\left(\tfrac{22}{7}\right) = -0.0001 \times 10^1,$$

which converts to the floating-point approximation of this calculation:

$$p^* = fl\left(fl(\pi) - fl\left(\tfrac{22}{7}\right)\right) = -0.1000 \times 10^{-2}.$$

Although the relative errors using the floating-point representations for π and $\frac{22}{7}$ are small,

$$\left|\frac{\pi - fl(\pi)}{\pi}\right| \leq 0.0002 \quad \text{and} \quad \left|\frac{\frac{22}{7} - fl(\frac{22}{7})}{\frac{22}{7}}\right| \leq 0.0003,$$

the relative error produced by subtracting the nearly equal numbers is about 700 times as large:

$$\left| \frac{\left(\pi - \frac{22}{7}\right) - p^*}{\left(\pi - \frac{22}{7}\right)} \right| \approx 0.2092.$$

■ ■ ■

EXERCISE SET 1.3

1. Using the IBM mainframe format, find the decimal equivalent of the following floating-point machine numbers:
   (a) 0   1000011   1010100100110000000000000
   (b) 1   1000011   1010100100110000000000000
   (c) 0   0111111   0100011110000000000000000
   (d) 0   0111111   0100011110000000000000001

2. Find the next largest and next smallest machine numbers in decimal form for the numbers given in Exercise 1.

3. Compute the absolute error and relative error in the following approximations of p by p^*:
   (a) $p = \pi,\ p^* = \frac{22}{7}$ (b) $p = \pi,\ p^* = 3.1416$
   (c) $p = e,\ p^* = 2.718$ (d) $p = \sqrt{2},\ p^* = 1.414$
   (e) $p = e^{10},\ p^* = 22000$ (f) $p = 10^\pi,\ p^* = 1400$
   (g) $p = 8!,\ p^* = 39900$ (h) $p = 9!,\ p^* = \sqrt{18\pi}(9/e)^9$

4. Perform the following computations (i) exactly, (ii) using three-digit chopping arithmetic, and (iii) using three-digit rounding arithmetic. (iv) Compare the relative errors in parts (ii) and (iii).
   (a) $\frac{4}{5} + \frac{1}{3}$ (b) $\frac{4}{5} \cdot \frac{1}{3}$
   (c) $\left(\frac{1}{3} - \frac{3}{11}\right) + \frac{3}{20}$ (d) $\left(\frac{1}{3} + \frac{3}{11}\right) - \frac{3}{20}$

5. Use three-digit rounding arithmetic to perform the following calculations. Compute the absolute error and relative error with the exact value determined to at least 5 digits.
   (a) $133 + 0.921$ (b) $133 - 0.499$
   (c) $(121 - 0.327) - 119$ (d) $(121 - 119) - 0.327$
   (e) $\left(\frac{13}{14} - \frac{6}{7}\right)/(2e - 5.4)$ (f) $-10\pi + 6e - \frac{3}{62}$
   (g) $\left(\frac{2}{9}\right)\left(\frac{9}{7}\right)$ (h) $\left(\pi - \frac{22}{7}\right)/\frac{1}{17}$

6. The first three nonzero terms of the Maclaurin series for the arctangent function produce the polynomial $x - \frac{1}{3}x^3 + \frac{1}{5}x^5$. Compute the absolute error and relative error in the following approximations of π using the polynomial in place of the arctangent.
   (a) $\pi \approx 4\left[\arctan \frac{1}{2} + \arctan \frac{1}{3}\right]$ (b) $\pi \approx 16 \arctan \frac{1}{5} - 4 \arctan \frac{1}{239}$

7. Repeat Exercise 5 using four-digit rounding arithmetic.

8. Repeat Exercise 5 using three-digit chopping arithmetic.

9. Repeat Exercise 5 using four-digit chopping arithmetic.

10. Repeat Exercise 6 using four-digit rounding arithmetic.

11. Repeat Exercise 6 using four-digit chopping arithmetic.

1.4 Errors in Scientific Computation

In the previous section we saw how computational devices represent and manipulate numbers using finite-digit arithmetic. We now examine how the problems with this arithmetic can compound and look at some ways that arithmetic calculations can be arranged to reduce the effects of this inaccuracy.

The loss of accuracy due to round-off error can often be avoided by a careful sequencing of operations or reformulation of the problem. This is most easily described by considering a common computational problem.

EXAMPLE 1 The quadratic formula states that the roots of $ax^2 + bx + c = 0$, when $a \neq 0$, are

$$x_1 = \frac{-b + \sqrt{b^2 - 4ac}}{2a} \quad \text{and} \quad x_2 = \frac{-b - \sqrt{b^2 - 4ac}}{2a}.$$

Consider this formula applied, using four-digit rounding arithmetic, to the equation $x^2 + 62.10x + 1 = 0$, whose roots are approximately $x_1 = -0.01610723$ and $x_2 = -62.08390$. In this equation, b^2 is much larger than $4ac$, so the numerator in the calculation for x_1 involves the *subtraction* of nearly equal numbers. Since

$$\sqrt{b^2 - 4ac} = \sqrt{(62.10)^2 - (4.000)(1.000)(1.000)}$$
$$= \sqrt{3856 - 4.000} = 62.06,$$

we have

$$fl(x_1) = \frac{-62.10 + 62.06}{2.000} = \frac{-0.04000}{2.000} = -0.02000,$$

a poor approximation to $x_1 = -0.01611$ with the large relative error

$$\frac{|-0.01611 + 0.02000|}{|-0.01611|} = 2.4 \times 10^{-1}.$$

On the other hand, the calculation for x_2 involves the *addition* of the nearly equal numbers $-b$ and $-\sqrt{b^2 - 4ac}$. This presents no problem, since

$$fl(x_2) = \frac{-62.10 - 62.06}{2.000} = \frac{-124.2}{2.000} = -62.10$$

has the small relative error

$$\frac{|-62.08 + 62.10|}{|-62.10|} = 3.2 \times 10^{-4}.$$

To obtain a more accurate four-digit rounding approximation for x_1, we can change the form of the quadratic formula by *rationalizing the numerator*:

$$x_1 = \left(\frac{-b + \sqrt{b^2 - 4ac}}{2a} \right) \left(\frac{-b - \sqrt{b^2 - 4ac}}{-b - \sqrt{b^2 - 4ac}} \right) = \frac{b^2 - (b^2 - 4ac)}{2a(-b - \sqrt{b^2 - 4ac})},$$

which simplifies to

$$x_1 = \frac{-2c}{b + \sqrt{b^2 - 4ac}}.$$

Using this form of the equation gives

$$fl(x_1) = \frac{-2.000}{62.10 + 62.06} = \frac{-2.000}{124.2} = -0.01600,$$

which has the small relative error 6.2×10^{-4}. ∎ ∎ ∎

The rationalization technique in Example 1 can also be applied to give an alternative formula for x_2:

$$x_2 = \frac{-2c}{b - \sqrt{b^2 - 4ac}}.$$

This is the form to use if b is negative. In Example 1, however, the use of this formula results in the subtraction of nearly equal numbers, which produces the result

$$fl(x_2) = \frac{-2.000}{62.10 - 62.06} = \frac{-2.000}{0.04000} = -50.00,$$

with the large relative error 1.9×10^{-1}.

EXAMPLE 2 Evaluate $f(x) = x^3 - 6x^2 + 3x - 0.149$ at $x = 4.71$ using three-digit arithmetic. Table 1.1 gives the intermediate results in the calculations.

Table 1.1

	x	x^2	x^3	$6x^2$	$3x$
Exact:	4.71	22.1841	104.487111	133.1046	14.13
Three-digit (chopping):	4.71	22.1	104	132	14.1
Three-digit (rounding):	4.71	22.2	105	133	14.1

Note that the three-digit chopping values simply retain the leading three digits, with no rounding involved, and differ significantly from the three-digit rounding values.

Exact: $f(4.71) = 104.487111 - 133.1046 + 14.13 - 0.149 = -14.636489;$

Three-digit (chopping): $f(4.71) = 104. - 132. + 14.1 - 0.149 = -14.0;$

Three-digit (rounding): $f(4.71) = 105. - 133. + 14.1 - 0.149 = -14.0.$

The relative error for both three-digit methods is

$$\left| \frac{-14.636489 + 14.0}{14.636489} \right| \approx 0.04.$$

As an alternative approach, $f(x)$ could be written in a *nested* manner as

$$f(x) = x^3 - 6x^2 + 3x - 0.149 = ((x - 6)x + 3)x - 0.149.$$

This gives a three-digit chopping answer of

$$f(4.71) = ((4.71 - 6)4.71 + 3)4.71 - 0.149 = -14.5$$

and a three-digit rounding answer of -14.6. The new relative errors are

$$\text{Three-digit (chopping):} \quad \left| \frac{-14.636489 + 14.5}{14.636489} \right| \approx 0.0093;$$

$$\text{Three-digit (rounding):} \quad \left| \frac{14.636489 + 14.6}{-14.636489} \right| \approx 0.0025.$$

The results produced using nested arithmetic are superior in each instance.

■ ■ ■

Nested multiplication should be performed whenever a polynomial is evaluated, since it minimizes the number of error producing computations.

We are interested in choosing methods that will produce dependably accurate results. One criterion we will impose whenever possible is that small changes in the initial data produce correspondingly small changes in the final results. A method that satisfies this property is called **stable**; it is **unstable** when this criterion is not fulfilled. Some methods are stable only for certain choices of initial data. These methods will be called *conditionally stable*. We will attempt to characterize stability properties whenever possible.

To further consider the subject of round-off error growth and its connection to stability, suppose an error with magnitude $E_0 > 0$ is introduced at some stage in the calculations and that the magnitude of the error after n subsequent operations is E_n. There are two distinct cases that often arise in practice.

If a constant C exists, independent of n, with $E_n \approx CnE_0$, the growth of error is **linear.** If a constant $C > 1$ exists, independent of n, with $E_n \approx C^n E_0$, the growth of error is **exponential**. (It would be unlikely to have $E_n \approx C^n E_0$, with $C < 1$, since this would imply that the error tends to zero.)

Linear growth of error is usually unavoidable and, when C and E_0 are small, the results are generally acceptable. Methods having exponential growth of error should be avoided, since the term C^n becomes large for even relatively small values of n and E_0. As a consequence, a method that exhibits linear error growth is stable, whereas one exhibiting exponential error growth is unstable. (See Figure 1.7 on page 18.)

To reduce the effects of round-off error, we can use high-order digit arithmetic, such as the double- or multiple-precision option available on most digital computers. A disadvantage in using double-precision arithmetic is that it takes more computational time, and the growth of round-off error is only postponed, not entirely eliminated.

Since iterative techniques involving sequences are often used, the section concludes with a brief discussion of some terminology used to describe the rate at which convergence occurs when employing a numerical technique. In general, we would like to choose techniques that converge as rapidly as possible. The following definition is used to compare the convergence rates of various methods.

Figure 1.7

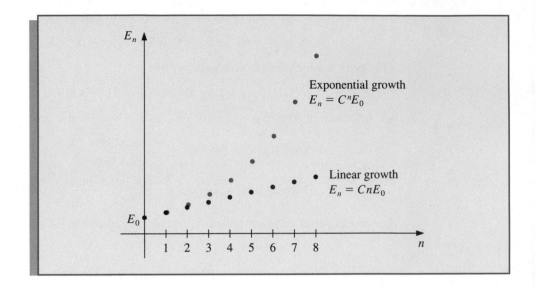

Suppose $\{\beta_n\}$ is a sequence known to converge to zero and K is a positive constant. If $\{\alpha_n\}_{n=1}^{\infty}$ converges to a number α and

$$|\alpha_n - \alpha| \le K|\beta_n| \quad \text{for large } n,$$

then $\{\alpha_n\}_{n=1}^{\infty}$ **converges to α with rate of convergence** $O(\beta_n)$ (read "big oh of β_n"). This is indicated by writing $\alpha_n = \alpha + O(\beta_n)$, or $\alpha_n \to \alpha$ with rate of convergence $O(\beta_n)$.

EXAMPLE 3   Suppose that the sequences $\{\alpha_n\}$ and $\{\hat{\alpha}_n\}$ are described by $\alpha_n = (n + 1)/n^2$ and $\hat{\alpha}_n = (n + 3)/n^3$. Although both $\lim_{n\to\infty} \alpha_n = 0$ and $\lim_{n\to\infty} \hat{\alpha}_n = 0$, the sequence $\{\hat{\alpha}_n\}$ converges to this limit much faster that does $\{\alpha_n\}$. This can be seen from the five-digit rounding entries for the sequences shown in Table 1.2.

Table 1.2

n	1	2	3	4	5	6	7
α_n	2.00000	0.75000	0.44444	0.31250	0.24000	0.19444	0.16327
$\hat{\alpha}_n$	4.00000	0.62500	0.22222	0.10938	0.064000	0.041667	0.029155

Since

$$|\alpha_n - 0| = \frac{n + 1}{n^2} \le \frac{n + n}{n^2} = 2 \cdot \frac{1}{n}$$

and

$$|\hat{\alpha}_n - 0| = \frac{n + 3}{n^3} \le \frac{n + 3n}{n^3} = 4 \cdot \frac{1}{n^2},$$

we have

$$\alpha_n = 0 + O\left(\frac{1}{n}\right) \quad \text{and} \quad \hat{\alpha}_n = 0 + O\left(\frac{1}{n^2}\right).$$

This implies that the convergence of the sequence $\{\alpha_n\}$ is similar to the convergence of $\{1/n\}$ to zero, but the sequence $\{\hat{\alpha}_n\}$ converges in a manner similar to the faster convergent sequence $\{1/n^2\}$. ■ ■ ■

We also use the "big oh" concept to describe the rate of convergence of functions, particularly when the independent variable approaches zero. If $G(h)$ is a function known to converge to zero as h approaches zero and if

$$|F(h) - L| \le K \; G(h), \quad \text{as } h \to 0,$$

then $F(h)$ **converges to L with rate of convergence $O(G(h))$.**

EXERCISE SET 1.4

1. Use four-digit rounding arithmetic and the formulas of Example 1 to find the most accurate approximations to the roots of the following quadratic equations. Compute the absolute errors and relative errors.
    (a) $\frac{1}{3}x^2 - \frac{123}{4}x + \frac{1}{6} = 0$ (b) $\frac{1}{3}x^2 + \frac{123}{4}x - \frac{1}{6} = 0$
    (c) $1.002x^2 - 11.01x + 0.01265 = 0$ (d) $1.002x^2 + 11.01x + 0.01265 = 0$

2. Repeat Exercise 1 using four-digit chopping arithmetic.

3. Let $f(x) = 1.013x^5 - 5.262x^3 - 0.01732x^2 + 0.8389x - 1.912$.
    (a) Evaluate $f(2.279)$ by first calculating $(2.279)^2, (2.279)^3, (2.279)^4$, and $(2.279)^5$ using four-digit rounding arithmetic.
    (b) Evaluate $f(2.279)$ using the formula

    $$f(x) = (((1.013x^2 - 5.262)x - 0.01732)x + 0.8389)x - 1.912$$

    and four-digit rounding arithmetic.
    (c) Compute the absolute and relative errors in parts (a) and (b).

4. Repeat Exercise 3 using four-digit chopping arithmetic.

5. The two by two linear system
    $$ax + by = e$$
    $$cx + dy = f$$

    where a, b, c, d, e, f are given can be solved for x and y as follows:

    $$\text{set } m = \frac{c}{a}, \quad \text{provided } a \ne 0;$$

    $$d_1 = d - mb;$$
    $$f_1 = f - me;$$
    $$y = \frac{f_1}{d_1};$$
    $$x = (e - by)/a.$$

Solve the following linear systems using this procedure with four-digit rounding arithmetic:

(a) $1.130x - 6.990y = 14.20$
$8.110x + 12.20y = -0.1370$

(b) $1.013x - 6.099y = 14.22$
$-18.11x + 112.2y = -0.1376$

6. Repeat Exercise 5 using four-digit chopping arithmetic.

7. The fifth Maclaurin polynomials for e^{2x} and e^{-2x} are

$$P_5(x) = \left(\left(\left(\left(\tfrac{4}{15}x + \tfrac{2}{3}\right)x + \tfrac{4}{3}\right)x + 2\right)x + 2\right)x + 1$$

and

$$\hat{P}_5(x) = \left(\left(\left(\left(-\tfrac{4}{15}x + \tfrac{2}{3}\right)x - \tfrac{4}{3}\right)x + 2\right)x - 2\right)x + 1$$

(a) Approximate $e^{-0.98}$ using $\hat{P}_5(0.49)$ and four-digit rounding arithmetic. Compute the absolute and relative error.

(b) Approximate $e^{-0.98}$ using $1/P_5(0.49)$ and four-digit rounding arithmetic. Compute the absolute and relative errors.

8. Repeat Exercise 7 using four-digit chopping arithmetic.

9. Find the rates of convergence of the following.

(a) $\lim\limits_{n\to\infty} \sin \dfrac{1}{n} = 0$ (b) $\lim\limits_{n\to\infty} \sin \dfrac{1}{n^2} = 0$

(c) $\lim\limits_{n\to\infty} \sin^2 \dfrac{1}{n} = 0$ (d) $\lim\limits_{n\to\infty} \ln(n+1) - \ln(n) = 0$

10. Find the rates of convergence of the following.

(a) $\lim\limits_{h\to0} \dfrac{\sin h - h\cos h}{h} = 0$ (b) $\lim\limits_{h\to0} \dfrac{1 - e^h}{h} = -1$

(c) $\lim\limits_{h\to0} \dfrac{\sin h}{h} = 1$ (d) $\lim\limits_{h\to0} \dfrac{1 - \cos h}{h} = 0$

1.5 Computer Software

Computer software for approximating the numerical solutions to problems is available in many forms. With this book we have provided programs written in Pascal and FORTRAN 77 that can be used to solve the problems given in the examples and exercises. These programs are dependable for nearly every problem that you are likely to need to solve, but they are what we call *special-purpose* programs. In this sense, the Numerical Toolkit from Borland, TK Solver from Universal Technical Systems, and packages available with various textbooks in numerical methods contain special-purpose software programs. We have adopted this term to distinguish these packages from those available from the standard mathematical subroutine libraries such as the International Mathematical and Statistics Library (IMSL), the Numerical Algorithms Group (NAG), and the like. The programs in these packages will be called *general-purpose* software.

The programs written for the general-purpose software packages such as the IMSL and NAG libraries differ in their intent from the algorithms and programs provided with this book. A general-purpose software program must take into consideration ways to reduce errors due to machine rounding, underflow, and overflow. It must also describe the range of input that will be expected to lead to results of a certain specified accuracy. Since these are machine-dependent characteristics, general-purpose software programs require input parameters that describe the floating-point characteristics of the machine being used for the calculations. Most of the early software was written for mainframe computers; a good reference for this is *Sources and Development of Mathematical Software*, edited by Wayne Crowell.* Now that the desktop (and portable) computer has become sufficiently powerful, much of the standard numerical software is available for personal computers and workstations. Most of this numerical software is written in FORTRAN 77, although some packages are written in C and Pascal.

In *Handbook for Automatic Computation, Vol. 2, Linear Algebra*, edited by Wilkinson and Reinsch, ALGOL procedures are presented for matrix computations. A package of FORTRAN subroutines based mainly on the ALGOL procedures was developed into the EISPACK routines. These routines are documented in the manuals published by Springer-Verlag as part of Lecture Notes in Computer Science series. (See *Matrix Eigensystem Routines: EISPACK Guide*, 2nd ed., by Smith, Boyle, Dongarra, Garbow, Ikebe, Klema, and Moler and *Matrix Eigensystem Routines: EISPACK Guide Extension* by Garbow, Boyle, Dongarra, and Moler.)

The FORTRAN subroutines compute eigenvalues and eigenvectors for a variety of types of matrices. EISPACK is available from Argonne National Laboratory free of charge. The EISPACK project was the first large scale numerical software package to be placed in the public domain.

LINPACK is a package of FORTRAN subroutines for analyzing and solving various systems of simultaneous linear algebraic equations and linear least squares problems. LINPACK is also available from Argonne National Laboratory. The documentation for this package is contained in *The LINPACK User's Guide* by Dongarra, Bunch, Moler, and Stewart. A step-by-step introduction to LINPACK, EISPACK, and BLAS (Basic Linear Algebra Subroutines) is given in *Handbook for Matrix Computations* by Coleman and Van Loan. The new package LAPACK is a transportable library of FORTRAN 77 subroutines that has been designed to supersede LINPACK and EISPACK by integrating these two sets of algorithms into a unified and improved package. The software has been restructured to achieve greater efficiency on vector processors and other high-performance or shared-memory multiprocessors. LAPACK is accessible through netlib for specified subroutines or from NAG for the complete library. *LAPACK User's Guide* by Anderson et al. should be available by the time this book is published.

Other packages for solving specific types of problems are available in the public domain. Information about these programs can be obtained through electronic mail by sending the line "help" to one of the following Internet addresses: netlib@research.att.com, netlib@ornl.gov, netlib@ nac.no, or netlib@draci.cs.uow.edu.au or to the uucp address: uunet!research!netlib. These packages are considered by experts to be highly efficient, accurate, and reliable. They are thoroughly tested and documentation is readily available. Although the packages are *portable*, it is a good idea to investigate the machine dependence and thoroughly read the documentation. The programs test for almost all special contingencies that might result in error and failures. At the end of each chapter we discuss some of the appropriate general-purpose packages.

---

*Complete references are given in the Bibliography.

Commercially available packages also represent the state of the art in numerical methods. Their contents are often based on the public-domain packages but include methods in libraries for almost every type of problem.

IMSL (International Mathematical Software Library) consists of libraries MATH, STAT, and SFUN for numerical mathematics, statistics, and special functions, respectively. These libraries contain over 700 subroutines written in FORTRAN 77, which solve most common numerical analysis problems. In 1970 IMSL became the first large-scale scientific library for mainframes. Since that time the libraries have been made available for computer systems ranging from supercomputers to personal computers. The libraries are available commercially from IMSL, 2500 Park West Tower One, 2500 City West Boulevard, Houston, TX 77042-3020. The packages are delivered in compiled form with extensive documentation. There is an example program for each routine as well as background reference information. IMSL contains methods for linear systems, eigensystem analysis, interpolation and approximation, integration and differentiation, differential equations, transforms, nonlinear equations, optimization, and basic matrix and vector operations. The library also contains extensive statistical routines.

The Numerical Algorithms Group (NAG) has been in existence in the United Kingdom since 1971. NAG offers over 600 subroutines in a FORTRAN 77 library for over 90 different computers. Subsets of their library are available for IBM personal computers (the PC50 Library consists of 50 of the most frequently used routines) and workstations (the Workstation Library contains 172 routines). The NAG user's manual includes instructions and examples, along with sample output for each of the routines. A useful introduction to the NAG routines is *The NAG Library: A Beginner's Guide* by Phillips. The NAG library contains routines to perform most standard numerical analysis tasks in a manner similar to those in the IMSL. It also includes are some statistical routines and a set of graphic routines. The library is commercially available from Numerical Algorithms Group, Inc., 1400 Opus Place, Suite 200, Downers Grove, IL 60515-5702.

The IMSL and NAG packages are designed for the mathematician, scientist, or engineer who wishes to call high-quality FORTRAN subroutines from within a program. The documentation available with the commercial packages illustrates the typical driver program required to use the library routines. The next two software packages are stand-alone environments. When activated, the user enters commands to cause the package to solve a problem. However, each package allows programming within the command language. The incorporated language resembles Pascal, but it is possible to call external routines that consist of compiled FORTRAN or C subprograms.

MATLAB is a matrix laboratory that was originally a Fortran program published as *Demonstration of a Matrix Library* by C. B. Moler. The laboratory is based mainly on the EISPACK and LINPACK subroutines, although functions such as nonlinear systems, numerical integration, cubic splines, curve fitting, optimization, ordinary differential equations, and graphical tools have been incorporated. MATLAB is currently written in C and assembler, and the PC version of this package requires a numeric coprocessor. There is, however, a student version that will use—but does not require—a numeric coprocessor. The basic structure of MATLAB is to perform matrix operations such as finding the eigenvalues of a matrix entered from the command line or from an external file via function calls. It is a powerful, self-contained system that is especially useful for instruction in an applied linear algebra course. A well-written book by David

Hill entitled *Experiments in Computational Matrix Algebra*, with a complete solutions manual, is also available. MATLAB has been available since 1985 and may be purchased from The MathWorks Inc., Cochituate Place, 24 Prime Park Way, Natick, MA 01760. The MATLAB software is designed to run on many computers, including IBM PC compatibles, APPLE Macintosh, and SUN workstations.

The second package is GAUSS, a mathematical and statistical system for IBM personal computers produced by Lee E. Edlefson and Samuel D. Jones in 1985. It is coded mainly in assembler and is based primarily on EISPACK and LINPACK. As in the case of MATLAB, integration and differentiation, nonlinear systems, fast Fourier transforms, and graphics are available. GAUSS is less oriented toward instruction in linear algebra and more oriented to statistical analysis of data. This package uses a numeric coprocessor if one is available. It can be purchased from Aptech Systems, Inc., 1914 N. 34th St., Suite 301, Seattle, WA 98103.

There are numerous packages available that can be classified as supercalculator packages for the PC. These should not be confused, however, with the general-purpose software listed previously. If you have an interest in one of these packages, you should read "Supercalculators on the PC," by Simon and Wilson.

The supercalculator and toolkit packages are generally much less expensive than the complete libraries of IMSL or NAG. However, the libraries have been thoroughly tested and are clearly of high quality. The former may be good packages but are less versatile.

Generally, packages available as supplements to numerical analysis and methods textbooks are designed to solve the problems presented in the textbooks. With this book we have provided programs written in Pascal and in FORTRAN 77 that can be used to solve the problems given in the examples and exercises. They will give satisfactory results for most problems that occur in practice for the engineer or scientist. However, they could fail to give sufficient accuracy for certain problems for one of the reasons described earlier in the section.

In the past decade a number of software packages have been developed to produce symbolic mathematical computations. Predominant among them are *MACSYMA, DERIVE, Maple,* and *Mathematica*. Versions of the software exist for most common computer systems, and student versions of some of the packages are available at reasonable prices. Although there are significant differences among the packages both in performance and price, they all can perform standard algebra and calculus operations.

Having a symbolic computation package available can be very useful in the study of approximation techniques. The results in most of our examples and exercises have been generated using problems for which exact values can be determined so that the performance of the approximation method can be monitored. Exact solutions can often be obtained quite easily using a symbolic computation package. In addition, many numerical techniques have error bounds that require bounding a higher ordinary or partial derivative of a function. This can be a very tedious task—one that is not particularly instructive once the techniques of calculus have been mastered. These derivatives can be quickly obtained symbolically, and a little insight will often permit the symbolic computation packages to aid in the bounding process as well.

Our purpose is to introduce the student to the methods available for solving problems. We have not attempted to write software libraries to rival the outstanding ones already available. We feel that the student can understand the methods and see when they can fail without having to program all the methods. Too much time can be spent programming and debugging when it is unlikely that the program will be reused. We have used our programs to solve all the problems in the textbook and we will illustrate the use of our programs within the text. However, because we believe in the importance of the professionally written libraries, we discuss some of the general-purpose software that is available for solving the problems we have considered in each chapter. We do not, however, consider further other special-purpose software such as supercalculators and toolkits.

Solutions of Equations of One Variable

2.1 Introduction

In this chapter we consider one of the most basic problems of numerical approximation, the root-finding problem, which involves finding a **root** x of an equation of the form $f(x) = 0$, that is, a **zero** of the function f. This is one of the oldest known approximation problems, yet research continues in this area at the present time.

The problem of finding an approximation to the root of an equation can be traced back at least as far as 1700 B.C. A cuneiform table in the Yale Babylonian Collection dating from that period gives approximations to $\sqrt{2}$, approximations that can essentially be found by applying a technique we will see in Section 2.4.

2.2 The Bisection Method

The first and most elementary technique we consider is the Bisection, or Binary-Search, method. The Bisection method is used to determine, to any specified accuracy that your computer will permit, a solution to $f(x) = 0$ on an interval $[a, b]$, provided that f is continuous on the interval and that $f(a)$ and $f(b)$ are of opposite sign. Although the method will work for the case when more than one root is contained in the interval $[a, b]$, we assume for simplicity of our discussion that the root in this interval is unique.

To begin the Bisection method, set $a_1 = a$ and $b_1 = b$, as shown in Figure 2.1 on page 26, and let p_1 be the midpoint of the interval $[a, b]$:

$$p_1 = a_1 + \frac{b_1 - a_1}{2}.$$

If $f(p_1) = 0$, then the root p is given by $p = p_1$; if not, then $f(p_1)$ has the same sign as either $f(a_1)$ or $f(b_1)$.

Figure 2.1

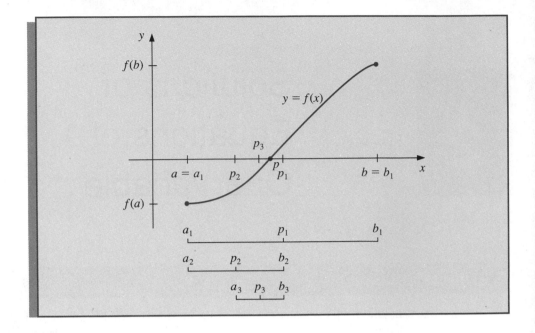

If $f(p_1)$ and $f(a_1)$ have the same sign, then p is in the interval (p_1, b_1), and we set

$$a_2 = p_1 \quad \text{and} \quad b_2 = b_1.$$

If, on the other hand, $f(p_1)$ and $f(a_1)$ have opposite signs, then p is in the interval (a_1, p_1), and we set

$$a_2 = a_1 \quad \text{and} \quad b_2 = p_1.$$

We reapply the process to the interval $[a_2, b_2]$ and continue forming $[a_3, b_3]$, $[a_4, b_4]$, ... Each new interval will contain p and will have one-half the length of the preceding interval.

Bisection Method

An interval $[a_{i+1}, b_{i+1}]$ containing an approximation to a root of $f(x) = 0$ is constructed from an interval $[a_i, b_i]$ containing the root by first letting

$$p_{i+1} = a_i + \frac{b_i - a_i}{2}.$$

Then

$$a_{i+1} = a_i \quad \text{and} \quad b_{i+1} = p_{i+1} \quad \text{if } f(a_i)f(p_{i+1}) < 0, \text{ and}$$
$$a_{i+1} = p_{i+1} \quad \text{and} \quad b_{i+1} = b_i \quad \text{otherwise.}$$

There are three stopping criteria commonly incorporated in the Bisection method. First, the method stops if one of the midpoints happens to coincide with the root. It also stops when the length of the search interval is less than some prescribed tolerance *TOL*. It is also good programming practice to impose stopping criteria independent of the given problem. For example, for a small tolerance, round-off error could cause the length of the search interval test always to be false, resulting in an infinite loop. To avoid this, a stopping technique involving a bound N_0 on the number of iterations should be imposed.

To start the Bisection method, an interval $[a, b]$ must be found with $f(a) \cdot f(b) < 0$. At each step the length of the interval known to contain a zero of f is reduced by a factor of 2. As a consequence, it is easy to determine a bound for the number of iterations needed to ensure a given tolerance. If the root needs to be determined within the tolerance *TOL*, we can be sure that, within the computational limits of the machine, the Bisection method will produce a result meeting this requirement in n iterations, provided that

$$n > \log_2 \left(\frac{b-a}{TOL} \right);$$

for in this case we have

$$2^n > \frac{b-a}{TOL}, \qquad \text{which implies that} \qquad |p_n - p| \leq \frac{b-a}{2^n} < TOL.$$

Since the number of required iterations to guarantee a given accuracy depends on the length of the initial interval $[a, b]$, it is advantageous to choose this interval as small as possible. For example, if $f(x) = 2x^3 - x^2 + x - 1$, then

$$f(-4) \cdot f(4) < 0 \qquad \text{and} \qquad f(0) \cdot f(1) < 0,$$

so the Bisection method could be used on $[-4, 4]$ or on $[0, 1]$. Starting the Bisection method on $[0, 1]$ instead of $[-4, 4]$ reduces by 3 the number of iterations required to achieve a specified accuracy.

The results in the following example were obtained by applying the Bisection method, as given in the program BISECT21.

EXAMPLE 1 The equation $f(x) = x^3 + 4x^2 - 10 = 0$ has a root in $[1, 2]$, since $f(1) = -5$ and $f(2) = 14$. It is easily seen from a sketch of the graph of f in Figure 2.2 on page 28 that there is only one root in $[1, 2]$. The program BISECT21, provided with the inputs $a = 1$, $b = 2$, $TOL = 0.0005$, and $N_0 = 20$ gives the values in Table 2.1. The actual root p, to ten decimal places, is $p = 1.3652300134$ and $|p - p_{11}| < 0.0005$. Since the expected number of iterations is $\log_2((2-1)/0.0005) \approx 10.96$, the bound N_0 was certainly sufficient. ■ ■ ■

The Bisection method, though conceptually clear, has significant drawbacks. It is slow to converge relative to the other techniques we will discuss, and a good intermediate approximation may be inadvertently discarded. This happened, for example, with p_9 in Example 1. However, the method has the important property that it always converges to a solution and it is easy to determine a bound for the number of iterations needed

Figure 2.2

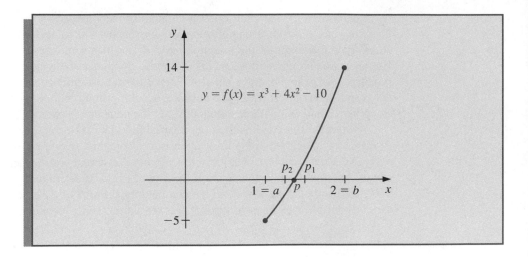

Table 2.1

n	a_n	b_n	p_n	$f(p_n)$
1	1.0000000000	2.0000000000	1.5000000000	2.3750000000
2	1.0000000000	1.5000000000	1.2500000000	−1.7968750000
3	1.2500000000	1.5000000000	1.3750000000	0.1621093750
4	1.2500000000	1.3750000000	1.3125000000	−0.8483886719
5	1.3125000000	1.3750000000	1.3437500000	−0.3509826660
6	1.3437500000	1.3750000000	1.3593750000	−0.0964088440
7	1.3593750000	1.3750000000	1.3671875000	0.0323557854
8	1.3593750000	1.3671875000	1.3632812500	−0.0321499705
9	1.3632812500	1.3671875000	1.3652343750	0.0000720248
10	1.3632812500	1.3652343750	1.3642578125	−0.0160466908
11	1.3642578125	1.3652343750	1.3647460938	−0.0079892628

to ensure a given accuracy. For these reasons, the Bisection method is frequently used as a dependable starting procedure for the more efficient methods presented later in this chapter.

The bound for the number of iterations for the Bisection method assumes that the calculations are performed using infinite-digit arithmetic. When implementing the method on a computer, consideration must be given to the effects of round-off error. For example, the computation of the midpoint of the interval $[a_n, b_n]$ should be found from the equation

$$p_n = a_n + \frac{b_n - a_n}{2}$$

instead of from the algebraically equivalent equation

$$p_n = \frac{a_n + b_n}{2}.$$

The first equation adds a small correction $(b_n - a_n)/2$ to the known value a_n. When $b_n - a_n$ is near the maximum precision of the machine, this correction might be in error, but the error would not significantly affect the computed value of p_n. In this situation, however, it is possible for p_n as defined in the second equation to return a midpoint that is not even in the interval $[a_n, b_n]$.

A number of tests can be used to see if a root has been found. We would normally use a test of the form

$$|f(p_n)| < \epsilon,$$

where $\epsilon > 0$ would be a small number related in some way to the tolerance. However, it is also possible for the value $f(p_n)$ to be very small when p_n is far from the root p.

As a final remark, to determine which subinterval of $[a_n, b_n]$ contains a root of f, it is better to use the condition

$$\text{sign } (f(a_n)) \text{ sign } (f(p_n)) > 0 \qquad \text{instead of} \qquad f(a_n)f(p_n) > 0.$$

This test avoids the possibility of overflow or underflow in the multiplication of $f(a_n)$ and $f(p_n)$.

EXERCISE SET 2.2

1. Use the Bisection method to find solutions accurate to within 10^{-2} for $x^3 - 7x^2 + 14x - 6 = 0$ on each of the following intervals:
   (a) [0, 1] (b) [1, 3.2] (c) [3.2, 4]

2. Use the Bisection method to find solutions accurate to within 10^{-2} for $x^4 - 2x^3 - 4x^2 + 4x + 4 = 0$ on each of the following intervals:
   (a) [−2, −1] (b) [0, 2] (c) [2, 3] (d) [−1, 0]

3. Use the Bisection method to find a solution accurate to within 10^{-3} for $x = \tan x$ on [4, 4.5].

4. Use the Bisection method to find a solution accurate to within 10^{-3} for $2 + \cos(e^x - 2) - e^x = 0$ on [0.5, 1.5].

5. Use the Bisection method to find solutions accurate to 10^{-5} for the following problems:
   (a) $x - 2^{-x} = 0$, for $0 \le x \le 1$,
   (b) $e^x - x^2 + 3x - 2 = 0$, for $0 \le x \le 1$,
   (c) $2x \cos(2x) - (x + 1)^2 = 0$, for $-3 \le x \le -2$ and $-1 \le x \le 0$
   (d) $x \cos x - 2x^2 + 3x - 1 = 0$, for $0.2 \le x \le 0.3$ and $1.2 \le x \le 1.3$

6. (a) Use the Bisection method to find a solution accurate to within 10^{-2} for $x + 0.5 + 2 \cos \pi x = 0$ on [0.5, 1.5].
   (b) Suppose the test

   If $f(b_i)f(p_i) > 0$, then set $b_{i+1} = p_i$, $a_{i+1} = a_i$; else set $a_{i+1} = p_i$, $b_{i+1} = b_i$

   is used in the Bisection method. Use this test to find a solution accurate to within 10^{-2} for $x + 0.5 + 2 \cos \pi x = 0$.
   (c) Discuss any discrepancy between parts (a) and (b).

7. Find a bound for the number of iterations needed to achieve an approximation with accuracy 10^{-4} to the solution of $x^3 - x - 1 = 0$ lying in the interval [1, 2]. Find an approximation to the root with this accuracy.

8. Find a bound for the number of iterations needed to achieve an approximation with accuracy 10^{-4} to the solution of $x^3 + x - 4 = 0$ lying in the interval $[1, 4]$. Find an approximation to the root with this accuracy.

9. Find a bound for the number of iterations needed to achieve an approximation with accuracy 0.5×10^{-2} to the solution of $x^3 + 4.001x^2 + 4.002x + 1.101$ lying in the interval $[-0.5, 0]$. Use three-digit rounding arithmetic to find an approximation to the root with this accuracy.

10. Find a bound for the number of iterations needed to achieve an approximation with accuracy 0.5×10^{-2} to the solution of $x - 0.5(\sin x + \cos x) = 0$ lying in the interval $[0,1]$. Use three-digit rounding arithmetic to find an approximation to the root with this accuracy.

11. Find the four zeros of $f(x) = 4x \cos(2x) - (x - 2)^2$ in the interval $[0, 8]$ accurate to within 10^{-4}.

12. Determine, to within 10^{-3}, the three smallest positive solutions to $\sin x = e^{-x}$.

13. A trough of length L has a cross section in the shape of a semicircle with radius r, as shown. When filled with water to within a distance h of the top, the volume V of the water is

$$V = L\left[0.5\pi r^2 - r^2 \arcsin\left(\frac{h}{r}\right) - h(r^2 - h^2)^{1/2}\right].$$

Suppose $L = 10$ ft, $r = 1$ ft, and $V = 12.4$ ft³. Find the depth of the water in the trough to within 0.01 ft.

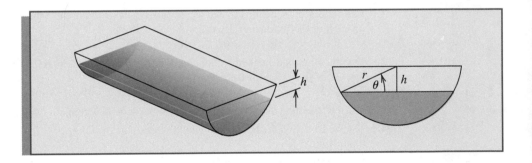

2.3 The Secant Method

Although the Bisection method always converges, the speed of convergence is usually too slow for general use. Figure 2.3 gives a graphical interpretation of the Bisection method that can be used to discover how improvements on this technique can be derived. This figure shows the graph of a continuous function that is negative at a_1 and positive at b_1. The first approximation p_1 to the root p is found by drawing the line joining the points $(a_1, \text{sign}(f(a_1))) = (a_1, -1)$ and $(b_1, \text{sign}(f(b_1))) = (b_1, 1)$ and letting p_1 be the point where this line intersects the x-axis. In essence, the line joining $(a_1, -1)$ and $(b_1, 1)$ has been used to approximate the graph of f on the interval $[a_1, b_1]$. Successive approximations apply this same process on subintervals of $[a_1, b_1]$, $[a_2, b_2]$, and so on. Notice that the Bisection method uses no information about the function f except the sign of $f(x)$ at certain values of x.

Figure 2.3

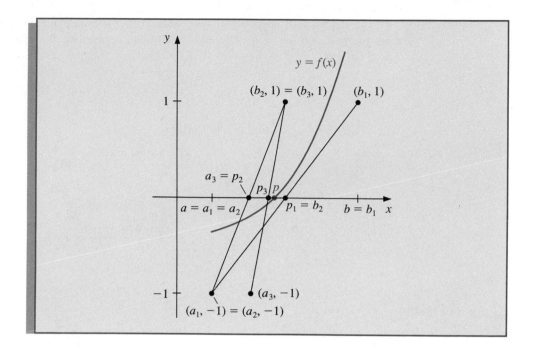

Suppose that in the initial step we know that $|f(a_1)| < |f(b_1)|$. Then we would expect the root p to be closer to a_1 than to b_1. Alternatively, if $|f(b_1)| < |f(a_1)|$, p is likely to be closer to b_1 than to a_1. Instead of choosing the intersection of the line through $(a_1, \text{sign}(f(a_1))) = (a_1, -1)$ and $(b_1, \text{sign}(f(b_1))) = (b_1, 1)$ as the approximation to the root p, the Secant method chooses the x-intercept of the secant line to the curve, the line through $(a_1, f(a_1))$ and $(b_1, f(b_1))$. This places the approximation closer to the endpoint of the interval for which f has smaller absolute value, as shown in Figure 2.4.

Figure 2.4

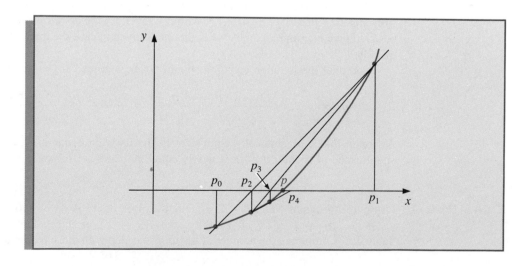

The sequence of approximations generated by the Secant method is started by setting $p_0 = a$ and $p_1 = b$. The equation of the secant line through $(p_0, f(p_0))$ and $(p_1, f(p_1))$ is

$$y = f(p_1) + \frac{f(p_1) - f(p_0)}{p_1 - p_0}(x - p_1).$$

The x-intercept $(p_2, 0)$ of this line satisfies

$$0 = f(p_1) + \frac{f(p_1) - f(p_0)}{p_1 - p_0}(p_2 - p_1),$$

and solving for p_2 gives

$$p_2 = p_1 - \frac{f(p_1)(p_1 - p_0)}{f(p_1) - f(p_0)}.$$

Secant Method

The approximation p_{i+1} to a root of $f(x) = 0$ is computed from the approximations p_i and p_{i-1} using the equation

$$p_{i+1} = p_i - \frac{f(p_i)(p_i - p_{i-1})}{f(p_i) - f(p_{i-1})}.$$

The Secant method does not have the root-bracketing property of the Bisection method, since there is no guarantee that the root is in the interval bounded by p_{i+1} and p_i. As a consequence, the method does not always converge, but when it does converge it generally does so much faster than the Bisection method.

We use two stopping conditions in the Secant method. The successful criterion results from iterating until $| p_i - p_{i-1} |$ is within a given tolerance. However, a safeguard exit based upon a maximum number of iterations is given in case the method fails to converge quickly.

Note that the iteration equation should not be algebraically simplified to

$$p_i = p_{i-1} - \frac{f(p_{i-1})(p_{i-1} - p_{i-2})}{f(p_{i-1}) - f(p_{i-2})} = \frac{f(p_{i-2})p_{i-1} - f(p_{i-1})p_{i-2}}{f(p_{i-2}) - f(p_{i-1})}.$$

Although this is algebraically equivalent to the iteration equation, it could increase the significance of round-off error if nearly equal numbers are subtracted.

EXAMPLE 1   Using the program SECANT22 to find the root of $x^3 + 4x^2 - 10 = 0$ with inputs $p_0 = 1$, $p_1 = 2$, $TOL = 0.0005$, and $N_0 = 20$ produces the results shown in Table 2.2. Comparing the number of iterations needed with those required in Example 1 of the previous section to solve this problem by the Bisection method, we see that about half the previous number are required. Further, $\mid p - p_6 \mid = \mid 1.3652300134 - 1.3652300011 \mid <$ 1.3×10^{-8}. ■ ■ ■

Table 2.2

n	p_n	$f(p_n)$
2	1.2631578947	-1.6022743840
3	1.3388278388	-0.4303647480
4	1.3666163947	0.0229094308
5	1.3652119026	-0.0002990679
6	1.3652300011	-0.0000002032

There are other reasonable choices for generating a sequence of approximations based on the intersection of an approximating line and the x-axis. The **method of False Position** (or *Regula Falsi*) is a hybrid Bisection-Secant method that constructs approximating lines similar to those of the Secant method but forms intervals to bracket the root in the manner of the Bisection method. As with the Bisection method, the method of False Position requires that an initial interval $[a, b]$ first be found with $f(a)$ and $f(b)$ of opposite sign. With $a_1 = a$ and $b_1 = b$, the approximation p_1 is given by

$$p_1 = a_1 - \frac{f(a_1)(b_1 - a_1)}{f(b_1) - f(a_1)}.$$

If $f(p_1)$ and $f(a_1)$ have the same sign, then set $a_2 = p_1$ and $b_2 = b_1$. Alternatively, if $f(p_1)$ and $f(b_1)$ have the same sign, set $a_2 = a_1$ and $b_2 = p_1$. (See Figure 2.5.)

Figure 2.5

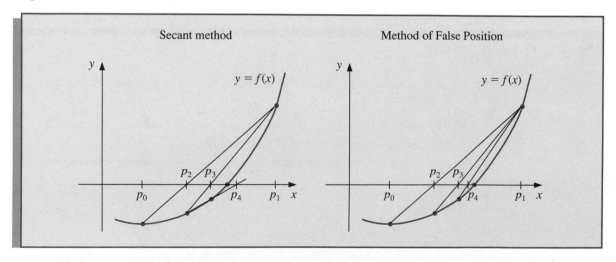

Method of False Position

An interval $[a_{i+1}, b_{i+1}]$ containing an approximation to a root of $f(x) = 0$ is found from an interval $[a_i, b_i]$ containing the root by first computing

$$p_{i+1} = a_i - \frac{f(a_i)(b_i - a_i)}{f(b_i) - f(a_i)}.$$

Then

$$a_{i+1} = a_i \quad \text{and} \quad b_{i+1} = p_{i+1} \quad \text{if } f(a_i)f(p_{i+1}) < 0, \text{ and}$$
$$a_{i+1} = p_{i+1} \quad \text{and} \quad b_{i+1} = b_i \quad \text{otherwise.}$$

Although the method of False Position may appear superior to the Secant method, it generally converges more slowly, as the results in Table 2.3 indicate for the problem we considered in Example 1. In fact, the method of False Position can converge even more slowly than the Bisection method (as the problem given in Exercise 12 shows), although this is not usually the case. The program FALPOS23 implements the method of False Position.

Table 2.3

n	a_n	b_n	p_n	$f(p_n)$
2	1.00000000	2.00000000	1.26315789	−1.60227438
3	1.26315789	2.00000000	1.33882784	−0.43036475
4	1.33882784	2.00000000	1.35854634	−0.11000879
5	1.35854634	2.00000000	1.36354744	−0.02776209
6	1.36354744	2.00000000	1.36480703	−0.00698342
7	1.36480703	2.00000000	1.36512372	−0.00175521
8	1.36512372	2.00000000	1.36520330	−0.00044106

EXERCISE SET 2.3

1. Use the Secant method to approximate to within 10^{-4} the roots of the following equations in the given intervals:
   (a) $x^3 - 2x^2 - 5 = 0$, $[1, 4]$ (b) $x^3 + 3x^2 - 1 = 0$, $[-3, -2]$
   (c) $x - \cos x = 0$, $[0, \pi/2]$ (d) $x - 0.8 - 0.2 \sin x = 0$, $[0, \pi/2]$

2. Repeat Exercise 1 using the method of False Position.

3. Use the Secant method to approximate the solutions of the following equations to within 10^{-5}:
   (a) $e^x + 2^{-x} + 2\cos x - 6 = 0$,   for $1 \le x \le 2$
   (b) $\ln(x - 1) + \cos(x - 1) = 0$,   for $1.3 \le x \le 2$
   (c) $3x^2 - e^x = 0$,   for $0 \le x \le 1$ and $3 \le x \le 5$
   (d) $(x - 2)^2 - \ln x = 0$ for $1 \le x \le 2$ and $e \le x \le 4$

4. Repeat Exercise 3 using the method of False Position.

5. Use the method of False Position and three-digit rounding arithmetic to find the solutions accurate to within 0.5×10^{-2} of the following equations in the given intervals:
   (a) $x^3 + 4.001x^2 + 4.002x + 1.101 = 0$, $[-0.5, 0]$
   (b) $x - 0.5(\sin x + \cos x) = 0$, $[0, 1]$

6. Repeat Exercise 5 using the Secant method.

7. Use the Secant method to approximate to within 10^{-5} the roots of the following equations in the given intervals:
   (a) $2x \cos(2x) - (x - 2)^2 = 0$, $[2, 3]$ and $[3, 4]$
   (b) $x \cos x - 2x^2 + 3x - 1 = 0$, $[0.2, 0.3]$ and $[1.2, 1.3]$

8. Use the Secant method to approximate to within 10^{-6} the roots of the following equations in the given intervals:
   (a) $\cos(x + \sqrt{2}) + x(x/\sqrt{2} + 2)/\sqrt{2} = 0$, $[-2, -1]$
   (b) $x^3 - 3x^2(2^{-x}) + 3x(4^{-x}) - 8^{-x} = 0$, $[0, 1]$

9. Repeat Exercise 8 using the method of False Position.

10. Find all four solutions of $4x \cos(2x) - (x - 2)^2 = 0$ in $[0, 8]$ accurate to within 10^{-5}.

11. Find all solutions of $x^2 + 10 \cos x = 0$ accurate to within 10^{-5}.

12. The equation $300x^4 - 275x^3 + 100x^2 - 10x - \frac{1}{2} = 0$ has a solution in $[0.1, 1]$. Find an approximation to the root accurate to 10^{-4} using each method:
    (a) Bisection method   (b) Secant method   (c) Method of False Position

13. A particle starts at rest on a smooth inclined plane whose angle θ is increasing at a constant rate

$$\frac{d\theta}{dt} = \omega,$$

as shown. At the end of t seconds, the position of the object is given by

$$x(t) = \frac{g}{2\omega^2} \left(\frac{e^{\omega t} - e^{-\omega t}}{2} - \sin \omega t \right).$$

Suppose the particle has moved 1.7 ft in 1 s. Find, to within 10^{-5}, the rate ω at which θ changes if $\omega_0 = -0.5$ and $\omega_1 = -0.1$. Assume that $g = -32.17$ ft/s^2.

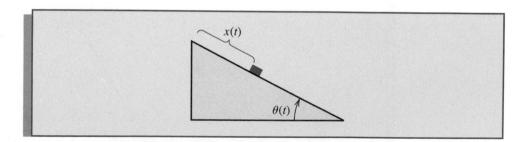

2.4 Newton's Method

The Bisection method and the Secant method both have geometric representations that involve finding the zero of an approximating line to the graph of a function f and using the zero of this line to approximate the solution to $f(x) = 0$. The increase in accuracy of the Secant method over the Bisection method is a consequence of the fact that the secant line to the curve better approximates the graph of f than does the line used to generate the approximations in the Bisection method.

The line that *best* approximates the graph of the function at a point on its graph is the tangent line to the graph at that point. Using this line instead of the secant line produces **Newton's method** (also called the *Newton–Raphson method*), the technique we consider in this section.

Suppose that p_0 is an initial approximation to the root p of the equation $f(x) = 0$. We will assume that the function f is differentiable sufficiently near this root so that f' exists in an interval containing all of our approximations to p. The slope of the tangent line to the graph of f at the point $(p_0, f(p_0))$ is $f'(p_0)$, so the equation of this tangent line is

$$y - f(p_0) = f'(p_0)(x - p_0).$$

Since this line crosses the x-axis when the y-coordinate of the point on the line is zero, the next approximation to p, p_1, satisfies

$$0 - f(p_0) = f'(p_0)(p_1 - p_0),$$

which implies that

$$p_1 = p_0 - \frac{f(p_0)}{f'(p_0)},$$

provided that $f'(p_0) \neq 0$. Subsequent approximations are found for p in a similar manner, as shown in Figure 2.6.

Figure 2.6

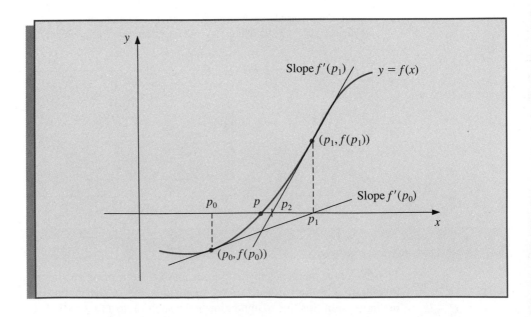

Newton's Method

The approximation p_{i+1} to a root of $f(x) = 0$ is computed from the approximation p_i using the equation

$$p_{i+1} = p_i - \frac{f(p_i)}{f'(p_i)}.$$

EXAMPLE 1 Using the Newton's method program NEWTON24 to find the root of $x^3 + 4x^2 - 10 = 0$ that lies in $[1, 2]$ gives the results in Table 2.4. To compare the convergence of this method with those applied to this problem previously, we use $p_0 = 1$, $TOL = 0.0005$, and $N_0 = 20$. The number of iterations needed to solve the problem by Newton's method is less, by one-third, than the number needed for the Secant method, which, in turn, required less than one-half the iterations needed for the Bisection method. In truth, $|p - p_4| \leq 10^{-10}$. ■ ■ ■

Table 2.4

n	p_n	$f(p_n)$
1	1.4545454545	1.5401953418
2	1.3689004011	0.0607196886
3	1.3652366002	0.0001087706
4	1.3652300134	0.0000000004

Newton's method generally produces accurate results with few iterations. With the aid of Taylor polynomials, we can see why this is true. Suppose p is the solution to $f(x) = 0$ and that f'' exists on an interval containing both p and the approximation p_i. Expanding f in the first Taylor polynomial at p_i and evaluating at $x = p$ gives

$$0 = f(p) = f(p_i) + f'(p_i)(p - p_i) + \frac{f''(\xi)}{2}(p - p_i)^2,$$

where ξ lies between p_i and p. Consequently, if $f'(p_i) \neq 0$,

$$p - p_i + \frac{f(p_i)}{f'(p_i)} = -\frac{f''(\xi)}{2f'(p_i)}(p - p_i)^2.$$

Since

$$p_{i+1} = p_i - \frac{f(p_i)}{f'(p_i)},$$

this implies that

$$p - p_{i+1} = -\frac{f''(\xi)}{2f'(p_i)}(p - p_i)^2.$$

If a bound M is known for the second derivative of f on an interval about p and if p_i is within this interval, then

$$| p - p_{i+1} | \le \frac{M}{2 | f'(p_i) |} | p - p_i |^2 .$$

The important feature of this inequality is that the error of the $(i + 1)$st approximation is bounded by approximately the square of the error of the ith approximation. This implies that Newton's method has the tendency to approximately double the number of digits of accuracy with each successive approximation. For example, if the error in approximating p by p_i is on the order of 10^{-k} for some integer k, then the error in approximating p by p_{i+1} is on the order of $C(10^{-k})^2 = C 10^{-2k}$, where $C = (M/ | 2f'(p_i) |)$. The smaller the value of C, the better the bound on the approximations. In particular, the bound will rapidly approach zero provided that the second derivative of f stays bounded and that the first derivative is bounded away from zero.

EXAMPLE 2   Find an approximation to the solution of the equation $x = 3^{-x}$ that is accurate to within 10^{-8}.

A solution to this equation corresponds to a solution to $f(x) = 0$, where

$$f(x) = x - 3^{-x}.$$

Since f is continuous and $f(0) = -1$ while $f(1) = \frac{2}{3}$, a solution to the equation lies in the interval $(0, 1)$. As an initial approximation we have chosen the midpoint of this interval, $x_0 = 0.5$. Succeeding approximations are generated by applying the formula

$$p_{i+1} = p_i - \frac{f(p_i)}{f'(p_i)} = p_i - \frac{p_i - 3^{-p_i}}{1 + 3^{-p_i} \ln 3}.$$

These approximations are listed in Table 2.5, together with differences between successive approximations. The difference in successive approximations leads to the correct conclusion that p_3 is accurate to the places listed and that the solution to $x = 3^{-x}$ accurate to within 10^{-8} is 0.54780862.   ▪ ▪ ▪

Table 2.5

n	p_n	$| p_n - p_{n-1} |$
0	0.500000000	
1	0.547329757	0.047329757
2	0.547808574	0.000478817
3	0.547808622	0.000000048

The success of Newton's method is predicated on the assumption that the derivative of f is nonzero at the approximations to the root p. If f' is continuous, this means that the technique will be satisfactory provided that $f'(p) \ne 0$ and that a sufficiently accurate initial approximation is used. The condition $f'(p) \ne 0$ is not trivial; it is true precisely when the root p of the equation is a simple root of f, which means that a function q exists with the property that for $x \neq p$,

$$f(x) = (x - p)q(x), \qquad \text{where} \qquad \lim_{x \to p} q(x) \neq 0.$$

When the root of the equation is not simple, Newton's method may converge, but not with the speed we have seen in our previous examples.

EXAMPLE 3 The root $p = 0$ of the equations $f(x) = e^x - x - 1 = 0$ is not simple, since both $f(0) = e^0 - 0 - 1 = 0$ and $f'(0) = e^0 - 1 = 0$. The terms generated by Newton's method with $p_0 = 0$ are shown in Table 2.6 and converge slowly to zero. ■ ■ ■

Table 2.6

n	p_n	n	p_n
0	1.0	8	0.005545
1	0.58198	9	2.7750×10^{-3}
2	0.31906	10	1.3881×10^{-3}
3	0.16800	11	6.9411×10^{-4}
4	0.08635	12	3.4703×10^{-4}
5	0.04380	13	1.7416×10^{-4}
6	0.02206	14	8.8041×10^{-5}
7	0.01107		

EXERCISE SET 2.4

1. Use Newton's method to approximate to within 10^{-4} the roots of the following equations in the given intervals:
   (a) $x^3 - 2x^2 - 5 = 0$, $\quad$ [1, 4] $\qquad$ (b) $x^3 + 3x^2 - 1 = 0$, $\quad$ [−3, −2]
   (c) $x - \cos x = 0$, $\quad$ $[0, \pi/2]$ $\qquad$ (d) $x - 0.8 - 0.2 \sin x = 0$, $\quad$ $[0, \pi/2]$

2. Use Newton's method to approximate the solutions of the following equations to within 10^{-5}:
   (a) $e^x + 2^{-x} + 2\cos x - 6 = 0$, $\quad$ for $1 \leq x \leq 2$
   (b) $\ln(x - 1) + \cos(x - 1) = 0$, $\quad$ for $1.2 \leq x \leq 2$
   (c) $3x^2 - e^x = 0$, $\quad$ for $0 \leq x \leq 1$ and $3 \leq x \leq 5$
   (d) $(x - 2)^2 - \ln x = 0$, $\quad$ for $1 \leq x \leq 2$ and $e \leq x \leq 4$

3. Solve $4\cos x = e^x$ with accuracy 10^{-4} by using Newton's method with $p_0 = 1$.

4. Use Newton's method to find solutions accurate to 10^{-5} for the following equations in the given intervals:
   (a) $2 + \cos(e^x - 2) - e^x = 0$, $\quad$ [0.5, 1.5]
   (b) $2x\cos(2x) - (x - 2)^2 = 0$, $\quad$ [2, 3] and [3, 4]
   (c) $2x\cos(2x) - (x + 1)^2 = 0$, $\quad$ [−3, −2] and [−1, 0]
   (d) $x\cos x - 2x^2 + 3x - 1 = 0$, $\quad$ [0.2, 0.3] and [1.2, 1.3]

5. Find all four solutions of $4x\cos(2x) - (x - 2)^2 = 0$ in [0, 8] accurate to within 10^{-5}.

6. Find all solutions of $x^2 + 10 \cos x = 0$ accurate to within 10^{-5}.

7. Use Newton's method to solve the equation

$$0 = 2 - 2\cos 2x - 4x \sin x + x^2.$$

Iterate $P_0 = \pi/2$ using Newton's method until an accuracy of 10^{-5} is obtained for the approximate root. Explain why the results seem unusual for Newton's method. Also, solve the equation with $p_0 = 5\pi$ and $p_0 = 10\pi$.

8. Use Newton's method to approximate the solutions of the following equations to within 10^{-5} in the given intervals. In all problems, the convergence will be slower than normal since the roots are not simple roots.
   (a) $x^2 - 2xe^{-x} + e^{-2x} = 0$,   [0, 1]
   (b) $\cos(x + \sqrt{2}) + x(x/\sqrt{2} + 2)/\sqrt{2} = 0$,   [-2, -1]
   (c) $x^3 - 3x^2(2^{-x}) + 3x(4^{-x}) + 8^{-x} = 0$,   [0, 1]
   (d) $e^{6x} + 3(\ln 2)^2 e^{2x} - (\ln 8)e^{4x} - (\ln 2)^3$,   [-1, 0]

9. The fourth-degree polynomial

$$f(x) = 230x^4 + 18x^3 + 9x^2 - 221x - 9$$

has two real zeros, one in $[-1, 0]$ and the other in $[0, 1]$. Attempt to find these zeros to within 10^{-6} using each method:
   (a) The Method of False Position   (b) The Secant method
   (c) Newton's method

10. The numerical method defined by

$$p_n = p_{n-1} - \frac{f(p_{n-1})f'(p_{n-1})}{[f'(p_{n-1})]^2 - f(p_{n-1})f''(p_{n-1})}$$

for $n = 1, 2, \ldots$, can be used instead of Newton's method for equations having multiple roots. Repeat Exercise 8 using this method and compare your results to those of Exercise 8.

11. The function described by $f(x) = \ln(x^2 + 1) - e^{0.4x} \cos \pi x$ has an infinite number of zeros.
    (a) Determine, within 10^{-6}, the only negative zero.
    (b) Determine, within 10^{-6}, the four smallest positive zeros.
    (c) Determine a reasonable initial approximation to find the nth smallest positive zero of f.
         [Hint: Sketch a rough graph of f.]
    (d) Use part (c) to determine, within 10^{-6}, the 25th smallest positive zero of f.

12. The sum of two numbers is 20. If each number is added to its square root, the product of the two sums equals 155.55. Determine the two numbers to within 10^{-4}.

13. Problems involving the amount of money needed to pay off a mortgage over a fixed period of time involve the formula

$$A = \frac{P}{i}[1 - (1 + i)^{-n}],$$

known as an *ordinary annuity equation*. In this equation A is the amount of each payment and i is the interest rate per period for the n payment periods.

Suppose that a 30-year mortgage in the amount of $75,000 is needed and that the borrower can afford house payments of at most $625 per month. What is the maximum interest rate that the borrower can afford to pay?

14. A drug administered to a patient produces a concentration in the blood stream given by $c(t) = Ate^{-t/3}$ mg/ml t hours after A units have been injected. The maximum safe concentration is 1 mg/ml.

    (a) What amount should be injected to reach this maximum safe concentration and when does this maximum occur?

    (b) An additional amount of this drug is to be administered to the patient after the concentration falls to 0.25 mg/ml. Determine, to the nearest minute, when this second injection should be given.

    (c) Assuming that the concentration from consecutive injections is additive and that 75% of the amount originally injected is administered in the second injection, when is it time for the third injection?

2.5 Error Analysis and Accelerating Convergence

In the previous section we found that Newton's method generally converges very rapidly if a sufficiently accurate initial approximation has been found. This rapid speed of convergence is not common in approximation techniques and is due to the fact that Newton's method produces *quadratically* convergent approximations.

A method that produces a sequence $\{p_n\}$ of approximations that converge to a number p is said to be **linearly** convergent to p if a constant M exists with

$$|p - p_{n+1}| < M\,|p - p_n| \qquad \text{for each} \quad n = 0, 1, \ldots.$$

The sequence is said to be **quadratically** convergent if a constant M exists with

$$|p - p_{n+1}| < M\,|p - p_n|^2 \qquad \text{for each} \quad n = 0, 1, \ldots.$$

The following example illustrates the advantage of quadratic over linear convergence.

EXAMPLE 1 Suppose that p_n converges linearly to $p = 0$ and $\hat{p}_n$ converges quadratically to $p = 0$. We assume, for sake of comparison, that the constant $M = 0.5$ is the same in each case. Then

$$|p_1| < M\,|p_0| \le (0.5) \cdot |p_0| \qquad \text{and} \qquad |\hat{p}_1| < M\,|\hat{p}_0|^2 \le (0.5) \cdot |\tilde{p}_0|^2.$$

Similarly,

$$|p_2| < M\,|p_1| \le 0.5(0.5) \cdot |p_0| = (0.5)^2\,|p_0|$$

and

$$|\hat{p}_2| < M\,|\hat{p}_1|^2 \le 0.5(0.5\,|\tilde{p}_0|^2)^2 = (0.5)^3\,|\tilde{p}_0|^4.$$

Continuing,

$$|p_3| < M\,|p_2| \le 0.5\big((0.5)^2\,|p_0|\big) = (0.5)^3\,|p_0|$$

and

$$|\hat{p}_3| < M\,|\hat{p}_2|^2 \le 0.5\big((0.5)^3\,|\tilde{p}_0|^4\big)^2 = (0.5)^7\,|\tilde{p}_0|^8.$$

In general,

$$| \, p_n \, | < 0.5^n \, | \, p_0 \, |, \qquad \text{whereas} \qquad | \, \hat{p}_n \, | < (0.5)^{2^n - 1} \, | \, \tilde{p}_0 \, |^{2^n}$$

for each $n = 1, 2, \ldots$. Table 2.7 illustrates the relative speed of convergence of these error bounds to zero, assuming that $| \, p_0 \, | = | \, \tilde{p}_0 \, | = 1$.

Table 2.7

n	Linear Convergence Sequence Bound $(0.5)^n$	Quadratic Convergence Sequence Bound $(0.5)^{2^n - 1}$
1	5.0000×10^{-1}	5.0000×10^{-1}
2	2.5000×10^{-1}	1.2500×10^{-1}
3	1.2500×10^{-1}	7.8125×10^{-3}
4	6.2500×10^{-2}	3.0518×10^{-5}
5	3.1250×10^{-2}	4.6566×10^{-10}
6	1.5625×10^{-2}	1.0842×10^{-19}
7	7.8125×10^{-3}	5.8775×10^{-39}

The quadratically convergent sequence is within 10^{-38} of zero by the seventh term. At least 126 terms are needed to ensure this accuracy for the linearly convergent sequence. If $| \, \tilde{p}_0 \, | < 1$, the bound on the sequence $\{\tilde{p}_n\}$ will decrease even more rapidly. No significant change will occur, however, if $| \, p_0 \, | < 1$. ■ ■ ■

Quadratically convergent sequences generally converge much more quickly than those that converge only linearly, but many techniques that generate convergent sequences do so linearly. We next consider a technique called **Aitken's Δ^2 method** that can be used to accelerate the convergence of a sequence that is linearly convergent, regardless of its origin or application.

Suppose $\{p_n\}_{n=0}^{\infty}$ is a linearly convergent sequence with limit p. To motivate the construction of a sequence $\{\hat{p}_n\}$ that converges more rapidly to p than does $\{p_n\}$, let us first assume that the signs of $p_n - p$, $p_{n+1} - p$, and $p_{n+2} - p$ agree and that n is sufficiently large that

$$\frac{p_{n+1} - p}{p_n - p} \approx \frac{p_{n+2} - p}{p_{n+1} - p}.$$

Then

$$(p_{n+1} - p)^2 \approx (p_{n+2} - p)(p_n - p),$$

so

$$p_{n+1}^2 - 2p_{n+1}p + p^2 \approx p_{n+2}p_n - (p_n + p_{n+2})p + p^2$$

and

$$(p_{n+2} + p_n - 2p_{n+1})p \approx p_{n+2}p_n - p_{n+1}^2.$$

Solving for p gives

$$
\begin{aligned}
p &\approx \frac{p_{n+2}\,p_n - p_{n+1}^2}{p_{n+2} - 2p_{n+1} + p_n} \\[2mm]
&= \frac{p_n^2 + p_n\,p_{n+2} + 2p_n\,p_{n+1} - 2p_n\,p_{n+1} - p_n^2 - p_{n+1}^2}{p_{n+2} - 2p_{n+1} + p_n} \\[2mm]
&= \frac{(p_n^2 + p_n\,p_{n+2} - 2p_n\,p_{n+1}) - (p_n^2 - 2p_n\,p_{n+1} + p_{n+1}^2)}{p_{n+2} - 2p_{n+1} + p_n} \\[2mm]
&= p_n - \frac{(p_{n+1} - p_n)^2}{p_{n+2} - 2p_{n+1} + p_n}.
\end{aligned}
$$

Aitken's Δ^2 method uses the sequence $\{\hat{p}_n\}_{n=0}^{\infty}$ defined next to approximate the limit p.

Aitken's Δ^2 Method

$$
\hat{p}_n = p_n - \frac{(p_{n+1} - p_n)^2}{p_{n+2} - 2p_{n+1} + p_n}.
$$

EXAMPLE 2 The sequence $\{p_n\}_{n=1}^{\infty}$, where $p_n = \cos(1/n)$, converges linearly to $p = 1$. The first few terms of the sequences $\{p_n\}_{n=1}^{\infty}$ and $\{\hat{p}_n\}_{n=1}^{\infty}$ are given in Table 2.8. It certainly appears that $\{\hat{p}_n\}_{n=1}^{\infty}$ converges more rapidly to $p = 1$ than does $\{p_n\}_{n=1}^{\infty}$. ∎ ∎ ∎

Table 2.8

n	p_n	$\hat{p}_n$
1	0.54030	0.96178
2	0.87758	0.98213
3	0.94496	0.98979
4	0.96891	0.99342
5	0.98007	0.99541
6	0.98614	
7	0.98981	

The Δ notation associated with this technique has its origin in the following definition.

Given the sequence $\{p_n\}_{n=0}^{\infty}$, the **forward difference** Δp_n is

$$
\Delta p_n = p_{n+1} - p_n, \qquad \text{for} \quad n \geq 0.
$$

Higher powers $\Delta^k p_n$ are defined recursively by

$$
\Delta^k p_n = \Delta(\Delta^{k-1} p_n), \qquad \text{for} \quad k \geq 2.
$$

Because of the definition,

$$\Delta^2 p_n = \Delta(p_{n+1} - p_n) = \Delta p_{n+1} - \Delta p_n = (p_{n+2} - p_{n+1}) - (p_{n+1} - p_n),$$

so

$$\Delta^2 p_n = p_{n+2} - 2p_{n+1} + p_n.$$

Thus, the formula for $\hat{p}_n$ given in Aitken's Δ^2 method can be written as

$$\hat{p}_n = p_n - \frac{(\Delta p_n)^2}{\Delta^2 p_n}, \qquad \text{for all} \quad n \geq 0.$$

The sequence $\{\hat{p}_n\}_{n=1}^{\infty}$ converges to p more rapidly than does the original sequence $\{p_n\}_{n=0}^{\infty}$. A more precise statement is that if $\{p_n\}$ is a sequence that converges linearly to the limit p and $p_n - p \neq 0$ for all $n \geq 0$, then

$$\lim_{n \to \infty} \frac{\hat{p}_n - p}{p_n - p} = 0.$$

We find occasion to apply this acceleration technique frequently in our study of approximation methods.

EXERCISE SET 2.5

1. The following sequences are linearly convergent. Generate the first five terms of the sequence $\{\hat{p}_n\}$ using Aitken's Δ^2 method.
   (a) $p_0 = 0.5$, $p_n = (2 - e^{p_{n-1}} + p_{n-1}^2)/3$, $n = 1, 2, \ldots$
   (b) $p_0 = 0.75$, $p_n = (e^{p_{n-1}}/3)^{1/2}$, $n = 1, 2, \ldots$
   (c) $p_0 = 0.5$, $p_n = 3^{-p_{n-1}}$, $n = 1, 2, \ldots$
   (d) $p_0 = 0.5$, $p_n = \cos p_{n-1}$, $n = 1, 2, \ldots$

2. Consider the function $f(x) = e^{6x} + 3(\ln 2)^2 e^{2x} - \ln 8 e^{4x} - (\ln 2)^3$. Use Newton's method with $p_0 = 0$ to determine the root of $f(x) = 0$. Generate terms until $|\,p_{n+1} - p_n\,| < 0.0002$. Construct the sequence $\{\hat{p}_n\}$. Is the convergence improved?

3. Repeat Exercise 2 with the constants in $f(x)$ replaced by their four-digit approximations — that is, with $f(x) = e^{6x} + 1.441 e^{2x} - 2.079 e^{4x} - 0.3330$. Compare the solutions to the results in Exercise 2.

4. Newton's method does not converge quadratically for the following problems. Accelerate the convergence using the Aitken's Δ^2 method. Iterate until $|\,\hat{p}_n - \hat{p}_{n-1}\,| < 10^{-4}$.
   (a) $x^2 - 2xe^{-x} + e^{-2x} = 0$, $[0, 1]$
   (b) $\cos(x + \sqrt{2}) + x(x/2 + \sqrt{2}) = 0$, $[-2, -1]$
   (c) $x^3 - 3x^2(2^{-x}) + 3x(4^{-x}) - 8^{-x} = 0$, $[0, 1]$
   (d) $e^{3x} + 3(\ln 1.5)^2 e^x - (\ln 3.375)e^{2x} - (\ln 1.5)^3 = 0$, $[-1, 0]$.

5. Show that the following sequences $\{p_n\}$ converge linearly to $p = 0$. How large must n be before $|\,p_n - p\,| \leq 5 \times 10^{-2}$? Use Aitken's Δ^2 method to generate a sequence $\{\hat{p}_n\}$ until $|\,p_n - p\,| \leq 5 \times 10^{-2}$.
   (a) $p_n = \dfrac{1}{n}$, $n \geq 1$ (b) $p_n = \dfrac{1}{n^2}$, $n \geq 1$

6. Show that the sequence defined by $p_n = 1/n^k, n \geq 1$, for any positive integer k, converges linearly to $p = 0$. For each pair of integers k and m, determine a number N for which $1/N^k < 10^{-m}$.

7. **(a)** Show that the sequence $p_n = 10^{-2^n}$ converges quadratically to zero.

   (b) Show that the sequence $p_n = 10^{-n^k}$ does not converge to zero quadratically, regardless of the size of the exponent k.

2.6 Müller's Method

There are a number of root-finding problems for which the Secant, False Position, and Newton's methods will not give satisfactory results. They will not give rapid convergence, for example, when the function and its derivative are simultaneously close to zero. In addition, these methods cannot be used to approximate complex roots unless the initial approximation is a complex number with nonzero imaginary part. This often makes them a poor choice for use in approximating the roots of polynomials which, even with real coefficients, commonly have complex roots occurring in conjugate pairs.

In this section we consider Müller's method, a technique that is similar to the Secant method. The Secant method constructs each new approximation by finding the zero of the line passing through points on the graph of the function corresponding to the two immediately previous approximations, as shown in Figure 2.7(a). Müller's method uses the zero of the parabola through the three immediately previous points on the graph as the new approximation, as shown in part (b) of Figure 2.7.

Figure 2.7

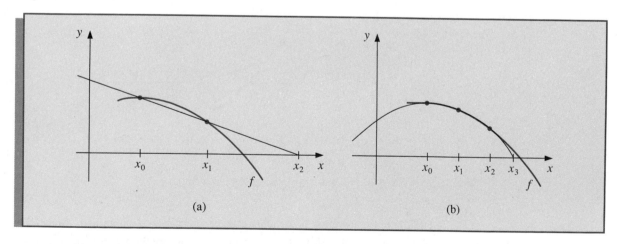

(a) (b)

Suppose that three initial approximations, p_0, p_1, and p_2, are given for a solution of $f(x) = 0$. The derivation of Müller's method for determining the next approximation p_3 begins by considering the quadratic polynomial

$$P(x) = a(x - p_2)^2 + b(x - p_2) + c$$

that passes through $(p_0, f(p_0))$, $(p_1, f(p_1))$, and $(p_2, f(p_2))$. The constants a, b, and c can be determined from the conditions

$$f(p_0) = a(p_0 - p_2)^2 + b(p_0 - p_2) + c,$$
$$f(p_1) = a(p_1 - p_2)^2 + b(p_1 - p_2) + c,$$

and

$$f(p_2) = a \cdot 0^2 + b \cdot 0 + c.$$

To determine p_3, the root of $P(x) = 0$, we apply the quadratic formula to $P(x)$. Because of round-off error problems caused by the subtraction of nearly equal numbers, however, we apply the formula in the manner prescribed in Example 1 of Section 1.4:

$$p_3 - p_2 = \frac{-2c}{b \pm \sqrt{b^2 - 4ac}}.$$

This gives two possibilities for p_3, depending on the sign preceding the radical term. In Müller's method, the sign is chosen to agree with the sign of b. Chosen in this manner, the denominator will be the largest in magnitude and will result in p_3 being selected as the closest root of $P(x) = 0$ to p_2.

Müller's Method

Given initial approximations p_0, p_1, and p_2, generate

$$p_3 = p_2 - \frac{2c}{b + \text{sign}(b)\sqrt{b^2 - 4ac}},$$

where

$$c = f(p_2),$$
$$b = \frac{(p_0 - p_2)^2[f(p_1) - f(p_2)] - (p_1 - p_2)^2[f(p_0) - f(p_2)]}{(p_0 - p_2)(p_1 - p_2)(p_0 - p_1)},$$

and

$$a = \frac{(p_1 - p_2)[f(p_0) - f(p_2)] - (p_0 - p_2)[f(p_1) - f(p_2)]}{(p_0 - p_2)(p_1 - p_2)(p_0 - p_1)}.$$

Then reiterate with p_1, p_2, and p_3 replacing p_0, p_1, and p_2.

The method continues until a satisfactory approximation is obtained. Since the method involves the radical $\sqrt{b^2 - 4ac}$ at each step, the method approximates complex roots when it is appropriate to do so.

EXAMPLE 1   Consider the polynomial $f(x) = 16x^4 - 40x^3 + 5x^2 + 20x + 6$. Using the program MULLER25 with accuracy tolerance 10^{-5} and various inputs for p_0, p_1, and p_2 produces the results in Tables 2.9, 2.10, and 2.11.

The actual values for the roots of the equation are 1.241677, 1.970446, and $-0.356062 \pm 0.162758i$, which demonstrates the accuracy of the approximations from Müller's method.
∎ ∎ ∎

Table 2.9

$p_0 = 0.5, \qquad p_1 = -0.5, \qquad p_2 = 0$

i	p_i	$f(p_i)$
3	$-0.555556 + 0.598352i$	$-29.4007 - 3.89872i$
4	$-0.435450 + 0.102101i$	$1.33223 - 1.19309i$
5	$-0.390631 + 0.141852i$	$0.375057 - 0.670164i$
6	$-0.357699 + 0.169926i$	$-0.146746 - 0.00744629i$
7	$-0.356051 + 0.162856i$	$-0.183868 \times 10^{-2} + 0.539780 \times 10^{-3}i$
8	$-0.356062 + 0.162758i$	$0.286102 \times 10^{-5} + 0.953674 \times 10^{-6}i$

Table 2.10

$p_0 = 0.5, \qquad p_1 = 1.0, \qquad p_2 = 1.5$

i	p_i	$f(p_i)$
3	1.28785	-1.37624
4	1.23746	0.126941
5	1.24160	0.219440×10^{-2}
6	1.24168	0.257492×10^{-4}
7	1.24168	0.257492×10^{-4}

Table 2.11

$p_0 = 2.5, \qquad p_1 = 2.0, \qquad p_2 = 2.25$

i	p_i	$f(p_i)$
3	1.96059	-0.611255
4	1.97056	0.748825×10^{-2}
5	1.97044	-0.295639×10^{-4}
6	1.97044	-0.295639×10^{-4}

Example 1 illustrates that Müller's method can approximate the roots of polynomials with a variety of starting values. In fact, the importance of Müller's method is that the technique generally converges to the root of a polynomial for any initial approximation choice, although problems can be constructed for which convergence fails for certain choices of initial approximations. General-purpose software packages using Müller's method request only one initial approximation per root and, as an option, may even supply this approximation.

Although Müller's method is not quite as efficient as Newton's method, it is generally better than the Secant method. The relative efficiency, however, is not as important as the ease of implementation and the likelihood that a root will be found. Any of these methods will converge quite rapidly once a reasonable initial approximation is determined.

Once a sufficiently accurate approximation p^* to a root has been found, $f(x)$ is divided by $x - p^*$ to produce what is called a *deflated* equation. If $f(x)$ is a polynomial of degree n, the deflated polynomial will be of degree $n - 1$, so the computations are simplified. After an approximation to the root of the deflated equation has been determined, use either Müller's method or Newton's method in the original polynomial with this root as the initial approximation. This will ensure that the root being approximated is a solution to the true equation, not to the less accurate deflated equation.

EXERCISE SET 2.6

1. Use Newton's method to find approximations to within 10^{-4} to all the real roots of $P(x) = 0$ for the following polynomials:
   (a) $P(x) = x^3 - 2x^2 - 5$
   (b) $P(x) = x^3 + 3x^2 - 1$
   (c) $P(x) = x^3 - x - 1$
   (d) $P(x) = x^4 + 2x^2 - x - 3$
   (e) $P(x) = x^3 + 4.001x^2 + 4.002x + 1.101$
   (f) $P(x) = x^5 - x^4 + 2x^3 - 3x^2 + x - 4$

2. Use Newton's method to find approximations to within 10^{-5} to all the roots of $P(x) = 0$. First find the real roots and then reduce to polynomials of lower degree to determine any complex roots.
   (a) $P(x) = x^4 + 5x^3 - 9x^2 - 85x - 136$
   (b) $P(x) = x^4 - 2x^3 - 12x^2 + 16x - 40$
   (c) $P(x) = x^4 + x^3 + 3x^2 + 2x + 2$
   (d) $P(x) = x^5 + 11x^4 - 21x^3 - 10x^2 - 21x - 5$
   (e) $P(x) = 16x^4 + 88x^3 + 159x^2 + 76x - 240$
   (f) $P(x) = x^4 - 4x^2 - 3x + 5$
   (g) $P(x) = x^4 - 2x^3 - 4x^2 + 4x + 4$
   (h) $P(x) = x^3 - 7x^2 + 14x - 6$

3. Repeat Exercise 1 using Müller's method.

4. Repeat Exercise 2 using Müller's method.

5. Use Müller's method to find, within 10^{-3}, the zeros and critical points of the following functions. Use this information to sketch the graph of f.
   (a) $f(x) = x^3 - 9x^2 + 12$ (b) $f(x) = x^4 - 2x^3 - 5x^2 + 12x - 5$

6. A can in the shape of a right circular cylinder is to be constructed to contain 1000 cm³, as shown. The circular top and bottom of the can must have a radius 0.25 cm more than the radius of the can so that the excess can be used to form a seal with the side. The sheet of material being formed into the side must be 0.25 cm longer than the circumference of the can so that a seal can be formed. Find, to within 10^{-4} cm², the minimal amount of material needed to construct the can.

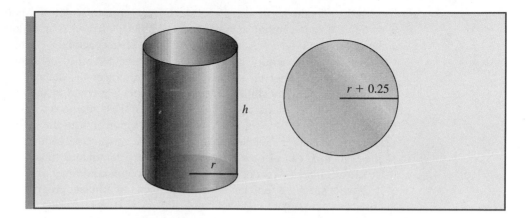

7. Two ladders crisscross an alley of width W, as shown. Each ladder reaches from the base of one wall to some point on the opposite wall. The ladders cross at a height H above the pavement. Given that the lengths of the ladders are $x_1 = 20$ ft and $x_2 = 30$ ft and that $H = 8$ ft, find W.

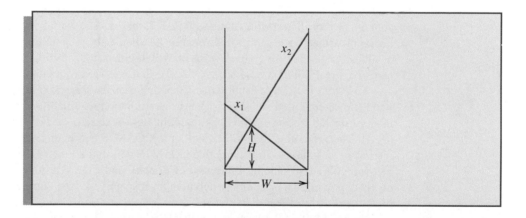

2.7 Survey of Methods and Software

In this chapter we have considered the problem of finding a root of the equation $f(x) = 0$, where f is a given continuous function. All the methods begin with an initial approximation and generate a sequence that converges to a root if the method is successful. If $[a, b]$ is an interval in which $f(a)$ and $f(b)$ are of opposite sign, then the Bisection method and the method of False Position will always converge. However, the convergence of these methods may be slow. Faster convergence is generally obtained using the Secant method or Newton's method. Good initial approximations are required for these methods, two for the Secant method and one for Newton's method, so the Bisection method or method of False Position can be used as starter methods for the Secant or Newton's methods.

Müller's method will also give rapid convergence to a root of a polynomial without a particularly good initial approximation. Müller's method is not quite as efficient as Newton's method; however it is better than the Secant method.

Deflation is generally used with Müller's method once an approximate root has been determined. After an approximation to the root of the deflated equation has been determined, use either Müller's method or Newton's method in the original polynomial with this root as the initial approximation. Use of these methods will ensure that the root being approximated is a solution to the true equation, not to the deflated equation. We recommended Müller's method for finding all the zeros of polynomials, real or complex. Müller's method can also be used for an arbitrary continuous function, but the program MULLER25 must be modified to incorporate complex arithmetic.

Given a specified function f and a tolerance, an efficient program should produce an approximation to one or more solutions of $f(x) = 0$, each having an absolute or relative error within the tolerance, and the results should be generated in a reasonable amount of time. If the program cannot accomplish this task, it should at least give meaningful explanations of why success was not obtained and an indication of how to remedy the cause of failure. Commercial software does precisely this.

The IMSL FORTRAN subroutine ZANLY uses Müller's method with deflation to approximate a number of roots of $f(x) = 0$. The routine ZBREN, due to R. P. Brent, uses a combination of linear interpolation, an inverse quadratic interpolation similar to Müller's method, and the Bisection method. It requires specifying an interval $[a, b]$ that contains a root. The IMSL routine ZREAL is based on a variation of Müller's method and approximates zeros of a real function f when only poor initial approximations are available. Routines for finding zeros of polynomials are ZPORC, ZPLRC for finding zeros of a real polynomial, and ZPOCC for finding zeros of a complex polynomial.

The NAG FORTRAN subroutine CO5ADF uses the Bisection method in conjunction with a method based on inverse linear interpolation, similar to the method of False Position or the Secant method, to approximate a real zero of $f(x) = 0$ in the interval $[a, b]$. The subroutine CO5AZF is similar to CO5ADF but requires a single starting value instead of an interval. The subroutine CO5AGF will find an interval containing a root on its own. NAG also supplies subroutines CO2AEF and CO2ADF to approximate all zeros of a real polynomial or complex polynomial, respectively. The subroutine CO2AGF also finds the roots of a real polynomial.

Within MATLAB, the function ROOTS is used to compute all the roots, both real and complex, of a polynomial. For an arbitrary function, FZERO computes a root within a specified tolerance that is near an initial approximation supplied by the user.

Notice that in spite of the diversity of methods, the professionally written packages are based primarily on the methods and principles discussed in this chapter. You should be able to use these packages by reading the manuals accompanying the packages to understand the parameters and the specifications of the results that are obtained.

CHAPTER 3

Interpolation and Polynomial Approximation

3.1 Introduction

Engineers and scientists commonly assume that relationships between variables in a physical problem have an analytic representation that can be approximately reproduced from data given by the problem. The ultimate goal of the representation might be to determine the values at intermediate points, to approximate the integral or derivative of the underlying function, or to simply give a smooth or continuous representation of the phenomena of interest.

Interpolation refers to the problem of determining a function that exactly represents a collection of data. The most elementary type of interpolation consists of fitting a polynomial to a collection of data points. Polynomials have derivatives and integrals that are themselves polynomials, so they are a natural choice for approximating derivatives and integrals. We will see in this chapter that approximating polynomials are easily constructed, and the result given next implies that the set of polynomials is sufficiently large to allow any continuous function to have an arbitrarily accurate representation.

Weierstrass Approximation Theorem

If f is defined and continuous on $[a, b]$ and $\varepsilon > 0$ is given, then there exists a polynomial P defined on $[a, b]$ with the property that

$$|f(x) - P(x)| < \varepsilon, \quad \text{for all } x \in [a, b].$$

Taylor polynomials were introduced in Section 1.2, where they were described as one of the fundamental building blocks of numerical analysis. Given this prominence, it would be reasonable to assume that polynomial interpolation would make heavy use of these functions. This is not, however, the case. Taylor polynomials agree as closely as possible with a given function at a specific point, but they concentrate their accuracy only near that point. A good interpolation polynomial needs to provide a relatively accurate approximation over an entire interval. Taylor polynomials do not generally satisfy this condition. For example, suppose we calculate the first six Taylor polynomials about $x_0 = 0$ for $f(x) = e^x$. Since the derivatives of f are all e^x, which evaluated at $x_0 = 0$ gives 1, the Taylor polynomials are

$$P_0(x) = 1, \quad P_1(x) = 1 + x, \quad P_2(x) = 1 + x + \frac{x^2}{2}, \quad P_3(x) = 1 + x + \frac{x^2}{2} + \frac{x^3}{6},$$

$$P_4(x) = 1 + x + \frac{x^2}{2} + \frac{x^3}{6} + \frac{x^4}{24}, \quad \text{and} \quad P_5(x) = 1 + x + \frac{x^2}{2} + \frac{x^3}{6} + \frac{x^4}{24} + \frac{x^5}{120}.$$

Table 3.1 lists the values of the Taylor polynomial for various values of x. Notice that even for the higher-degree polynomials, the error becomes progressively worse as we move away from zero.

Table 3.1

x	$P_0(x)$	$P_1(x)$	$P_2(x)$	$P_3(x)$	$P_4(x)$	$P_5(x)$	e^x
-2.0	1.00000	-1.00000	1.00000	-0.33333	0.33333	0.06667	0.13534
-1.5	1.00000	-0.50000	0.62500	0.06250	0.27344	0.21016	0.22313
-1.0	1.00000	0.00000	0.50000	0.33333	0.37500	0.36667	0.36788
-0.5	1.00000	0.50000	0.62500	0.60417	0.60677	0.60651	0.60653
0.0	1.00000	1.00000	1.00000	1.00000	1.00000	1.00000	1.00000
0.5	1.00000	1.50000	1.62500	1.64583	1.64844	1.64870	1.64872
1.0	1.00000	2.00000	2.50000	2.66667	2.70833	2.71667	2.71828
1.5	1.00000	2.50000	3.62500	4.18750	4.39844	4.46172	4.48169
2.0	1.00000	3.00000	5.00000	6.33333	7.00000	7.26667	7.38906

Although better approximations are obtained for this problem if higher-degree Taylor polynomials are used, this is not always true. Consider, as an extreme example, using Taylor polynomials of various degrees for $f(x) = 1/x$ expanded about $x_0 = 1$ to approximate $f(3) = \frac{1}{3}$. Since $f(x) = x^{-1}, f'(x) = -x^{-2}, f'(x) = (-1)^2 2 \cdot x^{-3}$, and, in general, $f^{(n)}(x) = (-1)^n n! x^{-n-1}$, the Taylor polynomials for $n \geq 0$ are

$$P_n(x) = \sum_{k=0}^{n} \frac{f^{(k)}(1)}{k!} (x-1)^k = \sum_{k=0}^{n} (-1)^k (x-1)^k.$$

To approximate $f(3) = \frac{1}{3}$ by $P_n(3)$ for increasing values of n, we obtain the values in Table 3.2—rather a dramatic failure.

Table 3.2

n	0	1	2	3	4	5	6	7
$P_n(3)$	1	-1	3	-5	11	-21	43	-85

Since the Taylor polynomials have the property that all the information used in the approximation is concentrated at the single point x_0, the type of difficulty that occurs here is quite common. This generally limits Taylor polynomial approximation to the situation in which approximations are needed only at points close to x_0. For ordinary computational purposes, it is more efficient to use methods that include information at various points, which we consider in the remainder of this chapter. The primary use of Taylor polynomials in numerical analysis is *not* for approximation purposes, but for the derivation of numerical techniques.

3.2 Lagrange Polynomials

In the previous section we discussed the general unsuitability of Taylor polynomials for approximation. These polynomials are useful only over small intervals for functions whose derivatives exist and are easily evaluated. In this section we find approximating polynomials that can be determined simply by specifying certain points on the plane through which they must pass.

The problem of determining a polynomial of degree 1 that passes through the distinct points (x_0, y_0) and (x_1, y_1) is the same as approximating a function f for which $f(x_0) = y_0$ and $f(x_1) = y_1$ by means of a first-degree polynomial interpolating, or agreeing with, the values of f at the given points. Consider the linear polynomial

$$P(x) = \frac{(x - x_1)}{(x_0 - x_1)} f(x_0) + \frac{(x - x_0)}{(x_1 - x_0)} f(x_1).$$

When $x = x_0$,

$$P(x_0) = 1 \cdot f(x_0) + 0 \cdot f(x_1) = f(x_0) = y_0,$$

and when $x = x_1$,

$$P(x_1) = 0 \cdot f(x_0) + 1 \cdot f(x_1) = f(x_1) = y_1,$$

so P has the required properties. (See Figure 3.1.)

Figure 3.1

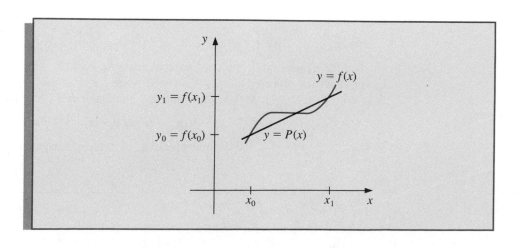

To generalize the concept of linear interpolation to higher-degree polynomials, consider the construction of a polynomial of degree at most n that passes through the $n + 1$ points

$$(x_0, f(x_0)), (x_1, f(x_1)), \ldots, (x_n, f(x_n)).$$

(See Figure 3.2.)

Figure 3.2

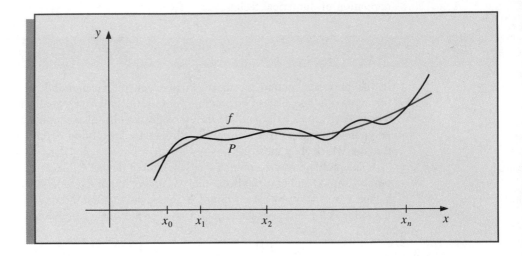

The linear polynomial passing through $(x_0, f(x_0))$ and $(x_1, f(x_1))$ uses the quotients

$$L_0(x) = \frac{(x - x_1)}{(x_0 - x_1)} \quad \text{and} \quad L_1(x) = \frac{(x - x_0)}{(x_1 - x_0)}.$$

This pair of linear functions is used to select the value of f that is needed for the approximation. When

$$x = x_0: \quad L_0(x_0) = 1 \quad \text{and} \quad L_1(x_0) = 0, \quad \text{so} \quad P_1(x_0) = f(x_0),$$

and when

$$x = x_1: \quad L_0(x_1) = 0 \quad \text{and} \quad L_1(x_1) = 1, \quad \text{so} \quad P_1(x_1) = f(x_1).$$

For the general case, we need to construct, for each $k = 0, 1, \ldots, n$, a quotient $L_{n,k}(x)$ with the property that $L_{n,k}(x_i) = 0$ when $i \neq k$ and $L_{n,k}(x_k) = 1$. To satisfy $L_{n,k}(x_i) = 0$ for each $i \neq k$ requires that the numerator of $L_{n,k}$ contain the term

$$(x - x_0)(x - x_1) \cdots (x - x_{k-1})(x - x_{k+1}) \cdots (x - x_n)$$

To satisfy $L_{n,k}(x_k) = 1$, the denominator of $L_{n,k}(x)$ must be equal to this term evaluated at $x = x_k$. Thus,

$$L_{n,k}(x) = \frac{(x - x_0) \cdots (x - x_{k-1})(x - x_{k+1}) \cdots (x - x_n)}{(x_k - x_0) \cdots (x_k - x_{k-1})(x_k - x_{k+1}) \cdots (x_k - x_n)}.$$

A sketch of the graph of a typical $L_{n,k}$ is shown in Figure 3.3.

Figure 3.3

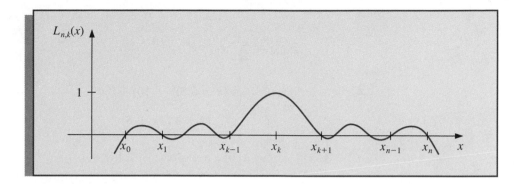

The interpolating polynomial is easily described now that the form of $L_{n,k}$ is known. This polynomial, called the nth Lagrange interpolating polynomial, is defined as follows:

---

nth Lagrange Interpolating Polynomial

$$P_n(x) = f(x_0)L_{n,0}(x) + \cdots + f(x_n)L_{n,n}(x) = \sum_{k=0}^{n} f(x_k)L_{n,k}(x),$$

where

$$L_{n,k}(x) = \frac{(x - x_0)(x - x_1)\cdots(x - x_{k-1})(x - x_{k+1})\cdots(x - x_n)}{(x_k - x_0)(x_k - x_1)\cdots(x_k - x_{k-1})(x_k - x_{k+1})\cdots(x_k - x_n)}$$

for each $k = 0, 1, \ldots, n$.

If $x_0, x_1, \ldots, x_n$ are $(n + 1)$ distinct numbers and f is a function whose values are given at these numbers, then $P_n(x)$ is the unique polynomial of degree at most n that agrees with $f(x)$ at $x_0, x_1, \ldots x_n$. The notation for describing the Lagrange interpolation polynomial $P_n(x)$ is rather complicated. To reduce this somewhat, we will write $L_{n,k}(x)$ simply as $L_k(x)$ when there is no confusion as to its degree.

EXAMPLE 1   Using the numbers, or *nodes*, $x_0 = 2$, $x_1 = 2.5$, and $x_2 = 4$ to find the second interpolating polynomial for $f(x) = 1/x$ requires that we first determine the coefficient polynomials L_0, L_1, and L_2. In nested form they are

$$L_0(x) = \frac{(x - 2.5)(x - 4)}{(2 - 2.5)(2 - 4)} = (x - 6.5)x + 10,$$

$$L_1(x) = \frac{(x - 2)(x - 4)}{(2.5 - 2)(2.5 - 4)} = \frac{(-4x + 24)x - 32}{3},$$

and

$$L_2(x) = \frac{(x - 2)(x - 2.5)}{(4 - 2)(4 - 2.5)} = \frac{(x - 4.5)x + 5}{3}.$$

Since $f(x_0) = f(2) = 0.5$, $f(x_1) = f(2.5) = 0.4$, and $f(x_2) = f(4) = 0.25$,

$$P_2(x) = \sum_{k=0}^{2} f(x_k)L_k(x)$$

$$= 0.5((x - 6.5)x + 10) + 0.4\frac{(-4x + 24)x - 32}{3}$$

$$+ 0.25\frac{(x - 4.5)x + 5}{3}$$

$$= (0.05x - 0.425)x + 1.15.$$

An approximation to $f(3) = \frac{1}{3}$ is

$$f(3) \approx P(3) = 0.325.$$

Compare this to Table 3.2 on page 52, where no Taylor polynomial expanded about $x_0 = 1$ can be used to reasonably approximate $f(3) = \frac{1}{3}$. (See Figure 3.4.)   ■ ■ ■

Figure 3.4

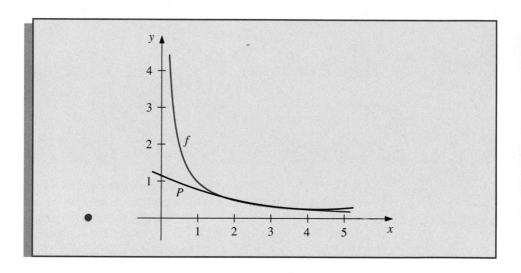

The Lagrange polynomials have remainder terms that are reminiscent of those for the Taylor polynomials. The nth Taylor polynomial about x_0 concentrates all the known information at x_0 and has an error term of the form

$$\frac{f^{(n+1)}(\xi(x))}{(n + 1)!}(x - x_0)^{n+1},$$

where $\xi(x)$ is between x and x_0. The nth Lagrange polynomial uses information at the distinct numbers $x_0, x_1, \ldots, x_n$. In place of $(x - x_0)^{n+1}$ its error formula uses a product of the $n + 1$ terms $(x - x_0), (x - x_1), \ldots, (x - x_n)$, and the number $\xi(x)$ can lie anywhere in the interval that contains the points $x, x_0, x_1, \ldots, x_n$. Otherwise it has the same form as the error formula for the Taylor polynomials.

Lagrange Polynomial Error Formula

$$f(x) = P_n(x) + \frac{f^{(n+1)}(\xi(x))}{(n+1)!}(x - x_0)(x - x_1)\cdots(x - x_n),$$

for some $\xi(x)$ between $x_0, x_1, \ldots, x_n$ and x.

This error formula is an important theoretical result because Lagrange polynomials are used extensively for deriving numerical differentiation and integration methods. Error bounds for these techniques are obtained from the Lagrange error formula. The specific use of this error formula, however, is restricted to those functions whose derivatives have known bounds. The next example illustrates interpolation techniques for a situation in which the Lagrange error formula cannot be used. This example also serves to illustrate that we should look for a more efficient way to obtain approximations via interpolation.

EXAMPLE 2 Table 3.3 lists values of a function at various points. The approximations to $f(1.5)$ obtained by various Lagrange polynomials will be compared.

Table 3.3

x	$f(x)$
1.0	0.7651977
1.3	0.6200860
1.6	0.4554022
1.9	0.2818186
2.2	0.1103623

Since 1.5 is between 1.3 and 1.6, the most appropriate linear polynomial uses $x_0 = 1.3$ and $x_1 = 1.6$. The value of the interpolating polynomial at 1.5 is

$$P_1(1.5) = \frac{(1.5 - 1.6)}{(1.3 - 1.6)}(0.6200860) + \frac{(1.5 - 1.3)}{(1.6 - 1.3)}(0.4554022) = 0.5102968.$$

Two polynomials of degree two could reasonably be used, one by letting $x_0 = 1.3$, $x_1 = 1.6$, and $x_2 = 1.9$, which gives

$$P_2(1.5) = \frac{(1.5 - 1.6)(1.5 - 1.9)}{(1.3 - 1.6)(1.3 - 1.9)}(0.6200860) + \frac{(1.5 - 1.3)(1.5 - 1.9)}{(1.6 - 1.3)(1.6 - 1.9)}(0.4554022)$$

$$+ \frac{(1.5 - 1.3)(1.5 - 1.6)}{(1.9 - 1.3)(1.9 - 1.6)}(0.2818186)$$

$$= 0.5112857,$$

and the other by letting $x_0 = 1.0$, $x_1 = 1.3$, and $x_2 = 1.6$, in which case

$$\hat{P}_2(1.5) = 0.5124715.$$

In the third-degree case there are also two reasonable choices for the polynomial. One uses $x_0 = 1.3$, $x_1 = 1.6$, $x_2 = 1.9$, and $x_4 = 2.2$, which gives $P_3(1.5) = 0.5118302$. The other is obtained by letting $x_0 = 1.0$, $x_1 = 1.3$, $x_2 = 1.6$, and $x_3 = 1.9$, giving $\hat{P}_3(1.5) = 0.5118127$.

The fourth Lagrange polynomial uses all the entries in the table. With $x_0 = 1.0$, $x_1 = 1.3$, $x_2 = 1.6$, $x_3 = 1.9$, and $x_4 = 2.2$, the approximation is $P_4(1.5) = 0.5118200$.

Since $P_3(1.5)$, $\hat{P}_3(1.5)$, and $P_4(1.5)$ all agree to within 2×10^{-5} units, we expect $P_4(1.5)$ to be the most accurate approximation and to be correct to within 2×10^{-5} units. The actual value of $f(1.5)$ is known to be 0.5118277, so the true accuracies of the approximations are as follows:

$$|P_1(1.5) - f(1.5)| \approx 1.53 \times 10^{-3}, \qquad |P_2(1.5) - f(1.5)| \approx 5.42 \times 10^{-4},$$
$$|\hat{P}_2(1.5) - f(1.5)| \approx 6.44 \times 10^{-4}, \qquad |P_3(1.5) - f(1.5)| \approx 2.5 \times 10^{-6},$$
$$|\hat{P}_3(1.5) - f(1.5)| \approx 1.50 \times 10^{-5}, \qquad |P_4(1.5) - f(1.5)| \approx 7.7 \times 10^{-6}.$$

Although P_3 is the most accurate approximation, if we had no knowledge of the actual value of $f(1.5)$, P_4 would be accepted as the best approximation, since it includes the most data about the function. The Lagrange error term cannot be applied here, since no knowledge of the fourth derivative of f is available. Unfortunately, this is generally the case. ■ ■ ■

A practical difficulty with Lagrange interpolation is that, since the error term is difficult to apply, the degree of the polynomial needed for the desired accuracy is generally not known until the computations are determined. The usual practice is to compute the results given from various polynomials until appropriate agreement is obtained, as was done in the previous example. However, the work done in calculating the approximation by the second polynomial does not lessen the work needed to calculate the third approximation; nor is the fourth approximation easier to obtain once the third approximation is known, and so on. We will now derive these approximating polynomials in a manner that uses the previous calculations to advantage. First we need some new notation.

Let f be a function defined at $x_0, x_1, x_2, \ldots, x_n$ and suppose that $m_1, m_2, \ldots, m_k$ are k distinct integers with $0 \leq m_i \leq n$ for each i. The Lagrange polynomial that agrees with $f(x)$ at the k points $x_{m_1}, x_{m_2}, \ldots, x_{m_k}$ is denoted $P_{m_1,m_2,\ldots,m_k}(x)$.

EXAMPLE 3 Let $x_0 = 1$, $x_1 = 2$, $x_2 = 3$, $x_3 = 4$, $x_4 = 6$, and $f(x) = x^3$. Then $P_{1,2,4}(x)$ is the polynomial that agrees with $f(x)$ at $x_1 = 2$, $x_2 = 3$, and $x_4 = 6$, and

$$P_{1,2,4}(x) = \frac{(x-3)(x-6)}{(2-3)(2-6)}(8) + \frac{(x-2)(2-6)}{(3-2)(3-6)}(27) + \frac{(x-2)(x-3)}{(6-2)(6-3)}(216).$$

■ ■ ■

The next result describes a method for recursively generating Lagrange polynomial approximations.

Recursively Generated Lagrange Polynomials

Let f be defined at $x_0, x_1, \ldots, x_k$ and x_j, x_i be two distinct numbers in this set. If

$$P(x) = \frac{(x - x_j)P_{0,1,\ldots,j-1,j+1,\ldots,k}(x) - (x - x_i)P_{0,1,\ldots,i-1,i+1,\ldots,k}(x)}{(x_i - x_j)},$$

then P is the kth Lagrange polynomial that interpolates f at the $k + 1$ points $x_0, x_1, \ldots, x_k$.

To see why this is true, first let $Q \equiv P_{0,1,\ldots,i-1,i+1,\ldots,k}$ and $\hat{Q} \equiv P_{0,1,\ldots,j-1,j+1,\ldots,k}$. Since $Q(x)$ and $\hat{Q}(x)$ are polynomials of degree $k - 1$ or less, $P(x)$ must be of degree at most k. If $0 \le r \le k$ and $r \ne i, j$, then $Q(x_r) = \hat{Q}(x_r) = f(x_r)$, so

$$P(x_r) = \frac{(x_r - x_j)\hat{Q}(x_r) - (x_r - x_i)Q(x_r)}{x_i - x_j} = \frac{(x_i - x_j)}{(x_i - x_j)}f(x_r) = f(x_r).$$

Moreover,

$$P(x_i) = \frac{(x_i - x_j)\hat{Q}(x_i) - (x_i - x_i)Q(x_i)}{x_i - x_j} = \frac{(x_i - x_j)}{(x_i - x_j)}f(x_i) = f(x_i);$$

similarly, $P(x_j) = f(x_j)$. But there is only one polynomial of degree at most k that agrees with $f(x)$ at $x_0, x_1, \ldots, x_k$ and this polynomial by definition is $P_{0,1,\ldots,k}(x)$.

This result implies that the approximations from the interpolating polynomials can be generated recursively in the manner shown in Table 3.4. The row-by-row generation is performed to move across the rows as rapidly as possible, since these entries are given by successively higher-degree interpolating polynomials.

Table 3.4

x_0	$P_0 = Q_{0,0}$				
x_1	$P_1 = Q_{1,0}$	$P_{0,1} = Q_{1,1}$			
x_2	$P_2 = Q_{2,0}$	$P_{1,2} = Q_{2,1}$	$P_{0,1,2} = Q_{2,2}$		
x_3	$P_3 = Q_{3,0}$	$P_{2,3} = Q_{3,1}$	$P_{1,2,3} = Q_{3,2}$	$P_{0,1,2,3} = Q_{3,3}$	
x_4	$P_4 = Q_{4,0}$	$P_{3,4} = Q_{4,1}$	$P_{2,3,4} = Q_{4,2}$	$P_{1,2,3,4} = Q_{4,3}$	$P_{0,1,2,3,4} = Q_{4,4}$

This procedure is called **Neville's method** and is implemented with the program NEVLLE31. The P notation used in Table 3.4 is cumbersome because of the number of subscripts used to represent the entries. Note, however, that as an array is being constructed, only two subscripts are needed. Proceeding down the table corresponds to using consecutive points x_i with larger i, and proceeding to the right corresponds to increasing the degree of the interpolating polynomial. Since the points appear consecutively in each entry, we need to describe only a starting point and the number of additional points used in constructing the approximation. To avoid the cumbersome subscripts we let $Q_{i,j}(x)$, for $0 \le i \le j$, denote the jth interpolating polynomial on the $(j + 1)$ numbers $x_{i-j}, x_{i-j+1}, \ldots, x_{i-1}, x_i$; that is,

$$Q_{i,j} = P_{i-j,i-j+1,\ldots,i-1,i}.$$

Using this notation for Neville's method provides the Q notation in Table 3.4.

EXAMPLE 4   In Example 2, values of various interpolating polynomials at $x = 1.5$ were obtained using the data shown in the first two columns of Table 3.3. In this example, we use Neville's method to calculate the approximation to $f(1.5)$. If $x_0 = 1.0$, $x_1 = 1.3$, $x_2 = 1.6$, $x_3 = 1.9$, and $x_4 = 2.2$, then $f(1.0) = Q_{0,0}$, $f(1.3) = Q_{1,0}$, $f(1.6) = Q_{2,0}$, $f(1.9) = Q_{3,0}$, and $f(2.2) = Q_{4,0}$; so these are the five polynomials of degree zero (constants) that approximate $f(1.5)$. Calculating the approximation $Q_{1,1}(1.5)$ gives

$$Q_{1,1}(1.5) = \frac{(1.5 - 1.0)Q_{1,0} - (1.5 - 1.3)Q_{0,0}}{(1.3 - 1.0)}$$

$$= \frac{0.5(0.6200860) - 0.2(0.7651977)}{0.3} = 0.5233449.$$

Similarly,

$$Q_{2,1}(1.5) = \frac{(1.5 - 1.3)(0.4554022) - (1.5 - 1.6)(0.6200860)}{(1.6 - 1.3)} = 0.5102968,$$

$$Q_{3,1}(1.5) = 0.5132634, \qquad \text{and} \qquad Q_{4,1}(1.5) = 0.5104270.$$

The best linear approximation is expected to be $Q_{2,1}$, since 1.5 is between $x_1 = 1.3$ and $x_2 = 1.6$.

In a similar manner, the approximations using quadratic polynomials are given by

$$Q_{2,2}(1.5) = \frac{(1.5 - 1.0)(0.5102968) - (1.5 - 1.6)(0.5233449)}{(1.6 - 1.0)} = 0.5124715,$$

$$Q_{3,2}(1.5) = 0.5112857, \qquad \text{and} \qquad Q_{4,2}(1.5) = 0.5137361.$$

The higher-degree approximations are generated in a similar manner and are shown in Table 3.5.   ■ ■ ■

Table 3.5

1.0	0.7651977			
1.3	0.6200860	0.5233449		
1.6	0.4554022	0.5102968	0.5124715	
1.9	0.2818186	0.5132634	0.5112857	0.5118127
2.2	0.1103623	0.5104270	0.5137361	0.5118302	0.5118200

If the latest approximation, $Q_{4,4}$, is not as accurate as desired, another node x_5 can be selected and another row can be added to the table:

$$x_5 \qquad Q_{5,0} \qquad Q_{5,1} \qquad Q_{5,2} \qquad Q_{5,3} \qquad Q_{5,4} \qquad Q_{5,5},$$

Now $Q_{4,4}$, $Q_{5,4}$, and $Q_{5,5}$ can be compared to determine further accuracy.

In Example 4, the function is the Bessel function of the first kind of order zero, whose value at 2.5 is -0.0483838. Using this we can construct a new row of approximations to $f(1.5)$:

2.5 -0.0483838 0.4807699   0.5301984   0.5119070   0.5118430   0.5118277.

The final new entry is correct to seven decimal places.

EXERCISE SET 3.2

1. Use appropriate Lagrange interpolating polynomials of degree one, two, three, and four to approximate each of the following:

   (a) $f(8.4)$ if $f(8) = 16.63553$, $f(8.1) = 17.61549$, $f(8.3) = 17.56492$, $f(8.6) = 18.50515$, $f(8.7) = 18.82091$

   (b) $f(-\frac{1}{3})$ if $f(-1) = 0.1$, $f(-0.75) = -0.0718125$, $f(-0.5) = -0.02475000$, $f(-0.25) = 0.33493750$, $f(0) = 1.10100000$

   (c) $f(0.25)$ if $f(0) = -1$, $f(0.1) = -0.62049958$, $f(0.2) = -0.28398668$, $f(0.3) = 0.00660095$, $f(0.4) = 0.24842440$

   (d) $f(0.9)$ if $f(0.5) = -0.34409873$, $f(0.6) = -0.17694460$, $f(0.7) = 0.01375227$, $f(0.8) = 0.22363362$, $f(1.0) = 0.65809197$

   (e) $f(\pi)$ if $f(2.9) = -4.827866$, $f(3.0) = -4.240058$, $f(3.1) = -3.496909$, $f(3.2) = -2.596792$, $f(3.4) = -0.3330587$

   (f) $f(\pi/2)$ if $f(1.1) = 1.964760$, $f(1.2) = 2.572152$, $f(1.3) = 3.602102$, $f(1.4) = 5.797884$, $f(1.5) = 14.10142$

   (g) $f(1.15)$ if $f(1) = 1.684370$, $f(1.1) = 1.949477$, $f(1.2) = 2.199796$, $f(1.3) = 2.439189$, $f(1.4) = 2.670324$

   (h) $f(4.1)$ if $f(3.6) = 1.16164956$, $f(3.8) = 0.80201036$, $f(4) = 0.30663842$, $f(4.2) = -0.35916618$, $f(4.4) = -1.23926000$

2. Use Neville's method to obtain the approximations for Exercise 1.

3. Use Neville's method with the function $f(x) = 3^x$ and the values $x_0 = -2$, $x_1 = -1$, $x_2 = 0$, $x_3 = 1$, and $x_4 = 2$ to approximate $\sqrt{3}$.

4. Use Neville's method with the function $f(x) = x^{1/2}$ and the values $x_0 = 0$, $x_1 = 1$, $x_2 = 2$, $x_3 = 4$, and $x_4 = 5$ to approximate $\sqrt{3}$. Compare this result with the result in Exercise 3.

5. The data for Exercise 1, parts (a), (b), (c), and (d), were generated using the following functions. Use the error formula to find a bound for the error and compare the bound to the actual error.

   (a) $f(x) = x \ln x$

   (b) $f(x) = x^3 + 4.001x^2 + 4.002x + 1.101$

   (c) $f(x) = x \cos x - 2x^2 + 3x - 1$

   (d) $f(x) = \sin(e^x - 2)$;   use $n = 1$ and $n = 2$ only.

6. Use the following values and four-digit rounding arithmetic to construct a third Lagrange polynomial approximation to $f(1.09)$. The function being approximated is $f(x) = \log_{10} \tan x$. Use this knowledge to find a bound for the error in the approximation.

$$f(1.00) = 0.1924 \qquad f(1.05) = 0.2414$$
$$f(1.10) = 0.2933 \qquad f(1.15) = 0.3492$$

7. Use Neville's method and four-digit rounding arithmetic to approximate $f(1.09)$ for the data given in Exercise 6.

8. Use the third Lagrange interpolating polynomial and four-digit chopping arithmetic to approximate $\cos 0.750$ using the following values. Find an error bound for the approximation.

$$\cos 0.698 = 0.7661 \qquad \cos 0.768 = 0.7193$$
$$\cos 0.733 = 0.7432 \qquad \cos 0.803 = 0.6946$$

The actual value of $\cos 0.750$ is 0.7317 (to four decimal places). If there is a discrepancy between the actual error and your error bound, explain why this occurred.

9. (a) Use Neville's method to approximate $f(1.03)$ with $P_{0,1,2}$ for the function $f(x) = 3xe^x - e^{2x}$ using $x_0 = 1$, $x_1 = 1.05$, and $x_2 = 1.07$.
    (b) Suppose the approximation of part (a) is not sufficiently accurate. Compute $P_{0,1,2,3}$ where $x_3 = 1.04$.

10. Repeat Exercise 9 using four-digit arithmetic. Do you think Neville's method is sensitive to round-off error?

11. Construct the Lagrange interpolating polynomials for the following functions and find a bound for the absolute error on the interval $[x_0, x_n]$:
    (a) $f(x) = e^{2x} \cos 3x$, $x_0 = 0$, $x_1 = 0.3$, $x_2 = 0.6$, $n = 2$
    (b) $f(x) = \sin(\ln x)$, $x_0 = 2.0$, $x_1 = 2.4$, $x_2 = 2.6$, $n = 2$
    (c) $f(x) = \ln x$, $x_0 = 1$, $x_1 = 1.1$, $x_2 = 1.3$, $x_3 = 1.4$, $n = 3$
    (d) $f(x) = \cos x + \sin x$, $x_0 = 0$, $x_1 = 0.25$, $x_2 = 0.5$, $x_3 = 1.0$, $n = 3$

12. Let $f(x) = e^x, 0 \le x \le 2$. Using the values given, perform the following computation:
    (a) Approximate $f(0.25)$ using linear interpolation with $x_0 = 0$ and $x_1 = 0.5$.
    (b) Approximate $f(0.75)$ using linear interpolation with $x_0 = 0.5$ and $x_1 = 1$.
    (c) Approximate $f(0.25)$ and $f(0.75)$ by using the second interpolating polynomial with $x_0 = 0$, $x_1 = 1$, and $x_2 = 2$.
    (d) Which approximations are better and why?

x	0.0	0.5	1.0	2.0
$f(x)$	1.00000	1.64872	2.71828	7.38906

13. Construct a sequence of interpolating values, $\{y_n\}$, to $f(1 + \sqrt{10})$, where $f(x) = (1+x^2)^{-1}$ for $-5 \le x \le 5$, as follows: For each $n = 1, 2, \ldots, 10$, let $h = 10/n$ and $y_n = P_n(1 + \sqrt{10})$, where $P_n(x)$ is the interpolating polynomial for $f(x)$ at the nodes $x_j^{(n)} = -5 + jh$ for each $j = 0, 1, 2, \ldots, n$. Does the sequence $\{y_n\}$ appear to converge to $f(1 + \sqrt{10})$?

14. The following table lists the population of the United States from 1940 to 1990:

Year	1940	1950	1960	1970	1980	1990
Population (in thousands)	132, 165	151, 326	179, 323	203, 302	226, 542	249, 633

Find the Lagrange polynomial of degree 5 fitting this data, and use this polynomial to estimate the population in the years 1930, 1965, and 2000. The population in 1930 was approximately 123,203,000. How accurate do you think your 1965 and 2000 figures are?

3.3 Divided Differences

Iterated interpolation was used in the previous section to generate successively higher-degree polynomial approximations at a specific point. Divided-difference methods introduced in this section are used to successively generate the polynomials themselves.

We first need to introduce the divided-difference notation, which should remind you of the Aitken's Δ^2 notation used in Section 2.5. The **zeroth divided difference** of the function f with respect to x_i, $f[x_i]$, is simply the value of f at x_i:

$$f[x_i] = f(x_i).$$

The remaining divided differences are defined inductively. The **first divided difference** of f with respect to x_i and x_{i+1} is denoted $f[x_i, x_{i+1}]$ and is defined as

$$f[x_i, x_{i+1}] = \frac{f[x_{i+1}] - f[x_i]}{x_{i+1} - x_i}.$$

After the $(k-1)$st divided differences

$$f[x_i, x_{i+1}, x_{i+2}, \ldots, x_{i+k-1}] \qquad \text{and} \qquad f[x_{i+1}, x_{i+2}, \ldots, x_{i+k-1}, x_{i+k}]$$

have been determined, the **kth divided difference** relative to $x_i, x_{i+1}, x_{i+2}, \ldots, x_{i+k}$ is given by

$$f[x_i, x_{i+1}, \ldots, x_{i+k-1}, x_{i+k}] = \frac{f[x_{i+1}, x_{i+2}, \ldots, x_{i+k}] - f[x_i, x_{i+1}, \ldots, x_{i+k-1}]}{x_{i+k} - x_i}.$$

With this notation, it can be shown that

$$P_n(x) = f[x_0] + f[x_0, x_1](x - x_0)$$
$$+ f[x_0, x_1, x_2](x - x_0)(x - x_1) + \cdots$$
$$+ f[x_0, x_1, \ldots, x_n](x - x_0)(x - x_1) \cdots (x - x_{n-1}).$$

In simplified form we have the following:

Newton's Interpolatory Divided-Difference Formula

$$P(x) = f[x_0] + \sum_{k=1}^{n} f[x_0, x_1, \ldots, x_k](x - x_0) \cdots (x - x_{k-1}).$$

The determination of the divided differences from tabulated data points is outlined in Table 3.6. Two fourth differences and one fifth difference could also be determined from these data, but these differences have not been recorded in the table.

The program DIVDIF32 computes the interpolating polynomial for f at $x_0, x_1, \ldots, x_n$ by the Newton divided-difference procedure. The form of the output can be modified to produce all the divided differences, as is done in Example 1.

EXAMPLE 1   In the previous section we approximated the value at 1.5 given the data shown in the second and third columns of Table 3.7. The remaining entries of this table contain the divided differences computed using the program DIVDIF32.

Table 3.6

x	$f(x)$	First Divided Differences	Second Divided Differences	Third Divided Differences
x_0	$f[x_0]$			
		$f[x_0, x_1] = \frac{f[x_1]-f[x_0]}{x_1-x_0}$		
x_1	$f[x_1]$		$f[x_0, x_1, x_2] = \frac{f[x_1,x_2]-f[x_0,x_1]}{x_2-x_0}$	
		$f[x_1, x_2] = \frac{f[x_2]-f[x_1]}{x_2-x_1}$		$f[x_0, x_1, x_2, x_3] = \frac{f[x_1,x_2,x_3]-f[x_0,x_1,x_2]}{x_3-x_0}$
x_2	$f[x_2]$		$f[x_1, x_2, x_3] = \frac{f[x_2,x_3]-f[x_1,x_2]}{x_3-x_1}$	
		$f[x_2, x_3] = \frac{f[x_3]-f[x_2]}{x_3-x_2}$		$f[x_1, x_2, x_3, x_4] = \frac{f[x_2,x_3,x_4]-f[x_1,x_2,x_3]}{x_4-x_1}$
x_3	$f[x_3]$		$f[x_2, x_3, x_4] = \frac{f[x_3,x_4]-f[x_2,x_3]}{x_4-x_2}$	
		$f[x_3, x_4] = \frac{f[x_4]-f[x_3]}{x_4-x_3}$		$f[x_2, x_3, x_4, x_5] = \frac{f[x_3,x_4,x_5]-f[x_2,x_3,x_4]}{x_5-x_2}$
x_4	$f[x_4]$		$f[x_3, x_4, x_5] = \frac{f[x_4,x_5]-f[x_3,x_4]}{x_5-x_3}$	
		$f[x_4, x_5] = \frac{f[x_5]-f[x_4]}{x_5-x_4}$		
x_5	$f[x_5]$			

Table 3.7

i	x_i	$f[x_i]$	$f[x_{i-1}, x_i]$	$f[x_{i-2}, x_{i-1}, x_i]$	$f[x_{i-3}, \dots, x_i]$	$f[x_{i-4}, \dots, x_i]$
0	1.0	0.7651977				
			-0.4837057			
1	1.3	0.6200860		-0.1087339		
			-0.5489460		0.0658784	
2	1.6	0.4554022		-0.0494433		0.0018251
			-0.5786120		0.0680685	
3	1.9	0.2818186		0.0118183		
			-0.5715210			
4	2.2	0.1103623				

The coefficients of the Newton forward divided-difference form of the interpolatory polynomial are along the upper diagonal in the table. The polynomial is

$$P_4(x) = 0.7651977 - 0.4837057(x - 1.0) - 0.1087339(x - 1.0)(x - 1.3)$$
$$+ 0.0658784(x - 1.0)(x - 1.3)(x - 1.6)$$
$$+ 0.0018251(x - 1.0)(x - 1.3)(x - 1.6)(x - 1.9).$$

It is easily verified that $P_4(1.5) = 0.5118200$, which agrees with the result determined in Example 4 of Section 3.2.   ■ · ■ ■

When $x_0, x_1, \dots, x_n$ are arranged consecutively with equal spacing, Newton's interpolatory divided-difference formula has a simpler form. We introduce the notation $h = x_{i+1} - x_i$ for each $i = 0, 1, \dots, n - 1$ and $x = x_0 + sh$. Then the difference

$x - x_i$ can be written as $x - x_i = (s - i)h$, and the formula becomes

$$P_n(x) = P_n(x_0 + sh) = f[x_0] + shf[x_0, x_1] + s(s - 1)h^2 f[x_0, x_1, x_2]$$
$$+ \cdots + s(s - 1) \cdots (s - n + 1)h^n f[x_0, x_1, \ldots, x_n]$$
$$= \sum_{k=0}^{n} s(s - 1) \cdots (s - k + 1)h^k f[x_0, x_1, \ldots, x_k].$$

Using a generalization of the binomial-coefficient notation

$$\binom{s}{k} = \frac{s(s - 1) \cdots (s - k + 1)}{k!},$$

where s need not be an integer, we can express $P_n(x)$ compactly as the **Newton forward divide-difference formula**.

Newton Forward Divided-Difference Formula

$$P_n(x) = P_n(x_0 + sh) = \sum_{k=0}^{n} \binom{s}{k} k! h^k f[x_0, x_1, \ldots, x_k].$$

Another form is constructed by making use of the forward-difference notation introduced in Section 2.5. With this notation

$$f[x_0, x_1] = \frac{f(x_1) - f(x_0)}{x_1 - x_0} = \frac{1}{h}\Delta f(x_0),$$

$$f[x_0, x_1, x_2] = \frac{1}{2h}\left[\frac{\Delta f(x_1) - \Delta f(x_0)}{h}\right] = \frac{1}{2h^2}\Delta^2 f(x_0),$$

and, in general,

$$f[x_0, x_1, \ldots, x_k] = \frac{1}{k!h^k}\Delta^k f(x_0).$$

Consequently, the Newton forward divided-difference formula can also be written as follows:

Newton Forward-Difference Formula

$$P_n(x) = \sum_{k=0}^{n} \binom{s}{k} \Delta^k f(x_0).$$

If the interpolating nodes are reordered as $x_n, x_{n-1}, \ldots, x_0$, we can write the interpolatory formula as

$$P_n(x) = f[x_n] + f[x_{n-1}, x_n](x - x_n) + f[x_{n-2}, x_{n-1}, x_n](x - x_n)(x - x_{n-1})$$
$$+ \cdots + f[x_0, \ldots, x_n](x - x_n)(x - x_{n-1}) \cdots (x - x_1).$$

Using equal spacing with $x = x_n + sh$ and $x = x_i + (s + n - i)h$ produces

$$P_n(x) = P_n(s_n + sh)$$
$$= f[x_n] + shf[x_{n-1}, x_n] + s(s + 1)h^2 f[x_{n-2}, x_{n-1}, x_n] + \cdots$$
$$+ s(s + 1) \cdots (s + n - 1)h^n f[x_0, x_1, \ldots, x_n].$$

This form is called the **Newton backward divided-difference formula**. It is used to derive a more commonly applied formula known as the **Newton backward-difference formula**. To discuss this formula, we first need to introduce some notation.

Given the sequence $\{p_n\}_{n=0}^{\infty}$, the **backward difference** ∇p_n is defined by

$$\nabla p_n \equiv p_n - p_{n-1} \qquad \text{for } n \geq 1$$

and higher powers are defined recursively by

$$\nabla^k p_n = \nabla(\nabla^{k-1} p_n) \qquad \text{for } k \geq 2.$$

This implies that

$$f[x_{n-1}, x_n] = \frac{1}{h}\nabla f(x_n), \qquad f[x_{n-2}, x_{n-1}, x_n] = \frac{1}{2h^2}\nabla^2 f(x_n),$$

and, in general,

$$f[x_{n-k}, \ldots, x_{n-1}, x_n] = \frac{1}{k!h^k}\nabla^k f(x_n).$$

Consequently,

$$P_n(x) = f[x_n] + s\nabla f(x_n) + \frac{s(s + 1)}{2}\nabla^2 f(x_n) + \cdots$$
$$+ \frac{s(s + 1) \cdots (s + n - 1)}{n!}\nabla^n f(x_n).$$

Extending the binomial coefficient notation to include all real values of s, we let

$$\binom{-s}{k} = \frac{-s(-s - 1) \cdots (-s - k + 1)}{k!}$$

so

$$\binom{-s}{k} = (-1)^k \frac{s(s + 1) \cdots (s + k - 1)}{k!}$$

and

$$P_n(x) = f(x_n) + (-1)^1 \binom{-s}{1}\nabla f(x_n) + (-1)^2 \binom{-s}{2}\nabla^2 f(x_n) + \cdots + (-1)^n \binom{-s}{n}\nabla^n f(x_n)$$

to obtain the following result:

Newton Backward-Difference Formula

$$P_n(x) = \sum_{k=0}^{n}(-1)^k \binom{-s}{k}\nabla^k f(x_n).$$

EXAMPLE 2 Consider the table of data given in Example 1 and reproduced in the first two columns of Table 3.8. If an approximation to $f(1.1)$ is required, the reasonable choice for $x_0, x_1, \ldots, x_4$ would be $x_0 = 1.0, x_1 = 1.3, x_2 = 1.6, x_3 = 1.9$, and $x_4 = 2.2$, since this choice makes the greatest possible use of the data points closest to $x = 1.1$ and also makes use of the fourth divided difference. This implies that $h = 0.3$ and $s = 1/3$, so the Newton forward divided-difference formula is used with the divided differences that have a solid underline in Table 3.8.

Table 3.8

		First Divided Differences	Second Divided Differences	Third Divided Differences	Fourth Divided Differences
1.0	0.7651977				
		−0.4837057			
1.3	0.6200860		−0.1087339		
		−0.5489460		0.0658784	
1.6	0.4554022		−0.0494433		0.0018251
		−0.5786120		0.0680685	
1.9	0.2818186		0.0118183		
		-0.5715210			
2.2	0.1103623				

$$P_4(1.1) = P_4(1.0 + \tfrac{1}{3}(0.3))$$
$$= 0.7651997 + \tfrac{1}{3}(0.3)(-0.4837057) + \tfrac{1}{3}\left(-\tfrac{2}{3}\right)(0.3)^2(-0.1087339)$$
$$+ \tfrac{1}{3}\left(-\tfrac{2}{3}\right)\left(-\tfrac{5}{3}\right)(0.3)^3(0.0658784) + \tfrac{1}{3}\left(-\tfrac{2}{3}\right)\left(-\tfrac{5}{3}\right)\left(-\tfrac{8}{3}\right)(0.3)^4(0.0018251)$$
$$= 0.7196480.$$

To approximate a value when x is close to the end of the tabulated values, say, $x = 2.0$, we would again like to make maximum use of the data points closest to x. This requires using the Newton backward divided-difference formula with $s = -\tfrac{2}{3}$ and the divided differences in Table 3.8 that have a dashed underline:

$$P_4(2.0) = P_4(2.2 - \tfrac{2}{3}(0.3))$$
$$= 0.1103623 - \tfrac{2}{3}(0.3)(-0.5715210) - \tfrac{2}{3}\left(\tfrac{1}{3}\right)(0.3)^2(0.0118183)$$
$$- \tfrac{2}{3}\left(\tfrac{1}{3}\right)\left(\tfrac{4}{3}\right)(0.3)^3(0.0680685) - \tfrac{2}{3}\left(\tfrac{1}{3}\right)\left(\tfrac{4}{3}\right)\left(\tfrac{7}{3}\right)(0.3)^4(0.0018251)$$
$$= 0.2238754.$$ ■ ■ ■

The Newton formulas are not appropriate for approximating a value, x, that lies near the center of the table, since employing either the backward or forward method in such a way that the highest-order difference is involved will not allow x_0 to be close to x. A number of divided-difference formulas are available in this instance, each of which has situations when it can be used to maximum advantage. These methods are known as *centered-difference formulas*. There are a number of such methods, but we will not discuss any of these techniques.

E X E R C I S E   S E T   3.3

1. Use Newton's interpolatory divided-difference formula to construct first, second, third, and fourth interpolating polynomials for the following data. Approximate the specified value using each of the polynomials.
   (a) $f(8.4)$ if $f(8) = 16.63553, f(8.1) = 17.61549, f(8.3) = 17.56492, f(8.6) = 18.50515, f(8.7) = 18.82091$
   (b) $f(0.9)$ if $f(0.5) = -0.34409873, f(0.6) = -0.17694460, f(0.7) = 0.01375227, f(0.8) = 0.22363362, f(1.0) = 0.65809197$
   (c) $f(\pi)$ if $f(2.9) = -4.827866, f(3.0) = -4.240058, f(3.1) = -3.496909, f(3.2) = -2.596792, f(3.4) = -0.3330587$

2. Use Newton's forward-difference formula to construct first, second, third, and fourth interpolating polynomials for the following data. Approximate the specified value using each of the polynomials.
   (a) $f(-\frac{1}{3})$ if $f(-1) = 0.1, f(-0.75) = -0.0718125, f(-0.5) = -0.02475000, f(-0.25) = 0.33493750, f(0) = 1.10100000$
   (b) $f(0.25)$ if $f(0) = -1, f(0.1) = -0.62049958, f(0.2) = -0.28398668, f(0.3) = 0.00660095, f(0.4) = 0.24842440$
   (c) $f(1.15)$ if $f(1) = 1.684370, f(1.1) = 1.949477, f(1.2) = 2.199796, f(1.3) = 2.439189, f(1.4) = 2.670324$

3. Use Newton's backward-difference formula to construct first, second, third, and fourth interpolating polynomials for the following data. Approximate the specified value using each of the polynomials.
   (a) $f(-\frac{1}{3})$ if $f(-1) = 0.1, f(-0.75) = -0.0718125, f(-0.5) = -0.02475000, f(-0.25) = 0.33493750, f(0) = 1.10100000$
   (b) $f(0.25)$ if $f(0) = -1, f(0.1) = -0.62049958, f(0.2) = -0.28398668, f(0.3) = 0.00660095, f(0.4) = 0.24842440$
   (c) $f(\frac{\pi}{2})$ if $f(1.1) = 1.964760, f(1.2) = 2.572152, f(1.3) = 3.602102, f(1.4) = 5.797884, f(1.5) = 14.10142$

4. (a) Construct the fourth interpolating polynomial for the unequally spaced points given in the following table:

x	$f(x)$
0.0	−6.00000
0.1	−5.89483
0.3	−5.65014
0.6	−5.17788
1.0	−4.28172

   (b) Suppose $f(1.1) = -3.99583$ is added to the table. Construct the fifth interpolating polynomial.

5. (a) Use the following data and the Newton forward divided-difference formula to approximate $f(0.05)$.

x	0.0	0.2	0.4	0.6	0.8
$f(x)$	1.00000	1.22140	1.49182	1.82212	2.22554

(b) Use the Newton backward divided-difference formula to approximate $f(0.65)$.

6. The following population table was given in Exercise 13 of Section 3.3.

Year	1940	1950	1960	1970	1980	1990
Population (in thousands)	132, 165	151, 326	179, 323	203, 302	226, 542	249, 633

Use an appropriate divided difference method to approximate:
(a) The population in the year 1930
(b) The population in the year 2000

3.4 Hermite Interpolation

The Lagrange polynomials agree with a function f at a specified number of points. The values of f are often determined from observation, and in some situations it is possible to determine the derivative of f as well. This is likely to be the case, for example, if the independent variable is time and the function describes the position of an object. The derivative of the function in this case is the velocity, which may be available.

In this section we consider Hermite interpolation, which determines a polynomial that agrees with the function and its first derivative at specified points. If $n + 1$ data points and values of the function f, $(x_0, f(x_0))$, $(x_1, f(x_1))$, ..., $(x_n, f(x_n))$, are given, the Lagrange polynomial agreeing with f at these points will generally have degree n. If we require, in addition, that the derivative of the Hermite polynomial agree with the derivative of f at $x_0, x_1, \ldots, x_n$, these additional $n + 1$ conditions raise the expected degree of the Hermite polynomial to $2n + 1$. The specific form of the Hermite polynomial is described next.

Hermite Polynomial

If $f \in C^1[a, b]$ and $x_0, \ldots, x_n \in [a, b]$ are distinct, the unique polynomial of least degree agreeing with f and f' at $x_0, \ldots, x_n$ is the polynomial of degree at most $2n + 1$ given by

$$H_{2n+1}(x) = \sum_{j=0}^{n} f(x_j) H_{n,j}(x) + \sum_{j=0}^{n} f'(x_j) \hat{H}_{n,j}(x),$$

where

$$H_{n,j}(x) = [1 - 2(x - x_j)L'_{n,j}(x_j)]L^2_{n,j}(x)$$

and

$$\hat{H}_{n,j}(x) = (x - x_j)L^2_{n,j}(x).$$

Here, $L_{n,j}$ denotes the jth Lagrange coefficient polynomial of degree n.

The error term for the Hermite polynomial is similar to that of the Lagrange polynomial, with the only modifications being those needed to accommodate the increased amount of data used in the Hermite polynomial.

Hermite Polynomial Error Formula

If $f \in C^{(2n+2)}[a, b]$, then

$$f(x) = H_{2n+1}(x) + \frac{f^{(2n+2)}(\xi(x))}{(2n + 2)!}(x - x_0)^2 \cdots (x - x_n)^2$$

for some $\xi(x)$ with $a < \xi(x) < b$.

Although the Hermite formula provides a complete description of the Hermite polynomials, the need to determine and evaluate the Lagrange polynomials and their derivatives makes the procedure tedious even for small values of n. An alternative method for generating Hermite approximations is based on the connection between the nth divided difference and the nth derivative of f:

Divided-Difference Relationship to the Derivative

If $f \in C^n[a, b]$ and $x_0, x_1, \ldots, x_n$ are distinct in $[a, b]$, then a number ξ in (a, b) exists with

$$f[x_0, x_1, \ldots, x_n] = \frac{f^{(n)}(\xi)}{n!}.$$

To use this result to generate the Hermite polynomial, first suppose that $n + 1$ distinct numbers $x_0, x_1, \ldots, x_n$ are given together with their values of f and f'. Define a new sequence $z_0, z_1, \ldots, z_{2n+1}$ by

$$z_{2i} = z_{2i+1} = x_i, \qquad \text{for each } i = 0, 1, \ldots, n.$$

Now construct the divided-difference table using the variables $z_0, z_1, \ldots, z_{2n+1}$.

Since $z_{2i} = z_{2i+1} = x_i$ for each i, $f[z_{2i}, z_{2i+1}]$ cannot be defined by the basic divided-difference relation:

$$f[z_{2i}, z_{2i+1}] = \frac{f[z_{2i+1}] - f[z_{2i}]}{z_{2i+1} - z_{2i}}.$$

But, $f[x_0, x_1] = f'(\xi)$ for some number ξ in (x_0, x_1) and $\lim_{x_1 \to x_0} f[x_0, x_1] = f'(x_0)$. So a reasonable substitution in this situation is $f[z_{2i}, z_{2i+1}] = f'(x_i)$, and we can use the entries

$$f'(x_0), f'(x_1), \ldots, f'(x_n)$$

in place of the undefined first divided differences

$$f[z_0, z_1], f[z_2, z_3], \ldots, f[z_{2n}, z_{2n+1}].$$

The remaining divided differences are produced as usual and the appropriate divided differences are employed in Newton's interpolatory divided-difference formula.

Divided-Difference Form of the Hermite Polynomial

If $f \in C^1[a, b]$ and $x_0, x_1, \ldots, x_n$ are distinct in $[a, b]$, then

$$H_{2n+1}(x) = f[z_0] + \sum_{k=1}^{2n+1} f[z_0, z_1, \ldots, z_k](x - z_0) \ldots, (x - z_{k-1}),$$

where $z_{2k} = z_{2k+1} = x_k$ and $f[z_{2k}, z_{2k+1}] = f'(x_k)$ for each $k = 1, 2 \ldots, n$.

Table 3.9 shows the entries that are used for the first three divided-difference columns when determining the Hermite polynomial H_5 for x_0, x_1, and x_2. The remaining entries are generated in the usual divided-difference manner.

Table 3.9

z	$f(z)$	First Divided Differences	Second Divided Differences
$z_0 = x_0$	$f[z_0] = f(x_0)$		
		$f[z_0, z_1] = f'(x_0)$	
$z_1 = x_0$	$f[z_1] = f(x_0)$		$f[z_0, z_1, z_2] = \frac{f[z_1,z_2]-f[z_0,z_1]}{z_2-z_0}$
		$f[z_1, z_2] = \frac{f[z_2]-f[z_1]}{z_2-z_1}$	
$z_2 = x_1$	$f[z_2] = f(x_1)$		$f[z_1, z_2, z_3] = \frac{f[z_2,z_3]-f[z_1,z_2]}{z_3-z_1}$
		$f[z_2, z_3] = f'(x_1)$	
$z_3 = x_1$	$f[z_3] = f(x_1)$		$f[z_2, z_3, z_4] = \frac{f[z_3,z_4]-f[z_2,z_3]}{z_4-z_2}$
		$f[z_3, z_4] = \frac{f[z_4]-f[z_3]}{z_4-z_3}$	
$z_4 = x_2$	$f[z_4] = f(x_2)$		$f[z_3, z_4, z_5] = \frac{f[z_4,z_5]-f[z_3,z_4]}{z_5-z_3}$
		$f[z_4, z_5] = f'(x_2)$	
$z_5 = x_2$	$f[z_5] = f(x_2)$		

EXAMPLE 1   The entries in Table 3.10 use the data in the examples we have previously considered together with known values of the derivative. The underlined entries are the given data; the remainder are generated by the standard divided-difference method. They produce the following approximation to $f(1.5)$.

Table 3.10

1.3	0.6200860				
		−0.5220232			
1.3	0.6200860		−0.0897427		
		−0.5489460		0.0663657	
1.6	0.4554022		−0.0698330		0.0026663
		−0.5698959		0.0679655		−0.0027738
1.6	0.4554022		−0.0290537		0.0010020
		−0.5786120		0.0685667	
1.9	0.2818186		−0.0084837		
		−0.5811571			
1.9	0.2818186				

$$H_5(1.5) = 0.6200860 + (1.5 - 1.3)(-0.5220232) + (1.5 - 1.3)^2(-0.0897427)$$
$$+ (1.5 - 1.3)^2(1.5 - 1.6)(0.0663657) + (1.5 - 1.3)^2(1.5 - 1.6)^2(0.0026663)$$
$$+ (1.5 - 1.3)^2(1.5 - 1.6)^2(1.5 - 1.9)(-0.0027738)$$
$$= 0.5118277.$$ ■ ■ ■

The program HERMIT33 generates the coefficients for the Hermite polynomials using this modified Newton interpolatory divided-difference formula. The structure of the program is slightly different from the discussion, to take advantage of efficiency of computation.

EXERCISE SET 3.4

1. Use Hermite interpolation to construct an approximating polynomial for the following data.

(a)	x	$f(x)$	$f'(x)$	(b)	x	$f(x)$	$f'(x)$
	8.3	17.56492	1.116256		0.8	0.22363362	2.1691753
	8.6	18.50515	1.151762		1.0	0.65809197	2.0466965

(c)	x	$f(x)$	$f'(x)$	(d)	x	$f(x)$	$f'(x)$
	−0.5	−0.0247500	0.7510000		3.0	−4.240058	6.649860
	−0.25	0.3349375	2.1890000		3.1	−3.496909	8.215853
	0	1.1010000	4.0020000		3.2	−2.596792	9.784636

(e) x	$f(x)$	$f'(x)$	(f) x	$f(x)$	$f'(x)$
0.1	−0.62049958	3.58502082	1.2	2.572152	7.615964
0.2	−0.28398668	3.14033271	1.3	3.602102	13.97514
0.3	0.00660095	2.66668043	1.4	5.797884	34.61546
0.4	0.24842440	2.16529366	1.5	14.10142	199.8500

(g) x	$f(x)$	$f'(x)$	(h) x	$f(x)$	$f'(x)$
1.0	1.684370	2.742245	3.6	1.16164956	−1.50728217
1.1	1.949477	2.569394	3.8	0.80201036	−2.11189875
1.2	2.199796	2.443303	4.0	0.30663842	−2.87057564
1.3	2.439189	2.348926	4.2	−0.35916618	−3.82381126
1.4	2.670324	2.276919	4.4	−1.23926000	−5.02312214

2. The data in Exercise 1 were generated using the following functions. For the given value of x, use the polynomials constructed in Exercise 1 to approximate $f(x)$, and calculate the actual error.
   (a) $f(x) = x \ln x$;   approximate $f(8.4)$.
   (b) $f(x) = \sin(e^x - 2)$;   approximate $f(0.9)$.
   (c) $f(x) = x^3 + 4.001x^2 + 4.002x + 1.101$;   approximate $f(-\frac{1}{3})$.
   (d) $f(x) = x \cos x - x^2 \sin x$;   approximate $f(\pi)$.
   (e) $f(x) = x \cos x - 2x^2 + 3x - 1$;   approximate $f(0.25)$.
   (f) $f(x) = \tan x$;   approximate $f(\frac{\pi}{2})$.
   (g) $f(x) = \ln(e^{2x} - 2)$;   approximate $f(1.15)$.
   (h) $f(x) = x - (\ln x)^x$;   approximate $f(4.1)$.

3. Use the Hermite polynomial error formula to find a bound for the errors in parts (a), (c), and (e) of Exercise 2.

4. Let $f(x) = 3xe^x - e^{2x}$.
   (a) Approximate $f(1.03)$ by the Hermite interpolating polynomial of degree at most three using $x_0 = 1$ and $x_1 = 1.05$. Compare the actual error to the error bound.
   (b) Repeat (a) with the Hermite interpolating polynomial of degree at most five, using $x_0 = 1$, $x_1 = 1.05$, and $x_2 = 1.07$.

5. (a) Use the following values and five-digit rounding arithmetic to construct a Hermite interpolatory polynomial to approximate $\sin 0.34$.

x	$\sin x$	$D_x \sin x = \cos x$
0.30	0.29552	0.95534
0.32	0.31457	0.94924
0.35	0.34290	0.93937

   (b) Determine an error bound for the approximation in part (a) and compare to the actual error.
   (c) Add $\sin 0.33 = 0.32404$ and $\cos 0.33 = 0.94604$ to the data and redo the calculations.

6. The following table lists data for the function described by $f(x) = e^{0.1x^2}$. Approximate $f(1.25)$, by using $H_5(1.25)$ and $H_3(1.25)$, where H_5 uses the nodes $x_0 = 1$, $x_1 = 2$, and $x_2 = 3$ and H_3 uses the nodes $\bar{x}_0 = 1$ and $\bar{x}_1 = 1.5$. Find error bounds for these approximations.

x	$f(x) = e^{0.1x^2}$	$f'(x) = 0.2xe^{0.1x^2}$
$x_0 = \bar{x}_0 = 1$	1.105170918	0.2210341836
$\bar{x}_1 = 1.5$	1.252322716	0.3756968148
$x_1 = 2$	1.491824698	0.5967298792
$x_2 = 3$	2.459603111	1.475761867

7. A car traveling along a straight road is clocked at a number of points. The data from the observations are given in the following table where the time is in seconds, the distance is in feet, and the speed is in feet per second.

Time	0	3	5	8	13
Distance	0	225	383	623	993
Speed	75	77	80	74	72

   (a) Use a Hermite polynomial to predict the position of the car and its speed when $t = 10$ s.
   (b) Use the derivative of the Hermite polynomial to determine whether the car ever exceeds a 55-mi/h speed limit on the road; if so, what is the first time the car exceeds this speed?
   (c) What is the predicted maximum speed for the car?

3.5 Spline Interpolation

The previous sections of this chapter use polynomials to approximate arbitrary functions on closed intervals. However, we have seen that relatively high-degree polynomials are needed for accurate approximation and that these have some serious disadvantages. They all have an oscillatory nature, and a fluctuation over a small portion of the interval can induce large fluctuations over the entire range.

An alternative approach is to divide the interval into a collection of subintervals and construct a different approximating polynomial on each subinterval. This is called **piecewise polynomial approximation**.

The simplest piecewise polynomial approximation consists of joining a set of data points $(x_0, f(x_0)), (x_1, f(x_1)), \ldots, (x_n, f(x_n))$ by a series of straight lines such as those shown in Figure 3.5. This is the interpolation used in elementary courses involving the study of trigonometric or logarithmic functions when intermediate values are required from a collection of tabulated values.

Figure 3.5

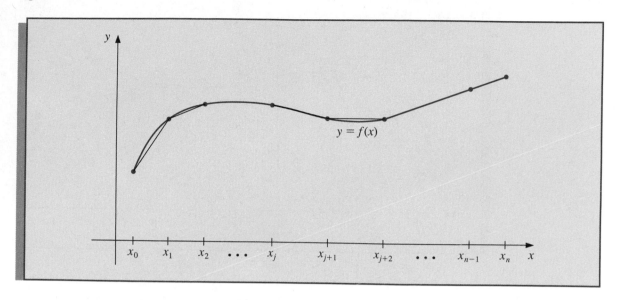

A disadvantage of linear approximation is that the approximation is generally not differentiable at the endpoints of the subintervals. In a geometric context, this means that the interpolating function is not "smooth" at these points. Often it is clear from physical conditions that smoothness is required, so the approximating function must be continuously differentiable.

One remedy for this problem is to use piecewise polynomials of Hermite type. For example, if the values of f and f' are known at each of the points $x_0 < x_1 < \cdots < x_n$, a Hermite polynomial of degree three can be used on each of the subintervals $[x_0, x_1], [x_1, x_2], \ldots, [x_{n-1}, x_n]$ to obtain an approximating function that is continuously differentiable on the interval $[x_0, x_n]$. To determine the appropriate Hermite cubic polynomial on a given interval, we simply compute the function $H_3(x)$ for that interval. The Lagrange interpolating polynomials needed to determine H_3 are of first degree, so this can be accomplished without great difficulty.

The Hermite cubic polynomials are commonly used in application problems to study the motion of particles moving in space. The difficulty with using Hermite piecewise polynomials for general interpolation problems concerns the need to know the derivative of the function being approximated. The remainder of this section considers approximation using piecewise polynomials that require no derivative information, except perhaps at the endpoints of the interval on which the function is being approximated.

The most common piecewise polynomial approximation uses cubic polynomials between pairs of nodes and is called **cubic spline** interpolation. A general cubic polynomial involves four constants, so there is sufficient flexibility in the cubic spline procedure to ensure that the interpolant has two continuous derivatives on the interval. The derivatives of the cubic spline do not, in general, however, agree with the derivatives of the function, even at the nodes. (See Figure 3.6.)

Figure 3.6

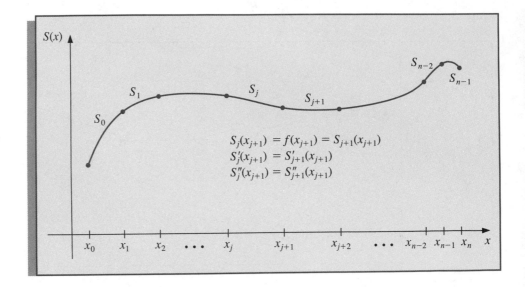

$$S_j(x_{j+1}) = f(x_{j+1}) = S_{j+1}(x_{j+1})$$
$$S_j'(x_{j+1}) = S_{j+1}'(x_{j+1})$$
$$S_j''(x_{j+1}) = S_{j+1}''(x_{j+1})$$

Cubic Spline Interpolation

Given a function f defined on $[a, b]$ and a set of nodes, $a = x_0 < x_1 < \cdots < x_n = b$, a cubic spline interpolant, S, for f is a function that satisfies the following conditions:

(a) S is a cubic polynomial, denoted S_j, on the subinterval $[x_j, x_{j+1}]$ for each $j = 0, 1, \ldots, n - 1$.

(b) $S(x_j) = f(x_j)$ for each $j = 0, 1, \ldots, n$.

(c) $S_{j+1}(x_{j+1}) = S_j(x_{j+1})$ for each $j = 0, 1, \ldots, n - 2$.

(d) $S_{j+1}'(x_{j+1}) = S_j'(x_{j+1})$ for each $j = 0, 1, \ldots, n - 2$.

(e) $S_{j+1}''(x_{j+1}) = S_j''(x_{j+1})$ for each $j = 0, 1, \ldots, n - 2$.

(f) One of the following sets of boundary conditions is satisfied:

   (i) $S''(x_0) = S''(x_n) = 0$ (natural or free boundary);

   (ii) $S'(x_0) = f'(x_0)$ and $S'(x_n) = f'(x_n)$ (clamped boundary).

Although cubic splines are defined with other boundary conditions, the conditions given here are used most frequently in practice. When the natural boundary conditions are used, the spline assumes the shape that a long flexible rod would take if forced to go through the points $\{(x_0, f(x_0)), (x_1, f(x_1)), \ldots, (x_n, f(x_n))\}$. This spline continues linearly when $x \leq x_0$ and when $x \geq x_n$.

In general, clamped boundary conditions lead to more accurate approximations, since they include more information about the function. However, for this type of boundary condition, we need values of the derivative at the endpoints or an accurate approximation to those values. This can be difficult to obtain.

To construct the cubic-spline interpolant for a given function f, the conditions in the definition are applied to the cubic polynomials

$$S_j(x) = a_j + b_j(x - x_j) + c_j(x - x_j)^2 + d_j(x - x_j)^3$$

for each $j = 0, 1, \ldots, n - 1$.

Since

$$S_j(x_j) = a_j = f(x_j),$$

condition (c) can be applied to obtain

$$a_{j+1} = S_{j+1}(x_{j+1}) = S_j(x_{j+1}) = a_j + b_j(x_{j+1} - x_j) + c_j(x_{j+1} - x_j)^2 + d_j(x_{j+1} - x_j)^3$$

for each $j = 0, 1, \ldots, n - 2$.

Since the term $(x_{j+1} - x_j)$ will be used repeatedly in this development, it is convenient to introduce the simpler notation

$$h_j = x_{j+1} - x_j,$$

for each $j = 0, 1, \ldots, n - 1$. If we also define $a_n = f(x_n)$, then the equation

(3.1) $$a_{j+1} = a_j + b_j h_j + c_j h_j^2 + d_j h_j^3$$

holds for each $j = 0, 1, \ldots, n - 1$.

In a similar manner, define $b_n = S'(x_n)$ and observe that

$$S_j'(x) = b_j + 2c_j(x - x_j) + 3d_j(x - x_j)^2$$

implies $S_j'(x_j) = b_j$ for each $j = 0, 1, \ldots, n - 1$. Applying condition (d),

(3.2) $$b_{j+1} = b_j + 2c_j h_j + 3d_j h_j^2,$$

for each $j = 0, 1, \ldots, n - 1$.

Another relation between the coefficients of S_i is obtained by defining $c_n = S''(x_n)/2$ and applying condition (e). In this case,

(3.3) $$c_{j+1} = c_j + 3d_j h_j,$$

for each $j = 0, 1, \ldots, n - 1$.

Solving for d_j in Eq. (3.3) and substituting this value into Eqs. (3.1) and (3.2) gives the new equations

(3.4) $$a_{j+1} = a_j + b_j h_j + \frac{h_j^2}{3}(2c_j + c_{j+1})$$

and

(3.5) $$b_{j+1} = b_j + h_j(c_j + c_{j+1}),$$

for each $j = 0, 1, \ldots, n - 1$.

The final relationship involving the coefficients is obtained by first solving the appropriate equation in the form of Eq. (3.4) for b_j,

(3.6) $$b_j = \frac{1}{h_j}(a_{j+1} - a_j) - \frac{h_j}{3}(2c_j + c_{j+1}),$$

and then, with a reduction of the index, for b_{j-1}:

$$b_{j-1} = \frac{1}{h_{j-1}}(a_j - a_{j-1}) - \frac{h_{j-1}}{3}(2c_{j-1} + c_j).$$

Substituting these values into the equation derived from Eq. (3.5), when the index is reduced by 1, gives the linear system of equations

(3.7) $h_{j-1}c_{j-1} + 2(h_{j-1} + h_j)c_j + h_jc_{j+1} = \dfrac{3}{h_j}(a_{j+1} - a_j) - \dfrac{3}{h_{j-1}}(a_j - a_{j-1})$

for each $j = 1, 2, \ldots, n - 1$. This system involves, as unknowns, only $\{c_j\}_{j=0}^n$, since the values of $\{h_j\}_{j=0}^{n-1}$ and $\{a_j\}_{j=0}^n$ are given by the spacing of the nodes $\{x_j\}_{j=0}^n$ and the values $\{f(x_j)\}_{j=0}^n$.

Once the values of $\{c_j\}_{j=0}^n$ are known, it is a simple matter to find the remainder of the constants $\{b_j\}_{j=0}^{n-1}$ from Eq. (3.6) and $\{d_j\}_{j=0}^{n-1}$ from Eq. (3.3) and to construct the cubic polynomials $\{S_j\}_{j=0}^{n-1}$.

The solution to the cubic spline problem with the natural boundary conditions $S''(x_0) = S''(x_n) = 0$ can be obtained by applying the program NCUBSP34. The program CCUBSP35 determines the cubic spline with the clamped boundary conditions $S'(x_0) = f'(x_0)$ and $S'(x_1) = f'(x_1)$.

EXAMPLE 1   Figure 3.7 shows a ruddy duck in flight.

Figure 3.7

To approximate the top profile of the duck, we have chosen points along the curve through which we want the approximating curve to pass. Table 3.11 lists the coordinates of 21 data points relative to the superimposed coordinate system shown in Figure 3.8. Notice that more points are used when the curve is changing rapidly than when it is changing more slowly.

Using NCUBSP34 to generate the natural cubic spline for these data produces the coefficients shown in Table 3.12. This spline curve is nearly identical to the profile, as shown in Figure 3.9 (page 80).

Table 3.11

x	0.9	1.3	1.9	2.1	2.6	3.0	3.9	4.4	4.7	5.0	6.0	7.0	8.0	9.2	10.5	11.3	11.6	12.0	12.6	13.0	13.3
$f(x)$	1.3	1.5	1.85	2.1	2.6	2.7	2.4	2.15	2.05	2.1	2.25	2.3	2.25	1.95	1.4	0.9	0.7	0.6	0.5	0.4	0.25

Figure 3.8

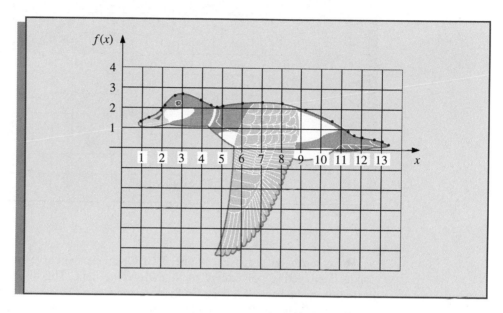

Table 3.12

j	x_j	a_j	b_j	c_j	d_j
0	0.9	1.3	0.54	0.00	−0.25
1	1.3	1.5	0.42	−0.30	0.95
2	1.9	1.85	1.09	1.41	−2.96
3	2.1	2.1	1.29	−0.37	−0.45
4	2.6	2.6	0.59	−1.04	0.45
5	3.0	2.7	−0.02	−0.50	0.17
6	3.9	2.4	−0.50	−0.03	0.08
7	4.4	2.15	−0.48	0.08	1.31
8	4.7	2.05	−0.07	1.27	−1.58
9	5.0	2.1	0.26	−0.16	0.04
10	6.0	2.25	0.08	−0.03	0.00
11	7.0	2.3	0.01	−0.04	−0.02
12	8.0	2.25	−0.14	−0.11	0.02
13	9.2	1.95	−0.34	−0.05	−0.01
14	10.5	1.4	−0.53	−0.10	−0.02
15	11.3	0.9	−0.73	−0.15	1.21
16	11.6	0.7	−0.49	0.94	−0.84
17	12.0	0.6	−0.14	−0.06	0.04
18	12.6	0.5	−0.18	0.00	−0.45
19	13.0	0.4	−0.39	−0.54	0.60
20	13.3	0.25			

Figure 3.9

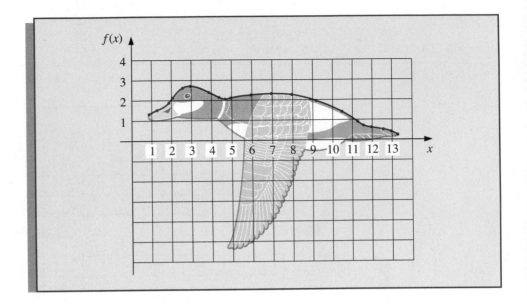

For comparison purposes, Figure 3.10 illustrates the curve generated using a Lagrange interpolating polynomial to fit these same data. This produces a very strange illustration of the back of a duck, in flight or otherwise. The interpolating polynomial in this case is of degree 20 and oscillates wildly, except near the middle of the curve.

Figure 3.10

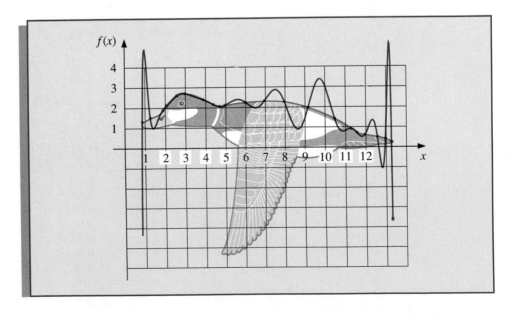

To use a clamped spline to approximate the curve in Example 1 we would need derivative approximations for the endpoints. Even if these approximations were available, we could expect little improvement, because of the close agreement of the natural cubic spline to the curve of the top profile.

Cubic splines generally agree quite well with the function being approximated, provided that the points are not too far apart and the fourth derivative of the function is well behaved. For example, suppose that f has four continuous derivatives on $[a, b]$ and that the fourth derivative on this interval has a magnitude bounded by M. Then the clamped cubic spline $S(x)$ agreeing with $f(x)$ at the points $a = x_0 < x_1 < \cdots < x_n = b$ has the property that for all x in $[a, b]$,

(3.8)
$$|f(x) - S(x)| \le \frac{5M}{384} \max_{0 \le j \le n-1} (x_{j+1} - x_j)^4.$$

A similar, but more complicated, error result holds for the free cubic splines.

EXERCISE SET 3.5

1. Construct the free cubic spline for the following data:

(a)

x	f(x)
8.3	17.56492
8.6	18.50515

(b)

x	f(x)
0.8	0.22363362
1.0	0.65809197

(c)

x	f(x)
-0.5	-0.0247500
-0.25	-0.3349375
0	1.1010000

(d)

x	f(x)
3.0	-4.240058
3.1	-3.496909
3.2	-2.596792

(e)

x	f(x)
0.1	-0.62049958
0.2	-0.28398668
0.3	0.00660095
0.4	0.24842440

(f)

x	f(x)
1.2	2.572152
1.3	3.602102
1.4	5.797884
1.5	14.10142

(g)

x	f(x)
1.0	1.684370
1.1	1.949477
1.2	2.199796
1.3	2.439189
1.4	2.670324

(h)

x	f(x)
3.6	1.16164956
3.8	0.80201036
4.0	0.30663842
4.2	-0.35916618
4.4	-1.23926000

2. The data in Exercise 1 were generated using the following functions. For the given value of x, use the cubic splines constructed in Exercise 1 to approximate $f(x)$ and $f'(x)$, and calculate the actual error.

   (a) $f(x) = x \ln x$;   approximate $f(8.4)$ and $f'(8.4)$.

   (b) $f(x) = \sin(e^x - 2)$;   approximate $f(0.9)$ and $f'(0.9)$.

   (c) $f(x) = x^3 + 4.001x^2 + 4.002x + 1.101$;   approximate $f(-\frac{1}{3})$ and $f'(-\frac{1}{3})$.

   (d) $f(x) = x \cos x - x^2 \sin x$;   approximate $f(\pi)$ and $f'(\pi)$.

   (e) $f(x) = x \cos x - 2x^2 + 3x - 1$;   approximate $f(0.25)$ and $f'(0.25)$.

   (f) $f(x) = \tan x$;   approximate $f(\pi/2)$ and $f'(\pi/2)$.

   (g) $f(x) = \ln(e^{2x} - 2)$;   approximate $f(1.15)$ and $f'(1.15)$.

   (h) $f(x) = x - (\ln x)^x$;   approximate $f(4.1)$ and $f'(4.1)$.

3. Construct the clamped cubic spline using the data of Exercise 1 and the fact that

   (a) $f'(8.3) = 1.116256$ and $f'(8.6) = 1.151762$

   (b) $f'(0.8) = 2.1691753$ and $f'(1.0) = 2.0466965$

   (c) $f'(-0.5) = 0.7510000$ and $f'(0) = 4.0020000$

   (d) $f'(3.0) = 6.649860$ and $f'(3.2) = 9.784636$

   (e) $f'(0.1) = 3.58502082$ and $f'(0.4) = 2.16529366$

   (f) $f'(1.2) = 7.615964$ and $f'(1.5) = 199.8500$

   (g) $f'(1.0) = 2.742245$ and $f'(1.4) = 2.276919$

   (h) $f'(3.6) = -1.50728217$ and $f'(4.4) = -5.02312214$

4. Repeat Exercise 2 using the cubic splines constructed in Exercise 3.

5. (a) Use the following values and five-digit rounding to construct a free cubic spline to approximate $\sin 0.34$.

   (b) Determine the error for the approximation in part (a).

   (c) Repeat part (a) using a clamped cubic spline.

   (d) Repeat part (b) for the clamped cubic spline constructed in part (c).

x	$\sin x$	$D_x(\sin x) = \cos x$
0.30	0.29552	0.95534
0.32	0.31457	0.94924
0.35	0.34290	0.93937

   (e) Use the spline constructed in part (a) to approximate $\cos 0.34$.

   (f) Use the spline constructed in part (c) to approximate $\cos 0.34$.

   (g) Use the spline constructed in part (a) to approximate

$$\int_{0.30}^{0.35} \sin x \, dx.$$

   (h) Use the spline constructed in part (c) to approximate

$$\int_{0.30}^{0.35} \sin x \, dx.$$

6. Add $\sin 0.33 = 0.32404$ and $\cos 0.33 = 0.94604$ to the data in Exercise 5 and redo the calculations in that exercise.

7. Use the cubic spline with clamped boundary conditions to approximate the function $f(x) = 3xe^x - e^{2x}$ at $x = 1.03$ using $f'(1.0) = 1.5315787$, $f'(1.06) = 1.1754977$, and the data in the following table:

x	1.0	1.02	1.04	1.06
$f(x)$	0.76578939	0.79536678	0.82268817	0.84752226

Estimate the error, using the inequality (3.8), and compare to the actual error.

8. Given the partition $x_0 = 0$, $x_1 = 0.05$, $x_2 = 0.1$ of $[0, 0.1]$ and $f(x) = e^{2x}$.
   (a) Find the cubic spline s with clamped boundary conditions that interpolates f.
   (b) Find an approximation for $\int_0^{0.1} e^{2x}\, dx$ by evaluating $\int_0^{0.1} s(x)\, dx$.
   (c) Use the inequality (3.8) to estimate $\max_{0 \le x \le 0.1} |f(x) - s(x)|$ and

$$\left| \int_0^{0.1} f(x)\, dx - \int_0^{0.1} s(x)\, dx \right|.$$

   (d) Determine the cubic spline S with free boundary conditions and compare $S(0.02)$, $s(0.02)$, and $e^{0.04} = 1.04081077$.

9. Given the partition $x_0 = 0$, $x_1 = 0.05$, $x_2 = 0.1$ of $[0, 0.1]$, find the piecewise linear interpolating function F for $f(x) = e^{2x}$. Approximate $\int_0^{0.1} e^{2x}\, dx$ with $\int_0^{0.1} F(x)\, dx$. Compare the results to those of Exercise 8.

10. Let $f \in C^2[a, b]$, and let the nodes $a = x_0 < x_1 < \cdots < x_n = b$ be given. Derive an error estimate similar to that in inequality (3.8) for the piecewise linear interpolating function F. Use this estimate to derive error bounds for Exercise 9.

11. Exercise 7 of Section 3.5 lists observed data concerning speed and distance of a car traveling along a straight road.
   (a) Use a clamped cubic spline to predict the position of the car and its speed at $t = 10$ s.
   (b) Use the spline to determine whether the car ever exceeds a 55–mi/h speed limit on the road, and if so, the first time the car exceeds this speed.
   (c) What is the predicted maximum speed for the car?

12. It is suspected that the high amounts of tannin in mature oak leaves inhibit the growth of the winter moth (*Operophtera bromata L.*, Geometridae) larvae that extensively damage these trees in certain years. The following table lists the average weight of two samples of larvae at times in the first 28 days after birth. The first sample was reared on young oak leaves, whereas the second sample was reared on mature leaves from the same tree.
   (a) Use a free cubic spline to approximate the average weight curve for each sample.
   (b) Find an approximate maximum average weight for each sample by determining the maximum of the spline.

Day	0	6	10	13	17	20	28
Sample 1 average weight (mg)	6.67	17.33	42.67	37.33	30.10	29.31	28.74
Sample 2 average weight (mg)	6.67	16.11	18.89	15.00	10.56	9.44	8.89

13. The 1979 Kentucky Derby was won by a horse named Spectacular Bid in a time of $2:02\frac{2}{5}$ (2 min $2\frac{2}{5}$ s) for the $1\frac{1}{4}$-mi race. Times at the quarter-mile, half-mile, and mile poles were $25\frac{2}{5}$, $49\frac{2}{5}$, and $1:37\frac{3}{5}$.
   - **(a)** Use these values together with the starting time to construct a free cubic spline for Spectacular Bid's race.
   - **(b)** Use the spline to predict the time at the three-quarter-mile pole, and compare this to the actual time of $1:12\frac{2}{5}$.
   - **(c)** Use the spline to approximate Spectacular Bid's starting speed and speed at the finish line.

14. The data in the following table list the population of the United States for the years 1940 to 1990 and were considered in Exercise 13 of Section 3.2.

Year	1940	1950	1960	1970	1980	1990
Population (in thousands)	132, 165	151, 326	179, 323	203, 302	226, 542	249, 633

Find a free cubic spline agreeing with these data, and use the spline to predict the population in the years 1930, 1965, and 2000. Compare your approximations with those previously obtained. If you had to make a choice, which interpolation procedure would you choose?

3.6 Parametric Curves

None of the techniques we have developed can be used to generate curves of the form shown in Figure 3.11, since this curve cannot be expressed as a function of one coordinate variable in terms of the other. In this section we will see how to represent general curves by using a parameter to express both the x- and y-coordinate variables. This technique can be extended to represent general curves and surfaces in space.

A straightforward parametric technique for determining a polynomial or piecewise polynomial to connect in order the points $(x_0, y_0), (x_1, y_1), \ldots, (x_n, y_n)$ is to use a parameter t on an interval $[t_0, t_n]$, with $t_0 < t_1 < \cdots < t_n$ and construct approximation functions with

$$x_i = x(t_i) \quad \text{and} \quad y_i = y(t_i) \qquad \text{for each} \quad i = 0, 1, \ldots, n.$$

The following example demonstrates the technique when both approximating functions are Lagrange interpolating polynomials.

Figure 3.11

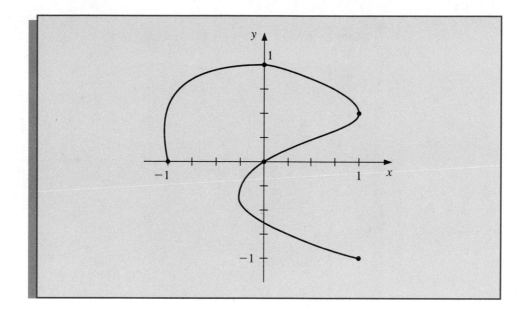

EXAMPLE 1   Construct a pair of Lagrange polynomials to approximate the curve shown in Figure 3.11, using the data points shown on the curve.

There is flexibility in choosing the parameter, and we will choose the points $\{t_i\}$ equally spaced in $[0, 1]$. In this case, we have the data in Table 3.13.

Table 3.13

i	0	1	2	3	4
t_i	0	0.25	0.5	0.75	1
x_i	-1	0	1	0	1
y_i	0	1	0.5	0	-1

This produces the Lagrange polynomials

$$x(t) = \left(\left(\left(64t - \tfrac{352}{3}\right)t + 60\right)t - \tfrac{14}{3}\right)t - 1$$

and

$$y(t) = \left(\left(\left(-\tfrac{64}{3}t + 48\right)t - \tfrac{116}{3}\right)t + 11\right)t.$$

Plotting this parametric system produces the graph in Figure 3.12 on page 86. Although it passes through the required points and has the same basic shape, it is quite a crude approximation to the original curve. A more accurate approximation would require additional nodes, with the accompanying increase in computation. ■ ■ ■

Figure 3.12

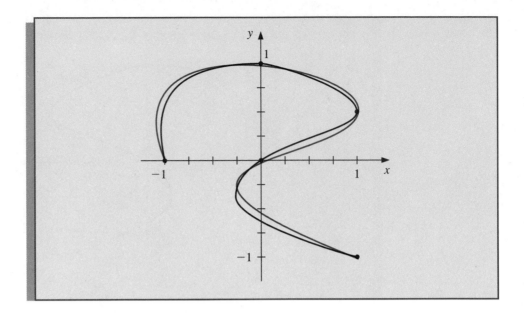

Hermite and spline curves can be generated in a similar manner, but again extensive computational effort is involved.

Applications in computer graphics require the rapid generation of smooth curves that can be easily and quickly modified. In addition, changing one portion of a curve should have little or no effect on other portions. This eliminates the use of interpolating polynomials and splines, since changing one portion of these curves affects the whole curve. This is undesirable both from an aesthetic and computational standpoint.

The choice of curve for use in computer graphics is generally a form of the piecewise cubic Hermite polynomial. Each portion of a cubic Hermite polynomial is completely determined by specifying its endpoints and the derivatives at these endpoints. As a consequence, one portion of the curve can be changed while leaving most of the curve the same. Only the adjacent portions must be modified if we want to ensure smoothness at the endpoints. The computations can be performed quickly and the curve can be modified a section at a time.

The problem with Hermite interpolation is the need to specify the derivatives at the endpoints of each section of the curve. Suppose the curve has $n + 1$ data points $(x_0, y_0), \ldots, (x_n, y_n)$, and we wish to parameterize the cubic to allow complex features. Then we must specify $x'(t_i)$ and $y'(t_i)$ for each $i = 0, 1, \ldots, n$, where $(x_i, y_i) = (x(t_i), y(t_i))$. This is not as difficult as it would first appear, however, since each portion can be generated independently. Essentially, then, we can simplify the process to one of determining a pair of cubic Hermite polynomials in the parameter t, where $t_0 = 0$, $t_1 = 1$, given the endpoint data $(x(0), y(0)), (x(1), y(1))$ and the derivatives dy/dx at $t = 0$ and dy/dx at $t = 1$.

Notice that we are specifying only six conditions, and each cubic polynomial has four parameters, for a total of eight. Thus we have considerable flexibility in choosing the pair of cubic Hermite polynomials to satisfy these conditions. This flexibility occurs because the natural form for determining $x(t)$ and $y(t)$ requires that we specify $x'(0)$, $x'(1)$, $y'(0)$, and $y'(1)$. The explicit Hermite curve in x and y requires specifying only the quotients

$$\frac{dy}{dx}(\text{at } t = 0) = \frac{x'(0)}{y'(0)} \qquad \text{and} \qquad \frac{dy}{dx}(\text{at } t = 1) = \frac{x'(1)}{y'(1)}.$$

By multiplying $x'(0)$ and $y'(0)$ by a common scaling factor, the tangent line to the curve at $(x(0), y(0))$ remains the same, but the shape of the curve varies. The larger the scaling factor, the closer the curve comes to approximating the tangent line near $(x(0), y(0))$. A similar situation exists at the other endpoint $(x(1), y(1))$.

To further simplify the process, the derivative at an endpoint is specified graphically by describing a second point, called a *guidepoint*, on the desired tangent line. The farther the guidepoint is from the node, the more closely the curve approximates the tangent line near the node.

In Figure 3.13, the nodes occur at (x_0, y_0) and (x_1, y_1), the guidepoint for (x_0, y_0) is $(x_0 + \alpha_0, y_0 + \beta_0)$, and the guidepoint for (x_1, y_1) is $(x_1 - \alpha_1, y_1 - \beta_1)$. The cubic Hermite polynomial $x(t)$ on $[0, 1]$ must satisfy

$$x(0) = x_0, \qquad x(1) = x_1, \qquad x'(0) = \alpha_0, \qquad \text{and} \qquad x'(1) = \alpha_1.$$

The unique cubic polynomial satisfying these conditions is

$$x(t) = [2(x_0 - x_1) + (\alpha_0 + \alpha_1)]t^3 + [3(x_1 - x_0) - (\alpha_1 + 2\alpha_0)]t^2 + \alpha_0 t + x_0.$$

In a similar manner, the unique cubic polynomial for y is

$$y(t) = [2(y_0 - y_1) + (\beta_0 + \beta_1)]t^3 + [3(y_1 - y_0) - (\beta_1 + 2\beta_0)]t^2 + \beta_0 t + y_0.$$

Figure 3.13

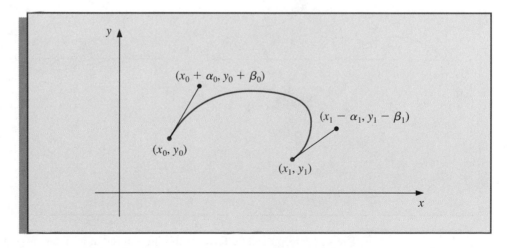

EXAMPLE 2 The graphs in Figure 3.14 on page 88 show some possibilities that occur when the nodes are $(0, 0)$ and $(1, 0)$, and the slopes at these nodes are 1 and -1, respectively. The specification of the slope at the endpoints requires only that $\alpha_0 = \beta_0$ and $\alpha_1 = -\beta_1$, since the ratios $\alpha_0 = \beta_0$ and $\alpha_1 = -\beta_1$ give the slopes at the left and right endpoints, respectively.

■ ■ ■

Figure 3.14

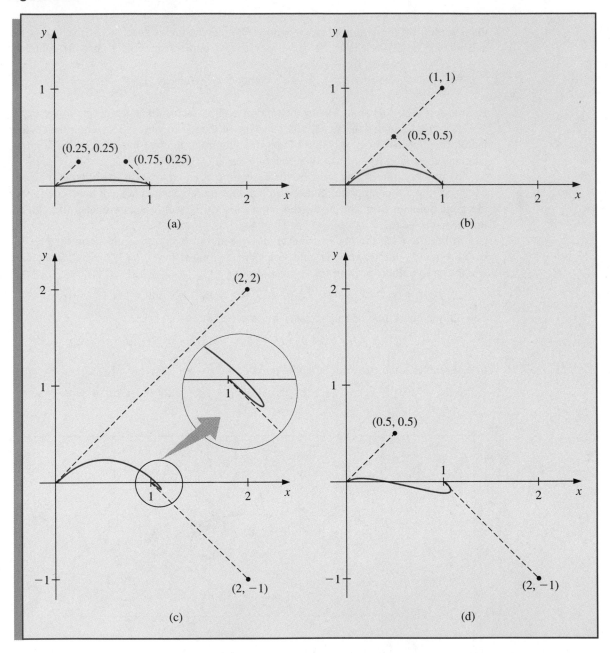

(a)

(b)

(c)

(d)

The standard procedure for determining curves in an interactive graphics mode is to first use an input device, such as a mouse or trackball, to set the nodes appropriately. The guidepoints are then placed to generate a first approximation to the desired curve. These can be set manually, but most graphics systems permit you to use your input device to draw the curve on the screen freehand and will select appropriate nodes and guidepoints for your freehand curve.

The nodes and guidepoints can then be manipulated into a position that produces an aesthetically satisfying curve. Since the computation is minimal, the curve is determined so quickly that the resulting change can be seen almost immediately. Moreover, all the data needed to compute the curves are imbedded in coordinates of the nodes and guidepoints, so no analytical knowledge is required of the user of the system.

Popular graphics programs use this type of system for their freehand graphic representations but in a slightly modified form. The Hermite cubics are described as Bézier polynomials, which incorporate a scaling factor of 3 when computing the derivatives at the endpoints. This modifies the parametric equations to

$$x(t) = [2(x_0 - x_1) + 3(\alpha_0 + \alpha_1)]t^3 + [3(x_1 - x_0) - 3(\alpha_1 + 2\alpha_0)]t^2 + 3\alpha_0 t + x_0$$
$$y(t) = [2(y_0 - y_1) + 3(\beta_0 + \beta_1)]t^3 + [3(y_1 - y_0) - 3(\beta_1 + 2\beta_0)]t^2 + 3\beta_0 t + y_0,$$

for $0 \le t \le 1$, but this change is transparent to the user of the system.

Three–dimensional curves can be generated in a similar manner by additionally specifying third components z_0 and z_1 for the nodes and $z_0 - \gamma_0$ and $z_1 - \gamma_1$ for the guidepoints. The more difficult problem involving the representation of three-dimensional curves concerns the loss of the third dimension when the curve is projected onto a two-dimensional medium such as a computer screen or printer paper. Various projection techniques are used, but this topic lies within the realm of computer graphics. For an introduction to this topic and ways that the technique can be modified for surface representations, see one of the many books on computer graphics methods.

Bézier curves can be generated using the program BEZIER36.

E X E R C I S E   S E T   3.6

1. Let $(x_0, y_0) = (0, 0)$ and $(x_1, y_1) = (5, 2)$ be the endpoints of a curve. Construct parametric cubic Hermite approximations $(x(t), y(t))$ to the curve using the given guidepoints and graph the approximations.
    (a) (1, 1) and (6, 1)   (b) (0.5, 0.5) and (5.5, 1.5)
    (c) (1, 1) and (6, 3)   (d) (2, 2) and (7, 0)

2. Repeat Exercise 1 using cubic Bézier polynomials.

3. Given the following points and guidepoints, construct and graph the cubic Bézier polynomials.
    (a) Point (1, 1) with guidepoint (1.5, 1.25) to point (6, 2) with guidepoint (7, 3)
    (b) Point (1, 1) with guidepoint (1.25, 1.5) to point (6, 2) with guidepoint (5, 3)
    (c) Point (0, 0) with guidepoint (0.5, 0.5) to point (4, 6) with entering guidepoint (3.5, 7) and exiting guidepoint (4.5, 5) to point (6, 1) with guidepoint (7, 2)
    (d) Point (0, 0) with guidepoint (0.5, 0.25) to point (2, 1) with entering guidepoint (3, 1) and exiting guidepoint (3, 1) to point (4, 0) with entering guidepoint (5, 1) and exiting guidepoint (3, −1) to point (6, −1) with guidepoint (6.5, −0.25)

4. Use the data in the following table to approximate the letter η.

i	x_i	y_i	α_i	β_i	α_i'	β_i'
0	3	6	3.3	6.5		
1	2	2	2.8	3.0	2.5	2.5
2	6	6	5.8	5.0	5.0	5.8
3	5	2	5.5	2.2	4.5	2.5
4	6.5	3			6.4	2.8

3.7   Survey of Methods and Software

In this chapter we have considered approximating a function using both polynomials and piecewise polynomials. The function can be specified by a given defining equation or by providing points in the plane through which the graph of the function passes. A set of nodes $x_0, x_1, \ldots, x_n$ is given in each case, and more information, such as the value of various derivatives, may also be required. We assume that the data are accurate, and we need to find an approximating function that interpolates the given function at the nodes. The interpolating polynomial P is the polynomial of least degree that satisfies, for a given function f,

$$P(x_i) = f(x_i), \qquad \text{for } i = 0, 1, \ldots, n.$$

Although there is a unique interpolating polynomial, it can take many different forms. The Lagrange form is most often used for interpolating tables when n is small and for deriving formulas for approximating derivatives and integrals. Neville's method is used for evaluating several interpolating polynomials at the same value of x. Newton's forms of the polynomial are more appropriate for computation and are also used for deriving formulas for solving differential equations. However, polynomial interpolation has the inherent weaknesses of oscillation, particularly if the number of nodes is large. In this case there are other methods that can be better applied.

The Hermite polynomials interpolate a function and its derivative at the nodes. These polynomials can be very accurate, but they require more information about the function being approximated. For large numbers of nodes, the Hermite polynomials also exhibit oscillation weaknesses.

The most commonly used form of polynomial interpolation is piecewise polynomial interpolation. If function and derivative values are available, piecewise cubic Hermite interpolation is recommended. This would be a preferred method for interpolating values of a function that is the solution to a differential equation. Free cubic spline interpolation is recommended when only the function values are available. This spline forces the second derivative of the spline to be zero at the endpoints. Other cubic splines require additional data. For example, the clamped cubic spline needs values of the derivative of the function at the endpoints of the interval.

There are other methods of interpolation that are commonly used. Trigonometric interpolation is used with large amounts of data when the function has a periodic nature. In particular, the fast Fourier transform discussed in Chapter 8 is employed. Interpolation by rational functions is also used. If the data are suspected to be inaccurate, smoothing techniques can be applied. In particular, some form of least squares fit of data is recommended. Polynomials, trigonometric functions, rational functions, and splines can be used in least squares fitting of data. We consider all these topics in Chapter 8.

Interpolation routines included in the IMSL Library are based on the book *A Practical Guide to Splines* by Carl de Boor. These subroutines use interpolation by cubic splines. The subroutine CSDEC is for interpolation by cubic splines with user-supplied end conditions, CSPER is for interpolation by cubic splines with periodic end conditions, and CSHER is for interpolation by quasi-Hermite piecewise polynomials. There are also cubic splines to minimize oscillations or to preserve concavity. B-spline interpolation routines are also included, as are methods for two-dimensional interpolation by bicubic splines.

The NAG library contains the subroutines EO1AEF for polynomial and Hermite interpolation, EO1BAF for cubic spline interpolation, and EO1BEF for piecewise cubic Hermite interpolation. To interpolate data at equally spaced points, the subroutine EO1ABF is used. The routine EO1AAF is applied if the data are given at unequally spaced points. NAG also contains subroutines for interpolating functions of two variables.

The MATLAB function POLYFIT can be used to find an interpolating function of degree at most n that passes through $n + 1$ specified points. Cubic splines can also be produced with the function SPLINE.

CHAPTER 4

Numerical Integration and Differentiation

4.1 Introduction

One of the most frequently used numerical procedures involves the approximation of definite integrals. Numerous techniques are described in calculus courses for the exact evaluation of integrals, but these techniques are given more for illustration of algebraic manipulation and simplification than for use in the evaluation of integrals that occur in real-life problems. Exact techniques are inadequate for many problems that arise in the physical world.

This chapter describes methods that are commonly applied to approximate definite integrals. The basic techniques are discussed in Section 4.2 and refinements and special applications of these procedures are given in the next six sections. The final section in the chapter considers approximating derivatives of functions.

You might wonder why there is so much more emphasis on approximating integrals than on approximating derivatives. Determining the derivative of a function is a constructive process that leads to straightforward rules for evaluation. Although the definition of the integral is also constructive, the principle tool for evaluating a definite integral is the Fundamental Theorem of Calculus. To apply this theorem we must determine the antiderivative of the function we wish to evaluate. This is not generally a constructive process, and it leads to the need for accurate approximation procedures.

In this chapter we will discover one of the more interesting facts in the study of numerical methods. The approximation of integrals—a task that is frequently needed—can usually be accomplished very accurately and often with little effort, whereas the accurate approximation of derivatives—which is needed far less frequently—is a much more difficult problem. There is something satisfying about a subject that provides good approximation methods for problems that need them but is less successful for problems that don't.

4.2 Basic Quadrature Rules

The basic procedure for approximating the definite integral of a function f on the interval $[a, b]$ is to determine an interpolating polynomial that approximates f and then integrate this polynomial. In this section we will determine approximations that arise when some basic polynomials are used for the approximation and determine error bounds for these approximations.

The approximations we consider use interpolating polynomials at equally spaced points in the interval $[a, b]$. The first of these is the Midpoint rule, which uses as its only interpolation point the midpoint of $[a, b]$, that is, the number $x_0 = (a + b)/2$. The Midpoint rule approximation can be easily determined geometrically, as shown in Figure 4.1, but to establish the pattern for the higher-order methods and to determine an error formula for the technique we will use the basic tool for these derivations, the Newton Forward Divided-Difference formula.

Figure 4.1

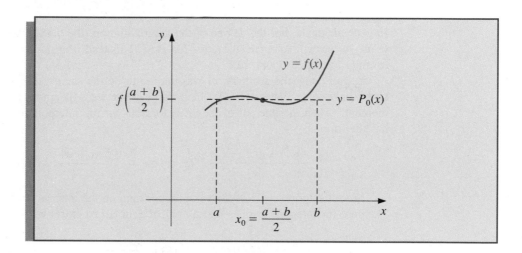

The Newton Forward Divided-Difference formula states that the interpolating polynomial for the function f using the nodes $x_0, x_1, \ldots, x_n$ can be expressed in the form

$$P_{0,1,\ldots,n}(x) = f[x_0] + f[x_0, x_1](x - x_0) + f[x_0, x_1, x_2](x - x_0)(x - x_1) + \cdots$$
$$+ f[x_0, x_1, \ldots, x_n](x - x_0)(x - x_1)\cdots(x - x_{n-1}).$$

Since this is also the nth Lagrange polynomial, the error formula has the form

$$f(x) - P_{0,1,\ldots,n}(x) = \frac{f^{(n+1)}(\xi(x))}{(n + 1)!}(x - x_0)(x - x_1)\cdots(x - x_n),$$

where $\xi(x)$ lies between $x, x_0, x_1, \ldots, x_n$.

To derive the Midpoint formula we could use the constant interpolating polynomial with $x_0 = (a + b)/2$ to produce

$$\int_a^b f(x)\, dx \approx \int_a^b f[x_0]\, dx = f[x_0](b - a) = f\left(\frac{a + b}{2}\right)(b - a).$$

We could also use a linear interpolating polynomial with this value of x_0 and an arbitrary value of x_1. This is due to the fact that the integral of the second term in the Newton Forward Divided-Difference formula is zero for our choice of x_0, independent of the value of x_1, and as such does not contribute to the approximation:

$$\int_a^b f[x_0, x_1](x - x_0)\, dx = \frac{f[x_0, x_1]}{2}(x - x_0)^2 \Big|_a^b$$

$$= \frac{f[x_0, x_1]}{2}\left(x - \frac{a+b}{2}\right)^2 \Big|_a^b$$

$$= \frac{f[x_0, x_1]}{2}\left[\left(b - \frac{a-b}{2}\right)^2 - \left(a - \frac{a+b}{2}\right)^2\right]$$

$$= \frac{f[x_0, x_1]}{2}\left[\left(\frac{b-a}{2}\right)^2 - \left(\frac{a-b}{2}\right)^2\right] = 0.$$

In general, the higher the degree of the approximation, the higher the order of the error term, so we will integrate the error for $P_{0,1}(x)$ instead of $P_0(x)$ to determine an error formula for the Midpoint rule.

Suppose that the arbitrary x_1 was chosen to be the same value as x_0. (In fact, this is the only value that we *cannot* have for x_1, but we will ignore this problem for the moment.) Then the integral of the error formula for the interpolating polynomial $P_0(x)$ has the form

$$\int_a^b \frac{(x - x_0)(x - x_1)}{2} f''(\xi(x))\, dx = \int_a^b \frac{(x - x_0)^2}{2} f''(\xi(x))\, dx.$$

Since the term $(x - x_0)^2$ does not change sign on the interval (a, b), the Mean Value Theorem for Integrals implies that a number ξ in (a, b) exists with

$$\int_a^b \frac{(x - x_0)^2}{2} f''(\xi(x))\, dx = f''(\xi) \int_a^b \frac{(x - x_0)^2}{2}\, dx = \frac{f''(\xi)}{6}(x - x_0)^3 \Big|_a^b$$

$$= \frac{f''(\xi)}{6}\left[\left(b - \frac{b+a}{2}\right)^3 - \left(a - \frac{b+a}{2}\right)^3\right]$$

$$= \frac{f''(\xi)}{6}\frac{(b-a)^3}{4} = \frac{f''(\xi)}{24}(b - a)^3.$$

As a consequence, the Midpoint rule with its error formula has the following form:

Midpoint Rule

$$\int_a^b f(x)\, dx = (b - a)f\left(\frac{a+b}{2}\right) + \frac{f''(\xi)}{24}(b - a)^3,$$

for some number ξ in (a, b).

The invalid assumption $x_1 = x_0$ that led to this result can be avoided by taking x_1 close, but not equal, to x_0 and using limits to show that the error formula is valid.

The Midpoint rule uses a constant interpolating polynomial disguised as a linear interpolating polynomial. The next method we consider uses a true linear interpolating polynomial, one with the distinct nodes $x_0 = a$ and $x_1 = b$. This approximation is also easy to generate geometrically, as is shown in Figure 4.2, and is aptly called the Trapezoidal rule. If we integrate the linear interpolating polynomial with $x_0 = a$ and $x_1 = b$ we also produce this formula:

$$\int_a^b \{f[x_0] + f[x_0, x_1](x - x_0)\}\, dx = \left[f[a]x + f[a, b]\frac{(x - a)^2}{2} \right]_a^b$$

$$= f(a)(b - a) + \frac{f(b) - f(a)}{b - a}\left[\frac{(b - a)^2}{2} - \frac{(a - a)^2}{2} \right]$$

$$= (b - a)\frac{f(a) + f(b)}{2}.$$

Figure 4.2

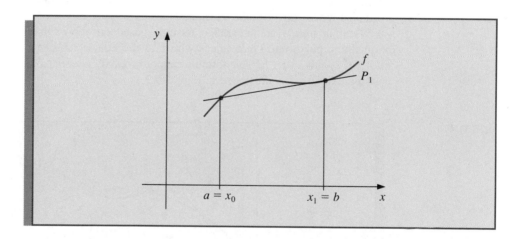

The error for the Trapezoidal rule follows from integrating the error term for $P_{0,1}(x)$ when $x_0 = a$ and $x_1 = b$. Since $(x - x_0)(x - x_1) = (x - a)(x - b)$ does not change sign in the interval (a, b), we can again apply the Mean Value Theorem for Integrals. In this case it implies that a number ξ in (a, b) exists with

$$\int_a^b \frac{(x - a)(x - b)}{2}f''(\xi(x))\, dx = \frac{f''(\xi)}{2}\int_a^b (x - a)[(x - a) + (a - b)]\, dx$$

$$= \frac{f''(\xi)}{2}\left[\frac{(x - a)^3}{3} + \frac{(x - a)^2}{2}(a - b) \right]_a^b$$

$$= \frac{f''(\xi)}{2}\left[\frac{(b - a)^3}{3} + \frac{(b - a)^2}{2}(a - b) \right].$$

Simplifying this equation gives the Trapezoidal rule with its error formula.

Trapezoidal Rule

$$\int_a^b f(x)\, dx = (b-a)\frac{f(a)+f(b)}{2} - \frac{f''(\xi)}{12}(b-a)^3,$$

for some ξ in (a,b).

We cannot improve on the order of this error formula as we did in the case of the Midpoint rule, because the integral of the next higher term in the Newton Forward Divided-Difference formula is

$$\int_a^b f[x_0,x_1,x_2](x-x_0)(x-x_1)\, dx = f[x_0,x_1,x_2]\int_a^b (x-a)(x-b)\, dx.$$

Since $(x-a)(x-b) < 0$ for all x in (a,b), this error term will not be zero unless $f[x_0,x_1,x_2] = 0$. As a consequence, the error formulas for the Midpoint and the Trapezoidal rules are both of order 3, even though they are derived from interpolation formulas with error formulas of order 1 and 2, respectively.

Next in line is an integration formula based on approximating the function f by a quadratic polynomial that agrees with f at the equally spaced points $x_0 = a$, $x_1 = (a+b)/2$, and $x_2 = b$. This formula cannot be easily generated geometrically, although the approximation is illustrated in Figure 4.3.

Figure 4.3

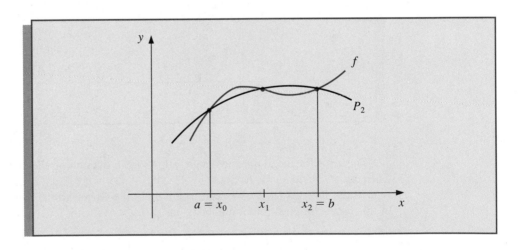

To derive the formula we will integrate $P_{0,1,2}(x)$.

$$\int_a^b P_{0,1,2}(x)\, dx = \int_a^b \left\{ f(a) + f\left[a, \frac{a+b}{2}\right](x-a) + f\left[a, \frac{a+b}{2}, b\right](x-a)\left(x - \frac{a+b}{2}\right) \right\} dx$$

$$= \left[f(a)x + f\left[a, \frac{a+b}{2}\right]\frac{(x-a)^2}{2} \right]_a^b$$

$$+ f\left[a, \frac{a+b}{2}, b\right]\int_a^b (x-a)\left[(x-a) + \left(a - \frac{a+b}{2}\right)\right] dx$$

$$= f(a)(b-a) + \frac{f\left(\frac{a+b}{2}\right) - f(a)}{\frac{a+b}{2} - a} \frac{(b-a)^2}{2}$$

$$+ \frac{f\left[\frac{a+b}{2}, b\right] - f\left[a, \frac{a+b}{2}\right]}{b-a}\left[\frac{(x-a)^3}{3} + \frac{(x-a)^2}{2}\left(\frac{a-b}{2}\right)\right]_a^b$$

$$= (b-a)\left[f(a) + f\left(\frac{a+b}{2}\right) - f(a)\right]$$

$$+ \left(\frac{1}{b-a}\right)\left[\frac{f(b) - f\left(\frac{a+b}{2}\right)}{\frac{b-a}{2}} - \frac{f\left(\frac{a+b}{2}\right) - f(a)}{\frac{b-a}{2}}\right]\left[\frac{(b-a)^3}{3} - \frac{(b-a)^3}{4}\right]$$

$$= (b-a)f\left(\frac{a+b}{2}\right) + \frac{2}{(b-a)^2}\left[f(b) - 2f\left(\frac{a+b}{2}\right) + f(a)\right]\frac{(b-a)^3}{12}.$$

Simplifying this equation gives the approximation method known as Simpson's rule:

$$\int_a^b f(x)\, dx \approx \frac{(b-a)}{6}\left[f(a) + 4f\left(\frac{a+b}{2}\right) + f(b)\right].$$

An order 4 error formula for Simpson's rule can be derived by using the error formula for the interpolating polynomial $P_{0,1,2}(x)$, but, similar to the case of the Midpoint rule, the integral of the next term in the Newton Forward Divided-Difference formula is zero. This implies that the next higher error formula can be used to produce an order 5 error formula. When simplified, Simpson's rule with this error formula is as follows:

Simpson's Rule

$$\int_a^b f(x)\, dx = \frac{(b-a)}{6}\left[f(a) + 4f\left(\frac{a+b}{2}\right) + f(b)\right] - \frac{f^{(4)}(\xi)}{2880}(b-a)^5,$$

for some ξ in (a, b).

This higher-order error term makes Simpson's rule significantly superior to the Midpoint and Trapezoidal rules in almost all situations if the width of the interval is small. This is illustrated in the following example.

EXAMPLE 1   Tables 4.1 and 4.2 (page 98) show results of the Midpoint, Trapezoidal, and Simpson's rules applied to a variety of functions integrated over the intervals $[1, 1.2]$ and $[0, 2]$.

■   ■   ■

Notice that the results are quite accurate for all the rules when the interval of integration is small, but the Midpoint and Trapezoidal rules give inaccurate approximations on the larger interval. Although the approximations from Simpson's rule are much better for the larger interval, they are still accurate only to within one or two decimal places.

Table 4.1   Integrals on the Interval [1,1.2]

$f(x)$	x^2	x^4	$1/(x+1)$	$\sqrt{1+x^2}$	$\sin x$	e^x
Exact value	0.24267	0.29766	0.09531	0.29742	0.17794	0.60184
Midpoint	0.24200	0.29282	0.09524	0.29732	0.17824	0.60083
Trapezoidal	0.24400	0.30736	0.09545	0.29626	0.17735	0.60384
Simpson's	0.24267	0.29767	0.09531	0.29742	0.17794	0.60184

Table 4.2   Integrals on the Interval [0, 2]

$f(x)$	x^2	x^4	$1/(x+1)$	$\sqrt{1+x^2}$	$\sin x$	e^x
Exact value	2.667	6.400	1.099	2.958	1.416	6.389
Midpoint	2.000	2.000	1.000	2.818	1.682	5.436
Trapezoidal	4.000	16.000	1.333	3.326	0.909	8.389
Simpson's	2.667	6.667	1.111	2.964	1.425	6.421

There are higher-order formulas that can be used to improve the accuracy when integrating over large intervals, some of which are considered in the exercises, but a better solution to the problem is considered in the next section.

EXERCISE SET 4.2

1. Approximate the following integrals using the Midpoint rule.

   (a) $\int_1^2 x \ln x\, dx$

   (b) $\int_{-2}^2 x^3 e^x\, dx$

   (c) $\int_0^2 \dfrac{2\,dx}{x^2+4}$

   (d) $\int_0^\pi x^2 \cos x\, dx$

   (e) $\int_0^2 e^{2x} \sin 3x\, dx$

   (f) $\int_1^3 \dfrac{x\,dx}{x^2+4}$

   (g) $\int_3^5 \dfrac{dx}{\sqrt{x^2-4}}$

   (h) $\int_0^{3\pi/8} \tan x\, dx$

2. Find a bound for the error in Exercise 1 using the error formula. Compute the actual error.

3. Repeat Exercise 1 using the Trapezoidal rule.

4. Repeat Exercise 2 using the Trapezoidal rule and the results of Exercise 3.

5. Repeat Exercise 1 using Simpson's rule.

6. Repeat Exercise 2 using Simpson's rule and the results of Exercise 5.

7. Approximate the following integrals using the Midpoint rule.

   (a) $\int_1^{1.5} x^2 \ln x\, dx$

   (b) $\int_0^1 x^2 e^{-x}\, dx$

   (c) $\int_0^{0.35} \dfrac{2\,dx}{x^2-4}$

   (d) $\int_0^{\pi/4} x^2 \sin x\, dx$

   (e) $\int_0^{\pi/4} e^{3x} \sin 2x\, dx$

   (f) $\int_1^{1.6} \dfrac{2x\,dx}{x^2-4}$

   (g) $\int_3^{3.5} \dfrac{x\,dx}{\sqrt{x^2-4}}$

   (h) $\int_0^{\pi/4} \cos^2 x\, dx$

8. Find a bound for the error in Exercise 7 using the error formula and compare to the actual error.

9. Repeat Exercise 7 using the Trapezoidal rule.

10. Repeat Exercise 8 using the Trapezoidal rule and the results of Exercise 9.

11. Repeat Exercise 7 using Simpson's rule.

12. Repeat Exercise 8 using Simpson's rule and the results of Exercise 11.

Other quadrature formulas with error terms are as follows:

1. $\int_a^b f(x)\, dx = \dfrac{b-a}{8}[f(a) + 3f(a+h) + 3f(a+2h) + f(b)] - \dfrac{3h^5}{80}f^{(4)}(\xi),$

   where $h = \dfrac{b-a}{3}$

2. $\int_a^b f(x)\, dx = \dfrac{b-a}{90}[7f(a) + 32f(a+h) + 12f(a+2h) + 32f(a+3h) + 7f(b)]$

   $- \dfrac{8h^7}{945}f^{(6)}(\xi),$ where $h = \dfrac{b-a}{4}$

3. $\int_a^b f(x)\, dx = \dfrac{b-a}{2}[f(a+h) + f(a+2h)] + \dfrac{3h^3}{4}f''(\xi),$ where $h = \dfrac{b-a}{3}$

4. $\int_a^b f(x)\, dx = \dfrac{b-a}{3}[2f(a+h) - f(a+2h) + 2f(a+3h)] + \dfrac{14h^5}{45}f^{(4)}(\xi),$

   where $h = \dfrac{b-a}{4}$

5. $\int_a^b f(x)\, dx = \dfrac{b-a}{24}[11f(a+h) + f(a+2h) + f(a+3h) + 11f(a+4h)]$

   $+ \dfrac{95h^5}{144}f^{(4)}(\xi),$ where $h = \dfrac{b-a}{5}$

13. Repeat Exercises 1 and 2 using formula (1).

14. Repeat Exercises 1 and 2 using formula (2).

15. Repeat Exercises 1 and 2 using formula (3).

16. Repeat Exercises 1 and 2 using formula (4).

17. Repeat Exercises 1 and 2 using formula (5).

4.3 Composite Quadrature Rules

The basic notions underlying numerical integration were derived in the previous section, but the techniques given there are not satisfactory for most problems. We saw an example of this in the example at the end of that section, where the approximations were all rather poor for integrals of functions on the interval $[0, 2]$. To see why this occurs, let us consider Simpson's method, generally the most accurate of these techniques. With its error formula it is given by

$$\int_a^b f(x)\, dx = \frac{b-a}{6}\left[f(a) + 4f\left(\frac{a+b}{2}\right) + f(b)\right] - \frac{(b-a)^5}{2880}f^{(4)}(\xi)$$

$$= \frac{h}{3}[f(a) + 4f(a+h) + f(b)] - \frac{h^5}{90}f^{(4)}(\xi)$$

where $h = (b - a)/2$ and ξ lies in (a, b). If the fourth derivative of f is bounded on (a, b), the error term in this formula is $O(h^5)$. That is, the error term behaves in a manner similar to h^5 as h approaches zero. As a consequence, we expect the error to be small provided that the fourth derivative of f is well behaved and the interval $[a, b]$ is small. The first assumption we can live with, but the second is quite unreasonable. There is no reason, in general, to expect that the interval over which the integration is performed is small, and if it is not, the $O(h^5)$ portion in the error term can dominate the calculations.

We circumvent the problem of a large interval of integration by subdividing the interval $[a, b]$ into a collection of smaller intervals. Each of these will be sufficiently small so that the error over that part is kept under control. In this section we see how this is accomplished while keeping the maximum total error bounded.

EXAMPLE 1 Consider finding an approximation to $\int_0^2 e^x \, dx$. If Simpson's rule is used with $h = 1$,

$$\int_0^2 e^x \, dx \approx \frac{1}{3}(e^0 + 4e^1 + e^2) = 6.4207278.$$

Since the exact answer in this case is $e^2 - e^0 = 6.3890561$, the error of magnitude 0.0316717 is larger than would generally be regarded as acceptable. To apply a piecewise technique to this problem, subdivide $[0, 2]$ into $[0, 1]$ and $[1, 2]$ and use Simpson's rule twice with $h = \frac{1}{2}$, giving

$$\int_0^2 e^x \, dx = \int_0^1 e^x \, dx + \int_1^2 e^x \, dx \approx \frac{1}{6}[e^0 + 4e^{0.5} + e^1] + \frac{1}{6}[e^1 + 4e^{1.5} + e^2]$$

$$= \frac{1}{6}[e^0 + 4e^{0.5} + 2e^1 + 4e^{1.5} + e^2] = 6.3912102.$$

The magnitude of the error now has been reduced to 0.0021541. Encouraged by our results, we subdivide each of the intervals $[0, 1]$ and $[1, 2]$ and use Simpson's rule with $h = \frac{1}{4}$, giving

$$\int_0^2 e^x \, dx = \int_0^{0.5} e^x \, dx + \int_{0.5}^1 e^x \, dx + \int_1^{1.5} e^x \, dx + \int_{1.5}^2 e^x \, dx$$

$$\approx \frac{1}{12}[e^0 + 4e^{0.25} + e^{0.5}] + \frac{1}{12}[e^{0.5} + 4e^{0.75} + e^1]$$

$$+ \frac{1}{12}[e^1 + 4e^{1.25} + e^{1.5}] + \frac{1}{12}[e^{1.5} + 4e^{1.75} + e^2]$$

$$= \frac{1}{12}[e^0 + 4e^{0.25} + 2e^{0.5} + 4e^{0.75} + 2e^1 + 4e^{1.25} + 2e^{1.5} + 4e^{1.75} + e^2]$$

$$= 6.3891937.$$

The magnitude of the error for this approximation is 0.0001376. Each subdivision has produced a result that improved the accuracy by more than one decimal place. ■ ■ ■

The generalization of the procedure considered in this example is called the Composite Simpson's rule and is described as follows: Subdivide the interval $[a, b]$ into n

subintervals and use Simpson's rule on each consecutive pair of subintervals. Since each application of Simpson's rule requires two subintervals, n must be an even integer. With $h = (b - a)/n$ and $a = x_0 < x_1 < \cdots < x_n = b$, where $x_j = x_0 + jh$ for each $j = 0, 1, \ldots, n$, we have

$$\int_a^b f(x)\, dx = \sum_{j=1}^{\frac{n}{2}} \int_{x_{2j-2}}^{x_{2j}} f(x)\, dx$$

$$= \sum_{j=1}^{\frac{n}{2}} \left\{ \frac{h}{3}[f(x_{2j-2}) + 4f(x_{2j-1}) + f(x_{2j})] - \frac{h^5}{90} f^{(4)}(\xi_j) \right\}$$

for some ξ_j with $x_{2j-2} < \xi_j < x_{2j}$, provided that $f \in C^4[a, b]$. To simplify this formula, first note that for each $j = 1, 2, \ldots, \frac{n}{2} - 1$, $f(x_{2j})$ appears in the term corresponding to the interval $[x_{2j-2}, x_{2j}]$ and also in the term corresponding to the interval $[x_{2j-2}, x_{2j}]$. This simplifies the formula to (see Figure 4.4)

$$\int_a^b f(x)\, dx = \frac{h}{3} \left[f(x_0) + 2 \sum_{j=1}^{\frac{n}{2}-1} f(x_{2j}) + 4 \sum_{j=1}^{\frac{n}{2}} f(x_{2j-1}) + f(x_n) \right] - \frac{h^5}{90} \sum_{j=1}^{\frac{n}{2}} f^{(4)}(\xi_j).$$

Figure 4.4

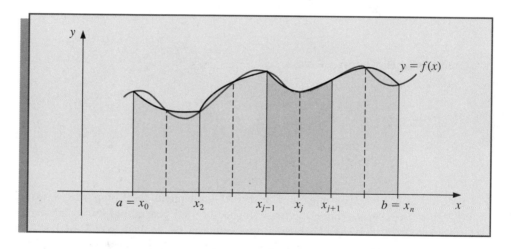

The error associated with the approximation is

$$E(f) = -\frac{h^5}{90} \sum_{j=1}^{\frac{n}{2}} f^{(4)}(\xi_j),$$

where $x_{2j-2} < \xi_j < x_{2j}$ for each $j = 1, 2, \ldots, \frac{n}{2}$. If $f \in C^4[a, b]$, the Extreme Value Theorem implies that $f^{(4)}$ assumes its maximum and minimum in $[a, b]$. Since

$$\min_{x \in [a,b]} f^{(4)}(x) \le f^{(4)}(\xi_j) \le \max_{x \in [a,b]} f^{(4)}(x),$$

we have

$$\frac{n}{2} \min_{x \in [a,b]} f^{(4)}(x) \le \sum_{j=1}^{\frac{n}{2}} f^{(4)}(\xi_j) \le \frac{n}{2} \max_{x \in [a,b]} f^{(4)}(x),$$

and

$$\min_{x \in [a,b]} f^{(4)}(x) \leq \frac{2}{n} \sum_{j=1}^{\frac{n}{2}} f^{(4)}(\xi_j) \leq \max_{x \in [a,b]} f^{(4)}(x).$$

Since the term in the middle lies between values of $f^{(4)}$, the Intermediate Value Theorem implies that a number μ in (a, b) exists with

$$f^{(4)}(\mu) = \frac{2}{n} \sum_{j=1}^{\frac{n}{2}} f^{(4)}(\xi_j).$$

Since $h = (b - a)/n$, this simplifies the error formula to

$$E(f) = -\frac{h^5}{90} \sum_{j=1}^{\frac{n}{2}} f^{(4)}(\xi_j) = -\frac{h^4(b - a)}{180} f^{(4)}(\mu).$$

Summarizing these results we have the following: If $f \in C^4[a, b]$, there exists a $\mu \in (a, b)$ for which the Composite Simpson's rule for n subintervals of $[a, b]$ can be expressed with error term as follows:

Composite Simpson's Rule

$$\int_a^b f(x)\, dx = \frac{h}{3}\left[f(a) + 2\sum_{j=1}^{\frac{n}{2}-1} f(x_{2j}) + 4\sum_{j=1}^{\frac{n}{2}} f(x_{2j-1}) + f(b) \right] - \frac{(b - a)h^4}{180} f^{(4)}(\mu),$$

for some μ in (a, b).

The program CSIMPR41 implements the Composite Simpson's rule on n subintervals. This is the most frequently used general-purpose quadrature technique.

The subdivision approach can be applied to any of the formulas we saw in the preceding section. Since the Trapezoidal rule (see Figure 4.5) uses only interval endpoint values to determine its approximation, there is no restriction on the integer n in this case.

Let $f \in C^2[a, b]$. With $h = (b - a)/n$ and $x_j = a + jh$ for each $j = 0, 1, \ldots, n$, the Composite Trapezoidal rule for n subintervals is as follows:

Composite Trapezoidal Rule

$$\int_a^b f(x)\, dx = \frac{h}{2}\left[f(a) + f(b) + 2\sum_{j=1}^{n-1} f(x_j) \right] - \frac{(b - a)h^2}{12} f''(\mu),$$

for some $\mu \in (a, b)$.

Figure 4.5

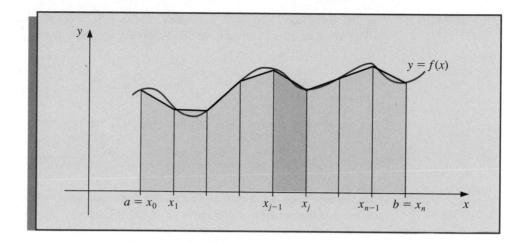

Let $f \in C^2[a, b]$ and n be an even integer. With $h = (b - a)/(n + 2)$ and $x_j = a + (j + 1)h$ for each $j = -1, 0, \ldots, n + 1$, the Composite Midpoint rule for $\frac{n}{2} + 1$ subintervals is as follows (see Figure 4.6):

Composite Midpoint Rule

$$\int_a^b f(x)\,dx = 2h \sum_{j=0}^{\frac{n}{2}} f(x_{2j}) + \frac{(b - a)h^2}{6} f''(\mu),$$

for some $\mu \in (a, b)$.

Figure 4.6

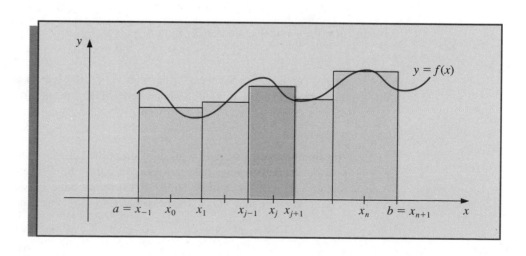

EXAMPLE 2   Consider using the Composite Simpson's rule to approximate the integral $\int_0^\pi \sin x \, dx$ with an absolute error at most 0.00002. Applying the formula to $\int_0^\pi \sin x \, dx$ gives

$$\int_0^\pi \sin x \, dx = \frac{h}{3}\left[2\sum_{j=1}^{\frac{n}{2}-1}\sin x_{2j} + 4\sum_{j=1}^{\frac{n}{2}}\sin x_{2j-1}\right] - \frac{\pi h^4}{180}\sin\mu.$$

Since the absolute error is to be less than 0.00002, the inequality

$$\left|\frac{\pi h^4}{180}\sin\mu\right| \le \frac{\pi h^4}{180}\cdot 1 = \frac{\pi^5}{180n^4} \le 0.00002$$

is used to determine n and then h. Completing these calculations gives $n \ge 18$. Verifying this, we find that when $n = 20$ and $h = \pi/20$, Composite Simpson's rule gives

$$\int_0^\pi \sin x \, dx \approx \frac{\pi}{60}\left[2\sum_{j=1}^{9}\sin\left(\frac{j\pi}{10}\right) + 4\sum_{j=1}^{10}\sin\left(\frac{(2j-1)\pi}{20}\right)\right] = 2.000006.$$

To be assured of this degree of accuracy using the Composite Trapezoidal rule requires that

$$\left|\frac{\pi h^2}{12}\sin\mu\right| \le \frac{\pi h^2}{12} = \frac{\pi^3}{12n^2} < 0.00002$$

which implies that $n \ge 360$. The Composite Midpoint rule would require approximately twice this number of function evaluations.

For comparison purposes, the Composite Midpoint rule with $n = 19$ and $h = \pi/40$ gives

$$\int_0^\pi \sin x \, dx \approx \frac{\pi}{20}\sum_{j=0}^{19}\sin\left(\frac{(2j+1)\pi}{40}\right) = 2.0020577$$

and the Composite Trapezoidal rule with $n = 20$ and $h = \pi/20$ gives

$$\int_0^\pi \sin x \, dx \approx \frac{\pi}{40}\left[\sin 0 + \sin\pi + 2\sum_{j=1}^{19}\sin\left(\frac{j\pi}{20}\right)\right] = \frac{\pi}{40}\left[2\sum_{j=1}^{19}\sin\left(\frac{j\pi}{20}\right)\right] = 1.9958860.$$

The exact answer is 2; so Simpson's rule with $n = 20$ gave an answer well within the required error bound of 0.00002, whereas the Midpoint and Trapezoidal rules with $n = 20$ clearly do not.   ■ ■ ■

An important property shared by all the Composite rules is stability with respect to round-off error. To demonstrate this, suppose we apply the Composite Simpson's rule with n subintervals to a function on $[a, b]$ and determine the maximum bound for the round-off error. Assume that $f(x_i)$ is approximated by $\tilde{f}(x_i)$ and that

$$f(x_i) = \tilde{f}(x_i) + e_i, \qquad \text{for each} \quad i = 0, 1, \ldots, n,$$

where e_i denotes the round-off error associated with using $\tilde{f}(x_i)$ to approximate $f(x_i)$. Then the accumulated round-off error, $e(h)$, in the Composite Simpson's rule is

$$e(h) = \left| \frac{h}{3} \left[e_0 + 2 \sum_{j=1}^{\frac{n}{2}-1} e_{2j} + 4 \sum_{j=1}^{\frac{n}{2}} e_{2j-1} + e_n \right] \right|$$

$$\leq \frac{h}{3} \left[|e_0| + 2 \sum_{j=1}^{\frac{n}{2}-1} |e_{2j}| + 4 \sum_{j=1}^{\frac{n}{2}} |e_{2j-1}| + |e_n| \right].$$

If the round-off errors are uniformly bounded by some known tolerance ϵ, then

$$e(h) \leq \frac{h}{3} \left[\epsilon + 2 \left(\frac{n}{2} - 1 \right) \epsilon + 4 \left(\frac{n}{2} \right) \epsilon + \epsilon \right] = \frac{h}{3} 3n\epsilon = nh\epsilon.$$

But $nh = b - a$, so $e(h) \leq (b-a)\varepsilon$, a bound independent of h and n.

EXERCISE SET 4.3

1. Use the Composite Trapezoidal rule with the indicated values of n to approximate the following integrals.

   (a) $\int_1^2 x \ln x \, dx$, $n = 4$ (b) $\int_{-2}^2 x^3 e^x \, dx$, $n = 4$

   (c) $\int_0^2 \frac{2}{x^2 + 4} \, dx$, $n = 6$ (d) $\int_0^\pi x^2 \cos x \, dx$, $n = 6$

   (e) $\int_0^2 e^{2x} \sin 3x \, dx$, $n = 8$ (f) $\int_1^3 \frac{x}{x^2 + 4} \, dx$, $n = 8$

   (g) $\int_3^5 \frac{1}{\sqrt{x^2 - 4}} \, dx$, $n = 8$ (h) $\int_0^{3\pi/8} \tan x \, dx$, $n = 8$

2. Use the Composite Midpoint rule with $n/2 + 1$ subintervals to approximate the integrals in Exercise 1.

3. Use the Composite Simpson's rule to approximate the integrals in Exercise 1.

4. To approximate $\int_0^2 x^2 e^{-x^2} \, dx$,
   (a) use the Composite Trapezoidal rule with $n = 8$.
   (b) use the Composite Midpoint rule with $n = 8$.
   (c) use the Composite Simpson's rule with $n = 8$.

5. Determine the values of n and h required to approximate

$$\int_0^2 e^{2x} \sin 3x \, dx$$

   to within 10^{-4}.
   (a) Use the Composite Trapezoidal rule.   (b) Use the Composite Midpoint rule.
   (c) Use the Composite Simpson's rule.

6. Repeat Exercise 5 for the integral $\int_0^\pi x^2 \cos x \, dx$.

7. Determine the values of n and h required to approximate

$$\int_0^2 \frac{1}{x + 4} \, dx$$

to within 10^{-5} and compute the approximation.
(a) Use the Composite Trapezoidal rule.   (b) Use the Composite Midpoint rule.
(c) Use the Composite Simpson's rule.

8. Repeat Exercise 7 for the integral $\int_1^2 x \ln x \, dx$.

9. Use the Composite Simpson's rule to find an approximation to within 10^{-4} of the value of the integral

$$\int_0^{48} \sqrt{1 + (\cos x)^2} \, dx.$$

10. Find an approximation of the area of the region bounded by the normal curve

$$y = \frac{1}{\sigma \sqrt{2\pi}} e^{\frac{-(x/\sigma)^2}{2}}$$

and the x-axis on the interval $[-\sigma, \sigma]$ using the Composite Trapezoidal rule with $n = 8$.

11. Repeat Exercise 10 using the Composite Simpson's rule with $n = 8$.

12. A car laps a race track in 84 s. The speed of the car at each 6-s interval is determined using a radar gun and given from the beginning of the lap, in feet/second, by the entries in the following table:

Time	0	6	12	18	24	30	36	42	48	54	60	66	72	78	84
Speed	124	134	148	156	147	133	121	109	99	85	78	89	104	116	123

How long is the track?

13. A particle of mass m moving through a fluid is subjected to a viscous resistance R, which is a function of the velocity v. The relationship between the resistance R, velocity v, and time t is given by the equation

$$t = \int_{v(t_0)}^{v(t)} \frac{m}{R(u)} \, du.$$

Suppose that $R(v) = -v\sqrt{v}$ for a particular fluid, where R is in newtons and v is in meters/second. If $m = 10$ kg and $v(0) = 10$ m/s, approximate the time required for the particle to slow to $v = 5$ m/s.
(a) Use the Composite Simpson's rule with $h = 0.25$.
(b) Use the Composite Trapezoidal rule with $h = 0.25$.
(c) Compare these approximations to the actual value.

14. The equation

$$\int_0^x \frac{1}{\sqrt{2\pi}} e^{-\frac{t^2}{2}} \, dt = 0.45$$

can be solved using Newton's method with

$$f(x) = \int_0^x \frac{1}{\sqrt{2\pi}} e^{-\frac{t^2}{2}} \, dt - 0.45$$

and

$$f'(x) = \frac{1}{\sqrt{2\pi}} e^{-\frac{x^2}{2}}.$$

To evaluate f at the approximation p_k, we need a quadrature formula to approximate

$$\int_0^{p_k} \frac{1}{\sqrt{2\pi}} e^{-\frac{t^2}{2}} \, dt.$$

(a) Find a solution to $f(x) = 0$ accurate to within 10^{-5} using Newton's method with $p_0 = 0.5$ and the Composite Simpson's method.

(b) Repeat (a) using the Composite Trapezoidal method in place of the Composite Simpson's method.

4.4 Gaussian Quadrature

The quadrature formulas for approximating the integral of a function were derived in Section 4.2 by integrating successively higher-order interpolating polynomials. The error term in the interpolating polynomial of degree n involves the $(n + 1)$st derivative of the function being approximated. Since every polynomial of degree less than or equal to n has its $(n + 1)$st derivative zero, applying a formula of this type to such polynomials gives an exact result.

All the formulas in Section 4.2 use values of the function at equally spaced points. This is convenient when the formulas are combined to form the composite rules we considered in the preceding section, but this restriction can significantly decrease the accuracy of the approximation. Consider, for example, the Trapezoidal rule applied to determine the integrals of the functions shown in Figure 4.7.

Figure 4.7

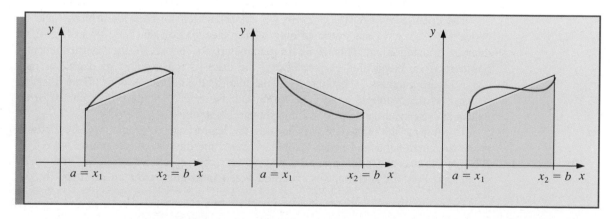

The Trapezoidal rule approximates the integral of the function by integrating the linear function that joins the endpoints of the graph of the function. But this is not likely to be the best line for approximating the integral. Lines such at those shown in Figure 4.8 would give better approximations in most cases.

Figure 4.8

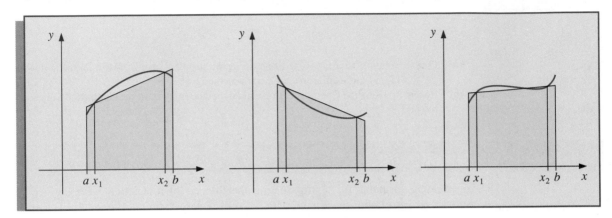

Gaussian quadrature chooses the points for evaluation in an optimal, rather than equally spaced, manner. The nodes $x_1, x_2, \ldots, x_n$ in the interval $[a, b]$ and coefficients $c_1, c_2, \ldots, c_n$ are chosen to minimize the expected error obtained in performing the approximation

$$\int_a^b f(x)\, dx \approx \sum_{i=1}^n c_i f(x_i)$$

for an arbitrary function f. To measure this accuracy, we assume that the best choice of these values is that producing the exact result for the largest class of polynomials.

The coefficients $c_1, c_2, \ldots, c_n$ in the approximation formula are arbitrary and the nodes $x_1, x_2, \ldots, x_n$ are restricted only by the specification that they lie in $[a, b]$, the interval of integration. This gives $2n$ parameters to choose. If the coefficients of a polynomial are considered as parameters, the class of polynomials of degree at most $(2n - 1)$ also contains $2n$ parameters. This, then, is the largest class of polynomials for which it is reasonable to expect the formula to be exact. For the proper choice of the values and constants, exactness on this set can be obtained.

To illustrate the procedure for choosing the appropriate constants, we show how to select the coefficients and nodes when $n = 2$ and the interval of integration is $[-1, 1]$. We then discuss the more general situation for arbitrary choice of nodes and coefficients and how the technique is modified when integrating over an arbitrary interval $[a, b]$.

EXAMPLE 1   Suppose we want to determine c_1, c_2, x_1, and x_2 so that the integration formula

$$\int_{-1}^1 f(x)\, dx \approx c_1 f(x_1) + c_2 f(x_2)$$

gives the exact result whenever $f(x)$ is a polynomial of degree $2(2)-1 = 3$ or less—that is, when

$$f(x) = a_0 + a_1x + a_2x^2 + a_3x^3,$$

for some collection of constants, $a_0, a_1, a_2,$ and a_3. Because

$$\int (a_0 + a_1x + a_2x^2 + a_3x^3)\, dx = a_0 \int 1\, dx + a_1 \int x\, dx + a_2 \int x^2\, dx + a_3 \int x^3\, dx,$$

this is equivalent to showing that the formula gives exact results when $f(x)$ is $1, x, x^2,$ and x^3. This is the condition that we will satisfy. To ensure this, we need $c_1, c_2, x_1,$ and x_2 so that

$$c_1 \cdot 1 + c_2 \cdot 1 = \int_{-1}^{1} 1\, dx = 2, \qquad\qquad c_1 \cdot x_1 + c_2 \cdot x_2 = \int_{-1}^{1} x\, dx = 0,$$

$$c_1 \cdot x_1^2 + c_2 \cdot x_2^2 = \int_{-1}^{1} x^2\, dx = \tfrac{2}{3}, \qquad \text{and} \qquad c_1 \cdot x_1^3 + c_2 \cdot x_2^3 = \int_{-1}^{1} x^3\, dx = 0.$$

A little algebra shows that this system of equations has the unique solution

$$c_1 = 1, \qquad c_2 = 1, \qquad x_1 = -\frac{\sqrt{3}}{3}, \qquad \text{and} \qquad x_2 = \frac{\sqrt{3}}{3}.$$

This produces the following integral approximation formula:

$$\int_{-1}^{1} f(x)\, dx \approx f\left(-\frac{\sqrt{3}}{3}\right) + f\left(\frac{\sqrt{3}}{3}\right),$$

which gives the exact result for every polynomial of degree 3 or less. ■ ■ ■

The technique in Example 1 can be used to determine the nodes and coefficients for formulas that give exact results for higher-degree polynomials, but an alternative method can be used to obtain them more easily. In Section 8.3 we will consider various collections of orthogonal polynomials, functions with the property that a particular definite integral of the product of any two of them is zero. The set that is relevant to our problem is the set of Legendre polynomials, a collection $\{P_0, P_1, \ldots, P_n, \ldots\}$ that has the following properties:

1. For each n, P_n is a polynomial of degree n.

2. $\displaystyle\int_{-1}^{1} P_i(x)P_j(x)\, dx = 0$ whenever $i \neq j$.

The first few Legendre polynomials are

$$P_0(x) = 1, \qquad P_1(x) = x, \qquad P_2(x) = x^2 - \tfrac{1}{3},$$

$$P_3(x) = x^3 - \tfrac{3}{5}x, \qquad \text{and} \qquad P_4(x) = x^4 - \tfrac{6}{7}x^2 + \tfrac{3}{35}.$$

The roots of these polynomials are distinct, lie in the interval $(-1, 1)$, have a symmetry with respect to the origin, and, most importantly, are the correct choice for determining the nodes that solve our problem. The nodes $x_1, x_2, \ldots, x_n$ needed to produce an integral approximation formula that will give exact results for any polynomial of degree $2n-1$

or less are the roots of the nth-degree Legendre polynomial. In addition, once the roots are known, the appropriate coefficients for the function evaluations at these nodes can be found from the fact that for each $i = 1, 2, \ldots, n$,

$$c_i = \int_{-1}^{1} \prod_{\substack{j=1 \\ j \neq i}}^{n} \frac{(x - x_j)}{(x_i - x_j)} \, dx.$$

However, both the coefficients and the roots of the Legendre polynomials are extensively tabulated, so it is not necessary to perform these evaluations. A small sample is given in Table 4.3, but listings for higher-degree polynomials are given, for example, in the book by Stroud and Secrest.

Table 4.3

n	Roots $r_{n,i}$	Coefficients $c_{n,i}$
2	0.5773502692	1.0000000000
	−0.5773502692	1.0000000000
3	0.7745966692	0.5555555556
	0.0000000000	0.8888888889
	−0.7745966692	0.5555555556
4	0.8611363116	0.3478548451
	0.3399810436	0.6521451549
	−0.3399810436	0.6521451549
	−0.8611363116	0.3478548451
5	0.9061798459	0.2369268850
	0.5384693101	0.4786286705
	0.0000000000	0.5688888889
	−0.5384693101	0.4786286705
	−0.9061798459	0.2369268850

This completes the solution to the approximation problem for definite integrals of functions on the interval $[-1, 1]$. But this solution is sufficient for any closed interval, since the simple linear transformation $t = (2x - a - b)/(b - a)$ translates the interval $[a, b]$ into $[-1, 1]$. Then the Legendre polynomials can be used to approximate

$$\int_{a}^{b} f(x) \, dx = \int_{-1}^{1} f\left(\frac{(b - a)t + b + a}{2}\right)\frac{(b - a)}{2} \, dt.$$

Using the roots $r_{n,1}, r_{n,2}, \ldots, r_{n,n}$ and the coefficients $c_{n,1}, c_{n,2}, \ldots, c_{n,n}$ given in Table 4.3 transforms the approximation to the following:

Gaussian Quadrature

$$\int_{a}^{b} f(x) \, dx = \frac{b - a}{2} \sum_{j=1}^{n} c_{n,j} f\left(\frac{(b - a)r_{n,j} + b + a}{2}\right).$$

The next example illustrates the technique.

EXAMPLE 2 Consider the problem of finding approximations to $\int_1^{1.5} e^{-x^2}\,dx$, whose value to seven decimal places is 0.1093643.

Gaussian quadrature applied to this problem requires that the integral be translated into one whose interval of integration is $[-1, 1]$:

$$\int_1^{1.5} e^{-x^2}\,dx = \frac{1}{4}\int_{-1}^{1} e^{-\frac{(t+5)^2}{16}}\,dt.$$

Using the values in Table 4.3, we can find the Gaussian quadrature approximations. $n = 2$:

$$\int_1^{1.5} e^{-x^2}\,dx \approx \tfrac{1}{4}[e^{-(5+0.5773502692)^2/16} + e^{-(5-0.5773502692)^2/16}] = 0.1094003,$$

$n = 3$:

$$\int_1^{1.5} e^{-x^2}\,dx \approx \tfrac{1}{4}[(0.5555555556)e^{-(5+0.7745966692)^2/16} + (0.8888888889)e^{-(5)^2/16}$$

$$+ (0.5555555556)e^{-(5-0.7745966692)^2/16}]$$

$$= 0.1093642.$$

Using Gaussian quadrature with $n = 3$ requires three function evaluations and produces an approximation that is accurate to within 10^{-7}. The same number of function evaluations is needed if Simpson's rule is applied to the original integral using $h = (1.5 - 1)/2 = 0.25$. This application of Simpson's rule gives the approximation

$$\int_1^{1.5} e^{-x^2}\,dx \approx \frac{0.25}{3}[e^1 + 4e^{1.25} + e^{1.5}] = 0.1093104,$$

a result that is accurate only to within 5×10^{-5}. ■ ■ ■

For small problems, Composite Simpson's rule may be acceptable to avoid the computational complexity of Gaussian quadrature, but for problems requiring expensive function evaluations, the Gaussian procedure should certainly be given consideration. Gaussian quadrature is particularly important for approximating multiple integrals, since the number of function evaluations increases as a power of the number of integrals being evaluated. This topic is considered in Section 4.7.

EXERCISE SET 4.4

1. Approximate the following integrals using Gaussian quadrature with $n = 2$ and compare the approximations to the actual values.

(a) $\int_1^{1.5} x^2 \ln x\,dx$ (b) $\int_0^1 x^2 e^{-x}\,dx$ (c) $\int_0^{0.35} \frac{2}{x^2 - 4}\,dx$

(d) $\int_0^{\pi/4} x^2 \sin x\,dx$ (e) $\int_0^{\pi/4} e^{3x} \sin 2x\,dx$ (f) $\int_1^{1.6} \frac{2x}{x^2 - 4}\,dx$

(g) $\int_3^{3.5} \frac{x}{\sqrt{x^2 - 4}}\,dx$ (h) $\int_0^{\pi/4} \cos^2 x\,dx$

2. Repeat Exercise 1 with $n = 3$.

3. Repeat Exercise 1 with $n = 4$.

4. Repeat Exercise 1 with $n = 5$.

5. Verify the entries for the values of $n = 2$ and 3 in Table 4.3 by finding the roots of the respective Legendre polynomials and use the equations preceding this table to find the coefficients associated with the values.

6. Use Gaussian quadrature to approximate the integral

$$\int_0^{48} \sqrt{1 + (\cos x)^2}\, dx$$

with $n = 2, 3, 4$, and 5.

4.5 Romberg Integration

The Trapezoidal rule is one of the simplest of the integration formulas, but it generally lacks the degree of accuracy required. Romberg integration uses the Composite Trapezoidal rule to give preliminary approximations and then applies an acceleration technique called Richardson extrapolation to obtain improved approximations.

Extrapolation is used to accelerate the convergence of many approximation techniques. It can be applied whenever it is known that the approximation technique has an error term with a predictable form, one that depends on a parameter, usually the step size h.

Suppose that $N(h)$ is a formula involving a step size h that approximates an unknown value M and that it is known that the error for $N(h)$ has the form

(4.1)
$$M = N(h) + K_1 h + K_2 h^2 + K_3 h^3 + \cdots.$$

for some unspecified collection of constants, $K_1, K_2, K_3 \ldots$. We assume here that $h > 0$ can be arbitrarily chosen and that improved approximations occur as h becomes small. The objective of extrapolation is to improve the formula from one of order $O(h)$ to one of higher order. Do not be misled by the relative simplicity of this equation. It may be quite difficult to obtain the approximation $N(h)$, particularly for small values of h.

Since the formula is assumed to hold for all positive h, consider the result when we replace the parameter h by half its value. Then we have the formula

$$M = N\left(\frac{h}{2}\right) + K_1 \frac{h}{2} + K_2 \frac{h^2}{4} + K_3 \frac{h^3}{8} + \cdots.$$

Subtracting (4.1) from twice this equation eliminates the term involving K_1 and gives

$$M = \left[N\left(\frac{h}{2}\right) + N\left(\frac{h}{2}\right) - N(h) \right] + K_2 \left(\frac{h^2}{2} - h^2 \right) + K_3 \left(\frac{h^3}{4} - h^3 \right) + \cdots.$$

To facilitate the discussion, define $N_1(h) \equiv N(h)$ and

$$N_2(h) = N_1\left(\frac{h}{2}\right) + \left[N_1\left(\frac{h}{2}\right) - N_1(h) \right].$$

Then we have the $O(h^2)$ approximation formula for M:

(4.2)
$$M = N_2(h) - \frac{K_2}{2}h^2 - \frac{3K_3}{4}h^3 - \cdots.$$

If we now replace h by $h/2$ in this formula, we have

(4.3)
$$M = N_2\left(\frac{h}{2}\right) - \frac{K_2}{8}h^2 - \frac{3K_3}{32}h^3 - \cdots.$$

This can be combined with equation (4.2) to eliminate the h^2 term. Specifically, subtracting (4.2) from 4 times equation (4.3) gives

$$3M = 4N_2\left(\frac{h}{2}\right) - N_2(h) + \frac{3K_3}{4}\left(-\frac{h^3}{2} + h^3\right) + \cdots,$$

which simplifies to the $O(h^3)$ formula for approximating M:

$$M = \left[N_2\left(\frac{h}{2}\right) + \frac{N_2(\frac{h}{2}) - N_2(h)}{3}\right] + \frac{K_3}{8}h^3 + \cdots.$$

Defining

$$N_3(h) \equiv N_2\left(\frac{h}{2}\right) + \frac{N_2(\frac{h}{2}) - N_2(h)}{3}$$

simplifies to the $O(h^3)$ formula:

$$M = N_3(h) + \frac{K_3}{8}h^3 + \cdots.$$

The process is continued by constructing the $O(h^4)$ approximation

$$N_4(h) = N_3\left(\frac{h}{2}\right) + \frac{N_3(\frac{h}{2}) - N_3(h)}{7},$$

the $O(h^5)$ approximation

$$N_5(h) = N_4\left(\frac{h}{2}\right) + \frac{N_4(\frac{h}{2}) - N_4(h)}{15},$$

and so on. In general, if M can be written in the form

$$M = N(h) + \sum_{j=1}^{m-1} K_j h^j + O(h^m),$$

then for each $j = 2, 3, \ldots, m$, we have an $O(h^j)$ approximation of the form

$$N_j(h) = N_{j-1}\left(\frac{h}{2}\right) + \frac{N_{j-1}(\frac{h}{2}) - N_{j-1}(h)}{2^{j-1} - 1}.$$

These approximations are generated by rows to take advantage of the highest-order formulas. The first four rows are shown in Table 4.4.

Table 4.4

$O(h)$	$O(h^2)$	$O(h^3)$	$O(h^4)$
$N_1(h) \equiv N(h)$			
$N_1(\frac{h}{2}) \equiv N(\frac{h}{2})$	$N_2(h)$		
$N_1(\frac{h}{4}) \equiv N(\frac{h}{4})$	$N_2(\frac{h}{2})$	$N_3(h)$	
$N_1(\frac{h}{8}) \equiv N(\frac{h}{8})$	$N_2(\frac{h}{4})$	$N_3(\frac{h}{2})$	$N_4(h)$

To begin the presentation of the Romberg integration scheme, recall that the Composite Trapezoidal rule for approximating the integral of a function f on an interval $[a, b]$ using m subintervals is

$$\int_a^b f(x)\, dx = \frac{h}{2}\left[f(a) + f(b) + 2\sum_{j=1}^{m-1} f(x_j) \right] - \frac{(b-a)}{12}h^2 f''(\mu),$$

where $a < \mu < b$, $h = (b-a)/m$, and $x_j = a + jh$ for each $j = 0, 1, \ldots, m$ (see Figure 4.9).

Figure 4.9

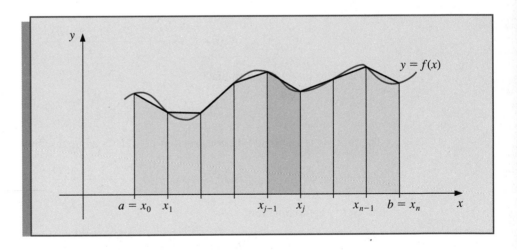

The first step in the Romberg process obtains the Composite Trapezoidal rule approximations with $m_1 = 1, m_2 = 2, m_3 = 4, \ldots$, and $m_n = 2^{n-1}$, where n is some positive integer. The values of the step size h_k corresponding to m_k are $h_k = (b-a)/m_k = (b-a)/2^{k-1}$. With this notation the Trapezoidal rule becomes

$$\int_a^b f(x)\, dx = \frac{h_k}{2}\left[f(a) + f(b) + 2\left(\sum_{i=1}^{2^{k-1}-1} f(a + ih_k) \right) \right] - \frac{b-a}{12}h_k^2 f''(\mu_k),$$

where μ_k is a number in (a, b). If the notation $R_{k,1}$ is introduced to denote the portion of this equation that is used for the trapezoidal approximation, then (see Figure 4.10):

$$R_{1,1} = \frac{h_1}{2} f(a) + f(b) = \frac{(b-a)}{2}[f(a) + f(b)];$$

$$R_{2,1} = \frac{1}{2}[R_{1,1} + h_1 f(a + h_2)];$$

$$R_{3,1} = \frac{1}{2}\{R_{2,1} + h_2[f(a + h_3) + f(a + 3h_3)]\}.$$

Figure 4.10

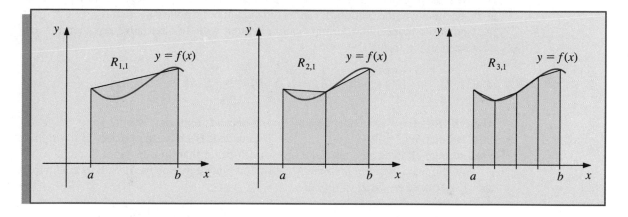

In general, the following holds:

Composite Trapezoidal Approximations

$$R_{k,1} = \frac{1}{2}\left[R_{k-1,1} + h_{k-1}\sum_{i=1}^{2^{k-2}} f(a + (2i-1)h_k)\right]$$

for each $k = 2, 3, \ldots, n$.

EXAMPLE 1   Using the Composite Trapezoidal formula to perform the first step of the Romberg integration scheme for approximating $\int_0^\pi \sin x \, dx$ with $n = 6$ leads to:

$$R_{1,1} = \frac{\pi}{2}[\sin 0 + \sin \pi] = 0;$$

$$R_{2,1} = \frac{1}{2}\left[R_{1,1} + \pi \sin \frac{\pi}{2}\right] = 1.57079633;$$

$$R_{3,1} = \frac{1}{2}\left[R_{2,1} + \frac{\pi}{2}\left(\sin \frac{\pi}{4} + \sin \frac{3\pi}{4}\right)\right] = 1.89611890;$$

$$R_{4,1} = \frac{1}{2}\left[R_{3,1} + \frac{\pi}{4}\left(\sin\frac{\pi}{8} + \sin\frac{3\pi}{8} + \sin\frac{5\pi}{8} + \sin\frac{7\pi}{8}\right)\right] = 1.97423160;$$

$$R_{5,1} = 1.99357034, \quad\text{and}\quad R_{6,1} = 1.99839336.\qquad \blacksquare\ \ \blacksquare\ \ \blacksquare$$

Since the correct value for the integral in Example 1 is 2, it is clear that, although the calculations involved are not difficult, the convergence is slow. To speed the convergence, the Richardson Extrapolation acceleration will be performed.

We must first have an approximation method with its error term in the form prescribed at the beginning of the section. It can be shown, although not easily, that if $f \in C^\infty[a, b]$, the Composite Trapezoidal formula can be written with an alternative error term. It can be written in the form:

(4.4)
$$\int_a^b f(x)\,dx - R_{k,1} = \sum_{i=1}^\infty K_i h_k^{2i} = K_1 h_k^2 + \sum_{i=2}^\infty K_i h_k^{2i},$$

where K_i for each i is independent of h_k and depends only on $f^{(2i-1)}(a)$ and $f^{(2i-1)}(b)$, provided that $h^i\left[f^{(2i-1)}(b) - f^{(2i-1)}(a)\right]$ is bounded. Because the powers of h_k in these equations are all even, the acceleration is much faster than in our initial discussion.

We eliminate the term involving h_k^2 by combining equation (4.4) with its counterpart with h_k replaced by $h_{k+1} = h_k/2$:

(4.5)
$$\int_a^b f(x)\,dx - R_{k+1,1} = \sum_{i=1}^\infty K_i h_{k+1}^{2i} = \sum_{i=1}^\infty \frac{K_i h_k^{2i}}{2^{2i}} = \frac{K_1 h_k^2}{4} + \sum_{i=2}^\infty \frac{K_i h_k^{2i}}{4^i}.$$

By subtracting equation (4.4) from 4 times (4.5) and simplifying, we have the $O(h_k^4)$ formula

$$\int_a^b f(x)\,dx - \left[R_{k+1,1} + \frac{R_{k+1,1} - R_{k,1}}{3}\right] = \sum_{i=2}^\infty \frac{K_i}{3}\left(\frac{h_k^{2i}}{4^{i-1}} - h_k^{2i}\right) = \sum_{i=2}^\infty \frac{K_i}{3}\left(\frac{1-4^{i-1}}{4^{i-1}}\right)h_k^{2i}.$$

Extrapolation can now be applied to this formula to eliminate the $O(h_k^4)$ term and obtain an $O(h_k^6)$ result, and so on. To simplify the notation, we define

$$R_{k,2} = R_{k,1} + \frac{R_{k,1} - R_{k-1,1}}{3}$$

for each $k = 2, 3, \ldots, n$, and apply the Richardson extrapolation procedure to these values. Continuing this notation, we have for each $k = 2, 3, 4, \ldots, n$ and $j = 2, \ldots, k$ an $O(h_k^{2j})$ approximation formula defined by

$$R_{k,j} = R_{k,j-1} + \frac{R_{k,j-1} - R_{k-1,j-1}}{4^{j-1} - 1}.$$

The results that are generated from these formulas are shown in Table 4.5.

The Romberg technique has the additional desirable feature that it allows an entire new row in the table to be calculated by simply doing one application of the Composite Trapezoidal rule and then using the previously calculated values to obtain the succeeding entries in the row. The method used to construct a table of this type calculates the entries row by row—that is, in the order $R_{1,1}, R_{2,1}, R_{2,2}, R_{3,1}, R_{3,2}, R_{3,3}$, etc. This is the process followed in the program ROMBERG42.

Table 4.5

$R_{1,1}$				
$R_{2,1}$	$R_{2,2}$			
$R_{3,1}$	$R_{3,2}$	$R_{3,3}$		
$R_{4,1}$	$R_{4,2}$	$R_{4,3}$	$R_{4,4}$	
$\vdots$	$\vdots$	$\vdots$	$\ddots$	
$R_{n,1}$	$R_{n,2}$	$R_{n,3}$	$\cdots$ $\cdots$	$R_{n,n}$

EXAMPLE 2 In Example 1, the values for $R_{1,1}$ through $R_{n,1}$ were obtained by approximating $\int_0^\pi \sin x \, dx$ with $n = 6$. The output from ROMBERG42 produces the Romberg table shown in Table 4.6. ■ ■ ■

Table 4.6

0					
1.57079633	2.09439511				
1.89611890	2.00455976	1.99857073			
1.97423160	2.00026917	1.99998313	2.00000555		
1.99357034	2.00001659	1.99999975	2.00000001	1.99999999	
1.99839336	2.00000103	2.00000000	2.00000000	2.00000000	2.00000000

ROMBRG42 requires a preset integer n to determine the number of rows to be generated. It is often also useful to prescribe an error tolerance for the approximation and generate n, within some upper bound, until consecutive diagonal entries $R_{n-1,n-1}$ and $R_{n,n}$ agree to within the tolerance. To guard against the possibility that two consecutive diagonal elements agree with each other but not with the value of the integral being approximated, we generate approximations until not only $|R_{n-1,n-1} - R_{n,n}|$ is within the tolerance, but also $|R_{n-2,n-2} - R_{n-1,n-1}|$. Although not a universal safeguard, it will ensure that two differently generated sets of approximations agree within the specified tolerance before $R_{n,n}$ is accepted as sufficiently accurate.

EXERCISE SET 4.5

1. Use Romberg integration to compute $R_{3,3}$ for the following integrals.

   (a) $\int_1^{1.5} x^2 \ln x \, dx$

   (b) $\int_0^1 x^2 e^{-x} \, dx$

   (c) $\int_0^{0.35} \frac{2}{x^2 - 4} \, dx$

   (d) $\int_0^{\pi/4} x^2 \sin x \, dx$

   (e) $\int_0^{\pi/4} e^{3x} \sin 2x \, dx$

   (f) $\int_1^{1.6} \frac{2x}{x^2 - 4} \, dx$

   (g) $\int_3^{3.5} \frac{x}{\sqrt{x^2 - 4}} \, dx$

   (h) $\int_0^{\pi/4} \cos^2 x \, dx$

2. Calculate $R_{4,4}$ for the integrals in Exercise 1.

3. Use Romberg integration to approximate the following integrals. Complete the Romberg table until $R_{n-1,n-1}$ and $R_{n,n}$ agree to within 10^{-6}, but do not continue if $n > 10$. Compare your results to the exact results.

(a) $\int_1^2 x \ln x \, dx$ (b) $\int_{-2}^2 x^3 e^x \, dx$ (c) $\int_0^2 \dfrac{2}{x^2 + 4} \, dx$

(d) $\int_0^\pi x^2 \cos x \, dx$ (e) $\int_0^2 e^{2x} \sin 3x \, dx$ (f) $\int_1^3 \dfrac{x}{x^2 + 4} \, dx$

(g) $\int_3^5 \dfrac{1}{\sqrt{x^2 - 4}} \, dx$ (h) $\int_0^{3\pi/8} \tan x \, dx$

4. Apply Romberg integration to the following integrals until $R_{n-1,n-1}$ and $R_{n,n}$ agree to within 10^{-5}.

(a) $\int_0^1 x^{1/3} \, dx$

(b) $\int_0^{0.3} f(x) \, dx$,   where

$$f(x) = \begin{cases} x^3 + 1, & 0 \le x \le 0.1, \\ 1.001 + 0.03(x - 0.1) + 0.3(x - 0.1)^2 + 2(x - 0.1)^3, & 0.1 < x \le 0.2, \\ 1.009 + 0.15(x - 0.2) + 0.9(x - 0.2)^2 + 2(x - 0.2)^3, & 0.2 < x \le 0.3. \end{cases}$$

5. Use Romberg integration to compute the following approximations to

$$\int_0^{48} \sqrt{1 + (\cos x)^2} \, dx.$$

[*Note:* The results in this exercise are most interesting if you are using a device with between seven- and nine-digit arithmetic.]

(a) Determine $R_{1,1}, R_{2,1}, R_{3,1}, R_{4,1}$, and $R_{5,1}$ and use these approximations to predict the value of the integral.

(b) Determine $R_{2,2}, R_{3,3}, R_{4,4}$, and $R_{5,5}$ and modify your prediction.

(c) Determine $R_{6,1}, R_{6,2}, R_{6,3}, R_{6,4}, R_{6,5}$, and $R_{6,6}$ and modify your prediction.

(d) Determine $R_{7,7}, R_{8,8}, R_{9,9}$, and $R_{10,10}$ and make a final prediction.

(e) Explain why this integral causes difficulty with Romberg integration and how it can be reformulated to determine an accurate approximation more easily.

6. Suppose the extrapolation table

$N_1(h)$

$N_1\left(\frac{h}{2}\right)$ $N_2(h)$

$N_1\left(\frac{h}{4}\right)$ $N_2\left(\frac{h}{2}\right)$ $N_3(h)$

has been constructed to approximate the number M with $M = N_1(h) + K_1 h^2 + K_2 h^4 + K_3 h^6$.

(a) Show that the linear interpolating polynomial $P_{0,1}(h)$ through $(h^2, N_1(h))$ and $(h^2/4, N_1(h/2))$ satisfies $P_{0,1}(0) = N_2(h)$. Similarly, show that $P_{1,2}(0) = N_2(h/2)$.

(b) Show that the linear interpolating polynomial $P_{0,2}(h)$ through $(h^4, N_2(h))$ and $(h^4/16, N_2(h/2))$ satisfies $P_{0,2}(0) = N_3(h)$.

4.6 Adaptive Quadrature

In Section 4.3 we used composite methods of approximation to break up an integral over a large interval into integrals over smaller subintervals. The approach used in that section involved equally sized subintervals, which permitted us to combine the individual approximations into a convenient form. Although this is satisfactory for most problems, it leads to increased computation when the function being integrated varies widely on some, but not all, parts of the interval of integration. In this case, techniques that adapt to the differing accuracy needs of the interval are superior. In this section we consider an Adaptive quadrature method and see how it can be used not only to reduce approximation error, but also to predict an error estimate for the approximation that does not rely on knowledge of higher derivatives of the function.

Suppose that we want to approximate $\int_a^b f(x)\,dx$ to within a specified tolerance $\epsilon > 0$. We first apply Simpson's rule with step size $h = (b - a)/2$, which gives (see Figure 4.11)

(4.6)
$$\int_a^b f(x)\,dx = S(a, b) - \frac{(b - a)^5}{2880}f^{(4)}(\xi) = S(a, b) - \frac{h^5}{90}f^{(4)}(\xi),$$

for some ξ in (a, b) where

$$S(a, b) = \frac{h}{3}[f(a) + 4f(a + h) + f(b)].$$

Figure 4.11

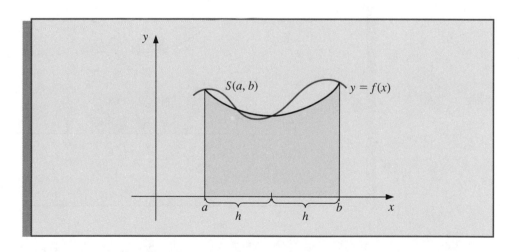

Next we determine an estimate for the accuracy of our approximation, in particular one that does not require determining $f^{(4)}(\xi)$. To do this, we first apply the Composite Simpson's rule to the problem with $n = 4$ and step size $(b - a)/4 = h/2$. Thus

(4.7)
$$\int_a^b f(x)\,dx = \frac{h}{6}\left[f(a) + 4f\left(a + \frac{h}{2}\right) + 2f(a + h) + 4f\left(a + \frac{3h}{2}\right) + f(b)\right]$$
$$- \left(\frac{h}{2}\right)^5 \frac{1}{90}f^{(4)}(\tilde{\xi}),$$

for some $\tilde{\xi}$ in (a, b). To simplify notation, let

$$S\left(a, \frac{a+b}{2}\right) = \frac{h}{6}\left[f(a) + 4f\left(a + \frac{h}{2}\right) + f(a + h)\right]$$

and

$$S\left(\frac{a+b}{2}, b\right) = \frac{h}{6}\left[f(a+h) + 4f\left(a + \frac{3h}{2}\right) + f(b)\right].$$

Then Eq. (4.7) can be rewritten (see Figure 4.12) as

(4.8) $$\int_a^b f(x)\,dx = S\left(a, \frac{a+b}{2}\right) + S\left(\frac{a+b}{2}, b\right) - \frac{1}{16}\left(\frac{h^5}{90}\right)f^{(4)}(\tilde{\xi}).$$

Figure 4.12

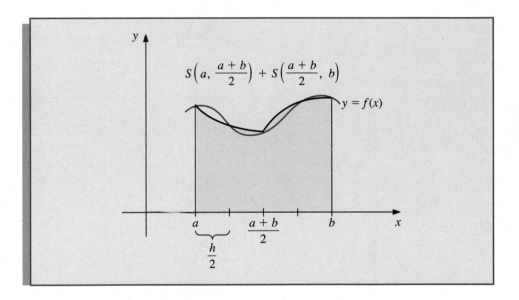

The error estimate is derived by assuming that $f^{(4)}(\xi) \approx f^{(4)}(\tilde{\xi})$. The success of the technique depends on the accuracy of this assumption. If it is accurate, then equating the integrals in Eqs. (4.6) and (4.8) implies that

$$S\left(a, \frac{a+b}{2}\right) + S\left(\frac{a+b}{2}, b\right) - \frac{1}{16}\left(\frac{h^5}{90}\right)f^{(4)}(\xi) \approx S(a, b) - \frac{h^5}{90}f^{(4)}(\xi),$$

so

$$\frac{h^5}{90}f^{(4)}(\xi) \approx \frac{16}{15}\left[S(a, b) - S\left(a, \frac{a+b}{2}\right) - S\left(\frac{a+b}{2}, b\right)\right].$$

Using this estimate in Eq. (4.8) produces the error estimation

$$\left| \int_a^b f(x)\,dx - S\left(a, \frac{a+b}{2}\right) - S\left(\frac{a+b}{2},\, b\right) \right|$$

$$\approx \frac{1}{15}\left| S(a, b) - S\left(a, \frac{a+b}{2}\right) - S\left(\frac{a+b}{2},\, b\right) \right|.$$

This implies that $S(a, (a+b)/2) + S((a+b)/2, b)$ approximates $\int_a^b f(x)\,dx$ fifteen times better than it agrees with the known value $S(a, b)$. Consequently, $S(a, (a+b)/2) + S((a+b)/2, b)$ approximates $\int_a^b f(x)\,dx$ to within ϵ provided that the two approximations $S(a, (a+b)/2) + S((a+b)/2, b)$ and $S(a, b)$ differ by less than 15ϵ.

Adaptive Quadrature Error Estimate

If

$$\left| S(a, b) - S\left(a, \frac{a+b}{2}\right) - S\left(\frac{a+b}{2},\, b\right) \right| < 15\epsilon,$$

then

$$\left| \int_a^b f(x)\,dx - S\left(a, \frac{a+b}{2}\right) - S\left(\frac{a+b}{2},\, b\right) \right| < \epsilon.$$

In this case,

$$S\left(a, \frac{a+b}{2}\right) + S\left(\frac{a+b}{2},\, b\right)$$

is assumed to be a sufficiently accurate approximation to $\int_a^b f(x)\,dx$.

EXAMPLE 1 To check the accuracy of the error estimate just given, consider its application to the integral

$$\int_0^{\pi/2} \sin x \, dx = 1.$$

In this case,

$$S\left(0, \frac{\pi}{2}\right) = \frac{(\pi/4)}{3}\left[\sin 0 + 4\sin\frac{\pi}{4} + \sin\frac{\pi}{2}\right] = \frac{\pi}{12}(2\sqrt{2} + 1) = 1.002279878$$

and

$$S\left(0, \frac{\pi}{4}\right) + S\left(\frac{\pi}{4}, \frac{\pi}{2}\right) = \frac{(\pi/8)}{3}\left[\sin 0 + 4\sin\frac{\pi}{8} + 2\sin\frac{\pi}{4} + 4\sin\frac{3\pi}{8} + \sin\frac{\pi}{2}\right]$$

$$= 1.000134585.$$

So,

$$\frac{1}{15}\left| S\left(0, \frac{\pi}{2}\right) - S\left(0, \frac{\pi}{4}\right) - S\left(\frac{\pi}{4}, \frac{\pi}{2}\right) \right| = 0.000143020.$$

This closely approximates the actual error,

$$\left| \int_0^{\pi/2} \sin x \, dx - 1.000134585 \right| = 0.000134585,$$

even though $D_x^4 \sin x = \sin x$ varies significantly in the interval $(0, \pi/2)$.

■ ■ ■

When the approximation is not sufficiently accurate (that is, when the two approximations differ by more than 15ϵ), the error estimation procedure can be applied individually to the subintervals $[a, (a + b)/2]$ and $[(a + b)/2, b]$ to determine if the approximation to the integral on each subinterval is within a tolerance of $\epsilon/2$. If so, the sum of the approximations agrees with $\int_a^b f(x) \, dx$ to within the tolerance ϵ. If the approximation on one of the subintervals fails to be within the tolerance $\epsilon/2$, that subinterval is itself subdivided and each of its subintervals analyzed to determine if the integral approximation on that subinterval is accurate to within $\epsilon/4$.

The halving procedure is continued until each portion is within the required tolerance. Although problems can be constructed for which this tolerance will never be met, the technique is successful for most problems, because each subdivision increases the accuracy of the approximation by a factor of approximately 15 while requiring an increased accuracy factor of only 2.

The program ADAPQR43 implements the adaptive quadrature procedure for Simpson's rule. Some technical difficulties require the implementation of the method to differ slightly from the preceding discussion. The tolerance between successive approximations has been set at 10ϵ rather than the derived value of 15ϵ. This bound is chosen conservatively to compensate for error in the assumption $f^{(4)}(\xi) \approx f^{(4)}(\tilde{\xi})$. In problems when $f^{(4)}$ is known to be widely varying, it would be reasonable to lower this bound even further.

EXAMPLE 2   The graph of the function $f(x) = (100/x^2)\sin(10/x)$ for x in $[1, 3]$ is shown in Figure 4.13. Using the program ADAPQR43 with tolerance 10^{-4} to approximate $\int_1^3 f(x) \, dx$ produces -1.476014, a result that is accurate to within 1.4×10^{-6}. The approximation required that Simpson's rule with $n = 4$ be performed on the 73 subintervals whose endpoints are shown on the horizontal axis in Figure 4.13. The total number of functional evaluations required for this approximation is 93.

For comparison purposes, suppose that the Composite Simpson's rule is used to approximate this integral with $h = \frac{1}{64}$. This requires 179 functional evaluations and gives the approximation -1.426059, a result that differs from the actual value by 2.4×10^{-6}.

■ ■ ■

Figure 4.13

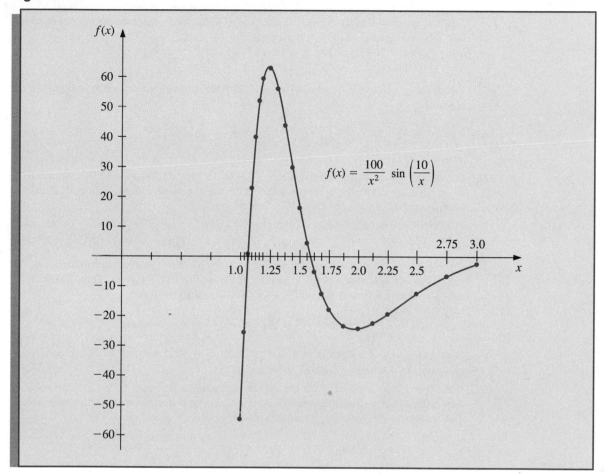

E X E R C I S E   S E T   4.6

1. Compute $S(a, b)$, $S(a, (a + b)/2)$, and $S((a + b)/2, b)$ for the following integrals and verify the estimate given in the approximation formula.

   (a) $\int_1^{1.5} x^2 \ln x \, dx$ (b) $\int_0^1 x^2 e^{-x} \, dx$ (c) $\int_0^{0.35} \dfrac{2}{x^2 - 4} \, dx$

   (d) $\int_0^{\pi/4} x^2 \sin x \, dx$ (e) $\int_0^{\pi/4} e^{3x} \sin 2x \, dx$ (f) $\int_1^{1.6} \dfrac{2x}{x^2 - 4} \, dx$

   (g) $\int_3^{3.5} \dfrac{x}{\sqrt{x^2 - 4}} \, dx$ (h) $\int_0^{\pi/4} \cos^2 x \, dx$

2. Use Adaptive quadrature to find approximations to within 10^{-3} for the integrals in Exercise 1. Do not use a computer program to generate these results.

3. Use Adaptive quadrature to approximate the following integrals to within 10^{-5}.

   (a) $\int_1^3 e^{2x} \sin 3x \, dx$ (b) $\int_1^3 e^{3x} \sin 2x \, dx$

   (c) $\int_0^5 [2x \cos(2x) - (x - 2)^2] \, dx$ (d) $\int_0^5 [4x \cos(2x) - (x - 2)^2] \, dx$

4. Use the Composite Simpson's rule with $n = 4, 6, 8, \ldots$ until successive approximations to the following integrals agree to within 10^{-6}. Determine the number of nodes required. Use Adaptive quadrature to approximate the integral to within 10^{-6} and count the number of nodes. Did Adaptive quadrature offer improvement?

   (a) $\int_0^\pi x \cos x^2 \, dx$ (b) $\int_0^\pi x \sin x^2 \, dx$
   (c) $\int_0^\pi x^2 \cos x \, dx$ (d) $\int_0^\pi x^2 \sin x \, dx$

5. Sketch the graphs of $\sin \frac{1}{x}$ and $\cos \frac{1}{x}$ on $[0.1, 2]$. Use Adaptive quadrature to approximate the integrals

$$\int_{0.1}^2 \sin \frac{1}{x} \, dx \quad \text{and} \quad \int_{0.1}^2 \cos \frac{1}{x} \, dx$$

   to within 10^{-3}.

6. Use Adaptive quadrature to approximate $\int_0^{48} \{1 + (\cos x)^2\}^{1/2} \, dx$ to within 10^{-4}. Does this technique have any advantage in this instance? (Compare the result and the number of functional evaluations with those of Exercise 9 in Section 4.3.)

7. The differential equation

$$m u''(t) + k u(t) = F_0 \cos \omega t$$

   describes a spring-mass system with mass m, spring constant k, and no applied damping. The term $F_0 \cos \omega t$ describes a periodic external force applied to the system. The solution to the equation when the system is initially at rest ($u'(0) = u(0) = 0$) is

$$u(t) = \frac{2F_0}{m(\omega_0^2 - \omega^2)} \sin \frac{(\omega_0 - \omega)}{2} t \sin \frac{(\omega_0 + \omega)}{2} t, \quad \text{where } \omega_0 = \sqrt{\frac{k}{m}} \neq \omega.$$

   Sketch the graph of u when $m = 1$, $k = 9$, $F_0 = 1$, $\omega = 2$, and $t \in [0, 2\pi]$. Use Adaptive quadrature to determine $\int_1^3 u(t) \, dt$ to within 10^{-4}.

8. If the term $c u'(t)$ is added to the left side of the motion equation in Exercise 7, the resulting differential equation describes a spring-mass system that is damped with damping constant c. The solution to this equation, when the solution is initially at rest, is

$$u(t) = c_1 e^{r_1 t} + c_2 e^{r_2 t} + \frac{F_0}{\sqrt{m^2(\omega_0^2 - \omega^2)^2 + c^2 \omega^2}} \cos(\omega t - \delta),$$

   where

$$\delta = \arctan\left(\frac{c\omega}{m(\omega_0^2 - \omega^2)}\right), \quad r_1 = \frac{-c + \sqrt{c^2 - 4\omega_0^2 m^2}}{2m},$$

   and

$$r_2 = \frac{-c - \sqrt{c^2 - 4\omega_0^2 m^2}}{2m}.$$

   Sketch the graph of u when $m = 1$, $k = 9$, $F_0 = 1$, $c = 10$, $\omega = 2$ and $t \in [0, 2\pi]$. Use Adaptive quadrature to determine $\int_1^3 u(t) \, dt$ to within 10^{-4}.

9. The study of light diffraction at a rectangular aperature involves the Fresnel integrals

$$c(t) = \int_0^t \cos \frac{\pi}{2} w^2 \, dw \quad \text{and} \quad s(t) = \int_0^t \sin \frac{\pi}{2} w^2 \, dw.$$

   Construct a table of values for $c(t)$ and $s(t)$ that is accurate to within 10^{-4} for values of $t = 0.1, 0.2, \ldots, 1.0$.

4.7 Multiple Integrals

The techniques discussed in the previous sections can be modified in a straightforward manner for use in the approximation of multiple integrals. Let us first consider the double integral

$$\iint_R f(x, y)\, dA,$$

where R is a rectangular region in the plane; that is,

$$R = \{(x, y)\,|\,a \le x \le b,\, c \le y \le d\},$$

for some constants $a, b, c,$ and d (see Figure 4.14). To illustrate the approximation technique, we employ the Composite Simpson's rule, although any other approximation formula could be used without major modifications.

Figure 4.14

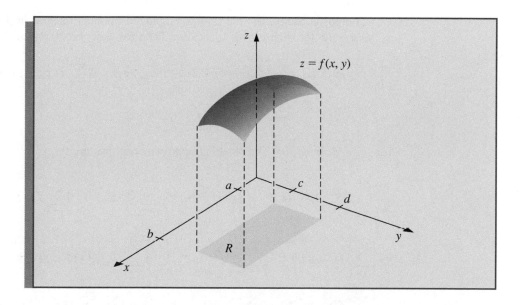

Suppose that integers n and m are chosen to determine the step sizes $h = (b-a)/2n$ and $k = (d - c)/2m$. Writing the double integral as an iterated integral

$$\iint_R f(x, y)\, dA = \int_a^b \left(\int_c^d f(x, y)\, dy \right) dx,$$

we first use the Composite Simpson's rule to evaluate

$$\int_c^d f(x, y)\, dy,$$

treating x as a constant. Let $y_j = c + jk$ for each $j = 0, 1, \ldots, 2m$. Then

$$\int_c^d f(x, y)\, dy = \frac{k}{3}\left[f(x, y_0)\, dx + 2\sum_{j=1}^{m-1} f(x, y_{2j}) + 4\sum_{j=1}^{m} f(x, y_{2j-1}) + f(x, y_{2m}) \right]$$
$$- \frac{(d-c)k^4}{180}\frac{\partial^4 f}{\partial y^4}(x, \mu)$$

for some μ in (c, d). Thus,

$$\int_a^b \int_c^d f(x, y)\, dy\, dx = \frac{k}{3}\int_a^b f(x, y_0)\, dx + \frac{2k}{3}\sum_{j=1}^{m-1}\int_a^b f(x, y_{2j})\, dx$$
$$+ \frac{4k}{3}\sum_{j=1}^{m}\int_a^b f(x, y_{2j-1})\, dx + \frac{k}{3}\int_a^b f(x, y_{2m})\, dx$$
$$- \frac{(d-c)k^4}{180}\int_a^b \frac{\partial^4 f}{\partial y^4}(x, \mu)\, dx.$$

Composite Simpson's rule is now employed on each integral in this equation. Let $x_i = a + ih$ for each $i = 0, 1, \ldots, 2n$. Then for each $j = 0, 1, \ldots, 2m$, we have

$$\int_a^b f(x, y_j)\, dx = \frac{h}{3}\left[f(x_0, y_j) + 2\sum_{i=1}^{n-1} f(x_{2i}, y_j) + 4\sum_{i=1}^{n} f(x_{2i-1}, y_j) + f(x_{2n}, y_j) \right]$$
$$- \frac{(b-a)h^4}{180}\frac{\partial^4 f}{\partial x^4}(\xi_j, y_j)$$

for some ξ_j in (a, b). The resulting approximation has the form

$$\int_a^b \int_c^d f(x, y)\, dy\, dx \approx \frac{hk}{9}\left\{ \left[f(x_0, y_0) + 2\sum_{i=1}^{n-1} f(x_{2i}, y_0) + 4\sum_{i=1}^{n} f(x_{2i-1}, y_0) + f(x_{2n}, y_0) \right] \right.$$

$$+ 2\left[\sum_{j=1}^{m-1} f(x_0, y_{2j}) + 2\sum_{j=1}^{m-1}\sum_{i=1}^{n-1} f(x_{2i}, y_{2j}) + 4\sum_{j=1}^{m-1}\sum_{i=1}^{n} f(x_{2i-1}, y_{2j}) + \sum_{j=1}^{m-1} f(x_{2n}, y_{2j}) \right]$$

$$+ 4\left[\sum_{j=1}^{m} f(x_0, y_{2j-1}) + 2\sum_{j=1}^{m}\sum_{i=1}^{n-1} f(x_{2i}, y_{2j-1}) + 4\sum_{j=1}^{m}\sum_{i=1}^{n} f(x_{2i-1}, y_{2j-1}) + \sum_{j=1}^{m} f(x_{2n}, y_{2j-1}) \right]$$

$$+ \left[f(x_0, y_{2m}) + 2\sum_{i=1}^{n-1} f(x_{2i}, y_{2m}) + 4\sum_{i=1}^{n} f(x_{2i-1}, y_{2m}) + f(x_{2n}, y_{2m}) \right] \right\}.$$

The error term E is given by

$$E = \frac{-k(b-a)h^4}{540}\left[\frac{\partial^4 f}{\partial x^4}(\xi_0, y_0) + 2\sum_{j=1}^{m-1}\frac{\partial^4 f}{\partial x^4}(\xi_{2j}, y_{2j}) + 4\sum_{j=1}^{m}\frac{\partial^4 f}{\partial x^4}(\xi_{2j-1}, y_{2j-1}) \right.$$

$$+\frac{\partial^4 f}{\partial x^4}(\xi_{2m}, y_{2m})\Bigg] - \frac{(d-c)k^4}{180}\int_a^b \frac{\partial^4 f}{\partial y^4}(x, \mu)\, dx.$$

If $\partial^4 f/\partial x^4$ and $\partial^4 f/\partial y^4$ are continuous, the Intermediate Value Theorem and Mean Value Theorem for Integrals can be used to show that the error formula can be simplified to

$$E = \frac{-(d-c)(b-a)}{180}\left[h^4\frac{\partial^4 f}{\partial x^4}(\bar{\eta}, \bar{\mu}) + k^4\frac{\partial^4 f}{\partial y^4}(\hat{\eta}, \hat{\mu})\right]$$

for some $(\bar{\eta}, \bar{\mu})$ and $(\hat{\eta}, \hat{\mu})$ in R.

EXAMPLE 1   The Composite Simpson's rule applied to approximate

$$\int_{1.4}^{2.0}\int_{1.0}^{1.5} \ln(x + 2y)\, dy\, dx$$

with $n = 2$ and $m = 1$ uses the step sizes $h = 0.15$ and $k = 0.25$. The region of integration R is shown in Figure 4.15, together with the nodes (x_i, y_j) for $i = 0, 1, 2, 3, 4$ and $j = 0, 1, 2$ and the coefficients $w_{i,j}$ of $f(x_i, y_i) = \ln(x_i + 2y_i)$ in the sum.

Figure 4.15

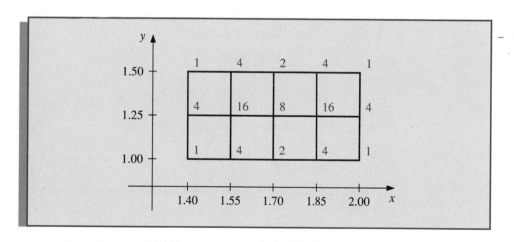

The approximation is

$$\int_{1.4}^{2.0}\int_{1.0}^{1.5} \ln(x + 2y)\, dy\, dx \approx \frac{(0.15)(0.25)}{9}\sum_{i=0}^{4}\sum_{j=0}^{2} w_{i,j}\ln(x_i + 2y_j) = 0.4295524387.$$

Since

$$\frac{\partial^4 f}{\partial x^4}(x, y) = \frac{-6}{(x+2y)^4} \qquad \text{and} \qquad \frac{\partial^4 f}{\partial y^4}(x, y) = \frac{-96}{(x^2+2y)^4},$$

the error is bounded by

$$|E| \leq \frac{(0.5)(0.6)}{180}\left[(0.15)^4 \max_{(x,y) \text{ in } R} \frac{6}{(x+2y)^4} + (0.25)^4 \max_{(x,y) \text{ in } R} \frac{96}{(x+2y)^4}\right] \leq 4.72 \times 10^{-6}.$$

The actual value of the integral to ten decimal places is

$$\int_{1.4}^{2.0} \int_{1.0}^{1.5} \ln(x+2y)\, dy\, dx = 0.4295545265,$$

so the approximation is accurate to within 2.1×10^{-6}. ■ ■ ■

The same techniques can be applied for the approximation of triple integrals as well as higher integrals for functions of more than three variables. The number of functional evaluations required for the approximation is the product of the number required when the method is applied to each variable. To reduce the number of functional evaluations, more efficient methods such as Gaussian quadrature, Romberg integration, or Adaptive quadrature can be incorporated in place of Simpson's formula. The following example illustrates the use of Gaussian quadrature for the integral considered in Example 1.

EXAMPLE 2 Consider the double integral given in Example 1. Before employing a Gaussian quadrature technique to approximate this integral, we must translate the region of integration

$$R = \{(x, y) | 1.4 \leq x \leq 2.0, 1.0 \leq y \leq 1.5\}$$

into

$$\hat{R} = \{(u, v) | -1 \leq u \leq 1, -1 \leq v \leq 1\}.$$

The linear transformations that accomplish this are

$$u = \frac{1}{2.0 - 1.4}(2x - 1.4 - 2.0) \qquad \text{and} \qquad v = \frac{1}{1.5 - 1.0}(2y - 1.0 - 1.5).$$

Employing this change of variables gives an integral on which Gaussian quadrature can be applied:

$$\int_{1.4}^{2.0} \int_{1.0}^{1.5} \ln(x+2y)\, dy\, dx = 0.075 \int_{-1}^{1} \int_{-1}^{1} \ln(0.3u + 0.5v + 4.2)\, dv\, du.$$

The Gaussian quadrature formula for $n = 3$ in both in u and v requires that we use the nodes

$$u_1 = v_1 = r_{3,2} = 0, \qquad u_0 = v_0 = r_{3,1} = -0.7745966692, \qquad \text{and}$$
$$u_2 = v_2 = r_{3,3,} = 0.7745966692.$$

The associated weights are found in Table 4.3 (Section 4.4) to be $c_{3,2} = 0.8888888889$ and $c_{3,1} = c_{3,3} = 0.5555555556$, so

$$\int_{1.4}^{2.0} \int_{1.0}^{1.5} \ln(x+2y)\, dy\, dx, \approx 0.075 \sum_{i=0}^{2} \sum_{j=0}^{2} c_{3,i} c_{3,j} \ln(0.3u_i + 0.5v_j + 4.2) = 0.4295545313.$$

Even though this result requires only 6 functional evaluations compared to 15 for the Composite Simpson's rule considered in Example 1, the result is accurate to within 4.8×10^{-9}, as compared to an accuracy of only 2×10^{-6} for Simpson's rule. ■ ■ ■

The use of approximation methods for double integrals is not limited to integrals with rectangular regions of integration. The techniques previously discussed can be modified to approximate double integrals with variable inner limits, integrals of the form

$$\int_{a}^{b} \int_{c(x)}^{d(x)} f(x, y)\, dy\, dx.$$

In fact, integrals over more regions can also be approximated by performing appropriate partitions of the region.

For this type of integral we begin as before by applying the Composite Simpson's rule to integrate with respect to both variables. The step size for the variable x is $h = (b - a)/2$, but the step size $k(x)$ for y varies with x (see Figure 4.16):

$$k(x) = \frac{d(x) - c(x)}{2}.$$

Figure 4.16

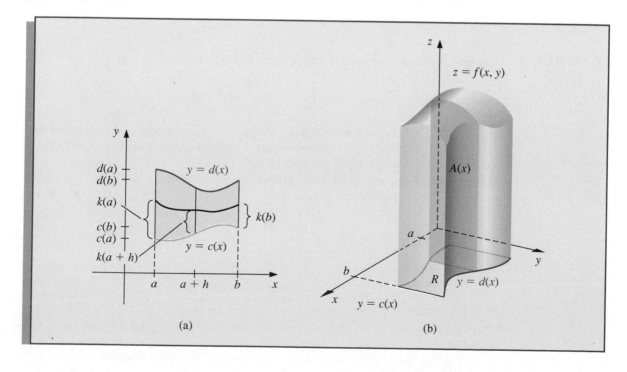

(a)

(b)

Consequently,

$$\int_a^b \int_{c(x)}^{d(x)} f(x, y)\, dy\, dx \approx \int_a^b \frac{k(x)}{3}[f(x, c(x)) + 4f(x, c(x) + k(x)) + f(x, d(x))]\, dx$$

$$\approx \frac{h}{3}\left\{ \frac{k(a)}{3}[f(a, c(a)) + 4f(a, c(a) + k(a)) + f(a, d(a))] \right.$$

$$+ \frac{4k(a + h)}{3}[f(a + h, c(a + h)) + 4f(a + h, c(a + h)$$

$$+ k(a + h)) + f(a + h, d(a + h))]$$

$$\left. + \frac{k(b)}{3}[f(b, c(b)) + 4f(b, c(b) + k(b)) + f(b, d(b))] \right\}.$$

The program DINTGL44 applies the Composite Simpson's rule to a double integral in this form.

To apply Gaussian quadrature to the double integral first requires translating, for each x in $[a, b]$, the interval $[c(x), d(x)]$ to $[-1, 1]$ and then applying Gaussian quadrature. This results in the formula

$$\int_a^b \int_{c(x)}^{d(x)} f(x, y)\, dy\, dx \approx \int_a^b \frac{d(x) - c(x)}{2} \sum_{j=1}^n c_{n,j} f\left(x, \frac{(d(x) - c(x))r_{n,j} + d(x) + c(x)}{2}\right) dx,$$

where, as before, the roots $r_{n,j}$ and coefficients $c_{n,j}$ come from Table 4.3. Now the interval $[a, b]$ is translated to $[-1, 1]$ and Gaussian quadrature is applied to approximate the integral on the right side of this equation. The program DGQINT45 uses this technique.

EXAMPLE 3   Applying Simpson's Double Integral program DINTGL44 with $n = m = 5$ to

$$\int_{0.1}^{0.5} \int_{x^3}^{x^2} e^{y/x}\, dy\, dx$$

requires 121 evaluations of the function $f(x, y) = e^{y/x}$ and produces the approximation 0.0333054, a result that is accurate to nearly 7 decimal places. (See Figure 4.17.) Applying the Gaussian quadrature program DGQINT45 with $n = m = 5$ requires only 25 function evaluations and, in addition, gives the approximation, 0.3330556611, which is accurate to 11 decimal places.   ■ ■ ■

Triple integrals of the form

$$\int_a^b \int_{c(x)}^{d(x)} \int_{\alpha(x,y)}^{\beta(x,y)} f(x, y, z)\, dz\, dy\, dx$$

are approximated in a similar manner. Because of the number of calculations involved, Gaussian quadrature is the method of choice. The program TINTGL46 implements this procedure.

The following example requires the evaluation of four triple integrals.

Figure 4.17

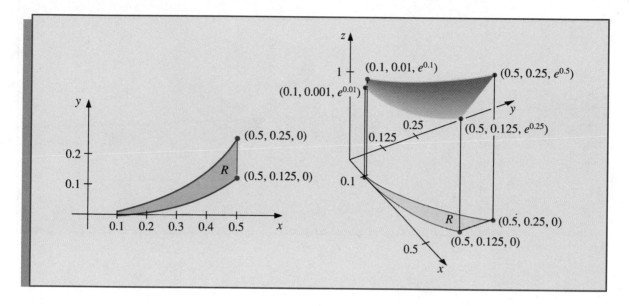

EXAMPLE 4 The center of mass of a solid region D with density function σ occurs at

$$(\bar{x}, \bar{y}, \bar{z}) = \left(\frac{M_{yz}}{M}, \frac{M_{xz}}{M}, \frac{M_{xy}}{M} \right),$$

where

$$M_{yz} = \iiint_D x\sigma(x, y, z) \, dV, \qquad M_{xz} = \iiint_D y\sigma(x, y, z) \, dV,$$

and

$$M_{xy} = \iiint_D z\sigma(x, y, z) \, dV$$

are the moments about the coordinate planes and

$$M = \iiint_D \sigma(x, y, z) \, dV$$

is the mass. The solid shown in Figure 4.18 on page 132 is bounded by the upper nappe of the cone $z^2 = x^2 + y^2$ and the plane $z = 2$ and has density function given by

$$\sigma(x, y, z) = \sqrt{x^2 + y^2}.$$

Applying TINTGL46 with $n = m = p = 5$ requires 125 function evaluations per integral and gives the following approximations:

$$M = \int_{-2}^{2} \int_{-\sqrt{4-x^2}}^{\sqrt{4-x^2}} \int_{\sqrt{x^2+y^2}}^{2} \sqrt{x^2 + y^2} \, dz \, dy \, dx \approx 8.37504476,$$

Figure 4.18

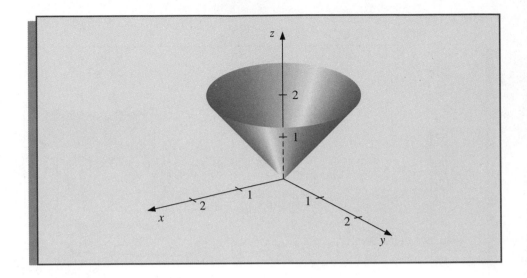

$$M_{yz} = \int_{-2}^{2} \int_{-\sqrt{4-x^2}}^{\sqrt{4-x^2}} \int_{\sqrt{x^2+y^2}}^{2} x\sqrt{x^2+y^2}\, dz\, dy\, dx \approx -5.55111512 \times 10^{-17},$$

$$M_{xz} = \int_{-2}^{2} \int_{-\sqrt{4-x^2}}^{\sqrt{4-x^2}} \int_{\sqrt{x^2+y^2}}^{2} y\sqrt{x^2+y^2}\, dz\, dy\, dx \approx -8.01513675 \times 10^{-17},$$

$$M_{xy} = \int_{-2}^{2} \int_{-\sqrt{4-x^2}}^{\sqrt{4-x^2}} \int_{\sqrt{x^2+y^2}}^{2} z\sqrt{x^2+y^2}\, dz\, dy\, dx \approx 13.40038156.$$

This implies that the approximate location of the center of mass is $\sqrt{(\bar{x}, \bar{y}, \bar{z})} = (0, 0, 1.60003701)$. By direct evaluation of the integrals, the center of mass can be shown to occur at $(0, 0, 1.6)$.   ■ ■ ■

EXERCISE SET 4.7

1. Use the Composite Simpson's rule for double integrals with $n = m = 2$ to approximate the following double integrals. Compare the results to the exact answer.

   (a) $\int_{2.1}^{2.5} \int_{1.2}^{1.4} xy^2\, dy\, dx$
   (b) $\int_{0}^{0.5} \int_{0}^{0.5} e^{y-x}\, dy\, dx$

   (c) $\int_{2}^{2.2} \int_{x}^{2x} (x^2 + y^3)\, dy\, dx$
   (d) $\int_{1}^{1.5} \int_{0}^{x} (x^2 + \sqrt{y})\, dy\, dx$

2. Find the smallest values for $n = m$ so that the Composite Simpson's rule for double integrals can be used to approximate the integrals in Exercise 1 to within 10^{-6} of the actual value.

3. Use the Composite Simpson's rule for double integrals with $n = 2, m = 4; n = 4, m = 2$; and $n = m = 3$ to approximate the following double integrals. Compare the results to the exact answer.

   (a) $\int_{0}^{\pi/4} \int_{\sin x}^{\cos x} (2y \sin x + \cos^2 x)\, dy\, dx$
   (b) $\int_{1}^{e} \int_{1}^{x} \ln xy\, dy\, dx$

   (c) $\int_{0}^{1} \int_{x}^{2x} (x^2 + y^3)\, dy\, dx$
   (d) $\int_{0}^{1} \int_{x}^{2x} (y^2 + x^3)\, dy\, dx$

(e) $\int_0^\pi \int_0^x \cos x \, dy \, dx$ (f) $\int_0^\pi \int_0^x \cos y \, dy \, dx$

(g) $\int_0^{\pi/4} \int_0^{\sin x} \dfrac{1}{\sqrt{1 - y^2}} \, dy \, dx$ (h) $\int_{-\pi}^{3\pi/2} \int_0^{2\pi} (y \sin x + x \cos y) \, dy \, dx$

4. Find the smallest values for $n = m$ so that the Composite Simpson's rule for double integrals can be used to approximate the integrals in Exercise 3 to within 10^{-6} of the actual value.

5. Use Gaussian quadrature for double integrals with $n = m = 2$ to approximate the integrals in Exercise 1 and compare the results to those obtained in Exercise 1.

6. Find the smallest values of $n = m$ so that Gaussian quadrature for double integrals may be used to approximate the integrals in Exercise 1 to within 10^{-6}. Do not continue beyond $n = m = 5$. Compare the number of functional evaluations required to the number required in Exercise 2.

7. Use Gaussian quadrature for double integrals with $n = m = 3$; $n = 3$, $m = 4$; $n = 4$, $m = 3$; and $n = m = 4$ to approximate the integrals in Exercise 3.

8. Use Gaussian quadrature for double integrals with $n = m = 5$ to approximate the integrals in Exercise 3. Compare the number of functional evaluations required to the number required in Exercise 4.

9. Use the Composite Simpson's rule for double integrals with $n = m = 7$ and Gaussian quadrature for double integrals with $n = m = 4$ to approximate

$$\iint_R e^{-(x+y)} \, dA$$

for the region R in the plane bounded by the curves $y = x^2$ and $y = \sqrt{x}$.

10. Use Gaussian quadrature for double integrals to approximate

$$\iint_R \sqrt{xy + y^2} \, dA,$$

where R is the region in the plane bounded by the lines $x + y = 6$, $3y - x = 2$, and $3x - y = 2$. First partition R into two regions R_1 and R_2 on which Gaussian quadrature for double integrals can be applied. Use $n = m = 3$ on both R_1 and R_2.

11. The area of the surface described by $z = f(x, y)$ for (x, y) in R is given by

$$\iint_R \sqrt{[f_x(x, y)]^2 + [f_y(x, y)]^2 + 1} \, dA.$$

Find an approximation to the area of the surface on the hemisphere $x^2 + y^2 + z^2 = 9$, $z \geq 0$ that lies above the region $R = \{(x, y) \mid 0 \leq x \leq 1, \ 0 \leq y \leq 1\}$, using
(a) the program DINTGL44 with $n = m = 4$;
(b) the program DGQINT45 with $n = m = 4$.

12. Use Gaussian quadrature for triple integrals with $n = m = p = 2$ to approximate the following triple integrals. Compare the results to the exact answer.
(a) $\int_0^1 \int_1^2 \int_0^{0.5} e^{x+y+z} \, dz \, dy \, dx$ (b) $\int_0^1 \int_x^1 \int_0^y y^2 z \, dz \, dy \, dx$
(c) $\int_0^1 \int_{x^2}^x \int_{x-y}^{x+y} y \, dz \, dy \, dx$ (d) $\int_0^1 \int_{x^2}^x \int_{x-y}^{x+y} z \, dz \, dy \, dx$
(e) $\int_0^\pi \int_0^x \int_0^{xy} \frac{1}{y} \sin \frac{z}{y} \, dz \, dy \, dx$ (f) $\int_0^1 \int_0^1 \int_{-xy}^{xy} e^{x^2+y^2} \, dz \, dy \, dx$

13. Repeat Exercise 12 using $n = m = p = 3$.

14. Repeat Exercise 12 using $n = m = p = 4$ and $n = m = p = 5$.

15. Use Gaussian quadrature for triple integrals with $n = m = p = 4$ to approximate

$$\iiint_S xy\sin(yz)\,dV,$$

where S is the solid bounded by the coordinate planes and the planes $x = \pi$, $y = \pi/2$, $z = \pi/3$. Compare this approximation to the exact result.

16. Use Gaussian quadrature for triple integrals with $n = m = p = 5$ to approximate

$$\iiint_S \sqrt{xyz}\,dV,$$

when S is the region in the first octant bounded by the cylinder $x^2 + y^2 = 4$, the sphere $x^2 + y^2 + z^2 = 4$, and the plane $x + y + z = 8$. Use eight nodes in each coordinate direction. How many functional evaluations are required for the approximation?

4.8 Improper Integrals

Improper integrals result when the notion of integration is extended either to an interval of integration on which the function is unbounded or to an interval with one or more infinite endpoints. In either circumstance, the normal rules of integral approximation must be modified.

We first handle the situation when the integrand is unbounded at the left endpoint of the interval of integration, as shown in Figure 4.19. We then show that, by a suitable manipulation, the other improper integrals can be reduced to problems of this form.

Figure 4.19

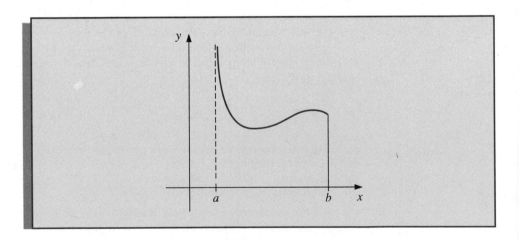

The improper integral with a singularity at the left endpoint,

$$\int_a^b \frac{1}{(x-a)^p}\,dx,$$

converges if and only if $p < 1$, and in this case,

$$\int_a^b \frac{1}{(x-a)^p}\, dx = \frac{(b-a)^{1-p}}{1-p}.$$

If f is a function that can be written in the form

$$f(x) = \frac{g(x)}{(x-a)^p},$$

where $0 < p < 1$ and g is continuous on $[a, b]$, then the improper integral

$$\int_a^b f(x)\, dx$$

also exists. We will approximate this integral using the Composite Simpson's rule. Assuming that $g \in C^5[a, b]$, we can construct the fourth Taylor polynomial, $P_4(x)$, for g about a,

$$P_4(x) = g(a) + g'(a)(x-a) + \frac{g''(a)}{2!}(x-a)^2 + \frac{g'''(a)}{3!}(x-a)^3 + \frac{g^{(4)}(a)}{4!}(x-a)^4,$$

and write

$$\int_a^b f(x)\, dx = \int_a^b \frac{g(x) - P_4(x)}{(x-a)^p}\, dx + \int_a^b \frac{P_4(x)}{(x-a)^p}\, dx.$$

We can exactly determine the value of

$$\int_a^b \frac{P_4(x)}{(x-a)^p}\, dx = \sum_{k=0}^4 \int_a^b \frac{g^{(k)}(a)}{k!}(x-a)^{k-p}\, dx = \sum_{k=0}^4 \frac{g^{(k)}(a)}{k!(k+1-p)}(b-a)^{k+1-p}.$$

This is generally the dominant portion of the approximation, especially when the Taylor polynomial $P_4(x)$ agrees closely with the function g throughout the interval $[a, b]$. To approximate the integral of f, then, we need to add this value to the approximation of

$$\int_a^b \frac{g(x) - P_4(x)}{(x-a)^p}\, dx.$$

To determine this we first define

$$G(x) = \begin{cases} \dfrac{g(x) - P_4(x)}{(x-a)^p}, & \text{if} \quad a < x \le b, \\ 0, & \text{if} \quad x = a. \end{cases}$$

Since $p < 1$ and $P_4^{(k)}(a)$ agrees with $g^{(k)}(a)$ for each $k = 0, 1, 2, 3, 4$, we have $G \in C^4[a, b]$. This implies that Composite Simpson's rule can be applied to approximate the integral of G on $[a, b]$ and the error term for this rule will be valid. Adding this approximation to the exact value gives an approximation to the improper integral of f on $[a, b]$ within the accuracy of the Composite Simpson's rule approximation.

EXAMPLE 1   To approximate the values of the improper integral

$$\int_0^1 \frac{e^x}{\sqrt{x}}\, dx,$$

we use the Composite Simpson's rule with $h = 0.25$. Since the fourth Taylor polynomial for e^x about $x = 0$ is

$$P_4(x) = 1 + x + \frac{x^2}{2} + \frac{x^3}{6} + \frac{x^4}{24},$$

we have

$$\int_0^1 \frac{P_4(x)}{\sqrt{x}}\, dx = \lim_{M \to 0^+} \left[2x^{\frac{1}{2}} + \frac{2}{3}x^{\frac{3}{2}} + \frac{1}{5}x^{\frac{5}{2}} + \frac{1}{21}x^{\frac{7}{2}} + \frac{1}{108}x^{\frac{9}{2}} \right]_M^1$$

$$= 2 + \frac{2}{3} + \frac{1}{5} + \frac{1}{21} + \frac{1}{108} \approx 2.9235450.$$

Table 4.7 lists the values needed for the Composite Simpson's rule for

$$G(x) = \begin{cases} \dfrac{e^x - P_4(x)}{\sqrt{x}}, & \text{when } 0 < x \le 1, \\ 0, & \text{when } x = 0. \end{cases}$$

Table 4.7

x	$f(x)$
0.00	0
0.25	0.0000170
0.50	0.0004013
0.75	0.0026026
1.00	0.0099485

Applying the Composite Simpson's rule to G using these data gives

$$\int_0^1 G(x)\, dx \approx \frac{0.25}{3}[0 + 4(0.0000170) + 2(0.0004013) + 4(0.0026026) + 0.0099485]$$

$$= 0.0017691.$$

Hence,

$$\int_0^1 \frac{e^x}{\sqrt{x}}\, dx \approx 2.9235450 + 0.0017691 = 2.9253141.$$

This result is accurate within the accuracy of the Composite Simpson's rule approximation for the function G. Since $|G^{(4)}(x)| < 1$ on $[0, 1]$, the error is bounded by

$$\frac{1 - 0}{180}(0.25)^4(1) = 0.0000217.$$

■ ■ ■

To approximate the improper integral with a singularity at the right endpoint, we simply apply the technique we used previously but expand in terms of the right endpoint b instead of the left endpoint a. Alternatively, we could make the substitution $z = -x$, $dz = -dx$ to change the improper integral into one of the form

$$\int_a^b f(x)\,dx = \int_{-b}^{-a} f(-z)\,dz,$$

which has its singularity at the left endpoint. (See Figure 4.20.)

Figure 4.20

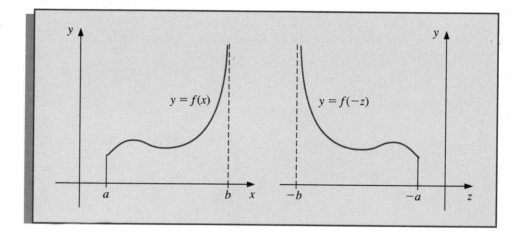

Improper integrals with interior singularities (for example, at c, where $a < c < b$) are treated as the sum of improper integrals with endpoint singularities, since

$$\int_a^b f(x)\,dx = \int_a^c f(x)\,dx + \int_c^b f(x)\,dx.$$

The other type of improper integrals involves infinite limits of integration. The basic integral of this type has the form

$$\int_a^\infty \frac{1}{x^p}\,dx,$$

which is converted to an integral with left endpoint singularity by making the integration substitution

$$t = x^{-1}, \qquad dt = -x^{-2}\,dx, \qquad \text{so} \qquad dx = -x^2\,dt = -t^{-2}\,dt.$$

Then

$$\int_a^\infty \frac{1}{x^p}\,dx = \int_{\frac{1}{a}}^0 -\frac{t^p}{t^2}\,dt = \int_0^{\frac{1}{a}} \frac{1}{t^{2-p}}\,dt.$$

In a similar manner, the variable change $t = x^{-1}$ converts the improper integral $\int_a^\infty f(x)\,dx$ into one that has a left endpoint singularity at zero:

$$\int_a^\infty f(x)\,dx = \int_0^{\frac{1}{a}} t^{-2} f\left(\frac{1}{t}\right) dt.$$

It can now be approximated using a quadrature formula of the type described earlier.

EXAMPLE 2   To approximate the value of the improper integral

$$I = \int_1^\infty x^{-\frac{3}{2}} \sin\frac{1}{x}\,dx$$

we make the change of variable $t = x^{-1}$ to obtain

$$I = \int_0^1 t^{-\frac{1}{2}} \sin t\,dt.$$

The fourth Taylor polynomial, $P_4(t)$, for $\sin t$ about 0 is

$$P_4(t) = t - \frac{1}{6}t^3,$$

so we have

$$
\begin{aligned}
I &= \int_0^1 \frac{\sin t - t + \frac{1}{6}t^3}{t^{\frac{1}{2}}}\,dt + \int_0^1 t^{\frac{1}{2}} - \frac{1}{6}t^{\frac{5}{2}}\,dt \\
&= \int_0^1 \frac{\sin t - t + \frac{1}{6}t^3}{t^{\frac{1}{2}}}\,dt + \left[\frac{2}{3}t^{\frac{3}{2}} - \frac{1}{21}t^{\frac{7}{2}}\right]\Big|_0^1 \\
&= \int_0^1 \frac{\sin t - t + \frac{1}{6}t^3}{t^{\frac{1}{2}}}\,dt + 0.61904761.
\end{aligned}
$$

The Composite Simpson's rule with $n = 8$ for the remaining integral is 0.0014890097. This gives a final approximation of

$$I = 0.0014890097 + 0.61904761 = 0.62053661,$$

which is accurate to within 4.0×10^{-8}.   ▪ ▪ ▪

EXERCISE SET 4.8

1. Use the Composite Simpson's rule and the given values of n to approximate the following improper integrals:

(a) $\int_0^1 x^{-\frac{1}{4}} \sin x\,dx;$ $n = 4$ (b) $\int_0^1 \frac{e^{2x}}{\sqrt[5]{x^2}}\,dx;$ $n = 6$

(c) $\int_1^2 \frac{\ln x}{(x-1)^{\frac{1}{5}}}\,dx;$ $n = 8$ (d) $\int_0^1 \frac{\cos 2x}{x^{\frac{1}{3}}}\,dx;$ $n = 6$

2. (a) $\int_0^1 \dfrac{e^{-x}}{\sqrt{1-x}}\,dx;$ $n = 6$ (b) $\int_0^{1/2} \dfrac{1}{(2x-1)^{\frac{1}{3}}}\,dx;$ $n = 4$

  (c) $\int_{-1}^0 \dfrac{1}{\sqrt[3]{3x+1}}\,dx;$ $n = 6$ (d) $\int_0^2 \dfrac{x e^x}{\sqrt[3]{(x-1)^2}}\,dx;$ $n = 8$

3. Make the transformation $t = x^{-1}$, and then use the Composite Simpson's rule and the given values of n to approximate the following improper integrals:

  (a) $\int_1^\infty \dfrac{1}{x^2+9}\,dx;$ $n = 4$ (b) $\int_1^\infty \dfrac{1}{x^4+1}\,dx;$ $n = 4$

  (c) $\int_1^\infty \dfrac{\cos x}{x^3}\,dx;$ $n = 6$ (d) $\int_1^\infty x^{-4}\sin x\,dx;$ $n = 6$

4. An improper integral of the form $\int_0^\infty f(x)\,dx$ cannot be converted into an integral with finite limits using the substitution $t = 1/x$ because the limit at zero becomes infinite. The problem is easily resolved by first writing $\int_0^\infty f(x)\,dx = \int_0^1 f(x)\,dx + \int_1^\infty f(x)\,dx$. Apply this technique to approximate the following improper integrals to within 10^{-6}:

  (a) $\int_0^\infty \dfrac{1}{1+x^4}\,dx$ (b) $\int_0^\infty \dfrac{1}{(1+x^2)^3}\,dx$

5. The improper integral $\int_{-\infty}^\infty f(x)\,dx$ can be written as $\int_{-\infty}^0 f(x)\,dx + \int_0^\infty f(x)\,dx$, where $\int_{-\infty}^0 f(x)\,dx = \int_0^\infty f(-x)\,dx$. Use the techniques of Exercises 1, 2, and 3 to approximate the following integrals:

  (a) $\int_{-\infty}^\infty \dfrac{1}{1+x^2}\,dx;$ $n = 4$ (b) $\int_{-\infty}^\infty \dfrac{x}{(x^2+1)^4}\,dx;$ $n = 4$

  (c) $\int_{-\infty}^\infty \dfrac{1}{(1+x^2)^3}\,dx;$ $n = 6$ (d) $\int_{-\infty}^\infty \dfrac{1}{1+x^4}\,dx;$ $n = 6$

6. Suppose a body of mass m is traveling vertically upward starting at the surface $x = R$ of the earth. If all resistance except gravity is neglected, then the escape velocity v is given by

$$v^2 = 2gR \int_1^\infty z^{-2}\,dz, \qquad \text{where} \quad z = \frac{x}{R},$$

and g is the gravitational field strength at the earth's surface. If $g = 0.00609$ mi/s^2 and $R = 3960$ mi, approximate the escape velocity v.

4.9 Numerical Differentiation

The derivative of the function f at x_0 is defined as

$$f'(x_0) = \lim_{h \to 0} \frac{f(x_0 + h) - f(x_0)}{h}.$$

To approximate this number, suppose first that $x_0 \in (a, b)$, where $f \in C^2[a, b]$, and that $x_1 = x_0 + h$ for some $h \neq 0$ that is sufficiently small to ensure that $x_1 \in [a, b]$. We construct the first Lagrange polynomial $P_{0,1}$ for f determined by x_0 and x_1, with its error term:

$$f(x) = P_{0,1}(x) + \frac{(x - x_0)(x - x_1)}{2!} f''(\xi(x))$$

$$= \frac{f(x_0)(x - x_0 - h)}{-h} + \frac{f(x_0 + h)(x - x_0)}{h} + \frac{(x - x_0)(x - x_0 - h)}{2} f''(\xi(x))$$

for some $\xi(x)$ in $[a, b]$. Differentiating this equation we have

$$f'(x) = \frac{f(x_0 + h) - f(x_0)}{h} + D_x\left[\frac{(x - x_0)(x - x_0 - h)}{2}f''(\xi(x))\right]$$

$$= \frac{f(x_0 + h) - f(x_0)}{h} + \frac{2(x - x_0) - h}{2}f''(\xi(x))$$

$$+ \frac{(x - x_0)(x - x_0 - h)}{2}D_x(f''(\xi(x))),$$

so

$$f'(x) \approx \frac{f(x_0 + h) - f(x_0)}{h}.$$

There are two terms for the error in this approximation. The first term involves $f''(\xi(x))$, which can be bounded if we have a bound for the second derivative of f. The second part of the truncation error involves $D_x f''(\xi(x)) = f'''(\xi(x)) \cdot \xi'(x)$, which generally cannot be estimated because it contains the unknown term $\xi'(x)$. However, when x is x_0, the coefficient of $D_x f''(\xi(x))$ is zero. In this case the formula simplifies to the following:

Two-Point Formula

$$f'(x_0) = \frac{f(x_0 + h) - f(x_0)}{h} - \frac{h}{2}f''(\xi),$$

where ξ lies between x_0 and $x_0 + h$.

For small values of h, the difference quotient $[f(x_0 + h) - f(x_0)]/h$ can be used to approximate $f'(x_0)$ with an error bounded by $Mh/2$, if M is a bound on $f''(x)$ for $x \in [a, b]$. This is a two-point formula known as the **forward-difference formula** if $h > 0$ (see Figure 4.21) and the **backward-difference formula** if $h < 0$.

Figure 4.21

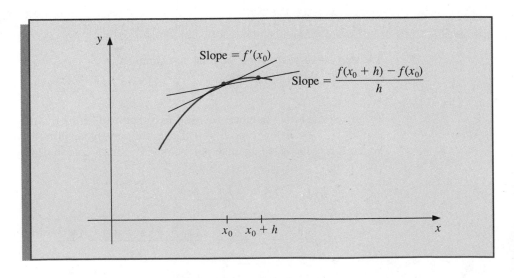

EXAMPLE 1 Let $f(x) = \ln x$ and $x_0 = 1.8$. The quotient

$$\frac{f(1.8 + h) - f(1.8)}{h}, \qquad h > 0,$$

can be used to approximate $f'(1.8)$ with error

$$\frac{|hf''(\xi)|}{2} = \frac{|h|}{2\xi^2} \le \frac{|h|}{2(1.8)^2}, \qquad \text{where } 1.8 < \xi < 1.8 + h.$$

The results in Table 4.8 are produced when $h = 0.1, 0.01$, and 0.001.

Table 4.8

h	$f(1.8 + h)$	$\dfrac{f(1.8 + h) - f(1.8)}{h}$	$\dfrac{\|h\|}{2(1.8)^2}$
0.1	0.64185389	0.5406722	0.0154321
0.01	0.59332685	0.5540180	0.0015432
0.001	0.58834207	0.5554013	0.0001543

Since $f'(x) = 1/x$, the exact value of $f'(1.8)$ is $0.\bar{5}$, and the error bounds are appropriate. ■ ■ ■

To obtain more general derivative approximation formulas, suppose that $x_0, x_1, \ldots, x_n$ are $(n + 1)$ distinct numbers in some interval I and that $f \in C^{n+1}(I)$. If we use the nth Lagrange polynomial to approximate f, then

$$f(x) = \sum_{k=0}^{n} f(x_k)L_k(x) + \frac{(x - x_0) \cdots (x - x_n)}{(n + 1)!} f^{(n+1)}(\xi(x))$$

for some $\xi(x)$ in I, where $L_k(x)$ denotes the kth Lagrange coefficient polynomial for f at $x_0, x_1, \ldots, x_n$. Differentiating this expression gives

$$f'(x) = \sum_{j=0}^{n} f(x_j)L_j'(x) + D_x\left[\frac{(x - x_0) \cdots (x - x_n)}{(n + 1)!}\right] f^{(n+1)}(\xi(x))$$

$$+ \frac{(x - x_0) \cdots (x - x_n)}{(n + 1)!} D_x[f^{(n+1)}(\xi(x))].$$

Again we have a problem with the second part of the truncation error unless x is one of the numbers x_k. In this case, the term involving $D_x[f^{(n+1)}(\xi(x))]$ is zero, and the formula becomes

$$f'(x_k) = \sum_{j=0}^{n} f(x_j)L_j'(x_k) + \frac{f^{(n+1)}(\xi(x_k))}{(n + 1)!} \prod_{\substack{j=0 \\ j \ne k}}^{n} (x_k - x_j).$$

Applying this technique using the second Lagrange polynomial at $x_0, x_0 + h$, and $x_0 + 2h$ produces the formula on page 142.

Three-Point Endpoint Formula

$$f'(x_0) = \frac{1}{2h}[-3f(x_0) + 4f(x_0 + h) - f(x_0 + 2h)] + \frac{h^2}{3}f^{(3)}(\xi),$$

where ξ lies between x_0 and $x_0 + 2h$.

This formula is most useful when approximating the derivative at the endpoint of an interval. This situation occurs, for example, when approximations are needed for the derivatives used for the clamped cubic splines discussed in Section 3.6. Left-endpoint approximations are found using $h > 0$ and right-endpoint approximations, when $h < 0$.

When approximating the derivative of a function at an interior point of an interval, it is better to use the formula produced by using the second Lagrange polynomial at $x_0 - h$, x_0, and $x_0 + h$.

Three-Point Midpoint Formula

$$f'(x_0) = \frac{1}{2h}[f(x_0 + h) - f(x_0 - h)] - \frac{h^2}{6}f^{(3)}(\xi),$$

where ξ lies between $x_0 - h$ and $x_0 + h$.

The error in the Midpoint formula is approximately half the error in the Endpoint formula and f needs to be evaluated at only two points, whereas in the Endpoint formula three evaluations are required. Figure 4.22 gives an illustration of the approximation produced from the Midpoint formula.

Figure 4.22

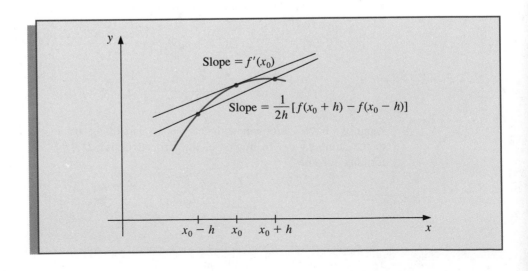

These methods are called *three-point formulas* (even though the "third point," $f(x_0)$, does not appear in the Midpoint formula). Similarly, there are methods known as *five-point formulas* that involve evaluating the function at two additional points, whose error term is of the form $O(h^4)$. These formulas are generated by differentiating fourth Lagrange polynomials that pass through the evaluation points. The most useful is the following interior-point formula:

Five-Point Midpoint Formula

$$f'(x_0) = \frac{1}{12h}[f(x_0 - 2h) - 8f(x_0 - h) + 8f(x_0 + h) - f(x_0 + 2h)] + \frac{h^4}{30}f^{(5)}(\xi),$$

where ξ lies between $x_0 - 2h$ and $x_0 + 2h$.

Another five-point formula that is useful, particularly with regard to the clamped cubic spline interpolation of Section 3.6, is the following endpoint formula:

Five-Point Endpoint Formula

$$f'(x_0) = \frac{1}{12h}[-25f(x_0) + 48f(x_0 + h) - 36f(x_0 + 2h)$$

$$+ 16f(x_0 + 3h) - 3f(x_0 + 4h)] + \frac{h^4}{5}f^{(5)}(\xi),$$

where ξ lies between x_0 and $x_0 + 4h$.

Left-endpoint approximations are found using $h > 0$ and right-endpoint approximations are found when $h < 0$.

EXAMPLE 2 Given in Table 4.9 are values for $f(x) = xe^x$.

Table 4.9

x	$f(x)$
1.8	10.889365
1.9	12.703199
2.0	14.778112
2.1	17.148957
2.2	19.855030

Since $f'(x) = (x + 1)e^x$, $f'(2.0) = 22.167168$. Approximating $f'(2.0)$ using the various three- and five-point formulas produces the results on page 144.

Three-Point formulas

Endpoint with $h = 0.1$: $\frac{1}{0.2}[-3f(2.0) + 4f(2.1) - f(2.2)]$ $= 22.032310.$

Endpoint with $h = -0.1$: $\frac{1}{-0.2}[-3f(2.0) + 4f(1.9) - f(1.8)]$ $= 22.054525.$

Midpoint with $h = 0.1$: $\frac{1}{0.2}[f(2.1) - f(1.9)]$ $= 22.228790.$

Midpoint with $h = 0.2$: $\frac{1}{0.4}[f(2.2) - f(1.8)]$ $= 22.414163.$

Five-Point formula

Endpoint with $h = 0.1$ (the only five-point formula applicable)

$$\frac{1}{1.2}[f(1.8) - 8f(1.9) + 8f(2.1) - f(2.2)] = 22.166999.$$

The errors are approximately 1.35×10^{-1}, 1.13×10^{-1}, -6.16×10^{-2}, -2.47×10^{-1}, and 1.69×10^{-4}, respectively. Clearly, the five-point formula gives a superior result. Note also that the error from the midpoint formula with $h = 0.1$ is approximately half of the magnitude of the error produced using the endpoint formula with either $h = 0.1$ or $h = -0.1$.   ■ ■ ■

A particularly important consideration in numerical differentiation is the effect of round-off error. In Section 4.3 we found that by reducing the step size in the Composite Simpson's rule, we could reduce the truncation error, and, even though the amount of calculation increased, total round-off error in the method was not effected. The truncation error in a numerical differentiation technique will also decrease if the step size is reduced, but only at the expense of increased round-off error. To see why this occurs, let us examine more closely the Three-Point Midpoint formula:

$$f'(x_0) = \frac{1}{2h}[f(x_0 + h) - f(x_0 - h)] - \frac{h^2}{6}f^{(3)}(\xi).$$

Suppose that, in evaluating $f(x_0 + h)$ and $f(x_0 - h)$, we encounter round-off errors $e(x_0 + h)$ and $e(x_0 - h)$; that is, our computed values $\tilde{f}(x_0 + h)$ and $\tilde{f}(x_0 - h)$ are related to the true values $f(x_0 + h)$ and $f(x_0 - h)$ by the formulas

$$f(x_0 + h) = \tilde{f}(x_0 + h) + e(x_0 + h) \quad \text{and} \quad f(x_0 - h) = \tilde{f}(x_0 - h) + e(x_0 - h).$$

In this case, the total error in the approximation,

$$f'(x_0) - \frac{\tilde{f}(x_0 + h) - \tilde{f}(x_0 - h)}{2h} = \frac{e(x_0 + h) - e(x_0 - h)}{2h} - \frac{h^2}{6}f^{(3)}(\xi),$$

is due partially to round-off and partially to truncating. If we assume that the round-off errors $e(x_0 \pm h)$ are bounded by some number $\epsilon > 0$ and that the third derivative of f is bounded by a number $M > 0$, then

$$\left| f'(x_0) - \frac{\tilde{f}(x_0 + h) - \tilde{f}(x_0 + h)}{2h} \right| \leq \frac{\epsilon}{h} + \frac{h^2}{6}M.$$

To reduce the truncation portion of the error, $h^2M/6$, we must reduce h. But as h is reduced, the round-off portion of the error, ϵ/h, grows. In practice, then, it is seldom advantageous to let h be too small, since the round-off error will dominate the calculations.

EXAMPLE 3 Consider approximating $f'(0.900)$ for $f(x) = \sin x$ using the values in Table 4.10. The true value is $\cos(0.900) = 0.62161$. Using the formula

$$f'(0.900) \approx \frac{f(0.900 + h) - f(0.900 - h)}{2h}$$

with different values of h gives the approximations in Table 4.11.

Table 4.10

x	$\sin x$	x	$\sin x$
0.800	0.71736	0.901	0.78395
0.850	0.75128	0.902	0.78457
0.880	0.77074	0.905	0.78643
0.890	0.77707	0.910	0.78950
0.895	0.78021	0.920	0.79560
0.898	0.78208	0.950	0.81342
0.899	0.78270	1.000	0.84147

Table 4.11

h	Approximation to $f'(0.900)$	Error
0.001	0.62500	0.00339
0.002	0.62250	0.00089
0.005	0.62200	0.00039
0.010	0.62150	−0.00011
0.020	0.62150	−0.00011
0.050	0.62140	−0.00021
0.100	0.62055	−0.00106

It appears that an optimal choice for h lies between 0.005 and 0.05. If we perform some analysis on the error term

$$e(h) = \frac{\epsilon}{h} + \frac{h^2}{6}M,$$

we can see that a minimum for e occurs when $e'(x) = 0$, that is, when $h = \sqrt[3]{3\epsilon/M}$. Note that

$$M = \max_{x \in [0.800, 1.00]} |f'''(x)| = \max_{x \in [0.800, 1.00]} |\cos x| \approx 0.69671.$$

Also, if the values of f are given to five decimal places, it is reasonable to assume that $\varepsilon = 0.000005$. Therefore, the optimal choice of h is

$$h = \sqrt[3]{\frac{3(0.000005)}{0.69671}} \approx 0.028,$$

which is consistent with our results. ■ ■ ■

In practice, we cannot compute an optimal h to use in approximating the derivative, since we have no knowledge of the third derivative of the function.

Although we have considered only the round-off error problems that are presented by the Three-Point Midpoint formula, similar difficulties occur with all the differentiation formulas. The reason for the problems can be traced to the need to divide by a power of h. As we found in Section 1.4 (see, in particular, Example 1), division by small numbers tends to exaggerate round-off error and should be avoided if possible. In the case of numerical differentiation, it is impossible to avoid the problem entirely, although the higher-order methods reduce the difficulty.

Keep in mind that as an approximation method, numerical differentiation is unstable, since the small values of h needed to reduce truncation error also cause the round-off error to grow. This is the first class of unstable methods we have encountered, and these techniques would be avoided if it were possible. However, in addition to being used for computational purposes, the formulas we have derived are needed for approximating the solutions of ordinary and partial-differential equations.

Methods for approximating higher derivatives of functions can be derived as was done for approximating the first derivative, or by using an averaging technique similar to that used for extrapolation. These techniques, of course, suffer from the same stability weaknesses as the approximation methods for first derivatives, but they are needed for approximating the solution to boundary value problems in differential equations. The only one we will need is a three-point midpoint formula, which has the following form:

Three-Point Midpoint Formula for Approximating f''

$$f''(x_0) = \frac{1}{h^2}[f(x_0 - h) - 2f(x_0) + f(x_0 + h)] - \frac{h^2}{12}f^{(4)}(\xi),$$

where ξ lies between $x_0 - h$ and $x_0 + h$.

EXERCISE SET 4.9

1. Use the forward-difference formulas and backward-difference formulas to determine approximations that will complete the following tables:

(a)

x	$f(x)$	$f'(x)$
0.5	-0.344099	
0.6	-0.176945	
0.7	0.0137523	

(b)

x	$f(x)$	$f'(x)$
0.0	-1.0000000	
0.2	-0.2839867	
0.4	0.2484244	

2. The data in Exercise 1 were taken from the following functions. Compute the actual errors in Exercise 1 and find error bounds using the error formulas.
   (a) $f(x) = \sin(e^x - 2)$
   (b) $f(x) = x \cos x - 2x^2 + 3x - 1$

3. Use the most appropriate Three-Point formula to determine approximations that will complete the following tables:

(a)

x	$f(x)$	$f'(x)$
1.1	1.949477	
1.2	2.199796	
1.3	2.439189	
1.4	2.670324	

(b)

x	$f(x)$	$f'(x)$
8.1	16.94410	
8.3	17.56492	
8.5	18.19056	
8.7	18.82091	

(c)

x	$f(x)$	$f'(x)$
2.9	-4.827866	
3.0	-4.240058	
3.1	-3.496909	
3.2	-2.596792	

(d)

x	$f(x)$	$f'(x)$
3.6	1.16164956	
3.8	0.80201036	
4.0	0.30663842	
4.2	-0.35916618	

4. The data in Exercise 3 were taken from the following functions. Compute the actual errors in Exercise 3 and find error bounds using the error formulas.
   (a) $f(x) = \ln(e^{2x} - 2)$
   (b) $f(x) = x \ln x$
   (c) $f(x) = x \cos x - x^2 \sin x$
   (d) $f(x) = x - (\ln x)^x$

5. Use the most accurate formula possible to determine approximations that will complete the following tables.

(a)

x	$f(x)$	$f'(x)$
2.1	-1.709847	
2.2	-1.373823	
2.3	-1.119214	
2.4	-0.9160143	
2.5	-0.7470223	
2.6	-0.6015966	

(b)

x	$f(x)$	$f'(x)$
-3.0	9.367879	
-2.8	8.233241	
-2.6	7.180350	
-2.4	6.209329	
-2.2	5.320305	
-2.0	4.513417	

6. The data in Exercise 5 were taken from the given functions. Compute the actual errors in Exercise 5 and find error bounds using the error formulas.
   (a) $f(x) = \tan x$
   (b) $f(x) = e^{\frac{1}{3}x} + x^2$

7. Repeat Exercise 1 using four-digit round-off arithmetic and compare the errors to those in Exercise 2.

8. Repeat Exercise 3 using four-digit chopping arithmetic and compare the errors to those in Exercise 4.

9. Repeat Exercise 5 using four-digit round-off arithmetic and compare the errors to those in Exercise 6.

10. Consider the table of data:

x	0.2	0.4	0.6	0.8	1.0
$f(x)$	0.9798652	0.9177710	0.808038	0.6386093	0.3843735

    (a) Use all appropriate formulas to approximate $f'(0.4)$ and $f''(0.4)$.
    (b) Use all appropriate formulas to approximate $f'(0.6)$ and $f''(0.6)$.

11. Let $f(x) = \cos \pi x$. Use the Three-Point Midpoint formula for approximating f'' and the values of $f(x)$ at $x = 0.25, 0.5$, and 0.75 to approximate $f''(0.5)$. Explain why this method is particularly accurate for this problem. Find a bound for the error.

12. Let $f(x) = 3xe^x - \cos x$. Use the following data and the Three-Point Midpoint formula for approximating f'' to approximate $f''(1.3)$ with $h = 0.1$ and $h = 0.01$.

x	1.20	1.29	1.30	1.31	1.40
$f(x)$	11.59006	13.78176	14.04276	14.30741	16.86187

Compare your results to $f''(1.3)$.

4.10 Survey of Methods and Software

In this chapter we have considered approximating integrals of functions of one, two, or three variables and approximating the first and second derivatives of a function of a single real variable.

The Midpoint method, Trapezoidal method, and Simpson's rule were studied to introduce the techniques and error analysis of quadrature methods. The Composite Simpson's rule is easy to use and produces accurate approximations unless the function oscillates in a subinterval of the interval of integration. Adaptive quadrature can be used if the function is suspected of oscillatory behavior. To minimize the number of nodes and increase the degree of precision, we studied Gaussian quadrature. Romberg integration was introduced to take advantage of extrapolation on the easily applied Composite Trapezoidal rule.

Most software for integrating a function of a single real variable is based on the adaptive approach or on extremely accurate Gaussian formulas. Cautious Romberg Integration is an adaptive technique that includes a check to make sure that the integrand is sufficiently smooth over subintervals of the interval of integration. This method has been successfully used in software libraries. Multiple integrals are generally approximated by extending adaptive methods to higher dimensions.

The main routines in both the IMSL and NAG Libraries are based on QUADPACK: A Subroutine Package for Automatic Integration by R. Piessens, E. de Doncker-Kapenga,

C. W. Überhuber, and D. K. Kahaner published by Springer-Verlag in 1983. The routines are in the public domain.

The IMSL Library contains the function QDAGS, which is an adaptive integration scheme based on the 21-point Gaussian-Kronrod rule using the 10-point Gaussian rule for error estimation. The Gaussian rule uses 10 points $x_1, \ldots, x_{10}$ and weights $w_1, \ldots, w_{10}$ to give the quadrature formula $\sum_{i=1}^{10} w_i f(x_i)$ to approximate $\int_a^b f(x)\, dx$. The additional 11 points $x_{11}, \ldots, x_{21}$ and the new weights $v_1, \ldots, v_{21}$ are then used to form the Kronrod formula $\sum_{i=1}^{21} v_i f(x_i)$. The results of the two formulas are compared to estimate error. The advantage in using $x_1, \ldots, x_{10}$ in each formula is that f needs to be evaluated only at 21 points. If independent 10- and 21-point Gaussian rules were used, 31 function evaluations would be needed. This procedure permits endpoint singularities in the integrand function. Other IMSL subroutines are QDAGP, which allows user-specified singularities; QDAGI, which allows infinite intervals of integration; and QDNG, which is a nonadaptive procedure for smooth functions. The subroutine TWODQ uses the Gauss-Kronrod rules to integrate a function of two variables. There is also a subroutine QAND to use Gaussian Quadrature to integrate a function of n variables over n intervals of the form $[a_i, b_i]$.

The NAG Library includes the subroutine DO1AJF to compute the integral of $f(x)$ over the interval $[a, b]$ using an adaptive method based on Gaussian Quadrature using Gauss 10-point and Kronrod 21-point rules. The subroutine DO1AHF is used to approximate $\int_a^b f(x)\, dx$ using a family of Gaussian-type formulas based on 1, 3, 5, 7, 15, 31, 63, 127, and 255 nodes. These interlacing high-precision rules are also used in an adaptive manner. The subroutine DO1GBF is for multiple integrals and DO1GAF approximates an integral given only data points instead of the function f. NAG includes many other subroutines for approximating integrals.

Although numerical differentiation is unstable, derivative approximation formulas are needed for solving differential equations. The NAG Library includes the subroutine DO4AAF for the numerical differentiation of a function of one real variable with possible differentiation to the fourteenth derivative. The IMSL function DERIV uses an adaptive change in step size for finite differences to approximate a derivative of f at x to within a given tolerance. IMSL also includes the subroutine QDDER to compute the derivatives of a function defined on a set of points using quadratic interpolation. Both packages allow the differentiation and integration of interpolatory cubic splines constructed by the methods mentioned in Section 3.5.

Numerical Solution of Initial-Value Problems

5.1 Introduction

Differential equations are used to model problems that involve the change of some variable with respect to another. These problems require the solution to an initial-value problem, that is, the solution to a differential equation that satisfies a given initial condition.

In many real-life situations, the differential equation that models the problem is too complicated to solve exactly, and one of two approaches is taken to approximate the solution. The first is to simplify the differential equation to one that can be solved exactly and then use the solution of the simplified equation to approximate the solution to the original equation. The other approach, the one we will examine in this chapter, involves finding methods for directly approximating the solution of the original problem. This is the approach most commonly taken, since more accurate results and more realistic error information can be obtained.

The methods we consider in this chapter do not produce a continuous approximation to the solution of the initial-value problem. Rather, approximations are found at certain specified, and often equally spaced, points. Some method of interpolation, commonly Hermite, is used if intermediate values are needed.

The first part of the chapter concerns approximating the solution $y(t)$ to a problem of the form

$$\frac{dy}{dt} = f(t, y), \qquad a \leq t \leq b,$$

subject to an initial condition

$$y(a) = \alpha.$$

These techniques form the core of the study, since more general procedures use these as a base. Later in the chapter we deal with the extension of these methods to a system of first-order differential equations in the form

$$\frac{dy_1}{dt} = f_1(t, y_1, y_2, \ldots, y_n),$$

$$\frac{dy_2}{dt} = f_2(t, y_1, y_2, \ldots, y_n),$$

$$\vdots$$

$$\frac{dy_n}{dt} = f_n(t, y_1, y_2, \ldots, y_n),$$

for $a \leq t \leq b$, subject to the initial conditions

$$y_1(a) = \alpha_1, \qquad y_2(a) = \alpha_2, \ldots, y_n(a) = \alpha_n.$$

We also examine the relationship of a system of this type to the general nth-order initial-value problem of the form

$$y^{(n)} = f(t, y, y', y'', \ldots, y^{(n-1)})$$

for $a \leq t \leq b$, subject to the initial conditions

$$y(a) = \alpha_0, \qquad y'(a) = \alpha_1, \ldots, y^{(n-1)}(a) = \alpha_{n-1}.$$

Before describing the methods for approximating the solution to our basic problem, we consider some situations for which we know the solution will exist. In fact, since we will not be solving the given problem, only an approximation to the problem, we need to know when problems that are close to the given problem have solutions that accurately approximate the solution to the given problem. This property of an initial-value problem is called **well-posed**, and these are the problems for which numerical methods are appropriate. The following result shows that the class of well-posed problems is quite broad.

Well-Posed Condition

Suppose that f and f_y, its first partial derivative with respect to y, are continuous for t in $[a, b]$. Then the initial-value problem

$$y' = f(t, y), \qquad a \leq t \leq b, \qquad y(a) = \alpha,$$

has a unique solution $y(t)$ for $a \leq t \leq b$ and the problem is well-posed.

EXAMPLE 1   Consider the initial-value problem

$$y' = 1 + t\sin(ty), \qquad 0 \leq t \leq 2, \qquad y(0) = 0.$$

Since the functions

$$f(t, y) = 1 + t\sin(ty) \qquad \text{and} \qquad f_y(t, y) = t^2\cos(ty)$$

are both continuous on [0, 2], a unique solution exists to this well-posed initial-value problem.

▪ ▪ ▪

If you have taken a course in differential equations you might like to attempt to determine the solution to the problem in Example 1 by using one of the techniques you learned in that course.

5.2 Taylor Methods

Many of the numerical methods we saw in the first four chapters have an underlying derivation from Taylor's Theorem. The approximation of the solution to initial-value problems is no exception. In this case, the function we need to expand in a Taylor polynomial is the (unknown) solution to the problem, $y(t)$. In its most elementary form this leads to **Euler's method**. Although Euler's method is seldom used in practice, the simplicity of its derivation illustrates the technique used in the more advanced techniques, without the cumbersome algebra that accompanies these constructions.

The object of Euler's method is to find an approximation to the solution of a problem of the form

$$\frac{dy}{dt} = f(t, y), \qquad a \le t \le b, \qquad y(a) = \alpha$$

at the equally spaced **mesh points** $\{t_0, t_1, t_2, \ldots, t_N\}$ (see Figure 5.1), where

$$t_i = a + ih, \qquad \text{for each } i = 0, 1, \ldots, N.$$

The common distance between the points, $h = (b - a)/N$, is called the **step size**.

Figure 5.1

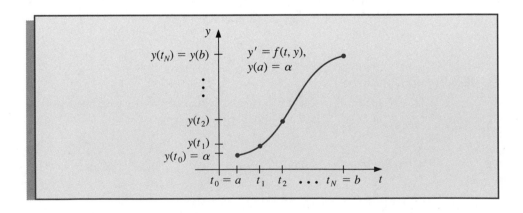

Suppose that $y(t)$, the solution to the problem, has two continuous derivatives on $[a, b]$, so that for each $i = 0, 1, 2, \ldots, N - 1$, Taylor's Theorem implies that

$$y(t_{i+1}) = y(t_i) + (t_{i+1} - t_i)y'(t_i) + \frac{(t_{i+1} - t_i)^2}{2}y''(\xi_i),$$

for some ξ_i in (t_i, t_{i+1}). Letting $h = t_{i+1} - t_i$, we have

$$y(t_{i+1}) = y(t_i) + hy'(t_i) + \frac{h^2}{2}y''(\xi_i)$$

and, since $y(t)$ satisfies the differential equation $y'(t) = f(t, y(t))$,

$$y(t_{i+1}) = y(t_i) + hf(t_i, y(t_i)) + \frac{h^2}{2}y''(\xi_i).$$

Euler's method constructs the approximation w_i to $y(t_i)$ for each $i = 1, 2, \ldots, N$, by deleting the error term in this equation. This produces a *difference equation* that approximates the differential equation. The program EULERM51 implements the following procedure:

Euler's Method

$$w_0 = \alpha,$$

$$w_{i+1} = w_i + hf(t_i, w_i),$$

Where $i = 0, 1, \ldots, N - 1$. The local error is $\frac{1}{2}y''(\xi)h^2$ for some ξ_i in (t_i, t_{i+1}).

To interpret Euler's method geometrically, note that when w_i is a close approximation to $y(t_i)$, the assumption that the problem is well-posed implies that

$$f(t_i, w_i) \approx y'(t_i) = f(t_i, y(t_i)).$$

The first step of Euler's method appears in part (a) of Figure 5.2, and a series of steps appears in part (b).

Figure 5.2

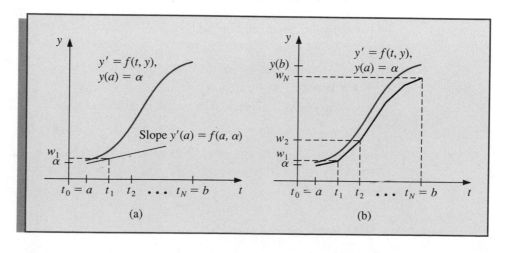

(a)   (b)

EXAMPLE 1 Suppose that Euler's method is used to approximate the solution to the initial-value problem

$$y' = y - t^2 + 1, \qquad 0 \le t \le 2, \qquad y(0) = 0.5,$$

assuming that $N = 10$. Then $h = 0.2$ and $t_i = 0.2i$. Using the fact that $f(t, y) = y - t^2 + 1$ gives $w_0 = 0.5$ and

$$w_{i+1} = w_i + h(w_i - t_i^2 + 1) = w_i + 0.2[w_i - 0.04i^2 + 1] = 1.2w_i - 0.008i^2 + 0.2,$$

for $i = 0, 1, \ldots, 9$. The exact solution is $y(t) = (t + 1)^2 - 0.5e^t$. Table 5.1 shows the comparison between the approximate values at t_i and the actual values.

■ ■ ■

Table 5.1

t_i	w_i	$y_i = y(t_i)$	$\|y_i - w_i\|$
0.0	0.5000000	0.5000000	0.0000000
0.2	0.8000000	0.8292986	0.0292986
0.4	1.1520000	1.2140877	0.0620877
0.6	1.5504000	1.6489406	0.0985406
0.8	1.9884800	2.1272295	0.1387495
1.0	2.4581760	2.6408591	0.1826831
1.2	2.9498112	3.1799415	0.2301303
1.4	3.4517734	3.7324000	0.2806266
1.6	3.9501281	4.2834838	0.3333557
1.8	4.4281538	4.8151763	0.3870225
2.0	4.8657845	5.3054720	0.4396874

Since Euler's method is derived from a Taylor polynomial whose error term involves the square of the step size h, the error at each step, called the **local error**, is proportional to h^2. However, the total error, called the **global error**, accumulates these local errors, so it generally grows at a much faster rate. The following result relates the global error of Euler's method to its local error and is typical of the type of results that are known for initial-value approximation methods.

Euler's Method Error Bound

Let $y(t)$ denote the unique solution to the initial-value problem

$$y' = f(t, y), \qquad a \le t \le b, \qquad y(a) = \alpha,$$

and $w_0, w_1, \ldots, w_N$ be the approximations generated by Euler's method for some positive integer N. If f is continuous for all t in $[a, b]$ and all y in $(-\infty, \infty)$ and if constants L and M exist with

$$\left| \frac{\partial f(t, y(t))}{\partial t} \right| \le L \qquad \text{and} \qquad |y''(t)| \le M,$$

then, for each $i = 0, 1, 2, \ldots, N$,

$$|y(t_i) - w_i| \le \frac{hM}{2L}[e^{L(t_i - a)} - 1].$$

The important feature of this result is that the global error depends linearly on h, which is a reduction from quadratically for the local error. This reduction of one power of h from local to global error is typical of initial-value techniques. Even though we have a reduction in order from local to global errors, the formula shows that the error tends to zero with h, which is necessary for convergence.

EXAMPLE 2 Returning to the initial-value problem

$$y' = y - t^2 + 1, \qquad 0 \le t \le 2, \qquad y(0) = 0.5$$

considered in Example 1, we see that with $f(t, y) = y - t^2 + 1$, $\partial f(t, y)/\partial y = 1$ for all y and $L = 1$. For this problem we know that the exact solution is $y(t) = (t + 1)^2 - \frac{1}{2}e^t$, so $y''(t) = 2 - 0.5e^t$, and

$$|y''(t)| \le 0.5e^2 - 2 \qquad \text{for all } t \in [0, 2].$$

Using the inequality in the error bound for Euler's method with $h = 0.2, L = 1$, and $M = 0.5e^2 - 2$ gives the error bound

$$|y_i - w_i| \le 0.1(0.5e^2 - 2)(e^{t_i} - 1).$$

Table 5.2 lists the actual error found in Example 1, together with this error bound.

∎ ∎ ∎

Table 5.2

t_i	0.2	0.4	0.6	0.8	1.0	1.2	1.4	1.6	1.8	2.0
Actual error:	0.02930	0.06209	0.09854	0.13875	0.18268	0.23013	0.28063	0.33336	0.38702	0.43969
Error bound:	0.03752	0.08334	0.13931	0.20767	0.29117	0.39315	0.51771	0.66985	0.85568	1.08264

Neglected in the global error result for Euler's method is the effect that round-off error plays in the choice of step size. As h becomes smaller, more calculations are necessary, so more round-off error is expected. Taking this into consideration produces an error estimate that becomes unbounded as the step size approaches zero. However, the lower bound for an effective step size is sufficiently small that round-off error does not significantly effect the approximation.

Since Euler's method was derived using Taylor's Theorem with $n = 2$, the first attempt to find methods for improving the convergence properties of difference methods is to extend this technique of derivation to larger values of n. Suppose the solution $y(t)$ to the initial-value problem

$$y' = f(t, y), \qquad a \le t \le b, \qquad y(a) = \alpha$$

has $(n + 1)$ continuous derivatives. If we expand the solution, $y(t)$, in terms of its nth Taylor polynomial about t_i, we obtain:

$$y(t_{i+1}) = y(t_i) + hy'(t_i) + \frac{h^2}{2}y''(t_i) + \cdots + \frac{h^n}{n!}y^{(n)}(t_i) + \frac{h^{n+1}}{(n + 1)!}y^{(n+1)}(\xi_i)$$

for some ξ in (t_i, t_{i+1}).

Successive differentiation of the solution, $y(t)$, gives

$$y'(t) = f(t, y(t)), \quad y''(t) = f'(t, y(t)), \quad \text{and, generally,} \quad y^{(k)}(t) = f^{(k-1)}(t, y(t)).$$

Substituting these results into the Taylor expansion gives

$$y(t_{i+1}) = y(t_i) + hf(t_i, y(t_i)) + \frac{h^2}{2}f'(t_i, y(t_i)) + \cdots$$

$$+ \frac{h^n}{n!}f^{(n-1)}(t_i, y(t_i)) + \frac{h^{n+1}}{(n+1)!}f^{(n)}(\xi_i, y(\xi_i)).$$

The difference-equation method corresponding to this equation is obtained by deleting the remainder term involving ξ_i.

Taylor Method of Order n

$$w_0 = \alpha,$$

$$w_{i+1} = w_i + hT^{(n)}(t_i, w_i)$$

for each $i = 0, 1, \ldots, N - 1$, where

$$T^{(n)}(t_i, w_i) = f(t_i, w_i) + \frac{h}{2}f'(t_i, w_i) + \cdots + \frac{h^{n-1}}{n!}f^{(n-1)}(t_i, w_i).$$

The local error is $\dfrac{1}{(n+1)!}y^{(n+1)}(\xi_i)h^{n+1}$ for some ξ_i in $(t_i, t_i + 1)$.

Although the formula for $T^{(n)}$ is easily expressed, it is difficult to use because it requires the derivatives of f with respect to t. Since f is described as a multivariable function of both t and y, the chain rule implies that the total derivative of f with respect to t, which we denoted $f'(t, y(t))$, is obtained by

$$f'(t, y(t)) = \frac{\partial f}{\partial t}(t, y(t)) \cdot \frac{dt}{dt} + \frac{\partial f}{\partial y}(t, y(t))\frac{dy(t)}{dt},$$

or, since $dt/dt = 1$ and $dy(t)/dt = f(t, y(t))$,

$$f'(t, y(t)) = \frac{\partial f}{\partial t}(t, y(t)) + f(t, y(t))\frac{\partial f}{\partial y}(t, y(t)).$$

Subsequent derivatives increase in complication. For example, $f''(t, y(t))$ involves the partial derivatives of all the terms on the right side of this equation with respect to both t and y.

EXAMPLE 3   To apply Taylor's method of orders two and four to the initial-value problem

$$y' = y - t^2 + 1, \quad 0 \le t \le 2, \quad y(0) = 0.5,$$

which was studied in Examples 1 and 2, we must find the first three derivatives of $f(t, y(t)) = y(t) - t^2 + 1$ with respect to the variable t:

$$f'(t, y(t)) = \frac{d}{dt}(y - t^2 + 1) = y' - 2t = y - t^2 + 1 - 2t$$

$$f''(t, y(t)) = \frac{d}{dt}(y - t^2 + 1 - 2t) = y' - 2t - 2$$
$$= y - t^2 + 1 - 2t - 2 = y - t^2 - 2t - 1$$

and

$$f'''(t, y(t)) = \frac{d}{dt}(y - t^2 - 2t - 1) = y' - 2t - 2 = y - t^2 - 2t - 1.$$

So

$$T^{(2)}(t_i, w_i) = f(t_i, w_i) + \frac{h}{2}f'(t_i, w_i) = w_i - t_i^2 + 1 + \frac{h}{2}(w_i - t_i^2 - 2t + 1)$$
$$= \left(1 + \frac{h}{2}\right)(w_i - t_i^2 + 1) - ht_i$$

and

$$T^{(4)}(t_i, w_i) = f(t_i, w_i) + \frac{h}{2}f'(t_i, w_i) + \frac{h^2}{6}f''(t_i, w_i) + \frac{h^3}{24}f'''(t_i, w_i)$$
$$= w_i - t_i^2 + 1 + \frac{h}{2}(w_i - t_i^2 - 2t_i + 1) + \frac{h^2}{6}(w_i - t_i^2 - 2t_i - 1)$$
$$+ \frac{h^3}{24}(w_i - t_i^2 - 2t_i - 1)$$
$$= \left(1 + \frac{h}{2} + \frac{h^2}{6} + \frac{h^3}{24}\right)(w_i - t_i^2) - \left(1 + \frac{h}{3} + \frac{h^2}{12}\right)(ht_i)$$
$$+ 1 + \frac{h}{2} - \frac{h^2}{6} - \frac{h^3}{24}.$$

The Taylor methods of orders two and four are, consequently,

$$w_0 = 0.5,$$
$$w_{i+1} = w_i + h\left[\left(1 + \frac{h}{2}\right)(w_i - t_i^2 + 1) - ht_i\right]$$

and

$$w_0 = 0.5,$$
$$w_{i+1} = w_i + h\left[\left(1 + \frac{h}{2} + \frac{h^2}{6} + \frac{h^3}{24}\right)(w_i - t_i^2)\right.$$
$$\left. - \left(1 + \frac{h}{3} + \frac{h^2}{12}\right)ht_i + 1 + \frac{h}{2} - \frac{h^2}{6} - \frac{h^3}{24}\right],$$

for $i = 0, 1, \ldots, N - 1$, respectively. If $h = 0.2$, then $N = 10$ and $t_i = 0.2i$ for each $i = 1, 2, \ldots, 10$, so the second-order method becomes

$$w_0 = 0.5,$$
$$w_{i+1} = w_i + 0.2\left[\left(1 + \frac{0.2}{2}\right)(w_i - 0.04i^2 + 1) - 0.04i\right] = 1.22w_i - 0.0088i^2 - 0.008i + 0.22,$$

and the fourth-order method becomes

$$w_{i+1} = w_i + 0.2\left[\left(1 + \frac{0.2}{2} + \frac{0.04}{6} + \frac{0.008}{24}\right)(w_i - 0.04i^2)\right.$$

$$\left. - \left(1 + \frac{0.2}{3} + \frac{0.04}{12}\right)(0.04i) + 1 + \frac{0.2}{2} - \frac{0.04}{6} - \frac{0.008}{24}\right]$$

$$= 1.2214w_i - 0.008856i^2 - 0.00856i + 0.2186,$$

for each $i = 0, 1, \ldots, 9$.

Table 5.3 lists the actual values of the solution $y(t) = (t + 1)^2 - 0.5e^t$, the results from the Taylor methods of orders two and four, and the actual errors involved with these methods.

Table 5.3

t_i	Taylor Order 2 w_i	Error $\|y(t_i) - w_i\|$	Taylor Order 4 w_i	Error $\|y(t_i) - w_i\|$	Exact $y(t_i)$
0.0	0.5000000	0	0.5000000	0	0.5000000
0.2	0.8300000	0.0007014	0.8293000	0.0000014	0.8292986
0.4	1.2158000	0.0017123	1.2140910	0.0000034	1.2140877
0.6	1.6520760	0.0031354	1.6489468	0.0000062	1.6489406
0.8	2.1323327	0.0051032	2.1272396	0.0000101	2.1272295
1.0	2.6486459	0.0077868	2.6408744	0.0000153	2.6408591
1.2	3.1913480	0.0114065	3.1799640	0.0000225	3.1799415
1.4	3.7486446	0.0162446	3.7324321	0.0000321	3.7324000
1.6	4.3061464	0.0226626	4.2835285	0.0000447	4.2834838
1.8	4.8462986	0.0311223	4.8152377	0.0000615	4.8151763
2.0	5.3476843	0.0422123	5.3055554	0.0000834	5.3054720

Suppose we need to determine an approximation to an intermediate point in the table—for example, at $t = 1.25$. If we use linear interpolation on the Taylor method of order four approximations at $t = 1.2$ and $t = 1.4$, we have

$$y(1.25) \approx \frac{1.25 - 1.4}{1.2 - 1.4} 3.1799640 + \frac{1.25 - 1.2}{1.4 - 1.2} 3.7324321 = 3.3180810.$$

Since $y(1.25) = 3.3173285$, this approximation has an error of 0.0007525, which is nearly 30 times the average of the approximation errors at 1.2 and 1.4.

To improve the approximation to $y(1.25)$, we can use cubic Hermite interpolation. This requires approximations to $y'(1.2)$ and $y'(1.4)$ as well as approximations to $y(1.2)$ and $y(1.4)$. But the derivative approximations are available from the differential equation, since $y'(t) = f(t, y(t))$. In our example this means that $y'(t) = y(t) - t^2 + 1$, so

$$y'(1.2) = y(1.2) - (1.2)^2 + 1 \approx 3.1799640 - 1.44 + 1 = 2.7399640$$

and

$$y'(1.4) = y(1.4) - (1.4)^2 + 1 \approx 3.7324327 - 1.96 + 1 = 2.7724321.$$

Following the divided-difference procedure in Section 3.3, we have the information in Table 5.4. The underlined entries come from the data and the other entries use the divided-difference formulas.

Table 5.4

1.2	3.1799640			
		2.7399640		
1.2	3.1799640		0.1118825	
		2.7623405		-0.3071225
1.4	3.7324321		0.0504580	
		2.7724321		
1.4	3.7324321			

The cubic Hermite polynomial is

$$y(t) \approx 3.1799640 + (t - 1.2)2.7399640 + (t - 1.2)^2 0.1118825$$
$$+ (t - 1.2)^2(t - 1.4)(-0.3071225),$$

so

$$y(1.25) \approx 3.1799640 + 0.1369982 + 0.0002797 + 0.0001152 = 3.3173571,$$

a result that is accurate to within 0.0000286. This is less than twice the average error at 1.2 and 1.4, or only about 4% of the error obtained using linear interpolation. ■ ■ ■

Error estimates for the Taylor methods are similar to those for Euler's method. If sufficient differentiability conditions are met, an nth-order Taylor method will have local error $O(h^{n+1})$ and global error $O(h^n)$.

EXERCISE SET 5.2

1. Use Euler's method to approximate the solutions for each of the following initial-value problems.
   (a) $y' = te^{3t} - 2y,$ $0 \le t \le 1,$ $y(0) = 0$, with $h = 0.5$
   (b) $y' = 1 + (t - y)^2,$ $2 \le t \le 3,$ $y(2) = 1$, with $h = 0.5$
   (c) $y' = 1 + y/t,$ $1 \le t \le 2,$ $y(1) = 2$, with $h = 0.25$
   (d) $y' = \cos 2t + \sin 3t,$ $0 \le t \le 1,$ $y(0) = 1$, with $h = 0.25$

2. The actual solutions to the initial-value problems in Exercise 1 are given here. Compare the actual error at each step to the error bound.
   (a) $y(t) = \frac{1}{5}te^{3t} - \frac{1}{25}e^{3t} + \frac{1}{25}e^{-2t}$
   (b) $y(t) = t + (1 - t)^{-1}$
   (c) $y(t) = t \ln t + 2t$
   (d) $y(t) = \frac{1}{2}\sin 2t - \frac{1}{3}\cos 3t + \frac{4}{3}$

3. Use Euler's method to approximate the solutions for each of the following initial-value problems.

   (a) $y' = \dfrac{y}{t} - \left(\dfrac{y}{t}\right)^2$, $1 \le t \le 2$, $y(1) = 1$, with $h = 0.1$

   (b) $y' = 1 + \dfrac{y}{t} + \left(\dfrac{y}{t}\right)^2$, $1 \le t \le 3$, $y(1) = 0$, with $h = 0.2$

   (c) $y' = -(y + 1)(y + 3)$, $0 \le t \le 2$, $y(0) = -2$, with $h = 0.2$
   (d) $y' = -5y + 5t^2 + 2t$, $0 \le t \le 1$, $y(0) = \frac{1}{3}$, with $h = 0.1$

4. The actual solutions to the initial-value problems in Exercise 2 are given here. Compute the actual error in the approximations of Exercise 3.

   (a) $y(t) = t(1 + \ln t)^{-1}$
   (b) $y(t) = t \tan(\ln t)$
   (c) $y(t) = -3 + 2(1 + e^{-2t})^{-1}$
   (d) $y(t) = t^2 + \frac{1}{3}e^{-5t}$

5. Repeat Exercise 1 using Taylor's method of order two.

6. Repeat Exercises 3 and 4 using Taylor's method of order two.

7. Repeat Exercises 3 and 4 using Taylor's method of order four.

8. Use the results of Exercise 3 and linear interpolation to approximate the following values of $y(t)$. Compare the approximations obtained to the actual values obtained using the functions given in Exercise 4.

   (a) $y(1.25)$ and $y(1.93)$ (b) $y(2.1)$ and $y(2.75)$
   (c) $y(1.3)$ and $y(1.93)$ (d) $y(0.54)$ and $y(0.94)$

9. Use the results of Exercise 6 and linear interpolation to approximate the following values of $y(t)$. Compare the approximations obtained to the actual values.

   (a) $y(1.25)$ and $y(1.93)$ (b) $y(2.1)$ and $y(2.75)$
   (c) $y(1.4)$ and $y(1.93)$ (d) $y(0.54)$ and $y(0.94)$

10. Repeat Exercise 9 using cubic Hermite interpolation.

11. In a circuit with impressed voltage $\mathscr{E}$, having resistance R, inductance L, and capacitance C in parallel, the current i satisfies the differential equation

$$\frac{di}{dt} = C\frac{d^2\mathscr{E}}{dt^2} + \frac{1}{R}\frac{d\mathscr{E}}{dt} + \frac{1}{L}\mathscr{E}.$$

Suppose $i(0) = 0$, $C = 0.3$ farads, $R = 1.4$ ohms, $L = 1.7$ henries, and the voltage is given by

$$\mathscr{E}(t) = e^{-0.06\pi t} \sin(2t - \pi).$$

Use Euler's method to find the current i for the values $t = 0.1j$, $j = 0, 1, \ldots, 100$.

12. In a book entitled *Looking at History Through Mathematics*, Rashevsky considers a model for a problem involving the production of nonconformists in society. Suppose that a society has a population of $x(t)$ individuals at time t, in years, and that all nonconformists who mate with other nonconformists have offspring who are also nonconformists, whereas a fixed proportion r of all other offspring are also nonconformist. If the birth and death rates for all individuals are assumed to be the constants b and d, respectively, and if conformists and nonconformists mate at random, the problem can be expressed by the differential equations

$$\frac{dx(t)}{dt} = (b - d)x(t) \qquad \text{and} \qquad \frac{dx_n(t)}{dt} = (b - d)x_n(t) + rb(x(t) - x_n(t)),$$

where $x_n(t)$ denotes the number of nonconformists in the population at time t.

(a) If the variable $p(t) = x_n(t)/x(t)$ is introduced to represent the proportion of non-conformists in the society at time t, show that these equations can be combined and simplified to the single differential equation

$$\frac{dp(t)}{dt} = rb(1 - p(t)).$$

(b) Assuming that $p(0) = 0.01, b = 0.02, d = 0.015$, and $r = 0.1$, use Euler's method to approximate the solution $p(t)$ from $t = 0$ to $t = 50$ when the step size is $h = 1$ year.

(c) Solve the differential equation for $p(t)$ exactly, and compare your result in part (b) when $t = 50$ with the exact value at that time.

13. A projectile of mass $m = 0.11$ kg shot vertically upward with initial velocity $v(0) = 8$ m/s is slowed due to the force of gravity $F_g = mg$ and due to air resistance $F_r = -kv|v|$, where $g = -9.8$ m/s^2 and $k = 0.002$ kg/m. The differential equation for the velocity v is given by

$$mv' = mg - kv|v|.$$

(a) Use the Taylor methods of orders two and four to find the velocity after $0.1, 0.2, \ldots, 1.0$ s.

(b) Use the Taylor method of order four to determine, to the nearest tenth of a second, when the projectile reaches its maximum height.

5.3 Runge–Kutta Methods

In the last section we saw that Taylor methods of arbitrary high order can easily be generated. The application of one of the higher-order methods to a specific problem is anything but easy, however, due to the need to determine and evaluate high-order derivatives on the right side of the differential equation with respect to t. In this section we consider Runge–Kutta methods, which modify the Taylor methods in such a way that the high-order error bounds are preserved, but the need to determine and evaluate the high-order partial derivatives is eliminated. The strategy behind these techniques involves approximating a Taylor method with a method that is easier to evaluate. This approximation increases the error, but the increase does not exceed the order of the error that was already present in the Taylor method. So, the new error does not dominate the calculations.

The Runge–Kutta techniques make use of the Taylor expansion of f, the function on the right side of the differential equation. Since f is a function of two variables, t and y, we must first consider the generalization of Taylor's Theorem to functions of this type. This generalization appears more complicated than the original form, but this is only because all the possible derivatives of the function are involved.

Taylor's Theorem for Two Variables

If f and all its partial derivatives of order less than or equal to n are continuous on $D = \{(t, y) | a \le t \le b, \; c \le y \le d\}$ and (t, y) and $(t + h, y + k)$ both belong to D, then

$$f(t + h, y + k) \approx f(t, y) + \left[h \frac{\partial f(t, y)}{\partial t} + k \frac{\partial f(t, y)}{\partial y} \right]$$

$$+ \left[\frac{h^2}{2} \frac{\partial^2 f(t, y)}{\partial t^2} + hk \frac{\partial^2 f(t, y)}{\partial t \partial y} + \frac{k^2}{2} \frac{\partial^2 f(t, y)}{\partial y^2} \right] + \cdots$$

$$+ \frac{1}{n!} \sum_{j=0}^{n} \binom{n}{j} h^{n-j} k^j \frac{\partial^n f(t, y)}{\partial t^{n-j} \partial y^j}.$$

The error term in this approximation is similar to that given in Taylor's Theorem, with the added complications that arise because of the incorporation of all the partial derivatives of order $n + 1$.

To illustrate the use of this formula in developing the Runge–Kutta methods, let us consider the most elementary Runge–Kutta method, the method of order two. We saw in the previous section that the Taylor method of order two comes from

$$y(t_{i+1}) = y(t_i) + h y'(t_i) + \frac{h^2}{2} y''(t_i) + \frac{h^3}{3!} y'''(\xi)$$

$$= y(t_i) + h f(t_i, y(t_i)) + \frac{h^2}{2} f'(t_i, y(t_i)) + \frac{h^3}{3!} y'''(\xi),$$

or, since

$$f'(t_i, y(t_i)) = \frac{\partial f}{\partial t}(t_i, y(t_i)) + \frac{\partial f}{\partial y}(t_i, y(t_i)) y'(t_i)$$

and $y'(t_i) = f(t_i, y(t_i))$,

$$y(t_{i+1}) = y(t_i) + h \left\{ f(t_i, y(t_i)) + \frac{h}{2} \frac{\partial f(t_i, y(t_i))}{\partial t} + \frac{h}{2} \frac{\partial f(t_i, y(t_i))}{\partial y} f(t_i, y(t_i)) \right\} + \frac{h^3}{3!} y'''(\xi).$$

Taylor's Theorem of two variables permits us to replace the term in the braces by a multiple of a function evaluation of f of the form $a f(t_i + \alpha, y(t_i) + \beta)$. If we expand this term using the theorem with $n = 1$, we have

$$a f(t_i + \alpha, y(t_i) + \beta) = a \left[f(t_i, y(t_i)) + \alpha \frac{\partial f(t_i, y(t_i))}{\partial t} + \beta \frac{\partial f(t_i, y(t_i))}{\partial y} \right]$$

$$= a f(t_i, y(t_i)) + a\alpha \frac{\partial f(t_i, y(t_i))}{\partial t} + a\beta \frac{\partial f(t_i, y(t_i))}{\partial y}.$$

Equating this expression with the terms enclosed in the braces in the preceding equation implies that a, α, and β should be chosen so that

$$1 = a, \qquad \frac{h}{2} = a\alpha, \qquad \text{and} \qquad \frac{h}{2} f(t_i, y(t_i)) = a\beta;$$

that is,

$$a = 1, \qquad \alpha = \frac{h}{2}, \qquad \text{and} \qquad \beta = \frac{h}{2} f(t_i, y(t_i)).$$

The error introduced by replacing the term in the Taylor method with its approximation has the same order as the error term for the method, so the Runge–Kutta method produced in this way, called the **Midpoint method**, is also a second-order method. As a consequence, the global error of the method is expected to be proportional to the square of the step size.

Midpoint Method

$$w_0 = \alpha$$
$$w_{i+1} = w_i + h[f(t_i + \frac{h}{2}, w_i + \frac{h}{2} f(t_i, w_i))],$$

where $i = 0, 1, \ldots, N - 1$. The local error is $O(h^3)$.

Using $af(t + \alpha, y + \beta)$ to replace the term in the Taylor method is the easiest, but not the only, choice. If we instead use a term of the form

$$a_1 f(t, y) + a_2 f(t + \alpha, y + \beta f(t, y)),$$

the extra parameter in this formula provides an infinite number of second-order Runge–Kutta formulas. When $a_1 = a_2 = \frac{1}{2}$ and $\alpha = \beta = h$, we have the **Modified Euler method**.

Modified Euler Method

$$w_0 = \alpha$$
$$w_{i+1} = w_i + \frac{h}{2}[f(t_i, w_i) + f(t_{i+1}, w_i + h f(t_i, w_i))]$$

where $i = 0, 1, \ldots, N - 1$. The local error is $O(h^3)$.

When $a_1 = \frac{1}{4}$, $a_2 = \frac{3}{4}$, and $\alpha = \beta = \frac{2}{3} h$, we have **Heun's method**.

Heun's Method

$$w_0 = \alpha$$
$$w_{i+1} = w_i + \frac{h}{4}[f(t_i, w_i) + 3f(t_i + \frac{2}{3} h, w_i + \frac{2}{3} h f(t_i, w_i))]$$

where $i = 0, 1, \ldots, N - 1$. The local error is $O(h^3)$.

EXAMPLE 1   Suppose we apply the Runge–Kutta methods of order two to our usual example,

$$y' = y - t^2 + 1, \qquad 0 \le t \le 2, \qquad y(0) = 0.5,$$

with $N = 10$, $h = 0.2$, $t_i = 0.2i$, and $w_0 = 0.5$ in each case. The difference equations are

Midpoint method : $w_{i+1} = 1.22w_i - 0.0088i^2 - 0.008i + 0.218;$

Modified Euler method : $w_{i+1} = 1.22w_i - 0.0088i^2 - 0.008i + 0.216;$

Heun's method : $w_{i+1} = 1.22w_i - 0.0088i^2 - 0.008i + 0.217\overline{3};$

for each $i = 0, 1, \ldots, 9$. Table 5.5 lists the results of these calculations.

■   ■   ■

Table 5.5

t_i	$y(t_i)$	Midpoint Method	Error	Modified Euler Method	Error	Heun's Method	Error
0.0	0.5000000	0.5000000	0	0.5000000	0	0.5000000	0
0.2	0.8292986	0.8280000	0.0012986	0.8260000	0.0032986	0.8273333	0.0019653
0.4	1.2140877	1.2113600	0.0027277	1.2069200	0.0071677	1.2098800	0.0042077
0.6	1.6489406	1.6446592	0.0042814	1.6372424	0.0116982	1.6421869	0.0067537
0.8	2.1272295	2.1212842	0.0059453	2.1102357	0.0169938	2.1176014	0.0096281
1.0	2.6408591	2.6331668	0.0076923	2.6176876	0.0231715	2.6280070	0.0128521
1.2	3.1799415	3.1704634	0.0094781	3.1495789	0.0303627	3.1635019	0.0164396
1.4	3.7324000	3.7211654	0.0112346	3.6936862	0.0387138	3.7120057	0.0203944
1.6	4.2834838	4.2706218	0.0128620	4.2350972	0.0483866	4.2587802	0.0247035
1.8	4.8151763	4.8009586	0.0142177	4.7556185	0.0595577	4.7858452	0.0293310
2.0	5.3054720	5.2903695	0.0151025	5.2330546	0.0724173	5.2712645	0.0342074

Higher-order Taylor formulas can be converted into Runge–Kutta techniques in a similar way, but the algebra becomes tedious. The most common Runge–Kutta method is of order 4 and is obtained by expanding an expression that involves only four function evaluations. Deriving this expression requires solving a system of equations involving 12 unknowns. Once the algebra has been performed the method has a simple representation.

Runge–Kutta Method of Order 4

$$w_0 = \alpha,$$

$$k_1 = hf(t_i, w_i),$$

$$k_2 = hf\left(t_i + \frac{h}{2}, w_i + \frac{1}{2}k_1\right),$$

$$k_3 = hf\left(t_i + \frac{h}{2}, w_i + \frac{1}{2}k_2\right),$$

$$k_4 = hf(t_{i+1}, w_i + k_3),$$

$$w_{i+1} = w_i + \tfrac{1}{6}(k_1 + 2k_2 + 2k_3 + k_4),$$

where $i = 0, 1, \ldots, N - 1$. The local error is $O(h^5)$.

The program RKOR4M52 implements this method.

EXAMPLE 2 The Runge–Kutta method of order 4 applied to the initial-value problem

$$y' = y - t^2 + 1, \qquad 0 \le t \le 2, \qquad y(0) = 0.5,$$

with $h = 0.2, N = 10$, and $t_i = 0.2i$ gives the results and errors listed in Table 5.6.

■ ■ ■

Table 5.6

t_i	Runge–Kutta Order 4 w_i	Exact $y_i = y(t_i)$	Error $|y_i - w_i|$
0.0	0.5000000	0.5000000	0
0.2	0.8292933	0.8292986	0.0000053
0.4	1.2140762	1.2140877	0.0000114
0.6	1.6489220	1.6489406	0.0000186
0.8	2.1272027	2.1272295	0.0000269
1.0	2.6408227	2.6408591	0.0000364
1.2	3.1798942	3.1799415	0.0000474
1.4	3.7323401	3.7324000	0.0000599
1.6	4.2834095	4.2834838	0.0000743
1.8	4.8150857	4.8151763	0.0000906
2.0	5.3053630	5.3054720	0.0001089

The main computational effort in applying the Runge–Kutta methods involves the function evaluations of f. In the second-order methods, the local error is $O(h^3)$ and the cost is two functional evaluations per step. The Runge–Kutta method of order 4 requires four evaluations per step, and the local error is $O(h^5)$. The relationship between the number of evaluations per step and the order of the local error is shown in Table 5.7. Because of the relative decrease in the order for n greater than four, the methods of order less than 5 with smaller step size are used in preference to the higher-order methods using a larger step size.

Table 5.7

Evaluations per step	2	3	4	$5 \le n \le 7$	$8 \le n \le 9$	$10 \le n$
Best possible local error	$O(h^3)$	$O(h^4)$	$O(h^5)$	$O(h^n)$	$O(h^{n-1})$	$O(h^{n-2})$

One way to compare the lower-order Runge–Kutta methods is described as follows: Since the Runge–Kutta method of order 4 requires four evaluations per step, to be superior it should give more accurate answers than Euler's method with one-quarter the mesh size; where by mesh size, we mean the difference between consecutive mesh points. Similarly, if the Runge–Kutta method of order 4 is to be superior to the second-order Runge–Kutta methods, it should give more accuracy with step size h than a second-order method with step size $\frac{1}{2}h$, because the fourth-order method requires twice as many evaluations per step. An illustration of the superiority of the Runge–Kutta fourth-order method by this measure is shown in the following example.

EXAMPLE 3   For the problem

$$y' = y - t^2 + 1, \qquad 0 \le t \le 2, \qquad y(0) = 0.5,$$

Euler's method with $h = 0.025$, the Modified Euler's method with $h = 0.05$, and the Runge–Kutta method of order 4 with $h = 0.1$ are compared at the mesh points 0.1, 0.2, 0.3, 0.4, and 0.5. Each of these techniques requires 20 functional evaluations to approximate $y(0.5)$. (See Table 5.8.) In this example, the fourth-order method is clearly superior.   ■ ■ ■

Table 5.8

t_i	Exact	Euler $h = 0.025$	Modified Euler $h = 0.05$	Runge–Kutta Order 4 $h = 0.1$
0.0	0.5000000	0.5000000	0.5000000	0.5000000
0.1	0.6574145	0.6554982	0.6573085	0.6574144
0.2	0.8292986	0.8253385	0.8290778	0.8292983
0.3	1.0150706	1.0089334	1.0147254	1.0150701
0.4	1.2140877	1.2056345	1.2136079	1.2140869
0.5	1.4256394	1.4147264	1.4250141	1.4256384

EXERCISE SET 5.3

1. Use the Modified Euler method to approximate the solutions to each of the following initial-value problems and compare the results to the actual values.

    (a) $y' = te^{3t} - 2y, \qquad 0 \le t \le 1, \qquad y(0) = 0$ with $h = 0.5$, actual solution $y(t) = \frac{1}{5}te^{3t} - \frac{1}{25}e^{3t} + \frac{1}{25}e^{-2t}$

    (b) $y' = 1 + (t - y)^2, \qquad 2 \le t \le 3, \qquad y(2) = 1$ with $h = 0.5$, actual solution $y(t) = t + (1 - t)^{-1}$

    (c) $y' = 1 + \frac{y}{t}, \qquad 1 \le t \le 2, \qquad y(1) = 2$ with $h = 0.25$, actual solution $y(t) = t \ln t + 2t$

    (d) $y' = \cos 2t + \sin 3t, \qquad 0 \le t \le 1, \qquad y(0) = 1$ with $h = 0.25$, actual solution $y(t) = \frac{1}{2}\sin 2t - \frac{1}{3}\cos 3t + \frac{4}{3}$

2. Repeat Exercise 1 using Heun's method.

3. Repeat Exercise 1 using the Midpoint method.

4. Use the Modified Euler method to approximate the solutions to each of the following initial-value problems and compare the results to the actual values.

   (a) $y' = \dfrac{y}{t} - \left(\dfrac{y}{t}\right)^2$, $\quad 1 \le t \le 2$, $\quad y(1) = 1$ with $h = 0.1$, actual solution $y(t) = t(1 + \ln t)^{-1}$

   (b) $y' = 1 + \dfrac{y}{t} + \left(\dfrac{y}{t}\right)^2$, $\quad 1 \le t \le 3$, $\quad y(1) = 0$ with $h = 0.2$, actual solution $y(t) = t \tan(\ln t)$

   (c) $y' = -(y + 1)(y + 3)$, $\quad 0 \le t \le 2$, $\quad y(0) = -2$ with $h = 0.2$, actual solution $y(t) = -3 + 2(1 + e^{-2t})^{-1}$

   (d) $y' = -5y + 5t^2 + 2t$, $\quad 0 \le t \le 1$, $\quad y(0) = \frac{1}{3}$ with $h = 0.1$, actual solution $y(t) = t^2 + \frac{1}{3}e^{-5t}$

5. Use the results of Exercise 4 and linear interpolation to approximate values of $y(t)$ and compare the results to the actual values.
   (a) $y(1.25)$ and $y(1.93)$
   (b) $y(2.1)$ and $y(2.75)$
   (c) $y(1.3)$ and $y(1.93)$
   (d) $y(0.54)$ and $y(0.94)$

6. Repeat Exercise 4 using Heun's method.

7. Repeat Exercise 5 using the results of Exercise 6.

8. Repeat Exercise 4 using the Midpoint method.

9. Repeat Exercise 5 using the results of Exercise 8.

10. Repeat Exercise 1 using the Runge–Kutta method of order 4.

11. Repeat Exercise 4 using the Runge–Kutta method of order 4.

12. Use the results of Exercise 11 and Cubic Hermite interpolation to approximate values of $y(t)$ and compare the approximations to the actual values.
    (a) $y(1.25)$ and $y(1.93)$
    (b) $y(2.1)$ and $y(2.75)$
    (c) $y(1.3)$ and $y(1.93)$
    (d) $y(0.54)$ and $y(0.94)$

13. Water flows from an inverted conical tank with circular orifice at the rate

    $$\frac{dx}{dt} = -0.6\pi r^2 \sqrt{-2g}\,\frac{\sqrt{x}}{A(x)},$$

    where r is the radius of the orifice, x is the height of the liquid level from the vertex of the cone, and $A(x)$ is the area of the cross section of the tank x units above the orifice. Suppose $r = 0.1$ ft, $g = -32.17$ ft/s^2, and the tank has an initial water level of 8 ft and initial volume of $512\frac{\pi}{3}$ ft^3.
    (a) Compute the water level after 10 min using the Runge–Kutta fourth-order method with $h = 20$ s.
    (b) Determine, to within 1 min, when the tank will be empty.

14. The irreversible chemical reaction in which two molecules of solid potassium dichromate ($K_2Cr_2O_7$), two molecules of water (H_2O), and three atoms of solid sulfur (S) combine to yield three molecules of the gas sulfur dioxide (SO_2), four molecules of solid potassium hydroxide (KOH), and two molecules of solid chromic oxide (Cr_2O_3) can be represented

symbolically by the stoichiometric equation

$$2 \text{ K}_2\text{Cr}_2\text{O}_7 + 2 \text{ H}_2\text{O} + 3 \text{ S} \longrightarrow 4 \text{ KOH} + 2 \text{ Cr}_2\text{O}_3 + 3 \text{ SO}_2.$$

If n_1 molecules of $\text{K}_2\text{Cr}_2\text{O}_7$, n_2 molecules of H_2O, and n_3 molecules of S are originally available, the following differential equation describes the amount $x(t)$ of KOH after time t:

$$\frac{dx}{dt} = k\left(n_1 - \frac{x}{2}\right)^2 \left(n_2 - \frac{x}{2}\right)^2 \left(n_3 - \frac{3x}{4}\right)^3,$$

where k is the velocity constant of the reaction. If $k = 6.22 \times 10^{-19}$, $n_1 = n_2 = 2 \times 10^3$, and $n_3 = 3 \times 10^3$, how many units of potassium hydroxide will have been formed after 0.2 s? Use the Runge–Kutta fourth-order method with $h = 0.01$.

5.4 Predictor–Corrector Methods

The Taylor and Runge–Kutta methods are examples of **one-step methods** for approximating the solution to initial-value problems. These methods are characterized by the fact that the approximation w_{i+1} to $y(t_{i+1})$ involves w_i but does not involve any of the prior approximations $w_0, w_1, \ldots, w_{i-1}$. Only the last approximation w_i to $y(t_i)$ and generally some functional evaluations of f at intermediate points are used, but the latter are discarded as soon as w_{i+1} is obtained. Since $|y(t_j) - w_j|$ decreases in accuracy as j increases, better approximation methods can be derived if, when approximating $y(t_{i+1})$, we include in the method some of the approximations prior to w_i. Methods developed using this philosophy are called **multistep methods**. In brief, one-step methods consider what occurred at only one previous step; multi-step methods consider what happened at more than one previous step.

To derive a multistep method, first note that the solution to the initial-value problem

$$\frac{dy}{dt} = f(t, y), \qquad a \leq t \leq b,$$

subject to an initial condition

$$y(a) = \alpha,$$

if integrated over the interval $[t_i, t_{i+1}]$, has the property that

$$y(t_{i+1}) - y(t_i) = \int_{t_i}^{t_{i+1}} y'(t)\, dt = \int_{t_i}^{t_{i+1}} f(t, y(t))\, dt.$$

Consequently,

$$y(t_{i+1}) = y(t_i) + \int_{t_i}^{t_{i+1}} f(t, y(t))\, dt.$$

Since we cannot integrate $f(t, y(t))$ without knowing $y(t)$, the solution to the problem, we instead integrate an interpolating polynomial, P, determined by some of the previously obtained data points $(t_0, w_0), (t_1, w_1), \ldots, (t_i, w_i)$. When we assume, in addition, that $y(t_i) \approx w_i$, we have

$$y(t_{i+1}) \approx w_i + \int_{t_i}^{t_{i+1}} P(t) \, dt.$$

Although any form of the interpolating polynomial can be used for the derivation, it is most convenient to use the Newton Backward-Difference formula, since this form gives the greatest prominence to the most recently computed approximations.

There are two distinct classes of multistep methods. An **explicit method** is one in which w_{i+1} does not depend on the function evaluation $f(t_{i+1}, w_{i+1})$. A method that does depend in part on $f(t_{i+1}, w_{i+1})$ is called an **implicit method**.

Some of the explicit multistep methods together with their required starting values and local error terms follow.

Adams–Bashforth Two-Step Method

$$w_0 = \alpha, w_1 = \alpha_1$$
$$w_{i+1} = w_i + \frac{h}{2}[3f(t_i, w_i) - f(t_{i-1}, w_{i-1})],$$

where $i = 1, 2, \ldots, N-1$. The local error is $\frac{5}{12} y'''(\mu_i) h^3$, for some μ_i in (t_{i-1}, t_{i+1}).

Adams–Bashforth Three-Step Method

$$w_0 = \alpha, w_1 = \alpha_1, w_2 = \alpha_2,$$
$$w_{i+1} = w_i + \frac{h}{12}[23f(t_i, w_i) - 16f(t_{i-1}, w_{i-1}) + 5f(t_{i-2}, w_{i-2})]$$

where $i = 2, 3, \ldots, N-1$. The local error is $\frac{3}{8} y^{(4)}(\mu_i) h^4$, for some μ_i in (t_{i-2}, t_{i+1}).

Adams–Bashforth Four-Step Method

$$w_0 = \alpha, w_1 = \alpha_1, w_2 = \alpha_2, w_3 = \alpha_3,$$
$$w_{i+1} = w_i + \frac{h}{24}[55f(t_i, w_i) - 59f(t_{i-1}, w_{i-1}) + 37f(t_{i-2}, w_{i-2})$$
$$- 9f(t_{i-3}, w_{i-3})]$$

where $i = 3, 4, \ldots, N-1$. The local error is $\frac{251}{720} y^{(5)}(\mu_i) h^5$, for some μ_i in (t_{i-3}, t_{i+1}).

Adams–Bashforth Five-Step Method

$$w_0 = \alpha, \qquad w_1 = \alpha_1, \qquad w_2 = \alpha_2, \qquad w_3 = \alpha_3, \qquad w_4 = \alpha_4$$

$$w_{i+1} = w_i + \frac{h}{720}[1901f(t_i, w_i) - 2774f(t_{i-1}, w_{i-1})$$

$$+ 2616f(t_{i-2}, w_{i-2}) - 1274f(t_{i-3}, w_{i-3}) + 251f(t_{i-4}, w_{i-4})]$$

where $i = 4, 5, \ldots, N-1$. The local error is $\frac{95}{288} y^{(6)}(\mu_i)h^6$, for some μ_i in (t_{i-4}, t_{i+1}).

Implicit methods use $(t_{i+1}, f(t_{i+1}, y(t_{i+1})))$ as an additional interpolation node in the approximation of the integral

$$\int_{t_i}^{t_{i+1}} f(t, y(t)) \, dt.$$

Some of the more common implicit methods are listed below. Notice that the local error of an $(m - 1)$-step implicit method is $O(h^{m+1})$, the same as that of an m-step explicit method. They both use m function evaluations, however, since the implicit methods use $f(t_{i+1}, w_{i+1})$, but the explicit methods do not.

Adams–Moulton Two-Step Method

$$w_0 = \alpha, \qquad w_1 = \alpha_1$$

$$w_{i+1} = w_i + \frac{h}{12}[5f(t_{i+1}, w_{i+1}) + 8f(t_i, w_i) - f(t_{i-1}, w_{i-1})]$$

where $i = 1, 2, \ldots, N - 1$. The local error is $-\frac{1}{24} y^{(4)}(\mu_i)h^4$, for some μ_i in (t_{i-1}, t_{i+1}).

Adams–Moulton Three-Step Method

$$w_0 = \alpha, \qquad w_1 = \alpha_1, \qquad w_2 = \alpha_2,$$

$$w_{i+1} = w_i + \frac{h}{24}[9f(t_{i+1}, w_{i+1}) + 19f(t_i, w_i) - 5f(t_{i-1}, w_{i-1}) + f(t_{i-2}, w_{i-2})],$$

where $i = 2, 3, \ldots, N - 1$. The local error is $-\frac{19}{720} y^{(5)}(\mu_i)h^5$, for some μ_i in (t_{i-2}, t_{i+1}).

Adams–Moulton Four-Step Method

$$w_0 = \alpha, \qquad w_1 = \alpha_1, \qquad w_2 = \alpha_2 \quad , w_3 = \alpha_3,$$

$$w_{i+1} = w_i + \frac{h}{720}[251f(t_{i+1}, w_{i+1}) + 646f(t_i, w_i) - 246f(t_{i-1}, w_{i-1})$$

$$+ 106f(t_{i-2}, w_{i-2}) - 19f(t_{i-3}, w_{i-3})]$$

where $i = 3, 4, \ldots, N - 1$. The local error is $-\frac{3}{160}y^{(6)}(\mu_i)h^6$, for some μ_i in (t_{i-3}, t_{i+1}).

It is interesting to compare an m-step Adams–Bashforth explicit method to an $(m-1)$-step Adams–Moulton implicit method. Both require m evaluations of f per step, and both have the terms $y^{(m+1)}(\mu_i)h^{m+1}$ in their local errors. In general, the coefficients of the terms involving f in the approximation and those in the local error are smaller for the implicit methods than for the explicit methods. This leads to greater stability and smaller round-off errors for the implicit methods.

EXAMPLE 1 Consider the initial-value problem

$$y' = y - t^2 + 1, \qquad 0 \le t \le 2, \qquad y(0) = 0.5,$$

and the approximations given by the Adams–Bashforth Four–Step method and the Adams–Moulton Three-Step method, both using $h = 0.2$. The explicit Adams–Bashforth method has the difference equation

$$w_{i+1} = w_i + \frac{h}{24}[55f(t_i, w_i) - 59f(t_{i-1}, w_{i-1}) + 37f(t_{i-2}, w_{i-2}) - 9f(t_{i-3}, w_{i-3})],$$

for $i = 3, 4, \ldots, 9$, which, when simplified using $f(t, y) = y - t^2 + 1, h = 0.2$, and $t_i = 0.2i$, becomes

$$w_{i+1} = \frac{1}{24}[35w_i - 11.8w_{i-1} + 7.4w_{i-2} - 1.8w_{i-3} - 0.192i^2 - 0.192i + 4.736].$$

The implicit Adams–Moulton method has the difference equation

$$w_{i+1} = w_i + \frac{h}{24}[9f(t_{i+1}, w_{i+1}) + 19f(t_i, w_i) - 5f(t_{i-1}, w_{i-1})] + f(t_{i-2}, w_{i-2})],$$

for $i = 2, 3, \ldots, 9$, which reduces to

$$w_{i+1} = \frac{1}{24}[1.8w_{i+1} + 27.8w_i - w_{i-1} + 0.2w_{i-2} - 0.192i^2 - 0.192i + 4.736].$$

To use this method explicitly, we solve for w_{i+1}, which gives

$$w_{i+1} = \frac{1}{22.2}[27.8w_i - w_{i-1} + 0.2w_{i-2} - 0.192i^2 - 0.192i + 4.736]$$

for $i = 2, 3, \ldots, 9$. The results in Table 5.9 were obtained using the exact values from $y(t) = (t + 1)^2 - 0.5e^t$ for $\alpha, \alpha_1,$ and α_2 in the explicit Adams–Bashforth case and for α and α_1 in the implicit Adams–Moulton case.   ■ ■ ■

Table 5.9

t_i	Adams– Bashforth w_i	Error	Adams– Moulton w_i	Error
0.0	0.5000000	0	0.5000000	0
0.2	0.8292986	0.0000000	0.8292986	0.0000000
0.4	1.2140877	0.0000000	1.2140877	0.0000000
0.6	1.6489406	0.0000000	1.6489341	0.0000065
0.8	2.1273124	0.0000828	2.1272136	0.0000160
1.0	2.6410810	0.0002219	2.6408298	0.0000293
1.2	3.1803480	0.0004065	3.1798937	0.0000478
1.4	3.7330601	0.0006601	3.7323270	0.0000731
1.6	4.2844931	0.0010093	4.2833767	0.0001071
1.8	4.8166575	0.0014812	4.8150236	0.0001527
2.0	5.3075838	0.0021119	5.3052587	0.0002132

In Example 1, the implicit Adams–Moulton method gave considerably better results than the explicit Adams–Bashforth method of the same order. Although this is generally the case, the implicit methods have the inherent weakness of first having to convert the method algebraically to an explicit representation for w_{i+1}. This procedure can become difficult, if not impossible, as can be seen by considering the elementary initial-value problem

$$y' = e^y, \qquad 0 \le t \le 0.25, \qquad y(0) = 1.$$

Since $f(t, y) = e^y$, the Adams–Moulton three-step method has

$$w_{i+1} = w_i + \frac{h}{24}[9e^{w_{i+1}} + 19e^{w_i} - 5e^{w_{i-1}} + e^{w_{i-2}}]$$

as its difference equation, and this equation cannot be solved explicitly for w_{i+1}. We could use Newton's method or the Secant method to approximate w_{i+1}, but this complicates the procedure considerably.

In practice, implicit multistep methods are not used alone. Rather, they are used to improve approximations obtained by explicit methods. The combination of an explicit and implicit technique is called a **predictor–corrector method**. The explicit method predicts an approximation, and the implicit method corrects this prediction.

Consider the following fourth-order method for solving an initial-value problem. The first step is to calculate the starting values $w_0, w_1, w_2,$ and w_3 for the Adams–Bashforth four-step method. To do this, we use a fourth-order one-step method, the Runge–Kutta method of order 4. The next step is to calculate an approximation, $w_4^{(0)}$, to $y(t_4)$ using the Adams–Bashforth four-step method as predictor:

$$w_4^{(0)} = w_3 + \frac{h}{24}[55f(t_3, w_3) - 59f(t_2, w_2) + 37f(t_1, w_1) - 9f(t_0, w_0)].$$

This approximation is improved by use of the Adams–Moulton three-step method as corrector:

$$w_4^{(1)} = w_3 + \frac{h}{24}[9f(t_4, w_4^{(0)}) + 19f(t_3, w_3) - 5f(t_2, w_2) + f(t_1, w_1)].$$

The value $w_4 \equiv w_4^{(1)}$ is then used as the approximation to $y(t_4)$ and the technique of using the Adams–Bashforth method as a predictor and the Adams–Moulton method as a corrector is repeated to find $w_5^{(0)}$ and $w_5^{(1)}$, the initial and final approximations to $y(t_5)$, and so on.

The program PRCORM53 is based on the Adams–Bashforth four-step method as predictor and one iteration of the Adams–Moulton three-step method as corrector, with the starting values obtained from the Runge–Kutta method of order 4.

EXAMPLE 2 Table 5.10 lists the results obtained by using the program PRCORM53 for the initial-value problem

$$y' = y - t^2 + 1, \qquad 0 \le t \le 2, \qquad y(0) = 0.5,$$

with $N = 10$. The results here are more accurate than those in Example 1, which used only the corrector (that is, the Adams–Moulton method). This would not generally be the case.   ■ ■ ■

Table 5.10

t_i	w_i	$y_i = y(t_i)$	Error $|y_i - w_i|$
0.0	0.5000000	0.5000000	0
0.2	0.8292933	0.8292986	0.0000053
0.4	1.2140762	1.2140877	0.0000114
0.6	1.6489220	1.6489406	0.0000186
0.8	2.1272056	2.1272295	0.0000239
1.0	2.6408286	2.6408591	0.0000305
1.2	3.1799026	3.1799415	0.0000389
1.4	3.7323505	3.7324000	0.0000495
1.6	4.2834208	4.2834838	0.0000630
1.8	4.8150964	4.8151763	0.0000799
2.0	5.3053707	5.3054720	0.0001013

The Adams methods are the most elementary multistep methods, since they use interpolation on the immediately previous approximation. However, other multistep methods can be derived using integration of interpolating polynomials over intervals of the form $[t_j, t_{i+1}]$ for $j \le i - 1$, where some of the data points are omitted. Milne's method is an explicit technique that results when a Newton Backward Polynomial is integrated over $[t_{i-3}, t_{i+1}]$.

Milne's Method

$$w_{i+1} = w_{i-3} + \frac{4h}{3}[2f(t_i, w_i) - f(t_{i-1}, w_{i-1}) + 2f(t_{i-2}, w_{i-2})],$$

where $i = 3, 4, \ldots, N-1$. The local error is $\frac{14}{45}h^5 y^{(5)}(\mu_i)$, for some μ_i in (t_{i-3}, t_{i+1}).

This method is used as a predictor for an implicit method called Simpson's method. Its name comes from the fact that it can be derived using Simpson's rule for approximating integrals.

Simpson's Method

$$w_{i+1} = w_{i-1} + \frac{h}{3}[f(t_{i+1}, w_{i+1}) + 4f(t_i, w_i) + f(t_{i-1}, w_{i-1})],$$

where $i = 1, 2, \ldots, N - 1$. The local error is $-\frac{1}{90}h^5 y^{(5)}(\mu_i)$, for some μ_i in (t_{i-1}, t_{i+1}).

Although the local error involved with a predictor-corrector method of the Milne–Simpson type is generally smaller than that of the Adams–Bashforth–Moulton method, the technique has limited use because of problems of stability, which do not occur with the Adams procedure.

EXERCISE SET 5.4

1. Use each of the Adams–Bashforth methods to approximate the solutions to the following initial-value problems. In each case use exact starting values and compare the results to the actual values.

   (a) $y' = te^{3t} - 2y$, $0 \le t \le 1$, $y(0) = 0$ with $h = 0.2$, actual solution $y(t) = \frac{1}{5}te^{3t} - \frac{1}{25}e^{3t} + \frac{1}{25}e^{-2t}$.

   (b) $y' = 1 + (t - y)^2$, $2 \le t \le 3$, $y(2) = 1$ with $h = 0.2$, actual solution $y(t) = t + (1 - t)^{-1}$

   (c) $y' = 1 + \frac{y}{t}$, $1 \le t \le 2$, $y(1) = 2$ with $h = 0.2$, actual solution $y(t) = t \ln t + 2t$

   (d) $y' = \cos 2t + \sin 3t$, $0 \le t \le 1$, $y(0) = 1$ with $h = 0.2$, actual solution $y(t) = \frac{1}{2}\sin 2t - \frac{1}{3}\cos 3t + \frac{4}{3}$.

2. Use each of the Adams–Moulton methods to approximate the solutions to the Exercises 1(a), 1(c), and 1(d). In each case use exact starting values and explicitly solve for w_{i+1}. Compare the results to the actual values.

3. Use each of the Adams–Bashforth methods to approximate the solutions to the following initial-value problems. In each case use starting values obtained from the Runge–Kutta method of

order 4. Compare the results to the actual values.

(a) $y' = \dfrac{y}{t} - \left(\dfrac{y}{t}\right)^2$, $\quad 1 \le t \le 2$, $\quad y(1) = 1$ with $h = 0.1$, actual solution $y(t) = t(1 + \ln t)^{-1}$

(b) $y' = 1 + \dfrac{y}{t} + \left(\dfrac{y}{t}\right)^2$, $\quad 1 \le t \le 3$, $\quad y(1) = 0$ with $h = 0.2$, actual solution $y(t) = t\tan(\ln t)$

(c) $y' = -(y + 1)(y + 3)$, $\quad 0 \le t \le 2$, $\quad y(0) = -2$ with $h = 0.1$, actual solution $y(t) = -3 + 2(1 + e^{-2t})^{-1}$

(d) $y' = -5y + 5t^2 + 2t$, $\quad 0 \le t \le 1$, $\quad y(0) = \frac{1}{3}$ with $h = 0.1$, actual solution $y(t) = t^2 + \frac{1}{3}e^{-5t}$

4. Use the predictor-corrector method based on the Adams–Bashforth four-step method and the Adams–Moulton three-step method to approximate the solutions to the initial-value problems in Exercise 1.

5. Use the predictor-corrector method based on the Adams–Bashforth four-step method and the Adams–Moulton three-step method to approximate the solutions to the initial-value problem in Exercise 3.

6. The initial-value problem
$$y' = e^y, \quad 0 \le t \le 0.20, \quad y(0) = 1$$
has solution
$$y(t) = 1 - \ln(1 - et).$$
Applying the three-step Adams–Moulton method to this problem is equivalent to finding the fixed point w_{i+1} of
$$g(w) = w_i + \frac{h}{24}[9e^w + 19e^{w_i} - 5e^{w_{i-1}} + e^{w_{i-2}}].$$
(a) With $h = 0.01$, obtain w_{i+1} by functional iteration for $i = 2, \ldots, 19$ using exact starting values w_0, w_1, and w_2. At each step use w_i as an initial approximation to w_{i+1}.
(b) Will Newton's method speed the convergence over functional iteration?

7. Use the Milne–Simpson predictor-corrector method to approximate the solutions to the initial-value problems in Exercise 3.

8. Use the Milne-Simpson predictor-corrector method to approximate the solution to
$$y' = -5y, \quad 0 \le t \le 2, \quad y(0) = e,$$
with $h = 0.1$. Repeat the procedure with $h = 0.05$. Are the answers consistent with the local truncation error?

5.5 Extrapolation Methods

Extrapolation was used in Section 4.6 for the approximation of definite integrals, where we found that by correctly averaging relatively inaccurate trapezoidal approximations we could produce new approximations that are exceedingly accurate. In this section we will apply extrapolation to increase the accuracy of approximations to the solution of initial-value problems. As we have previously seen, the original approximations must have an error expansion of a specific form for the procedure to be successful.

To apply extrapolation to solve initial-value problems, we use a technique based on the Midpoint method:

$$w_{i+1} = w_{i-1} + 2hf(t_i, w_i), \qquad \text{for} \qquad i \geq 1.$$

This requires two starting values, since both w_0 and w_1 are needed before the first midpoint approximation, w_2, can be determined. As usual, we use the initial condition for $w_0 = y(a) = \alpha$. To determine the second starting value, w_1, we apply Euler's method. Subsequent approximations are obtained from the Midpoint method. After a series of approximations of this type are generated, ending at a value t, an endpoint correction is performed that involves the final two midpoint approximations. This produces an approximation $w(t, h)$ to $y(t)$ that has the form

$$y(t) = w(t, h) + \sum_{k=1}^{\infty} \delta_k h^{2k},$$

where the δ_k are constants related to the derivatives of the solution $y(t)$. The important point is that the δ_k do not depend on the step size h.

To illustrate the extrapolation technique for solving

$$y'(t) = f(t, y), \qquad a \leq t \leq b, \qquad y(a) = \alpha,$$

let us assume that we have a fixed step size h and that we wish to approximate $y(a + h)$.

For the first extrapolation step we let $h_0 = h/2$ and use Euler's method with $w_0 = \alpha$ to approximate $y(a + h_0) = y(a + h/2)$ as

$$w_1 = w_0 + h_0 f(a, w_0).$$

We then apply the Midpoint method with $t_{i-1} = a$ and $t_i = a + h_0 = a + h/2$ to produce a first approximation

$$w_2 = w_0 + 2h_0 f(a + h, w_1)$$

to $y(a + h) = y(a + 2h_0)$. The endpoint correction is applied to obtain the final approximation to $y(a + h)$ for the step size h_0. This results in the $O(h_0^2)$ approximation

$$y_{1,1} = \tfrac{1}{2}[w_2 + w_1 + h_0 f(a + h, w_2)].$$

For the next approximation to $y(a + h)$ we let $h_1 = h/3$ and use Euler's method with $w_0 = \alpha$ to obtain an approximation

$$w_1 = w_0 + h_1 f(a, w_0)$$

to $y(a + h_1) = y(a + h/3)$. We now apply the Midpoint method twice. First, we apply $t_{i-1} = a$ and $t_i = a + h_1 = a + h/3$ to produce an approximation w_2 to $y(a + 2h_1) = y(a + 2h/3)$. Then, we apply $t_{i-1} = a + h_1 = a + h/3$ and $t_i = a + 2h_1 = a + 2h/3$ to produce a first approximation w_3 for $y(a + h) = y(a + 3h_1)$. The endpoint correction is applied to obtain the final $O(h_1^2)$ approximation to $y(a + h)$ for the step size h_1:

$$y_{2,1} = \tfrac{1}{2}[w_3 + w_2 + h_1 f(a + h, w_3)].$$

Because of the form of the error, the two approximations to $y(a + h)$ have the property that

$$y(a + h) = y_{1,1} + \delta_1 \left(\frac{h}{2}\right)^2 + \delta_2 \left(\frac{h}{2}\right)^4 + \ldots = y_{1,1} + \delta_1 \frac{h^2}{4} + \delta_2 \frac{h^4}{16} + \ldots$$

and

$$y(a + h) = y_{2,1} + \delta_1\left(\frac{h}{3}\right)^2 + \delta_2\left(\frac{h}{3}\right)^4 + \cdots = y_{2,1} + \delta_1\frac{h^2}{9} + \delta_2\frac{h^4}{81} + \cdots.$$

We can eliminate the $O(h^2)$ portion of this truncation error by averaging these two formulas appropriately. Specifically, if we subtract 4 times the first from 9 times the second and divide the result by 5, we have

$$y(a + h) = y_{2,1} + \frac{4}{5}(y_{2,1} - y_{1,1}) - \delta_2\frac{h^4}{36} + \cdots.$$

So the approximation

$$y_{2,2} = y_{2,1} + \tfrac{4}{5}(y_{2,1} - y_{1,1})$$

has error of order $O(h^4)$.

Continuing in this manner, we next let $h_2 = h/4$ and apply one application of Euler's method followed by three of the Midpoint method and the endpoint correction to determine another approximation, $y_{3,1}$, to $y(a + h)$. This approximation can be averaged with $y_{2,1}$ to produce a second $O(h^4)$ approximation that we denote $y_{3,2}$. Then $y_{3,2}$ and $y_{2,2}$ are averaged to eliminate the $O(h^4)$ error terms and produce an approximation with error of order $O(h^6)$. Higher-order formulas are generated by continuing the process.

The only significant difference between the extrapolation performed here and that used for Romberg Integration in Section 4.5 results from the way the subdivisions are chosen. In Romberg Integration there is a convenient formula for representing the Composite Trapezoidal Rule approximations that uses consecutive divisions of the step size by the integers $1, 2, 4, 8, 16, 32, 64, \ldots$. This permits the averaging process to proceed in an easily described manner.

We do not have a means for easily producing refined approximations for initial-value problems, so the divisions for the extrapolation technique used here are chosen to minimize the number of required function evaluations. The averaging procedure arising from this choice of subdivision is not as easily seen, but, with this exception, the derivation process is the same as that used for Romberg Integration.

The program EXTRAP54 uses the extrapolation technique with the sequence of integers $q_0 = 2$, $q_1 = 3$, $q_2 = 4$, $q_3 = 6$, $q_4 = 8$, $q_5 = 12$, $q_6 = 16$, $q_7 = 24$. A basic step size, h, is selected, and the method progresses by using $h_j = h/q_j$, for each $j = 0, \ldots, 7$, to approximate $y(t + h)$. The error is controlled by requiring that the approximations $y_{1,1}, y_{2,2}, \ldots$ be computed until $|y_{i,i} - y_{i-1,i-1}|$ is less than a given tolerance. If the tolerance is not achieved by $i = 8$, then h is reduced, and the process is reapplied. Minimum and maximum values of h, $hmin$ and $hmax$, respectively, are specified in the program to ensure control over the method.

If $y_{i,i}$ is found to be acceptable, then w_1 is set to $y_{i,i}$ and computations begin again to determine w_2, which will approximate $y(t_2) = y(a + 2h)$. The process is repeated until the approximation w_N to $y(b)$ is determined.

EXAMPLE 1 Consider the initial-value problem

$$y' = y - t^2 + 1, \qquad 0 \le t \le 2, \qquad y(0) = 0.5,$$

which has solution $y(t) = (t + 1)^2 - 0.5e^t$. The program EXTRAP54 applied to this problem with $h = 0.25, TOL = 10^{-5}, \ hmax = 0.25$, and $hmin = 0.01$ produces the approximations to w_1 shown in Table 5.11.   ■ ■ ■

Table 5.11

$y_{1,1} = 0.91870117$				
$y_{2,1} = 0.91969682$	$y_{2,2} = 0.92049334$			
$y_{3,1} = 0.92003798$	$y_{3,2} = 0.92047662$	$y_{3,3} = 0.92047105$		
$y_{4,1} = 0.92028737$	$y_{4,2} = 0.92048688$	$y_{4,3} = 0.92049029$	$y_{4,4} = 0.92049270$	
$y_{5,1} = 0.92037479$	$y_{5,2} = 0.92048719$	$y_{5,3} = 0.92048729$	$y_{5,4} = 0.92048680$	$y_{5,5} = 0.92048641$

The computations stopped with $w_1 = y_{5,5}$ because $|y_{5,5} - y_{4,4}| \le 10^{-5}$. The complete set of approximations is given in Table 5.12.

Table 5.12

t_i	w_i	h_i	$y_i = y(t_i)$	Error $\|y_i - w_i\|$	k
0.25	0.9204864	0.25	0.9204873	0.0000009	5
0.50	1.4256374	0.25	1.4256394	0.0000019	5
0.75	2.0039968	0.25	2.0040000	0.0000032	5
1.00	2.6408544	0.25	2.6408591	0.0000046	5
1.25	3.3173249	0.25	3.3173285	0.0000036	4
1.50	4.0091480	0.25	4.0091555	0.0000075	3
1.75	4.6851917	0.25	4.6851987	0.0000070	3
2.00	5.3054724	0.25	5.3054720	0.0000005	3

EXERCISE SET 5.5

1. Use the extrapolation program EXTRAP54 with tolerance $TOL = 10^{-4}$, $hmax = 0.25$, and $hmin = 0.05$ to approximate the solutions to the following initial-value problems. Compare the results to the actual values.

   (a) $y' = te^{3t} - 2y$, $0 \le t \le 1$, $y(0) = 0$, actual solution $y(t) = \frac{1}{5}te^{3t} - \frac{1}{25}e^{3t} + \frac{1}{25}e^{-2t}$

   (b) $y' = 1 + (t - y)^2$, $2 \le t \le 3$, $y(2) = 1$, actual solution $y(t) = t + 1(1 - t)^{-1}$

   (c) $y' = 1 + \frac{y}{t}$, $1 \le t \le 2$, $y(1) = 2$, actual solution $y(t) = t \ln t + 2t$

   (d) $y' = \cos 2t + \sin 3t$, $0 \le t \le 1$, $y(0) = 1$, actual solution $y(t) = \frac{1}{2} \sin 2t - \frac{1}{3} \cos 3t + \frac{4}{3}$

2. Use the extrapolation program EXTRAP54 with tolerance $TOL = 10^{-6}$, $hmax = 0.5$, and $hmin = 0.05$ to approximate the solutions to the following initial-value problems. Compare the results to the actual values.

   (a) $y' = \frac{y}{t} - \frac{y^2}{t^2}$, $1 \le t \le 4$, $y(1) = 1$, actual solution $y(t) = t/(1 + \ln t)$

   (b) $y' = 1 + \frac{y}{t} + \left(\frac{y}{t}\right)^2$, $1 \le t \le 3$, $y(1) = 0$, actual solution $y(t) = t \tan(\ln t)$

(c) $y' = -(y+1)(y+3)$, $0 \le t \le 3$, $y(0) = -2$, actual solution $y(t) = -3 + 2(1+e^{-2t})^{-1}$

(d) $y' = (t+2t^3)y^3 - ty$, $0 \le t \le 2$, $y(0) = \frac{1}{3}$, actual solution $y(t) = (3+2t^2 + 6e^{t^2})^{-1/2}$

3. Use the extrapolation program EXTRAP54 with tolerance $TOL = 10^{-6}$, $hmax = 0.2$, and $hmin = 0.05$ to approximate the solutions to the following initial-value problems. Compare the results to the actual values.

(a) $y' = -t^{-1}(\sin t + y) + \cos t^2$, $\sqrt{\pi} \le t \le \pi$, $y(\sqrt{\pi}) = 0$, actual solution $y(t) = t^{-1}(\cos t - \cos \sqrt{\pi} + 0.5 \sin t^2)$

(b) $y' = e^t \csc t - \left(\frac{1}{t} + \cot t\right) y$, $\frac{\pi}{2} \le t \le \frac{3\pi}{4}$, $y\left(\frac{\pi}{2}\right) = \frac{2}{\pi}$, actual solution $y(t) = (t \sin t)^{-1}((t-1)e^t - \left(\frac{\pi}{2} - 1\right)e^{\pi/2} + 1)$

(c) $y' = \dfrac{t + \ln t}{y^2 \ln t} - \dfrac{y}{t \ln t}$, $e \le t \le 2e$, $y(e) = 2$, actual solution $y(t) = \left[3t + (\ln t)^{-1}(1.5t^2 - 9t) + (\ln t)^{-2}(18t - 1.5t^2) + (\ln t)^{-3}(0.75t^2 - 18t + C)\right]^{1/3}$,

where $C = 8 + e(6 - 0.75e)$

(d) $y' = 1 + t(1 - y^2) + 2t^2 y - t^3$, $0 \le t \le 3$, $y(0) = 0$, actual solution $y(t) = t - 1 + 2(1 + e^{-t^2})^{-1}$

4. Use the extrapolation program EXTRAP54 with tolerance $TOL = 10^{-5}$, $hmax = 0.2$, and $hmin = 0.05$ to approximate the solutions to the following initial-value problems. Compare the results to the actual values.

(a) $y' = -2\dfrac{y}{t} + t^3 y^2 \ln t$, $1 \le t \le 1.9$, $y(1) = 2$, actual solution $y(t) = 4(t^4 - 2t^4 \ln t + t^2)^{-1}$

(b) $y'(t) = \sqrt{2 - y^2}e^t$, $0 \le t \le 0.5$, $y(0) = 0$, actual solution $y(t) = \sqrt{2} \sin(e^t - 1)$

(c) $y' = -\sin t(1 + y \sec t)$, $0 \le t \le 0.5$, $y(0) = 2$, actual solution $y(t) = (2 + \ln(\cos t)) \cos t$

(d) $y' = -5y + 5t^2 + 2t$, $0 \le t \le 1$, $y(0) = \frac{1}{3}$, actual solution $y(t) = t^2 + \frac{1}{3}e^{-5t}$

5. Let $P(t)$ be the number of individuals in a population at time t, measured in years. If the average birth rate b is constant and the average death rate d is proportional to the size of the population (due to overcrowding), then the growth rate of the population is given by the *logistic equation*

$$\frac{dP(t)}{dt} = bP(t) - k[P(t)]^2$$

where $d = kP(t)$. Suppose $P(0) = 50,976$, $b = 2.9 \times 10^{-2}$, and $k = 1.4 \times 10^{-7}$. Use the extrapolation program EXTRAP54 to find the population after 5 years.

5.6 Adaptive Techniques

The appropriate use of varying step size was seen in Section 4.6 to produce integral approximating methods that are efficient in the amount of computation required. This might not be sufficient to favor these methods due to the increased complication of applying them, but they have another important feature. They incorporate in the step-size procedure an estimate of the truncation error that does not require the approximation of the higher derivatives of the function. These methods are called *adaptive* because they adapt the number and position of the nodes used in the approximation to keep the truncation error within a specified bound.

There is a close connection between the problem of approximating the value of a definite integral and that of approximating the solution to an initial-value problem. It is not surprising, then, that there are adaptive methods for approximating the solutions to initial-value problems and that these methods are not only efficient but also incorporate the control of error.

An ideal difference-equation method

$$w_{i+1} = w_i + h_i \phi(t_i, w_i, h_i), \qquad \text{for each } i = 0, 1, \ldots, N - 1,$$

for approximating the solution, $y(t)$, to the initial-value problem

$$y' = f(t, y), \qquad a \le t \le b, \qquad y(a) = \alpha$$

would have the property that given a tolerance $\epsilon > 0$, the minimal number of mesh points would be used to ensure that the global error, $|y(t_i) - w_i|$, would not exceed ϵ for any $i = 0, 1, \ldots, N$. Having a minimal number of mesh points and also controlling the global error of a difference method is, not surprisingly, inconsistent with the points being equally spaced in the interval. In this section we examine techniques used to control the error of a difference-equation method in an efficient manner by the appropriate choice of mesh points.

Although we cannot generally determine the global error of a method, we saw in Section 5.2 that there is often a close connection between the local error and the global error. If a method has local error of order $n + 1$, then the global error of the method is of order n. By using methods of differing order we can predict the local error and, using this prediction, choose a step size that will keep the global error in check.

To illustrate the technique, suppose that we have two approximation techniques. The first is an nth-order method obtained from an nth-order Taylor method of the form

$$y(t_{i+1}) = y(t_i) + h\phi(t_i, y(t_i), h) + O(h^{n+1}),$$

producing approximations

$$w_0 = \alpha$$
$$w_{i+1} = w_i + h\phi(t_i, w_i, h) \qquad \text{for} \quad i > 0,$$

which satisfy, for some K and all relevant h and i,

$$|y(t_i) - w_i| < Kh^n.$$

In general, the method is generated by applying a Runge–Kutta modification to the Taylor method, but the specific derivation is unimportant.

The second method is similar, but of higher order. For example, let us suppose it comes from an $(n + 1)$st-order Taylor method of the form

$$y(t_{i+1}) = y(t_i) + h\tilde{\phi}(t_i, y(t_i), h) + O(h^{n+2}),$$

producing approximations

$$\tilde{w}_0 = \alpha$$
$$\tilde{w}_{i+1} = \tilde{w}_i + h\tilde{\phi}(t_i, \tilde{w}_i, h) \qquad \text{for } i > 0,$$

which satisfy, for some $\tilde{K}$ and all relevant h and i,

$$|y(t_i) - \tilde{w}_i| < \tilde{K}h^{n+1}.$$

We assume now that at the point t_i we have

$$w_i = \tilde{w}_i = z(t_i),$$

where $z(t)$ is the solution to the differential equation that does not satisfy the original initial condition, but instead satisfies the condition $z(t_i) = w_i$. The typical difference between $y(t)$ and $z(t)$ is shown in Figure 5.3.

Figure 5.3

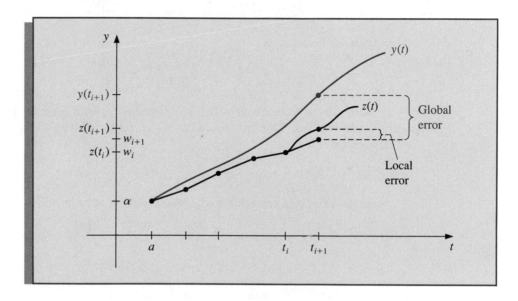

Applying the two methods to the differential equation with the fixed step size h produces two approximations, w_{i+1} and $\tilde{w}_{i+1}$, whose differences from $y(t_i + h)$ represent global errors but whose differences from $z(t_i + h)$ represent local errors.

Now consider

$$z(t_i + h) - w_{i+1} = (\tilde{w}_{i+1} - w_{i+1}) + (z(t_i + h) - \tilde{w}_{i+1}).$$

The term on the left side of this equation is $O(h^{n+1})$, the local error of the method, but the second term on the right side is $O(h^{n+2})$. This implies that the dominant portion on the right side comes from the first term; that is,

$$z(t_i + h) - w_{i+1} \approx \tilde{w}_{i+1} - w_{i+1}.$$

If we now assume that

$$Kh^{n+1} = |z(t_i + h) - w_{i+1}| \approx |\tilde{w}_{i+1} - w_{i+1}|,$$

we can approximate K by

(5.1)
$$K \approx \frac{|\tilde{w}_{i+1} - w_{i+1}|}{h^{n+1}}.$$

Let us now return to the global error associated with the problem we really want to solve,

$$y' = f(t, y), \quad a \le t \le b, \quad y(a) = \alpha,$$

and consider the adjustment to the step size needed if this global error is expected to be bounded by the tolerance ϵ. Using a multiple q of the original step size implies that we need to ensure that

$$|y(t_i + qh) - w_{i+1}(\text{using the new step size } qh)| < K(qh)^n < \epsilon.$$

This implies, from (5.1), that

$$Kq^n h^n \approx \frac{|\tilde{w}_{i+1} - w_{i+1}|}{h^{n+1}} q^n h^n = \frac{q^n |\tilde{w}_{i+1} - w_{i+1}|}{h} < \epsilon$$

and that, as an estimate,

$$q < \left[\frac{\epsilon h}{|\tilde{w}_{i+1} - w_{i+1}|} \right]^{1/n}.$$

One popular technique that uses this inequality for error control is the **Runge–Kutta–Fehlberg method**. This technique consists of using a Runge–Kutta method with local error of order five,

$$\tilde{w}_{i+1} = w_i + \frac{16}{135}k_1 + \frac{6656}{12825}k_3 + \frac{28561}{56430}k_4 - \frac{9}{50}k_5 + \frac{2}{55}k_6,$$

to estimate the local error in a Runge–Kutta method of order four,

$$w_{i+1} = w_i + \frac{25}{216}k_1 + \frac{1408}{2565}k_3 + \frac{2197}{4104}k_4 - \frac{1}{5}k_5,$$

where

$$k_1 = hf(t_i, w_i),$$

$$k_2 = hf\left(t_i + \frac{h}{4}, w_i + \frac{1}{4}k_1\right),$$

$$k_3 = hf\left(t_i + \frac{3h}{8}, w_i + \frac{3}{32}k_1 + \frac{9}{32}k_2\right),$$

$$k_4 = hf\left(t_i + \frac{12h}{13}, w_i + \frac{1932}{2197}k_1 - \frac{7200}{2197}k_2 + \frac{7296}{2197}k_3\right),$$

$$k_5 = hf\left(t_i + h, w_i + \frac{439}{216}k_1 - 8k_2 + \frac{3680}{513}k_3 - \frac{845}{4104}k_4\right),$$

$$k_6 = hf\left(t_i + \frac{h}{2}, w_i - \frac{8}{27}k_1 + 2k_2 - \frac{3544}{2565}k_3 + \frac{1859}{4104}k_4 - \frac{11}{40}k_5\right).$$

An advantage of this method is that only six evaluations of f are required per step, whereas arbitrary Runge–Kutta methods of order four and five used together would require (see Table 5.7 in Section 5.3) at least four evaluations of f for the fourth-order method and an additional six for the fifth-order method.

In the theory of error control, an initial value of h at the ith step was used to find the first values of w_{i+1} and $\tilde{w}_{i+1}$, which led to the determination of q for that step. Then the calculations were repeated with the step size h replaced by qh. This procedure requires twice the number of functional evaluations per step as without error control. In practice, the value of q to be used is chosen somewhat differently in order to make the increased functional-evaluation cost worthwhile. The value of q determined at the ith step is used for two purposes:

1. To reject the initial choice of h at the ith step and repeat the calculations using qh if the error is not within the required bound, and

2. To predict an appropriate initial choice of h for the $(i + 1)$st step.

Because of the penalty (in terms of functional evaluations) that must be paid if many of the steps are repeated, q tends to be chosen conservatively; in fact, for the Runge–Kutta–Fehlberg method with $n = 4$, the usual choice is

$$q = \left(\frac{\epsilon h}{2|\tilde{w}_{i+1} - w_{i+1}|}\right)^{1/4} \approx 0.84\left(\frac{\epsilon h}{|\tilde{w}_{i+1} - w_{i+1}|}\right)^{1/4}.$$

The program RKFVSM55, which implements the Runge–Kutta–Fehlberg method, incorporates a technique to eliminate large modifications in step size. This is done to avoid spending too much time with very small step sizes in regions with irregularities in the derivatives of y and to avoid large step sizes, which may result in skipping sensitive regions nearby. In some instances the step-size-increase procedure is omitted completely, and the step-size-decrease procedure is modified to be incorporated only when needed to bring the error under control.

EXAMPLE 1 The Runge–Kutta–Fehlberg method will be used to approximate the solution to the initial-value problem

$$y' = y - t^2 + 1, \qquad 0 \le t \le 2, \qquad y(0) = 0.5,$$

which has solution $y(t) = (t+1)^2 - 0.5e^t$. The input consists of tolerance $TOL = 10^{-5}$, a maximum step size $hmax = 0.25$, and a minimum step size $hmin = 0.01$. The results from the program RKFVSM55 are shown in Table 5.13. ■ ■ ■

Table 5.13

t_i	w_i	h_i	$y_i = y(t_i)$	$|y_i - w_i|$
0.0000000	0.5000000	0	0.5000000	0.0000000
0.2500000	0.9204886	0.2500000	0.9204873	0.0000013
0.4865522	1.3964910	0.2365522	1.3964884	0.0000026
0.7293332	1.9537488	0.2427810	1.9537446	0.0000042
0.9793332	2.5864260	0.2500000	2.5864198	0.0000062
1.2293332	3.2604605	0.2500000	3.2604520	0.0000085
1.4793332	3.9520955	0.2500000	3.9520844	0.0000111
1.7293332	4.6308268	0.2500000	4.6308127	0.0000141
1.9793332	5.2574861	0.2500000	5.2574687	0.0000173
2.0000000	5.3054896	0.0206668	5.3054720	0.0000177

The Runge–Kutta–Fehlberg method is popular for error control because at each step it provides, at little additional cost, *two* approximations that can be compared and related to the local error. Predictor-corrector techniques always generate two approximations at each step, so they are natural candidates for error-control adaptation.

To demonstrate the procedure, we construct a variable step-size predictor-corrector method using the explicit Adams–Bashforth four-step method as predictor and the implicit Adams–Moulton three-step method as corrector.

The Adams–Bashforth four-step method comes from the equation

$$y(t_{i+1}) = y(t_i) + \frac{h}{24}[55f(t_i, y(t_i)) - 59f(t_{i-1}y(t_{i-1}))$$
$$+ 37f(t_{i-2}, y(t_{i-2})) - 9f(t_{i-3}, y(t_{i-3}))] + \frac{251}{720}y^{(5)}(\hat{\mu}_i)h^5$$

for some μ_i in (t_{i-3}, t_{i+1}). Suppose we assume that the approximations $w_0, w_1, \ldots, w_i$ are all exact and, as in the case of the one-step methods, we let z represent the solution to the differential equation satisfying the initial condition $z(t_i) = w_i$. Then

(5.2)
$$z(t_{i+1}) - w_{i+1}^{(0)} = \frac{251}{720}z^{(5)}(\hat{\mu}_i)h^5.$$

A similar analysis of the Adams–Moulton three-step method leads to the local truncation error

(5.3) $$z(t_{i+1}) - w_{i+1} = -\frac{19}{720}z^{(5)}(\tilde{\mu}_i)h^5 \qquad \text{for some } \mu_i \text{ in } (t_{i-2}, t_{i+1}).$$

To proceed further, we must make the assumption that for small values of h,

$$z^{(5)}(\hat{\mu}_i) \approx z^{(5)}(\tilde{\mu}_i).$$

The effectiveness of the error-control technique depends directly on this assumption. If we subtract Eq. (5.3) from Eq. (5.2), we have

$$w_{i+1} - w_{i+1}^{(0)} = \frac{h^5}{720}[251z^{(5)}(\hat{\mu}_i) + 19z^{(5)}(\tilde{\mu}_i)] \approx \frac{3}{8}h^5z^{(5)}(\tilde{\mu}_i),$$

so

$$z^{(5)}(\tilde{\mu}_i) \approx \frac{8}{3h^5}(w_{i+1} - w_{i+1}^{(0)}).$$

Using this result to eliminate the term involving $h^5z^{(5)}$ from (5.2) gives the following approximation to the error:

$$|z(t_{i+1}) - w_{i+1}| \approx \frac{19h^5}{720} \cdot \frac{8}{3h^5}|w_{i+1} - w_{i+1}^{(0)}| = \frac{19|w_{i+1} - w_{i+1}^{(0)}|}{270}.$$

This expression was derived under the assumption that $w_0, w_1, \ldots, w_i$ are all exact, which means that this is an approximation to the local error. As in the case of the one-step methods, the global error is of order one degree less, so for the function y that is the solution to the original initial-value problem,

$$y' = f(t, y), \qquad a \le t \le b, \qquad y(a) = \alpha,$$

the global error can be estimated by

$$|y(t_{i+1}) - w_{i+1}| \approx \frac{|z(t_{i+1}) - w_{i+1}|}{h} \approx \frac{19|w_{i+1} - w_{i+1}^{(0)}|}{270h}.$$

Suppose we now reconsider the situation with a new step size qh generating new approximations $\hat{w}_{i+1}^{(0)}$ and $\hat{w}_{i+1}$. To control the global error to within ϵ, we want to choose q so that

$$\frac{|z(t_i + qh) - \hat{w}_{i+1}(\text{using the step size } qh)|}{qh} < \epsilon.$$

But from Eq. (5.3),

$$\frac{|z(t_i + qh) - \hat{w}_{i+1}(\text{using } qh)|}{qh} = \frac{19}{720}|z^{(5)}(\tilde{\mu}_i)|q^4h^4 \approx \frac{19}{720}\left[\frac{8}{3h^5}|w_{i+1} - w_{i+1}^{(0)}|\right]q^4h^4$$

so we need to choose q so that

$$\frac{|y(t_i + qh) - \hat{w}_{i+1}|}{qh} \approx \frac{19}{720}\left[\frac{8}{3h^5}|w_{i+1} - w_{i+1}^{(0)}|\right]q^4h^4 = \frac{19}{270}\frac{|w_{i+1} - w_{i+1}^{(0)}|}{h}q^4 < \epsilon.$$

Consequently,

$$q < \left(\frac{270}{19}\frac{h\epsilon}{|w_{i+1} - w_{i+1}^{(0)}|}\right)^{1/4} \approx 2\left(\frac{h\epsilon}{|w_{i+1} - w_{i+1}^{(0)}|}\right)^{1/4}.$$

A number of approximation assumptions have been made in this development, so in actual practice q is chosen conservatively, usually as

$$q = 1.5\left(\frac{h\epsilon}{|w_{i+1} - w_{i+1}^{(0)}|}\right)^{1/4}.$$

A change in step size for a multistep method is more costly in terms of functional evaluations than for a one-step method, since new equally spaced starting values must be computed. As a consequence, it is also common practice to ignore the step-size change whenever the global error is between $\epsilon/10$ and ϵ; that is, when

$$\frac{\epsilon}{10} < \frac{|y(t_{i+1}) - w_{i+1}|}{h} \approx \frac{19|w_{i+1} - w_{i+1}^{(0)}|}{270h} < \epsilon.$$

In addition, q is generally given an upper bound to ensure that a single unusually accurate approximation does not result in too large a step size. The program VPRCOR56 incorporates this safeguard with an upper bound of 4.

It should be emphasized that since the multistep methods require equal step sizes for the starting values, any change in step size necessitates recalculating new starting values at that point. In the program VPRCOR56 this is done by incorporating RKOR4M52, the Runge–Kutta method of order 4, as a subroutine.

EXAMPLE 2 Table 5.14 lists the results obtained using the program VPRCOR56 to find approximations to the solution of the initial-value problem

$$y' = y - t^2 + 1, \qquad 0 \le t \le 2, \qquad y(0) = 0.5,$$

which has solution $y(t) = (t+1)^2 - 0.5e^t$. Included in the input is the tolerance $TOL = 10^{-5}$, maximum step size $hmax = 0.25$, and minimum step size $hmin = 0.01$.

■ ■   ■

Table 5.14

t_i	w_i	h_i	σ_i	$y_i = y(t_i)$	$\|y_i - w_i\|$
0.1257017	0.7002318	0.1257017	4.051×10^{-6}	0.7002323	0.0000005
0.2514033	0.9230949	0.1257017	4.051×10^{-6}	0.9230960	0.0000011
0.3771050	1.1673877	0.1257017	4.051×10^{-6}	1.1673894	0.0000017
0.5028066	1.4317480	0.1257017	4.051×10^{-6}	1.4317502	0.0000022
0.6285083	1.7146306	0.1257017	4.610×10^{-6}	1.7146334	0.0000028
0.7542100	2.0142834	0.1257017	5.210×10^{-6}	2.0142869	0.0000035
0.8799116	2.3287200	0.1257017	5.913×10^{-6}	2.3287244	0.0000043
1.0056133	2.6556877	0.1257017	6.706×10^{-6}	2.6556930	0.0000054
1.1313149	2.9926319	0.1257017	7.604×10^{-6}	2.9926385	0.0000066
1.2570166	3.3366562	0.1257017	8.622×10^{-6}	3.3366642	0.0000080
1.3827183	3.6844761	0.1257017	9.777×10^{-6}	3.6844857	0.0000097
1.4857283	3.9697433	0.1030100	7.029×10^{-6}	3.9697541	0.0000108
1.5887383	4.2527711	0.1030100	7.029×10^{-6}	4.2527830	0.0000120
1.6917483	4.5310137	0.1030100	7.029×10^{-6}	4.5310269	0.0000133
1.7947583	4.8016488	0.1030100	7.029×10^{-6}	4.8016639	0.0000151
1.8977683	5.0615488	0.1030100	7.760×10^{-6}	5.0615660	0.0000172
1.9233262	5.1239764	0.0255579	3.918×10^{-8}	5.1239941	0.0000177
1.9488841	5.1854751	0.0255579	3.918×10^{-8}	5.1854932	0.0000181
1.9744421	5.2459870	0.0255579	3.918×10^{-8}	5.2460056	0.0000186
2.0000000	5.3054529	0.0255579	3.918×10^{-8}	5.3054720	0.0000191

EXERCISE SET 5.6

1. Use the Runge–Kutta–Fehlberg method with tolerance $TOL = 10^{-4}$, $hmax = 0.25$, and $hmin = 0.05$ to approximate the solutions to the following initial-value problems. Compare the results to the actual values.

(a) $y' = te^{3t} - 2y$, $\quad 0 \le t \le 1$, $\quad y(0) = 0$, actual solution $y(t) = \frac{1}{5}te^{3t} - \frac{1}{25}e^{3t} + \frac{1}{25}e^{-2t}$

(b) $y' = 1 + (t - y)^2$, $\quad 2 \le t \le 3$, $\quad y(2) = 1$, actual solution $y(t) = t + (1 - t)^{-1}$

(c) $y' = 1 + \frac{y}{t}$, $\quad 1 \le t \le 2$, $\quad y(1) = 2$, actual solution $y(t) = t \ln t + 2t$

(d) $y' = \cos 2t + \sin 3t$, $\quad 0 \le t \le 1$, $\quad y(0) = 1$, actual solution $y(t) = \frac{1}{2} \sin 2t - \frac{1}{3} \cos 3t + \frac{4}{3}$.

2. Use the Runge–Kutta–Fehlberg method with tolerance $TOL = 10^{-6}$, $hmax = 0.5$, and $hmin = 0.05$ to approximate the solutions to the following initial-value problems. Compare the results to the actual values.

(a) $y' = \frac{y}{t} - \frac{y^2}{t^2}$, $\quad 1 \le t \le 4$, $\quad y(1) = 1$, actual solution $y(t) = t(1 + \ln t)^{-1}$

(b) $y' = 1 + \frac{y}{t} + \left(\frac{y}{t}\right)^2$, $\quad 1 \le t \le 3$, $\quad y(1) = 0$, actual solution $y(t) = t \tan(\ln t)$

(c) $y' = -(y+1)(y+3)$, $\quad 0 \le t \le 3$, $\quad y(0) = -2$, actual solution $y(t) = -3 +$ $2(1 + e^{-2t})^{-1}$

(d) $y' = (t+2t^3)y^3 - ty$, $\quad 0 \le t \le 2$, $\quad y(0) = \frac{1}{3}$, actual solution $y(t) = (3+2t^2 + 6e^{t^2})^{-\frac{1}{2}}$

3. Use the Runge–Kutta–Fehlberg method with tolerance $TOL = 10^{-5}$, $hmax = 0.2$, and $hmin = 0.02$ to approximate the solutions to the following initial-value problems. Compare the results to the actual values.

(a) $y' = -2y/t + t^3y^2 \ln t$, $\quad 1 \le t \le 1.9$, $\quad y(1) = 2$, actual solution $y(t) =$ $4(t^4 - 2t^4 \ln t + t^2)^{-1}$

(b) $y'(t) = \sqrt{2 - y^2}e^t$, $\quad 0 \le t \le 0.5$, $\quad y(0) = 0$, actual solution $y(t) =$ $\sqrt{2}\sin(e^t - 1)$

(c) $y' = -\sin t(1 + y\sec t)$, $\quad 0 \le t \le 0.5$, $\quad y(0) = 2$, actual solution $y(t) =$ $(2 + \ln(\cos t))\cos t$

(d) $y' = -5y + 5t^2 + 2t$, $\quad 0 \le t \le 1$, $\quad y(0) = \frac{1}{3}$, actual solution $y(t) = t^2 + \frac{1}{3}e^{-5t}$

4. Use the Adams Variable Step-Size Predictor-Corrector method with tolerance $TOL = 10^{-4}$, $hmax = 0.25$, and $hmin = 0.025$ to approximate the solutions to the given initial-value problems. Compare the results to the actual values.

(a) $y' = te^{3t} - 2y$, $\quad 0 \le t \le 1$, $\quad y(0) = 0$, actual solution $y(t) = \frac{1}{5}te^{3t} - \frac{1}{25}e^{3t} + \frac{1}{25}e^{-2t}$

(b) $y' = 1 + (t-y)^2$, $\quad 2 \le t \le 3$, $\quad y(2) = 1$, actual solution $y(t) = t + 1(1-t)^{-1}$

(c) $y' = 1 + y/t$, $\quad 1 \le t \le 2$, $\quad y(1) = 2$, actual solution $y(t) = t \ln t + 2t$

(d) $y' = \cos 2t + \sin 3t$, $\quad 0 \le t \le 1$, $\quad y(0) = 1$, actual solution $y(t) = \frac{1}{2}\sin 2t - \frac{1}{3}\cos 3t + \frac{4}{3}$

5. Use the Adams Variable Step-Size Predictor-Corrector method with tolerance $TOL = 10^{-6}$, $hmax = 0.5$, and $hmin = 0.02$ to approximate the solutions to the given initial-value problems. Compare the results to the actual values.

(a) $y' = \frac{y}{t} - \frac{y^2}{t^2}$, $\quad 1 \le t \le 4$, $\quad y(1) = 1$, actual solution $y(t) = t(1 + \ln t)^{-1}$

(b) $y' = 1 + \frac{y}{t} + \left(\frac{y}{t}\right)^2$, $\quad 1 \le t \le 3$, $\quad y(1) = 0$, actual solution $y(t) = t \tan(\ln t)$

(c) $y' = -(y+1)(y+3)$, $\quad 0 \le t \le 3$, $\quad y(0) = -2$, actual solution $y(t) = -3 +$ $2(1 + e^{-2t})^{-1}$

(d) $y' = (t + 2t^3)y^3 - ty$, $\quad 0 \le t \le 2$, $\quad y(0) = \frac{1}{3}$, actual solution $y(t) =$ $(3 + 2t^2 + 6e^{t^2})^{-1/2}$

6. Use the Adams Variable Step-Size Predictor-Corrector method with tolerance $TOL = 10^{-5}$, $hmax = 0.2$, and $hmin = 0.02$ to approximate the solutions to the given initial-value problems. Compare the results to the actual values.

(a) $y' = -\frac{1}{t}(\sin t + y) + \cos t^2$, $\quad \sqrt{\pi} \le t \le \pi$, $\quad y(\sqrt{\pi}) = 0$, actual solution $y(t) = t^{-1}(\cos t - \cos\sqrt{\pi} + 0.5\sin t^2)$

(b) $y' = e^t \csc t - \left(\frac{1}{t} + \cot t\right)y$, $\quad \frac{\pi}{2} \le t \le \frac{3\pi}{4}$, $\quad y\left(\frac{\pi}{2}\right) = \frac{2}{\pi}$, actual solution $y(t) = (t \sin t)^{-1}((t-1)e^t - \left(\frac{\pi}{2} - 1\right)e^{\pi/2} + 1)$

(c) $y' = \frac{t + \ln t}{y^2 \ln t} - \frac{y}{t \ln t}$, $\quad e \le t \le 2e$, $\quad y(e) = 2$, actual solution
$y(t) = \left(3t + (\ln t)^{-1}(1.5t^2 - 9t) + (\ln t)^{-2}(18t - 1.5t^2) + (\ln t)^{-3}(0.75t^2 - 18t + C)\right)^{1/3}$,
where $C = 8 + e(6 - 0.75e)$

(d) $y' = 1 + t(1 - y^2) + 2t^2 y - t^3$, $\quad 0 \le t \le 3$, $\quad y(0) = 0$, actual solution $y(t) = t - 1 + 2(1 + e^{-t^2})^{-1}$

7. An electrical circuit consists of a capacitor of constant capacitance $C = 1.1$ farads in series with a resistor of constant resistance $R_0 = 2.1$ ohms. A voltage $\mathcal{E}(t) = 110 \sin t$ is applied at time $t = 0$. When the resistor heats up, the resistance becomes a function of the current i:

$$R(t) = R_0 + ki \quad \text{where } k = 0.9,$$

and the differential equation for i becomes

$$\left(1 + \frac{2k}{R_0} i\right) \frac{di}{dt} + \frac{1}{R_0 C} i = \frac{1}{R_0} \frac{d\mathcal{E}}{dt}.$$

Use the Runge–Kutta–Fehlberg method with $TOL = 10^{-3}$ and $i(0) = 0$ to find the current i after 2 s.

5.7 Methods for Systems of Equations

The most common application of numerical methods for approximating the solution of initial-value problems concerns not a single problem, but a linked system of differential equations. Why, then, have we spent the majority of this chapter considering the solution of a single equation? The answer is simple: To approximate the solution of a system of initial-value problems, we successively apply the techniques that we used to solve a single problem. As is generally the case in mathematics, the key to the methods for systems can be found by examining the easier problem and then modifying it to treat the more complicated situation.

An ***m*th-order system** of first-order initial-value problems can be expressed in the form

$$\frac{du_1}{dt} = f_1(t, u_1, u_2, \ldots, u_m),$$

$$\frac{du_2}{dt} = f_2(t, u_1, u_2, \ldots, u_m),$$

$$\vdots$$

$$\frac{du_m}{dt} = f_m(t, u_1, u_2, \ldots, u_m),$$

for $a \le t \le b$, with the initial conditions

$$u_1(a) = \alpha_1, \quad u_2(a) = \alpha_2, \ldots, u_m(a) = \alpha_m.$$

The object is to find m functions $u_1, u_2, \ldots, u_m$ that satisfy the system of differential equations together with all the initial conditions.

Methods to solve systems of first-order differential equations are generalizations of the methods for a single first-order equation presented earlier in this chapter. For example, the classical Runge–Kutta method of order 4 given by

$$w_0 = \alpha,$$

$$k_1 = hf(t_i, w_i),$$

$$k_2 = hf\left(t_i + \frac{h}{2}, w_i + \frac{1}{2}k_1\right),$$

$$k_3 = hf\left(t_i + \frac{h}{2}, w_i + \frac{1}{2}k_2\right),$$

$$k_4 = hf(t_{i+1}, w_i + k_3),$$

and

$$w_{i+1} = w_i + \tfrac{1}{6}[k_1 + 2k_2 + 2k_3 + k_4],$$

for each $i = 0, 1, \ldots, N - 1$, used to solve the first-order initial-value problem

$$y' = f(t, y), \qquad a \leq t \leq b, \qquad y(a) = \alpha,$$

is generalized as follows. Let an integer $N > 0$ be chosen and set $h = (b - a)/N$. Partition the interval $[a, b]$ into N subintervals with the mesh points

$$t_j = a + jh, \qquad \text{for each } j = 0, 1, \ldots, N.$$

Use the notation w_{ij} to denote an approximation to $u_i(t_j)$ for each $j = 0, 1, \ldots, N$ and $i = 1, 2, \ldots, m$; that is, w_{ij} will approximate the ith solution $u_i(t)$ of the system at the jth mesh point t_j. For the initial conditions, set

$$w_{1,0} = \alpha_1, \qquad w_{2,0} = \alpha_2, \ldots, \qquad w_{m,0} = \alpha_m.$$

If we assume that the values $w_{1,j}, w_{2,j}, \ldots, w_{m,j}$ have been computed, we obtain $w_{1,j+1}, w_{2,j+1}, \ldots, w_{m,j+1}$ by first calculating, for each $i = 1, 2, \ldots, m$,

$$k_{1,i} = hf_i(t_j, w_{1,j}, w_{2,j}, \ldots, w_{m,j})$$

and then finding, for each i,

$$k_{2,i} = hf_i\left(t_j + \frac{h}{2}, w_{1,j} + \frac{1}{2}k_{1,1}, w_{2,j} + \frac{1}{2}k_{1,2}, \ldots, w_{m,j} + \frac{1}{2}k_{1,m}\right).$$

We next determine all the terms

$$k_{3,i} = hf_i\left(t_j + \frac{h}{2}, w_{1,j} + \frac{1}{2}k_{2,1}, w_{2,j} + \frac{1}{2}k_{2,2}, \ldots, w_{m,j} + \frac{1}{2}k_{2,m}\right)$$

and finally calculate all the terms

$$k_{4,i} = hf_i(t_j + h, w_{1,j} + k_{3,1}, w_{2,j} + k_{3,2}, \ldots, w_{m,j} + k_{3,m}).$$

Combining these values gives

$$w_{i,j+1} = w_{i,j} + \tfrac{1}{6}[k_{1,i} + 2k_{2,i} + 2k_{3,i} + k_{4,i}]$$

for each $i = 1, 2, \ldots m$.

Note that all the values $k_{1,1}, k_{1,2}, \ldots, k_{1,m}$ must be computed before any of the terms of the form $k_{2,i}$ can be determined. In general, each $k_{l,1}, k_{l,2}, \ldots, k_{l,m}$ must be computed before any of the expressions $k_{l+1,i}$. The program RKO4SY57 implements the Runge–Kutta fourth-order method for systems.

EXAMPLE 1   Kirchhoff's law states that the sum of all instantaneous voltage changes around a closed circuit is zero. This implies that the current $I(t)$ in a closed circuit containing a resistance of R ohms, a capacitance of C farads, an inductance of L henries, and a voltage source of $\mathcal{E}(t)$ volts must satisfy the equation

$$LI'(t) + RI(t) + \frac{1}{C} \int I(t)\, dt = \mathcal{E}(t).$$

Figure 5.4

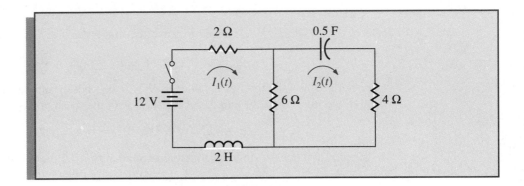

The currents $I_1(t)$ and $I_2(t)$ in the left and right loops, respectively, of the circuit shown in Figure 5.4 are the solutions to the following system of equations:

$$2I_1(t) + 6[I_1(t) - I_2(t)] + 2I_1'(t) = 12,$$

$$\frac{1}{0.5} \int I_2(t)\, dt + 4I_2(t) + 6[I_2(t) - I_1(t)] = 0.$$

If we assume that the switch in the circuit is closed at time $t = 0$, then differentiating the second equation and substituting the first equation into the resulting equation gives the following system of initial-value problems:

$$I_1' = f_1(t, I_1, I_2) = -4I_1 + 3I_2 + 6, \quad I_1(0) = 0,$$
$$I_2' = f_2(t, I_1, I_2) = 0.6I_1' - 0.2I_2 = -2.4I_1 + 1.6I_2 + 3.6, \quad I_2(0) = 0.$$

The exact solution to this system can be shown to be

$$I_1(t) = -3.375e^{-2t} + 1.875e^{-0.4t} + 1.5,$$
$$I_2(t) = -2.25e^{-2t} + 2.25e^{-0.4t}.$$

Suppose we apply the Runge–Kutta method of order 4 to this system with $h = 0.1$. Since $w_{1,0} = I_1(0) = 0$ and $w_{2,0} = I_2(0) = 0$,

$$k_{1,1} = hf_1(t_0, w_{1,0}, w_{2,0}) = 0.1\, f_1(0, 0, 0) = 0.1[-4(0) + 3(0) + 6] = 0.6,$$

$$k_{1,2} = hf_2(t_0, w_{1,0}, w_{2,0}) = 0.1\, f_2(0, 0, 0) = 0.1[-2.4(0) + 1.6(0) + 3.6] = 0.36,$$

$$k_{2,1} = hf_1(t_0 + \tfrac{1}{2}h, w_{1,0} + \tfrac{1}{2}k_{1,1}, w_{2,0} + \tfrac{1}{2}k_{1,2}) = 0.1\, f_1(0.05, 0.3, 0.18)$$
$$= 0.1[-4(0.3) + 3(0.18) + 6] = 0.534,$$

$$k_{2,2} = hf_2(t_0 + \tfrac{1}{2}h, w_{1,0} + \tfrac{1}{2}k_{1,1}, w_{2,0} + \tfrac{1}{2}k_{1,2}) = 0.1\, f_2(0.05, 0.3, 0.18)$$
$$= 0.1[-2.4(0.3) + 1.6(0.18) + 3.6] = 0.3168.$$

Generating the remaining entries in a similar manner produces

$$k_{3,1} = (0.1)f_1(0.05, 0.267, 0.1584) = 0.54072,$$
$$k_{3,2} = (0.1)f_2(0.05, 0.267, 0.1584) = 0.321264,$$
$$k_{4,1} = (0.1)f_1(0.1, 0.54072, 0.321264) = 0.4800912,$$

and

$$k_{4,2} = (0.1)f_1(0.1, 0.54072, 0.321264) = 0.28162944.$$

As a consequence,

$$I_1(0.1) \approx w_{1,1} = w_{1,0} + \tfrac{1}{6}[k_{1,1} + 2k_{2,1} + 2k_{3,1} + k_{4,1}]$$
$$= 0 + \tfrac{1}{6}[0.6 + 2(0.534) + 2(0.54072) + 0.4800912] = 0.5382552$$

and

$$I_2(0.1) \approx w_{2,1} = w_{2,0} + \tfrac{1}{6}[k_{1,2} + 2k_{2,2} + 2k_{3,2} + k_{4,2}] = 0.3196263.$$

The remaining entries in Table 5.15 are generated in a similar manner. ■ ■ ■

Table 5.15

t_j	$w_{1,j}$	$w_{2,j}$	$\|I_1(t_j) - w_{1,j}\|$	$\|I_2(t_j) - w_{2,j}\|$
0.0	0	0	0	0
0.1	0.5382550	0.3196263	0.8285×10^{-5}	0.5803×10^{-5}
0.2	0.9684983	0.5687817	0.1514×10^{-4}	0.9596×10^{-5}
0.3	1.310717	0.7607328	0.1907×10^{-4}	0.1216×10^{-4}
0.4	1.581263	0.9063208	0.2098×10^{-4}	0.1311×10^{-4}
0.5	1.793505	1.014402	0.2193×10^{-4}	0.1240×10^{-4}

Many important physical problems—for example, electrical circuits and vibrating systems—involve initial-value problems whose equations have degree higher than one. New techniques are not required for solving these problems, since by relabeling the variables we can reduce any higher-order differential equation into a system of first-order differential equations and then apply one of the methods we have already discussed.

A general mth-order initial-value problem has the form

$$y^{(m)}(t) = f(t, y, y', \ldots, y^{(m-1)}),$$

for $a \leq t \leq b$, with initial conditions

$$y(a) = \alpha_1, \qquad y'(a) = \alpha_2, \ldots, y^{(m-1)}(a) = \alpha_m.$$

To convert this into a system of first-order differential equations, define

$$u_1(t) = y(t), \qquad u_2(t) = y'(t), \ldots, \qquad u_m(t) = y^{(m-1)}(t).$$

Using this notation, we obtain the first-order system

$$\frac{du_1}{dt} = \frac{dy}{dt} = u_2,$$

$$\frac{du_2}{dt} = \frac{dy'}{dt} = u_3,$$

$$\vdots$$

$$\frac{du_{m-1}}{dt} = \frac{dy^{(m-2)}}{dt} = u_m,$$

and

$$\frac{du_m}{dt} = \frac{dy^{(m-1)}}{dt} = y^{(m)} = f(t, y, y', \ldots, y^{(m-1)}) = f(t, u_1, u_2, \ldots, u_m),$$

with initial conditions

$$u_1(a) = y(a) = \alpha_1, \qquad u_2(a) = y'(a) = \alpha_2, \ldots, \qquad u_m(a) = y^{(m-1)}(a) = \alpha_m.$$

EXAMPLE 2   Consider the second-order initial-value problem

$$y'' - 2y' + 2y = e^{2t} \sin t, \qquad \text{for } 0 \le t \le 1, \qquad \text{with } y(0) = -0.4, \ y'(0) = -0.6.$$

With $u_1(t) = y(t)$ and $u_2(t) = y'(t)$, this equation is transformed into the system

$$u_1'(t) = u_2(t),$$
$$u_2'(t) = e^{2t} \sin t - 2u_1(t) + 2u_2(t),$$

with initial conditions

$$u_1(0) = -0.4, \qquad u_2(0) = -0.6.$$

The set of values $w_{1,j}$ and $w_{2,j}$ for $j = 0, 1, \ldots, 10$, are presented in Table 5.16 and compared to the actual values of $u_1(t) = 0.2e^{2t}(\sin t - 2 \cos t)$ and $u_2(t) = u_1'(t) = 0.2e^{2t}(4 \sin t - 3 \cos t)$.   ■ ■ ■

Table 5.16

t_j	$w_{1,j}$	$w_{2,j}$	$y(t_j) = u_1(t_j)$	$|y(t_j) - w_{1,j}|$	$y'(t_j) = u_2(t_j)$	$|y'(t_j) - w_{2,j}|$
0.0	−0.40000000	−0.60000000	−0.40000000	0	−0.60000000	0
0.1	−0.46173334	−0.63163124	−0.46173297	3.7×10^{-7}	−0.6316304	7.75×10^{-7}
0.2	−0.52555988	−0.64014895	−0.52555905	8.3×10^{-7}	−0.6401478	1.01×10^{-6}
0.3	−0.58860144	−0.61366381	−0.58860005	1.39×10^{-6}	−0.6136630	8.34×10^{-7}
0.4	−0.64661231	−0.53658203	−0.64661028	2.03×10^{-6}	−0.5365821	1.79×10^{-7}
0.5	−0.69356666	−0.38873810	−0.69356395	2.71×10^{-6}	−0.3887395	5.96×10^{-7}
0.6	−0.72115190	−0.14438087	−0.72114849	3.41×10^{-6}	−0.1443834	7.75×10^{-7}
0.7	−0.71815295	0.22899702	−0.71814890	4.05×10^{-6}	0.2289917	2.03×10^{-6}
0.8	−0.66971133	0.77199180	−0.66970677	4.56×10^{-6}	0.7719815	5.30×10^{-6}
0.9	−0.55644290	0.15347815	−0.55643814	4.76×10^{-6}	1.534764	9.54×10^{-6}
1.0	−0.35339886	0.25787663	−0.35339436	4.50×10^{-6}	2.578741	1.34×10^{-5}

The other one-step methods can be similarly extended to systems. If the Runge–Kutta–Fehlberg method is extended, then each component of the numerical solution

$w_{1j}, w_{2j}, \ldots, w_{mj}$ must be examined for accuracy. If any of the components fail to be sufficiently accurate, the entire numerical solution must be recomputed.

The multistep methods and predictor-corrector techniques are also extended easily to systems. Again, if error control is used, each component must be accurate. The extension of the extrapolation technique to systems can also be done, but the notation becomes quite involved.

EXERCISE SET 5.7

1. Use the Runge–Kutta method for systems to approximate the solutions of the following systems of first-order differential equations and compare the results to the actual solutions.

   (a) $u_1' = 2u_1 + 2u_2,$ $\quad 0 \le t \le 2,$ $\quad u_1(0) = -1;$
   $u_2' = 3u_1 + u_2,$ $\quad 0 \le t \le 2,$ $\quad u_2(0) = 4;$
   with $h = 0.5$. Actual solutions $u_1(t) = e^{4t} - 2e^{-t}$ and $u_2(t) = e^{4t} + 3e^{-t}$.

   (b) $u_1' = 3u_1 + 2u_2 - (2t^2 + 1)e^{2t},$ $\quad 0 \le t \le 1,$ $\quad u_1(0) = 1;$
   $u_2' = 4u_1 + u_2 + (t^2 + 2t - 4)e^{2t},$ $\quad 0 \le t \le 1,$ $\quad u_2(0) = 1;$
   with $h = 0.2$. Actual solutions $u_1(t) = \frac{1}{3}e^{5t} - \frac{1}{3}e^{-t} + e^{2t}$ and $u_2(t) = \frac{1}{3}e^{5t} + \frac{2}{3}e^{-t} + t^2 e^{2t}$.

   (c) $u_1' = -4u_1 - 2u_2 + \cos t + 4 \sin t,$ $\quad 0 \le t \le 2,$ $\quad u_1(0) = 0;$
   $u_2' = 3u_1 + u_2 - 3 \sin t,$ $\quad 0 \le t \le 2,$ $\quad u_2(0) = -1;$
   with $h = 0.1$. Actual solutions $u_1(t) = 2e^{-t} - 2e^{-2t} + \sin t$ and $u_2(t) = -3e^{-t} + 2e^{-2t}$.

   (d) $u_1' = u_2,$ $\quad 0 \le t \le 2,$ $\quad u_1(0) = 1;$
   $u_2' = -u_1 - 2e^t + 1,$ $\quad 0 \le t \le 2,$ $\quad u_2(0) = 0;$
   $u_3' = -u_1 - e^t + 1,$ $\quad 0 \le t \le 2,$ $\quad u_3(0) = 1;$
   with $h = 0.5$. Actual solutions $u_1(t) = \cos t + \sin t - e^t + 1,$ $u_2(t) = -\sin t + \cos t - e^t$, and $u_3(t) = -\sin t + \cos t$.

   (e) $u_1' = u_2 - u_3 + t,$ $\quad 0 \le t \le 1,$ $\quad u_1(0) = 1;$
   $u_2' = 3t^2,$ $\quad 0 \le t \le 1,$ $\quad u_2(0) = 1;$
   $u_3' = u_2 + e^{-t},$ $\quad 0 \le t \le 1,$ $\quad u_3(0) = -1;$
   with $h = 0.1$. Actual solutions $u_1(t) = -0.05t^5 + 0.25t^4 + t + 2 - e^{-t}, u_2(t) = t^3 + 1,$ and $u_3(t) = 0.25t^4 + t - e^{-t}$.

   (f) $u_1' = 4u_1 - u_2 + u_3 - (t + 2)^2,$ $\quad 0 \le t \le 1,$ $\quad u_1(0) = 3;$
   $u_2' = -u_1 + 3u_2 - 2u_3 + 2t^2 + t + 15,$ $\quad 0 \le t \le 1,$ $\quad u_2(0) = -7;$
   $u_3' = u_1 - 2u_2 + 3u_3 - 3t^2 + t - 10,$ $\quad 0 \le t \le 1,$ $\quad u_3(0) = -2;$
   with $h = 0.1$. Actual solutions $u_1(t) = e^{6t} + 2e^{3t} + t, u_2(t) = -e^{6t} + e^{3t} - 2e^t - 5,$ and $u_3(t) = e^{6t} - e^{3t} - 2e^t + t^2.$

2. Use the Runge–Kutta method for systems to approximate the solutions of the following higher-order differential equations and compare the results to the actual solutions.

   (a) $y'' - 2y' + y = te^t - t,$ $\quad 0 \le t \le 1,$ $\quad y(0) = y'(0) = 0$ with $h = 0.1$. Actual solution $y(t) = \frac{1}{6}t^3 e^t - te^t + 2e^t - t - 2.$

   (b) $y'' + 2y' + y = e^t,$ $\quad 0 \le t \le 2,$ $\quad y(0) = 1,$ $\quad y'(0) = -1$ with $h = 0.1$. Actual solution $y(t) = \frac{3}{4}e^{-t} - \frac{1}{2}te^{-t} + \frac{1}{4}e^t.$

   (c) $t^2 y'' - 2ty' + 2y = t^3 \ln t,$ $\quad 1 \le t \le 2,$ $\quad y(1) = 1,$ $\quad y'(1) = 0$ with $h = 0.05$. Actual solution $y(t) = \frac{7}{4}t + \frac{1}{2}t^3 \ln t - \frac{3}{4}t^3.$

   (d) $t^2 y'' - ty' - 3y = t \ln t - t^2,$ $\quad 1 \le t \le 3,$ $\quad y(1) = 0,$ $\quad y'(1) = 5$ with $h = 0.2$. Actual solution $y(t) = -\frac{67}{48}t^{-1} + \frac{17}{16}t^3 - \frac{1}{4}t \ln t + \frac{t^2}{3}.$

   (e) $t^2 y'' + ty' - t(t + 1)y = 2te^t,$ $\quad 1 \le t \le 2,$ $\quad y(1) = 0,$ $\quad y'(1) = e$ with $h = 0.05$. Actual solution $y(t) = e^t \ln t.$

(f) $y''' + 2y'' - y' - 2y = e^t$, $\quad 0 \le t \le 3$, $\quad y(0) = 1$, $\quad y'(0) = 2$, $\quad y''(0) = 0$ with $h = 0.2$. Actual solution $y(t) = \frac{43}{36}e^t + \frac{1}{4}e^{-t} - \frac{4}{9}e^{-2t} + \frac{1}{6}te^t$.

(g) $y''' = -6(y)^4$, $\quad 1 \le t \le 1.9$, $\quad y(1) = -1$, $\quad y'(1) = -1$, $\quad y''(1) = 2$ with $h = 0.05$. Actual solution $y(t) = 1/(t-2)$.

(h) $t^3 y''' - t^2 y'' + 3ty' - 4y = 5t^3 \ln t + 9t^3$, $\quad 1 \le t \le 2$, $\quad y(1) = 0$, $\quad y'(1) = 1$, $y''(1) = 3$ with $h = 0.1$. Actual solution $y(t) = -t^2 + t\cos(\ln t) + t\sin(\ln t) + t^3 \ln t$.

3. Change the Adams fourth-order predictor-corrector method to obtain approximate solutions to systems of first-order equations.

4. Repeat Exercise 1 using the method developed in Exercise 3.

5. Repeat Exercise 2 using the method developed in Exercise 3.

6. The study of the mathematical models for predicting the population dynamics of competing species has its origin in independent works published in the early part of this century by A.J. Lotka and V. Volterra. Consider the problem of predicting the population of two species, one of which is the predator, whose population at time t is $x_2(t)$, feeding on the other, the prey, whose population is $x_i(t)$. We will assume that the prey always has an adequate food supply and that its birth rate at any time is proportional to the number of prey alive at that time; that is, birth rate (prey) $= k_1 x_1(t)$. The death rate of the prey depends on the number of both prey and predators alive at the time. For simplicity, we assume the death rate (prey) $= k_2 x_1(t)x_2(t)$. The birth rate of the predator, on the other hand, depends on its food supply, $x_1(t)$, as well as on the number of predators available for reproduction purposes. For this reason, we assume that the birth rate (predator) $= k_3 x_1(t)x_2(t)$. The death rate of the predator will be taken simply proportional to the number of predators alive at the time; that is, death rate (predator) $= k_4 x_2(t)$.

   Since $x_1'(t)$ and $x_2'(t)$ represent the change in the prey and predator populations, respectively, with respect to time, the problem is expressed by the system of nonlinear differential equations

   $$x_1'(t) = k_1 x_1(t) - k_2 x_1(t)x_2(t) \quad \text{and} \quad x_2'(t) = k_3 x_1(t)x_2(t) - k_4 x_2(t).$$

   Solve this system for $0 \le t \le 4$, assuming that the initial population of the prey is 1000 and of the predators is 200 and that the constants are $k_1(t) = 3, k_2 = 0.002, k_3 = 0.0006$, and $k_4 = 0.5$. Sketch a graph of the solutions to this problem, plotting both populations with time, and describe the physical phenomena represented. Is there a stable solution to this population model? If so, for what values of x_1 and x_2 is the solution stable?

7. In Exercise 6 we considered the problem of predicting the population in a predator–prey model. Another problem of this type is concerned with two species competing for the same food supply. If the numbers of the species alive at the time t are denoted by $x_1(t)$ and $x_2(t)$, it is often assumed that, while the birth rate of each of the species is simply proportional to the number of species alive at the time, the death rate of each species depends on the population of both species. We will assume the population of a particular pair of species is described by the equations

   $$\frac{dx_1(t)}{dt} = x_1(t)[4 - 0.0003x_1(t) - 0.004x_2(t)]$$

   and

   $$\frac{dx_2(t)}{dt} = x_2(t)[2 - 0.0002x_1(t) - 0.0001x_2(t)].$$

   If it is known that the initial population of each species is 10,000, find the solution to this system for $0 \le t \le 4$. Is there a stable solution to this proportional model? If so, for what values of x_1 and x_2 is the solution stable?

5.8   Stiff Differential Equations

All the methods for approximating the solution to initial-value problems have error terms that involve a higher derivative of the solution of the equation.If the derivative can be reasonably bounded, then the method will have a predictable error bound that can be used to estimate the accuracy of the approximation. Even if the derivative grows as the steps increase, the error can be kept in relative control, provided that the solution also grows in magnitude. Problems frequently arise, however, when the magnitude of the derivative increases, but the solution does not. In this situation, the error can grow so large that it dominates the calculations. Initial-value problems for which this is likely to occur are called **stiff equations** and are quite common, particularly in the study of vibrations, chemical reactions, and electrical circuits. Stiff systems derive their name from the motion of spring and mass systems that have large spring constants.

Stiff differential equations are characterized as those whose exact solution has a term of the form e^{-ct}, where c is a large positive constant. This is usually only a part of the solution, called the *transient* solution; the more important portion of the solution is called the *steady state* solution. A transient portion of a stiff equation will rapidly decay to zero as t increases, but since the nth derivative of this term has magnitude $c^n e^{-ct}$, the derivative does not decay nearly as rapidly. In fact, since the derivative in the error term is evaluated not at t, but at a number between zero and t, the derivative terms may increase as t increases—and very rapidly indeed. Fortunately, stiff equations can generally be predicted from the physical problem from which the equation is derived, and with care the error can be kept under control. The manner in which this is done is considered in this section.

EXAMPLE 1   The system of initial-value problems

$$u_1' = 9u_1 + 24u_2 + 5\cos t - \tfrac{1}{3}\sin t, \qquad u_1(0) = \tfrac{4}{3}$$
$$u_2' = -24u_1 - 51u_2 - 9\cos t + \tfrac{1}{3}\sin t, \qquad u_2(0) = \tfrac{2}{3}$$

has the unique solution

$$u_1(t) = 2e^{-3t} - e^{-39t} + \tfrac{1}{3}\cos t,$$
$$u_2(t) = -e^{-3t} + 2e^{-39t} - \tfrac{1}{3}\cos t.$$

The transient term e^{-39t} in the solution causes this system to be stiff. Applying the Runge–Kutta fourth-order method for systems gives results listed in Table 5.17 (page 196). Stability results, and accurate approximations occur, when $h = 0.05$. Increasing the step size to $h = 0.1$, however, leads to the disastrous results shown in the table.

■ ■ ■

Although stiffness is usually associated with systems of differential equations, the approximation characteristics of a particular numerical method applied to a stiff system can be predicted by examining the error produced when the method is applied to a simple *test equation*,

$$y' = \lambda y, \qquad y(0) = \alpha,$$

Table 5.17

t	$w_1(t)$ h = 0.05	$w_1(t)$ h = 0.1	$u_1(t)$	$w_2(t)$ h = 0.05	$w_2(t)$ h = 0.1	$u_2(t)$
0.1	1.712219	−2.645169	1.793061	−0.8703152	7.844527	−1.032001
0.2	1.414070	−18.45158	1.423901	−0.8550148	38.87631	−0.8746809
0.3	1.130523	−87.47221	1.131575	−0.7228910	176.4828	−0.7249984
0.4	0.9092763	−934.0722	0.9094086	−0.6079475	789.3540	−0.6082141
0.5	9.7387506	−1760.016	0.7387877	−0.5155810	3520.00	−0.5156575
0.6	0.6056833	−7848.550	0.6057094	−0.4403558	15697.84	−0.4404108
0.7	0.4998361	−34989.63	·0.4998603	−0.3773540	69979.87	−0.3774038
0.8	0.4136490	−155979.4	0.4136714	−0.3229078	311959.5	−0.3229535
0.9	0.3415939	−695332.0	0.3416143	−0.2743673	1390664.	−0.2744088
1.0	0.2796568	−3099671.	0.2796748	−0.2298511	6199352.	−0.2298877

where λ is a negative real number. The solution to this equation contains the transient $e^{\lambda t}$ and the steady-state form is zero, so the approximation characteristics of a method are easy to determine. (A more complete discussion of the round-off error associated with stiff systems requires examining the test equation when λ is a complex number with negative imaginary part.)

Suppose we apply Euler's method to the test equation. Letting $h = (b - a)/N$ and $t_j = jh$, for $j = 1, 2, \ldots, N$, implies that

$$w_0 = \alpha,$$
$$w_{j+1} = w_j + h(\lambda w_j) = (1 + h\lambda)w_j,$$

so

(5.4) $$w_{j+1} = (1 + h\lambda)^{j+1}w_0 = (1 + h\lambda)^{j+1}\alpha, \qquad \text{for } j = 0, 1, \ldots, N - 1.$$

Since the exact solution is $y(t) = \alpha e^{\lambda t}$, the absolute error is

$$|y(t_j) - w_j| = |e^{\lambda h j} - (1 + h\lambda)^j||\alpha|,$$

and the accuracy is determined by how well the term $1 + h\lambda$ approximates

$$e^{h\lambda} = 1 + h\lambda + \tfrac{1}{2!}(h\lambda)^2 + \tfrac{1}{3!}(h\lambda)^3 + \cdots.$$

When $\lambda < 0$, the exact solution $e^{\lambda h j}$ decays to zero, but by (5.4), the approximation w_i to $y(t_i)$ will have this property only if $|1 + h\lambda| < 1$. This effectively restricts the step size h for Euler's method to satisfy $|1 + h\lambda| < 0$, which simplifies to $h < 2/|\lambda|$.

Suppose now that a round-off error is introduced in the initial condition for Euler's method:

$$w_0 = \alpha + \delta_0.$$

At the jth step the round-off error is

$$\delta_j = (1 + h\lambda)^j \delta_0.$$

Since $\lambda < 0$, the condition for the control of the growth of round-off error is the same as the condition for controlling the absolute error, $h < 2/|\lambda|$.

The situation is similar for other one-step methods. In general, a function Q exists with the property that the difference method, when applied to the test equation, gives

$$w_{i+1} = Q(h\lambda)w_i.$$

The accuracy of the method depends upon how well $Q(h\lambda)$ approximates $e^{h\lambda}$, and the error grows intolerably unless $|Q(h\lambda)| \le 1$.

The problem is more complicated in the case of multistep methods due to the interplay of multiple previous approximations at each step. The problem tends to be particularly acute in the case of the explicit techniques, which carries over to predictor-corrector methods. In practice, the techniques used for stiff systems are implicit multistep methods. Generally, w_{i+1} is obtained by iteratively solving a nonlinear equation or nonlinear system, often by Newton's method. To illustrate the procedure, consider the following implicit technique:

Implicit Trapezoidal Method

$$w_0 = \alpha$$

$$w_{j+1} = w_j + \frac{h}{2}[f(t_{j+1}, w_{j+1}) + f(t_j, w_j)]$$

where $j = 0, 1, \ldots, N - 1$.

To determine w_1 using this technique, we apply Newton's method to find the root of the equation

$$0 = F(y) = y - w_0 - \frac{h}{2}[f(t_0, w_0) + f(t_1, y)] = y - \alpha - \frac{h}{2}[f(a, \alpha) + f(t_1, y)].$$

To approximate this solution, select $w_1^{(0)}$ (usually as w_0) and generate $w_1^{(k)}$ by applying Newton's method to obtain:

$$w_1^{(k)} = w_1^{(k-1)} - \frac{F\left(w_1^{(k-1)}\right)}{F'\left(w_1^{(k-1)}\right)}$$

$$= \frac{w_1^{(k-1)} - w_1^{(k-1)} - \alpha - \frac{h}{2}\left[f(a, \alpha) + f\left(t_1, w_1^{(k-1)}\right)\right]}{1 - \frac{h}{2}f_y\left(t_1, w_1^{(k-1)}\right)}$$

until $|w_1^{(k)} - w_1^{(k-1)}|$ is sufficiently small. Normally only three or four iterations are required.

Once a satisfactory approximation for w_1 has been determined, the method is repeated. In general, once w_j has been determined, Newton's method with initial approximation $w_{j+1}^{(0)} = w_j$ is applied to the equation

$$0 = F(y) = y - w_j - \frac{h}{2}[f(t_j, w_j) + f(t_{j+1}, y)]$$

to approximate the solution $y = w_{j+1}$. This is the procedure incorporated in TRAPNT58, which implements this technique.

The Secant method can be used as an alternative to Newton's method, but then two distinct initial approximations to w_{j+1} are required. To determine these, the usual practice is to let $w_{j+1}^{(0)} = w_j$ and obtain $w_{j+1}^{(1)}$ from some explicit multistep method. When a system of stiff equations is involved, a generalization is required for either Newton's or the Secant method. These topics are considered in Chapter 10.

EXAMPLE 2   The stiff initial-value problem

$$y' = 5e^{5t}(y - t)^2 + 1, \qquad 0 \le t \le 1, \qquad y(0) = -1$$

has solution $y(t) = t - e^{-5t}$. To show the effects of stiffness, the Trapezoidal method and the Runge–Kutta fourth-order method are applied both with $N = 4$, $h = 0.25$ and with $N = 5$, $h = 0.20$. The Trapezoidal method performs well in both cases using $M = 10$ and $TOL = 10^{-6}$, as does Runge–Kutta with $h = 0.2$. However, with $h = 0.25$ the Runge–Kutta method is unstable, which is evident from the results in Table 5.18.   ■ ■ ■

Table 5.18

	Runge–Kutta Method		Trapezoidal Method	
	$h = 0.2$		$h = 0.2$	
t_i	w_i	$|y(t_i) - w_i|$	w_i	$|y(t_i) - w_i|$
0.0	−1.0000000	0	−1.0000000	0
0.2	−0.1488521	1.9027×10^{-2}	−0.1414969	2.6383×10^{-2}
0.4	0.2684884	3.8237×10^{-3}	0.2748614	1.0197×10^{-2}
0.6	0.5519927	1.7798×10^{-3}	0.5539828	3.7700×10^{-3}
0.8	0.7822857	6.0131×10^{-4}	0.7830720	1.3876×10^{-3}
1.0	0.9934905	2.2845×10^{-4}	0.9937726	5.1050×10^{-4}
	$h = 0.25$		$h = 0.25$	
t_i	w_i	$|y(t_i) - w_i|$	w_i	$|y(t_i) - w_i|$
0.0	−1.0000000	0	−1.0000000	0
0.25	0.4014315	4.37936×10^{-1}	0.0054557	4.1961×10^{-2}
0.5	3.4374753	3.01956×10^{0}	0.4267572	8.8422×10^{-3}
0.75	1.44639×10^{23}	1.44639×10^{23}	0.7291528	2.6706×10^{-3}
1.0	Overflow		0.9940199	7.5790×10^{-4}

EXERCISE SET 5.8

1. Solve the following stiff initial-value problems using Euler's method and compare the results with the actual solution:

    (a) $y' = -9y$, $\quad 0 \le t \le 1$, $\quad y(0) = e$ with $h = 0.1$, actual solution $y(t) = e^{1-9t}$.

    (b) $y' = -8(y - t) + 1$, $\quad 0 \le t \le 2$, $\quad y(0) = 2$ with $h = 0.2$, actual solution $y(t) = t + 2e^{-8t}$.

(c) $y' = -20(y - t^2) + 2t$, $\quad 0 \le t \le 1$, $\quad y(0) = \frac{1}{3}$ with $h = 0.1$, actual solution
$y(t) = t^2 + \frac{1}{3}e^{-20t}$.

(d) $y' = -20y + 20 \sin t + \cos t$, $\quad 0 \le t \le 2$, $\quad y(0) = 1$ with $h = 0.25$, actual
solution $y(t) = \sin t + e^{-20t}$.

(e) $y' = -5y + \frac{1}{t}e^{-5t} + 5\cos(t - 1) - \sin(t - 1)$, $\quad 1 \le t \le 2$, $\quad y(1) = 1$ with $h = 0.1$, actual solution $y(t) = \cos(t - 1) + e^{-5t}\ln t$.

(f) $y' = \frac{50}{y} - 50y$, $\quad 0 \le t \le 1$, $\quad y(0) = \sqrt{2}$ with $h = 0.1$, actual solution $y(t) = (1 + e^{-100t})^{1/2}$.

2. Repeat Exercise 1 using the Runge–Kutta fourth-order method.

3. Repeat Exercise 1 using the Adams fourth-order predictor-corrector method.

4. Repeat Exercise 1 using the Implicit Trapezoidal method incorporating Newton's method to solve for w_{i+1}.

5. In Exercise 12 of Section 5.2, the differential equation

$$\frac{dp(t)}{dt} = rb(1 - p(t))$$

was obtained as a model for studying the proportion $p(t)$ of nonconformists in a society whose birth rate was b and where r represented the rate at which offspring would become nonconformists when at least one of their parents was a conformist. That exercise required that an approximation for $p(t)$ be found by using Euler's method for integral values of t when given $p(0) = 0.01$, $b = 0.02$, and $r = 0.1$, and that the approximation for $p(50)$ be compared with the actual value. Use the Trapezoidal method to obtain another approximation for $p(50)$, again assuming that $h = 1$ year.

5.9   Survey of Methods and Software

In this chapter we have considered methods for approximating the solutions to initial value problems for ordinary differential equations. We began with a discussion of the most elementary numerical technique, Euler's method. This procedure is not sufficiently accurate to be of use in applications, but it illustrates the general behavior of the more powerful techniques, without the accompanying algebraic difficulties. The Taylor methods were then considered as generalizations of Euler's method. They were found to be accurate but cumbersome because of the need to determine extensive partial derivatives of the differential equation. The Runge–Kutta formulas simplify the Taylor methods and do not significantly increase the error. To this point we had considered only one-step methods, techniques that use only data at the most recently computed point.

Multistep methods were discussed next. Explicit methods of the Adams–Bashforth type and implicit methods of the Adams–Moulton type were considered. These culminate in Predictor-Corrector methods, which use an explicit method, such as an Adams–Bashforth, to predict the solution and then apply a corresponding implicit method, like an Adams–Moulton, to correct the approximation.

These one- and multistep methods serve as an introduction to numerical methods for ordinary differential equations, since the more accurate adaptive methods are based on these relatively uncomplicated techniques. In particular, we saw that the Runge–Kutta–

Fehlberg method is a one-step procedure that seeks to select mesh spacing to keep the local error of the approximation under control. Our Variable Step-Size Predictor-Corrector method, based on the four-step Adams–Bashforth method and three-step Adams–Moulton method, also changes the step size in order to keep the local error within a given tolerance. Finally, the Extrapolation method is based on a modification of the Midpoint method and incorporates extrapolation to maintain a desired accuracy of approximation.

The final topic in the chapter concerns the difficulty that is inherent in the approximation of the solution to a stiff equation, a differential equation whose exact solution contains a portion of the form $e^{-\lambda t}$, where λ is a large positive constant. Special caution must be taken with problems of this type or the results can be overwhelmed by round-off error.

Methods of Runge–Kutta–Fehlberg type are generally sufficient for nonstiff problems where moderate accuracy is required. The extrapolation procedures are recommended for nonstiff problems where high accuracy is required. Finally, variable-order, variable step-size implicit Adams-type methods are used for stiff initial-value problems.

The IMSL Library includes three subroutines for approximating the solutions of initial-value problems. Each of the methods solves a system of m first-order equations in m variables. The equations are of the form

$$\frac{du_i}{dt} = f_i(t, u_1, u_2, \ldots, u_m), \qquad \text{for } i = 1, 2, \ldots, m,$$

where $u_i(x_0)$ is given for each i. The variable step-size subroutine IVPRK is based on the Runge–Kutta–Verner fifth- and sixth-order methods. The subroutine IVPBS is an extrapolation procedure based on the Bulirsch–Stoer extrapolation method. This technique uses extrapolation by rational functions instead of polynomial extrapolation, which we considered in Section 5.5. A subroutine of Adams type to be used for stiff equations is given in IVPAG. This method uses implicit multistep methods of order up to 12 and backward differentiation formulas of order up to 5.

The Runge–Kutta-type procedures contained in the NAG Library are called DO2BAF and DO2BBF and are based on the Merson form of the Runge–Kutta method. A variable-order and variable-step Adams method is contained in the procedures DO2CAF and DO2CBF. Variable-order, variable-step backward-difference formula methods for stiff systems are contained in the procedures DO2EAF and DO2EBF. The NAG Library contains numerous other subroutines for the numerical solution of initial-value problems.

Direct Methods for Solving Linear Systems

6.1 Introduction

Systems of equations are used to analytically represent physical problems that involve the interaction of various properties. The variables in the system quantitatively represent the properties being studied and the equations describe the interaction between the variables. The system is easiest to study when the equations are all linear. Often the number of equations is the same as the number of variables, for only in this case is it likely that a unique solution will exist.

Although not all physical problems can be reasonably represented using a linear system with the same number of equations as unknowns, the solutions to many problems either have this form or can be approximated by such a system. In fact, this is quite often the only approach that can give quantitative information about a physical problem.

In this chapter we consider direct methods for approximating the solution of a system of n linear equations in n unknowns. A direct method is one that gives the exact solution to the system, if it is assumed that all calculations can be performed without round-off error effects. This is an idealized assumption. We will need to consider quite carefully the role of finite-digit arithmetic error in the approximation to the solution to the system and how to arrange the calculations to minimize its effect.

6.2 Gaussian Elimination

If you have had any exposure to linear algebra or matrix theory, you have almost certainly been introduced to Gaussian elimination, the most elementary method for systematically determining the solution to a system of linear equations. Variables are eliminated from the various equations until one equation involves only one variable, another equation involves only that variable and one other, a third has only these two and one additional, and so on. The solution is found by solving for the variable in the single equation, using this to reduce the second equation to one that now contains a single variable, and so on, until values for all the variables are found.

Three operations are permitted on a system of equations (E_n):

Operations on Systems of Equations

1. Equation E_i can be multiplied by any nonzero constant λ and the resulting equation can be used in place of E_i. This operation is denoted $(\lambda E_i) \to (E_i)$.

2. Equation E_j can be multiplied by any constant λ, and added to equation E_i, and the resulting equation can be used in place of E_i. This operation is denoted $(E_i + \lambda E_j) \to (E_i)$.

3. Equations E_i and E_j can be transposed in order. This operation is denoted $(E_i) \leftrightarrow (E_j)$.

By a sequence of these operations, a linear system can be transformed to a more easily solved linear system with the same solutions. The sequence of operations is illustrated in the next example.

EXAMPLE 1   The four equations

$$
\begin{aligned}
E_1: \quad & x_1 + x_2 \quad\quad\quad + 3x_4 = 4, \\
E_2: \quad & 2x_1 + x_2 - x_3 + x_4 = 1, \\
E_3: \quad & 3x_1 - x_2 - x_3 + 2x_4 = -3, \\
E_4: \quad & -x_1 + 2x_2 + 3x_3 - x_4 = 4,
\end{aligned}
$$

will be solved for $x_1, x_2, x_3,$ and x_4. First use equation E_1 to eliminate the unknown x_1 from $E_2, E_3,$ and E_4 by performing $(E_2 - 2E_1) \to (E_2), (E_3 - 3E_1) \to (E_3),$ and $(E_4 + E_1) \to (E_4)$. The resulting system is:

$$
\begin{aligned}
E_1: \quad & x_1 + x_2 \quad\quad\quad + 3x_4 = 4, \\
E_2: \quad & - x_2 - x_3 - 5x_4 = -7, \\
E_3: \quad & - 4x_2 - x_3 - 7x_4 = -15, \\
E_4: \quad & 3x_2 + 3x_3 + 2x_4 = 8,
\end{aligned}
$$

where, for simplicity, the new equations are again labeled $E_1, E_2, E_3,$ and E_4.

In the new system, E_2 is used to eliminate x_2 from E_3 and E_4 by $(E_3 - 4E_2) \to (E_3)$ and $(E_4 + 3E_2) \to (E_4)$, resulting in

$$
\begin{aligned}
E_1: \quad & x_1 + x_2 \quad\quad\quad + 3x_4 = 4, \\
E_2: \quad & - x_2 - x_3 - 5x_4 = -7, \\
E_3: \quad & 3x_3 + 13x_4 = 13, \\
E_4: \quad & - 13x_4 = -13.
\end{aligned}
$$

The system of equations is now in *triangular* (or *reduced*) form and can be solved for the unknowns by a backward-substitution process. Noting that E_4 implies $x_4 = 1, E_3$ can be solved for x_3:

$$x_3 = \tfrac{1}{3}(13 - 13x_4) = \tfrac{1}{3}(13 - 13) = 0.$$

Continuing, E_2 gives

$$x_2 = -(-7 + 5x_4 + x_3) = -(-7 + 5 + 0) = 2;$$

and E_1 gives

$$x_1 = 4 - 3x_4 - x_2 = 4 - 3 - 2 = -1.$$

The solution is, therefore, $x_1 = -1$, $x_2 = 2$, $x_3 = 0$, and $x_4 = 1$. These values solve the original system of equations. ■ ■ ■

When performing the calculations of Example 1, we did not need to write out the full equations at each step or to carry the variables x_1, x_2, x_3, and x_4 through the calculations, since they always remained in the same column. The only variation from system to system occurred in the coefficients of the unknowns and in the values on the right side of the equations. For this reason, a linear system is often replaced by a **matrix**, a rectangular array of elements in which not only is the value of an element important, but also its position in the array. The matrix contains all the information about the system that is necessary to determine its solution, but in a compact form.

The notation for an $n \times m$ (n by m) matrix will be a capital letter, such as A, for the matrix and lowercase letters with double subscripts, such as a_{ij}, to refer to the entry at the intersection of the ith row and jth column; that is,

$$A = (a_{ij}) = \begin{bmatrix} a_{11} & a_{12} & \cdots & a_{1m} \\ a_{21} & a_{22} & \cdots & a_{2m} \\ \vdots & \vdots & & \vdots \\ a_{n1} & a_{n2} & \cdots & a_{nm} \end{bmatrix}.$$

EXAMPLE 2   The matrix

$$A = \begin{bmatrix} 2 & -1 & 7 \\ 3 & 1 & 0 \end{bmatrix}$$

is a 2×3 matrix with $a_{11} = 2$, $a_{12} = -1$, $a_{13} = 7$, $a_{21} = 3$, $a_{22} = 1$, and $a_{23} = 0$. ■ ■ ■

The $1 \times n$ matrix $A = [a_{11} \quad a_{12} \cdots a_{1n}]$ is called an ***n*-dimensional row vector**, and an $n \times 1$ matrix

$$A = \begin{bmatrix} a_{11} \\ a_{21} \\ \vdots \\ a_{n1} \end{bmatrix}$$

is called an **n-dimensional column vector**. Usually the unnecessary subscript is omitted for vectors and a boldface lowercase letter is used for notation. So,

$$\mathbf{x} = \begin{bmatrix} x_1 \\ x_2 \\ \vdots \\ x_n \end{bmatrix}$$

denotes a column vector, and $\mathbf{y} = [y_1 \; y_2 \; \cdots \; y_n]$, a row vector.

An n by $(n + 1)$ matrix can be used to represent the linear system

$$a_{11}x_1 + a_{12}x_2 + \cdots + a_{1n}x_n = b_1,$$

$$a_{21}x_1 + a_{22}x_2 + \cdots + a_{2n}x_n = b_2,$$

$$\vdots$$

$$a_{n1}x_1 + a_{n2}x_2 + \cdots + a_{nn}x_n = b_n,$$

by first constructing

$$A = (a_{ij}) = \begin{bmatrix} a_{11} & a_{12} & \cdots & a_{1n} \\ a_{21} & a_{22} & \cdots & a_{2n} \\ \vdots & \vdots & & \vdots \\ a_{n1} & a_{n2} & \cdots & a_{nn} \end{bmatrix} \quad \text{and} \quad \mathbf{b} = \begin{bmatrix} b_1 \\ b_2 \\ \vdots \\ b_n \end{bmatrix}$$

and then combining these matrices to form the *augmented matrix*:

$$[A, \mathbf{b}] = \begin{bmatrix} a_{11} & a_{12} & \cdots & a_{1n} & \vdots & b_1 \\ a_{21} & a_{22} & \cdots & a_{2n} & \vdots & b_2 \\ \vdots & \vdots & & \vdots & \vdots & \vdots \\ a_{n1} & a_{n2} & \cdots & a_{nn} & \vdots & b_n \end{bmatrix},$$

where the vertical dotted line is used to separate the coefficients of the unknowns from the values on the right-hand side of the equations.

Repeating the operations involved in Example 1 with the matrix notation results in first considering the augmented matrix

$$\begin{bmatrix} 1 & 1 & 0 & 3 & \vdots & 4 \\ 2 & 1 & -1 & 1 & \vdots & 1 \\ 3 & -1 & -1 & 2 & \vdots & -3 \\ -1 & 2 & 3 & -1 & \vdots & 4 \end{bmatrix}.$$

Performing the operations as described in that example produces the matrices

$$\begin{bmatrix} 1 & 1 & 0 & 3 & \vdots & 4 \\ 0 & -1 & -1 & -5 & \vdots & -7 \\ 0 & -4 & -1 & -7 & \vdots & -15 \\ 0 & 3 & 3 & 2 & \vdots & 8 \end{bmatrix} \quad \text{and} \quad \begin{bmatrix} 1 & 1 & 0 & 3 & \vdots & 4 \\ 0 & -1 & -1 & -5 & \vdots & -7 \\ 0 & 0 & 3 & 13 & \vdots & 13 \\ 0 & 0 & 0 & -13 & \vdots & -13 \end{bmatrix}.$$

The latter matrix can now be transformed into its corresponding linear system and solutions for x_1, x_2, x_3, and x_4 obtained. The procedure involved in this process is called **Gaussian Elimination with Backward Substitution**.

The general Gaussian elimination procedure applied to the linear system

$$E_1 : \quad a_{11}x_1 + a_{12}x_2 + \cdots + a_{1n}x_n = b_1,$$
$$E_2 : \quad a_{21}x_1 + a_{22}x_2 + \cdots + a_{2n}x_n = b_2,$$
$$\vdots$$
$$E_n : \quad a_{n1}x_1 + a_{n2}x_2 + \cdots + a_{nn}x_n = b_n,$$

is handled in a similar manner. First form the augmented matrix $\tilde{A}$:

$$\tilde{A} = [A, \mathbf{b}] = \begin{bmatrix} a_{11} & a_{12} & \cdots & a_{1n} & \vdots & a_{1,n+1} \\ a_{21} & a_{22} & \cdots & a_{2n} & \vdots & a_{2,n+1} \\ \vdots & \vdots & & \vdots & \vdots & \vdots \\ a_{n1} & a_{n2} & \cdots & a_{nn} & \vdots & a_{n,n+1} \end{bmatrix},$$

where A denotes the matrix formed by the coefficients and the entries in the $(n + 1)$st column are the values of $\mathbf{b}$; that is, $a_{i,n+1} = b_i$ for each $i = 1, 2, \ldots, n$.

Suppose the **pivot element** $a_{11} \neq 0$. To simplify the discussion we introduce the **multiplier** $m_{k1} = a_{k1}/a_{11}$ and perform the operations corresponding to $(E_k - m_{k1}E_1) \to (E_k)$ for each $k = 2, 3, \ldots, n$ to eliminate (that is, change to zero) the coefficient of x_1 in each of these rows. Although the entries in rows $2, 3, \ldots, n$ are expected to change, for ease of notation, we again denote the entry in the ith row and the jth column by a_{ij}.

We then follow this sequential procedure for the row $i = 2, 3 \ldots, n - 1$. Define the multiplier $m_{ki} = a_{ki}/a_{ii}$ and perform the operation

$$(E_k - m_{ki}E_i) \to (E_k)$$

for each $k = i + 1, i + 2, \ldots, n$, provided the pivot element $a_{ii} \neq 0$. This eliminates x_i in each row below the ith for all values of $i = 1, 2, \ldots, n - 1$. The resulting matrix has the form

$$\tilde{A} = \begin{bmatrix} a_{11} & a_{12} & \cdots & a_{1n} & \vdots & a_{1,n+1} \\ 0 & a_{22} & \cdots & a_{2n} & \vdots & a_{2,n+1} \\ \vdots & \ddots & \ddots & \vdots & \vdots & \vdots \\ 0 & \cdots & 0 & a_{n,n} & \vdots & a_{n,n+1} \end{bmatrix},$$

where, except in the first row, the values of a_{ij} are not expected to agree with those in the original matrix $\tilde{A}$. This matrix represents a linear system with the same solution set as the original system. Since the new linear system is triangular,

$$a_{11}x_1 + a_{12}x_2 + \cdots + a_{1n}x_n = a_{1,n+1},$$
$$a_{22}x_2 + \cdots + a_{2n}x_n = a_{2,n+1},$$
$$\vdots$$
$$a_{nn}x_n = a_{n,n+1},$$

backward substitution can be performed. Solving the nth equation for x_n gives

$$x_n = \frac{a_{n,n+1}}{a_{nn}}.$$

Then solving the $(n-1)$st equation for x_{n-1} and using x_n yields

$$x_{n-1} = \frac{a_{n-1,n+1} - a_{n-1,n}x_n}{a_{n-1,n-1}}.$$

Continuing this process, we obtain

$$x_i = \frac{a_{i,n+1} - a_{i,n}x_n - a_{i,n-1}x_{n-1} - \cdots - a_{i,i+1}x_{i+1}}{a_{ii}} = \frac{a_{i,n+1} - \sum_{j=i+1}^{n} a_{ij}x_j}{a_{ii}}$$

for each $i = n-1, n-2, \ldots, 2, 1$.

The procedure will not work if at the ith step the pivot element a_{ii} is zero, for then either the multipliers $m_{ki} = a_{ki}/a_{ii}$ are not defined (this occurs if $a_{ii} = 0$ for $1 \le i < n$), or the backward substitution cannot be performed (if $a_{nn} = 0$). This does not necessarily mean that the system has no solution, but rather that the technique for finding the solution must be altered. An illustration is given in the following example.

EXAMPLE 3 Consider the linear system

$$\begin{aligned}
E_1: \quad x_1 - x_2 + 2x_3 - x_4 &= -8, \\
E_2: \quad 2x_1 - 2x_2 + 3x_3 - 3x_4 &= -20, \\
E_3: \quad x_1 + x_2 + x_3 &= -2, \\
E_4: \quad x_1 - x_2 + 4x_3 + 3x_4 &= 4.
\end{aligned}$$

The augmented matrix is

$$\begin{bmatrix}
1 & -1 & 2 & -1 & \vdots & -8 \\
2 & -2 & 3 & -3 & \vdots & -20 \\
1 & 1 & 1 & 0 & \vdots & -2 \\
1 & -1 & 4 & 3 & \vdots & 4
\end{bmatrix}.$$

Performing the operations

$$(E_2 - 2E_1) \to (E_2), \qquad (E_3 - E_1) \to (E_3), \qquad \text{and} \qquad (E_4 - E_1) \to (E_4),$$

we have

$$\begin{bmatrix}
1 & -1 & 2 & -1 & \vdots & -8 \\
0 & 0 & -1 & -1 & \vdots & -4 \\
0 & 2 & -1 & 1 & \vdots & 6 \\
0 & 0 & 2 & 4 & \vdots & 12
\end{bmatrix}.$$

Since the element a_{22} in this matrix is zero, the procedure cannot continue in its present form. But the operation $(E_i) \leftrightarrow (E_p)$ is permitted, so a search is made of the elements

a_{32} and a_{42} for the first nonzero element. Since $a_{32} \neq 0$, the operation $(E_2) \leftrightarrow (E_3)$ is performed to obtain a new matrix:

$$\begin{bmatrix} 1 & -1 & 2 & -1 & \vdots & -8 \\ 0 & 2 & -1 & 1 & \vdots & 6 \\ 0 & 0 & -1 & -1 & \vdots & -4 \\ 0 & 0 & 2 & 4 & \vdots & 12 \end{bmatrix}.$$

The variable x_2 is already eliminated from E_3 and E_4, so the computations continue with the operation $(E_4 + 2E_3) \rightarrow (E_4)$, giving

$$\begin{bmatrix} 1 & -1 & 2 & -1 & \vdots & -8 \\ 0 & 2 & -1 & 1 & \vdots & 6 \\ 0 & 0 & -1 & -1 & \vdots & -4 \\ 0 & 0 & 0 & 2 & \vdots & 4 \end{bmatrix}.$$

Finally, the backward substitution is applied:

$$x_4 = \tfrac{4}{2} = 2, \qquad x_3 = \frac{[-4 - (-1)x_4]}{-1} = 2,$$

$$x_2 = \frac{[6 - x_4 - (-1)x_3]}{2} = 3, \qquad x_1 = \frac{[-8 - (-1)x_4 - 2x_3 - (-1)x_2]}{1} = -7.$$

■ ■ ■

Example 3 illustrates what is done if one of the pivot elements is zero. If the ith pivot element is zero, the ith column of the matrix is searched from the ith row downward for the first nonzero entry and then a row interchange is performed to obtain the new matrix. Then the procedure continues as before. If no nonzero entry is found, the procedure stops, and the linear system does not have a unique solution. The program GAUSEL61 implements Gaussian Elimination with Backward Substitution and incorporates pivoting when required.

The computations in the program are performed using only one n by $(n + 1)$ array for storage by replacing at each step the previous value of a_{ij} by the new one. In addition, the multipliers are stored in the locations of a_{ki} known to have zero values—that is, when $i < n$ and $k = i + 1, i + 2, \ldots, n$. Thus, the original matrix A is overwritten by the multipliers below the main diagonal and by the nonzero entries of the final reduced matrix on and above the main diagonal. These values can be used to solve other linear systems involving the original matrix A, as we will see in Section 6.5.

Both the amount of time required to complete the calculations and the subsequent round-off error depend on the number of floating-point arithmetic operations needed to solve a routine problem. In general, the amount of time required to perform a multiplication or division on a computer is approximately the same and is considerably greater than that required to perform an addition or subtraction. Even though the actual differences in execution time depend on the particular computing system being used, the count of the additions/subtractions are kept separate from the count of the multiplications/divisions because of the time differential. The total number of arithmetic operations depend on the size n, as follows:

Multiplications/divisions

$$\frac{n^3}{3} + n^2 - \frac{n}{3}.$$

Additions/subtractions

$$\frac{n^3}{3} + \frac{n^2}{2} - \frac{5n}{6}.$$

For large n, the total number of multiplications and divisions is approximately $n^3/3$, that is, $O(n^3)$, as is the total number of additions and subtractions. Thus, the amount of computation and the time required increases with n in proportion to n^3, as shown in Table 6.1.

Table 6.1

n	Multiplications/Divisions	Additions/Subtractions
3	17	11
10	430	375
50	44150	42875
100	343300	338250

EXERCISE SET 6.2

1. For each of the following linear systems, obtain a solution by graphical methods if possible. Explain the results from a geometrical standpoint.

(a) $x_1 + 2x_2 = 3$
    $x_1 - x_2 = 0$

(b) $x_1 + 2x_2 = 0$
    $x_1 - x_2 = 0$

(c) $x_1 + 2x_2 = 3$
    $2x_1 + 4x_2 = 6$

(d) $x_1 + 2x_2 = 3$
    $-2x_1 - 4x_2 = 6$

(e) $x_1 + 2x_2 = 0$
    $2x_1 + 4x_2 = 0$

(f) $0 \cdot x_1 + x_2 = 3$
    $2x_1 - x_2 = 7$

(g) $2x_1 + x_2 = -1$
    $x_1 + x_2 = 2$
    $x_1 - 3x_2 = 5$

(h) $2x_1 + x_2 = -1$
    $2x_1 + x_2 = 2$
    $x_1 - 3x_2 = 5$

(i) $2x_1 + x_2 = -1$
    $4x_1 + 2x_2 = -2$
    $x_1 - 3x_2 = 5$

(j) $2x_1 + x_2 + x_3 = 1$
    $2x_1 + 4x_2 - x_3 = -1$

2. Use Gaussian elimination with backward substitution and two-digit rounding arithmetic to solve the following linear systems. Do not reorder the equations. (The exact solution to each system is $x_1 = 1, x_2 = -1, x_3 = 3$.)

(a) $4x_1 - x_2 + x_3 = 8$
    $2x_1 + 5x_2 + 2x_3 = 3$
    $x_1 + 2x_2 + 4x_3 = 11$

(b) $x_1 + 2x_2 + 4x_3 = 11$
    $4x_1 - x_2 + x_3 = 8$
    $2x_1 + 5x_2 + 2x_3 = 3$

(c) $4x_1 + x_2 + 2x_3 = 9$
    $2x_1 + 4x_2 - x_3 = -5$
    $x_1 + x_2 - 3x_3 = -9$

(d) $2x_1 + 4x_2 - x_3 = -5$
    $x_1 + x_2 - 3x_3 = -9$
    $4x_1 + x_2 + 2x_3 = 9$

3. Use Gaussian elimination with backward substitution to solve the following linear systems, if possible, and determine whether row interchanges are necessary.

   (a) $x_1 - x_2 + 3x_3 = 2$
   $3x_1 - 3x_2 + x_3 = -1$
   $x_1 + x_2 = 3$

   (b) $x_2 + 4x_3 = 0$
   $x_1 - x_2 - x_3 = 0.375$
   $x_1 - x_2 + 2x_3 = 0$

   (c) $2x_1 - 1.5x_2 + 3x_3 = 1$
   $-x_1 + 2x_3 = 3$
   $4x_1 - 4.5x_2 + 5x_3 = 1$

   (d) $2x_1 - x_2 + x_3 = -1$
   $3x_1 + 3x_2 + 9x_3 = 0$
   $3x_1 + 3x_2 + 5x_3 = 4$

   (e) $2x_1 = 3$
   $x_1 + 1.5x_2 = 4.5$
   $- 3x_2 + 0.5x_3 = -6.6$
   $2x_1 - 2x_2 + x_3 + x_4 = 0.8$

   (f) $x_1 - \frac{1}{2}x_2 + x_3 = 4$
   $2x_1 - x_2 - x_3 + x_4 = 5$
   $x_1 + x_2 = 2$
   $x_1 - \frac{1}{2}x_2 + x_3 + x_4 = 5$

   (g) $x_1 + x_2 + x_4 = 2$
   $2x_1 + x_2 - x_3 + x_4 = 1$
   $4x_1 - x_2 - 2x_3 + 2x_4 = 0$
   $3x_1 - x_2 - x_3 + 2x_4 = -3$

   (h) $x_1 + x_2 + x_4 = 2$
   $2x_1 + x_2 - x_3 + x_4 = 1$
   $-x_1 + 2x_2 + 3x_3 - x_4 = 4$
   $3x_1 - x_2 - x_3 + 2x_4 = -3$

4. Use Gaussian elimination with backward substitution and, if available, single-precision arithmetic on a computer to solve the following linear systems.

   (a) $\frac{1}{4}x_1 + \frac{1}{5}x_2 + \frac{1}{6}x_3 = 9$
   $\frac{1}{3}x_1 + \frac{1}{4}x_2 + \frac{1}{5}x_3 = 8$
   $\frac{1}{2}x_1 + x_2 + 2x_3 = 8$

   (b) $3.333x_1 + 15920x_2 - 10.333x_3 = 15913$
   $2.222x_1 + 16.71x_2 + 9.612x_3 = 28.544$
   $1.5611x_1 + 5.1791x_2 + 1.6852x_3 = 8.4254$

   (c) $4.01x_1 + 1.23x_2 + 1.43x_3 - 0.73x_4 = 5.94$
   $1.23x_1 + 7.41x_2 + 2.41x_3 + 3.02x_4 = 14.07$
   $1.43x_1 + 2.41x_2 + 5.79x_3 - 1.11x_4 = 8.52$
   $-0.73x_1 + 3.02x_2 - 1.11x_3 + 6.41x_4 = 7.59$

   (d) $2x_1 + x_2 - x_3 + x_4 - 3x_5 = 7$
   $x_1 + 2x_3 - x_4 + x_5 = 2$
   $- 2x_2 - x_3 + x_4 - x_5 = -5$
   $3x_1 + x_2 - 4x_3 + 5x_5 = 6$
   $x_1 - x_2 - x_3 - x_4 + x_5 = 3$

   (e) $x_1 + \frac{1}{2}x_2 + \frac{1}{3}x_3 + \frac{1}{4}x_4 = \frac{1}{6}$
   $\frac{1}{2}x_1 + \frac{1}{3}x_2 + \frac{1}{4}x_3 + \frac{1}{5}x_4 = \frac{1}{7}$
   $\frac{1}{3}x_1 + \frac{1}{4}x_2 + \frac{1}{5}x_3 + \frac{1}{6}x_4 = \frac{1}{8}$
   $\frac{1}{4}x_1 + \frac{1}{5}x_2 + \frac{1}{6}x_3 + \frac{1}{7}x_4 = \frac{1}{9}$

   (f) $x_1 + \frac{1}{2}x_2 - \frac{1}{3}x_3 - \frac{1}{4}x_4 + \frac{1}{5}x_5 = 1$
   $\frac{1}{2}x_1 + x_2 - \frac{1}{3}x_3 + \frac{1}{4}x_4 - \frac{1}{5}x_5 = 0$
   $\frac{1}{3}x_1 + \frac{1}{4}x_2 - \frac{1}{5}x_3 + \frac{1}{6}x_4 - \frac{1}{7}x_5 = 1$
   $\frac{1}{4}x_1 + \frac{1}{5}x_2 - \frac{1}{6}x_3 - \frac{1}{7}x_4 + x_5 = 0$
   $\frac{1}{5}x_1 - \frac{1}{3}x_2 + \frac{1}{7}x_3 + \frac{1}{9}x_4 + x_5 = 1$

5. Suppose that in a biological system there are n species of animals and m sources of food. Let x_j represent the population of the jth species for each $j = 1, \ldots, n$; b_i represent the available daily supply of the ith food; and a_{ij} represent the amount of the ith food consumed

on the average by a member of the jth species. The linear system

$$a_{11}x_1 + a_{12}x_2 + \cdots + a_{1n}x_n = b_1,$$
$$a_{21}x_1 + a_{22}x_2 + \cdots + a_{2n}x_n = b_2,$$
$$\vdots$$
$$a_{m1}x_1 + a_{m2}x_2 + \cdots + a_{mn}x_n = b_m,$$

represents an equilibrium where there is a daily supply of food to precisely meet the average daily consumption of each species.

(a) Let

$$A = (a_{ij}) = \begin{bmatrix} 1 & 2 & 0 & 3 \\ 1 & 0 & 2 & 2 \\ 0 & 0 & 1 & 1 \end{bmatrix},$$

$\mathbf{x} = (x_j) = [1000, 500, 350, 400]$, and $\mathbf{b} = (b_i) = [3500, 2700, 900]$. Is there sufficient food to satisfy the average daily consumption?

(b) What is the maximum number of animals of each species that could be individually added to the system with the supply of food still meeting the consumption?

(c) If species 1 became extinct, how much individual increase of each of the remaining species could be supported?

(d) If species 2 became extinct, how much individual increase of each of the remaining species could be supported?

6. Consider the 2×2 linear system $(A + iB)(\mathbf{x} + i\mathbf{y}) = \mathbf{c} + i\mathbf{d}$ with complex entries in component form:

$$(a_{11} + ib_{11})(x_1 + iy_1) + (a_{12} + ib_{12})(x_2 + iy_2) = c_1 + id_1,$$
$$(a_{21} + ib_{21})(x_1 + iy_1) + (a_{22} + ib_{22})(x_2 + iy_2) = c_2 + id_2.$$

(a) Use the properties of complex numbers to convert this system to the equivalent 4×4 real linear system:

$$A\mathbf{x} - B\mathbf{y} = \mathbf{c}$$
$$B\mathbf{x} + A\mathbf{y} = \mathbf{d}.$$

(b) Solve the linear system

$$(1 - 2i)(x_1 + iy_1) + (3 + 2i)(x_2 + iy_2) = 5 + 2i,$$
$$(2 + i)(x_1 + iy_1) + (4 + 3i)(x_2 + iy_2) = 4 - i.$$

7. A Fredholm integral equation of the second kind is an equation of the form

$$u(x) = f(x) + \int_a^b K(x, t)u(t)\, dt,$$

where a and b and the functions f and K are given. To approximate the function u on the interval $[a, b]$, a partition $x_0 = a < x_1 < \cdots < x_{m-1} < x_m = b$ is selected and the equations

$$u(x_i) = f(x_i) + \int_a^b K(x_i, t)u(t)\, dt, \qquad \text{for each } i = 0, \ldots, m$$

are solved for $u(x_0), u(x_1), \ldots, u(x_m)$. The integrals are approximated using quadrature formulas based on the nodes $x_0, \ldots, x_m$. In our problem $a = 0, b = 1, f(x) = x^2$, and $K(x, t) = e^{|x-t|}$.

(a) Show that the linear system

$$u(0) = f(0) + \tfrac{1}{2}[K(0, 0)u(0) + K(0, 1)u(1)]$$
$$u(1) = f(1) + \tfrac{1}{2}[K(1, 0)u(0) + K(1, 1)u(1)]$$

must be solved when the Trapezoidal rule is used.

(b) Set up and solve the linear system that results when the Composite Trapezoidal rule is used with $n = 4$.

(c) Repeat part (b) using the Composite Simpson's rule.

6.3 Pivoting Strategies

If all the calculations could be done using exact arithmetic, we could nearly end the chapter with the previous section. We now know how many calculations are needed to perform Gaussian elimination on a system and from this we should be able to determine whether our computational device can solve our problem in reasonable time. In a practical situation, however, we do not have exact arithmetic, and the large number of arithmetic computations, on the order of $O(n^3)$, makes the consideration of computational round-off error necessary. In this section we see how the calculations in Gaussian elimination can be arranged to reduce the effect of this error.

In deriving the Gaussian elimination method, we found that a row interchange is needed when one of the pivot elements a_{ii} is zero. This row interchange has the form $(E_i) \leftrightarrow (E_p)$, where p is the smallest integer greater than i with $a_{pi} \neq 0$. To reduce the round-off error associated with finite-digit arithmetic, it is often necessary to perform row interchanges even when the pivot elements are not zero.

If a_{ii} is small in magnitude compared to a_{ki}, the magnitude of the multiplier,

$$m_{ki} = \frac{a_{ki}}{a_{ii}},$$

will be much larger than 1. A round-off error introduced in the computation of one of the terms a_{il} is multiplied by m_{ki} when computing a_{kl}, compounding the original error. Also, when performing the backward substitution for

$$x_i = \frac{a_{i,n+1} - \sum_{j=i+1}^{n} a_{ij}}{a_{ii}},$$

any round-off error in the numerator is dramatically increased when dividing by a_{ii}. An illustration of this difficulty is given in the following example.

EXAMPLE 1   The linear system

$$E_1 : \quad 0.003000x_1 + 59.14x_2 = 59.17,$$
$$E_2 : \quad 5.291x_1 - 6.130x_2 = 46.78,$$

has the solution $x_1 = 10.00$ and $x_2 = 1.000$. Suppose Gaussian elimination is performed on this system using four-digit arithmetic with rounding.

The first pivot element, $a_{11} = 0.003000$, is small, and its associated multiplier,

$$m_{21} = \frac{5.291}{0.003000} = 1763.\bar{6},$$

rounds to the large number 1764. Performing $(E_2 - m_{21}E_1) \rightarrow (E_2)$ and the appropriate rounding gives

$$0.003000x_1 + 59.14x_2 = 59.17$$
$$-104300x_2 \approx -104400,$$

instead of the precise values,

$$0.003000x_1 + 59.14x_2 = 59.17,$$
$$-104309.37\bar{6}x_2 = -104309.37\bar{6}.$$

The disparity in the magnitudes of $m_{21}a_{13}$ and a_{23} has introduced round-off error, but the error has not yet been propagated. Backward substitution yields

$$x_2 \approx 1.001,$$

which is a close approximation to the actual value, $x_2 = 1.000$. However, because of the small pivot $a_{11} = 0.003000$,

$$x_1 \approx \frac{59.17 - (59.14)(1.001)}{0.003000} = -10.00$$

contains the small error of 0.001 multiplied by $59.14/0.003000$. This ruins the approximation to the actual value $x_1 = 10.00$. (See Figure 6.1.)   ■   ■   ■

Figure 6.1

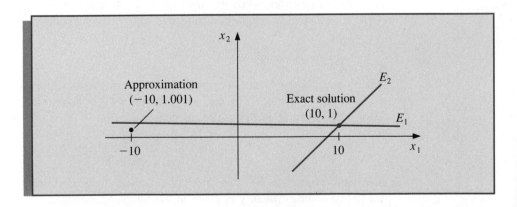

Example 1 shows how difficulties arise when the pivot element a_{ii} is small relative to the entries a_{kj} for $i \leq k \leq n$ and $i \leq j \leq n$. To avoid this problem, pivoting is performed by selecting a larger element a_{pq} for the pivot and interchanging the ith and pth rows, followed by the interchange of the ith and qth columns, if necessary. The simplest strategy is to select, at the ith step, the element in the same column that is below the diagonal and has the largest absolute value; that is, determine the smallest

$p \geq i$ such that

$$|a_{pi}| = \max_{i \leq k \leq n} |a_{ki}|,$$

and perform $(E_i) \leftrightarrow (E_p)$. In this case no interchange of columns is used.

EXAMPLE 2   Reconsider the system

$$E_1: \quad 0.003000x_1 + 59.14x_2 = 59.17,$$
$$E_2: \quad 5.291x_1 - 6.130x_2 = 46.78.$$

The pivoting procedure just described results in first finding

$$\max\{|a_{11}|, |a_{21}|\} = \max\{|0.003000|, |5.291|\} = |5.291| = |a_{21}|.$$

The operation $(E_2) \rightarrow (E_1)$ is then performed to give the system

$$E_1: \quad 5.291x_1 - 6.130x_2 = 46.78,$$
$$E_2: \quad 0.003000x_1 + 59.14x_2 = 59.17.$$

The multiplier for this system is

$$m_{21} = \frac{a_{21}}{a_{11}} = 0.0005670,$$

and the operation $(E_2 - m_{21}E_1) \rightarrow (E_2)$ reduces the system to

$$5.291x_1 - 6.130x_2 = 46.78,$$
$$59.14x_2 \approx 59.14.$$

The four-digit answers resulting from the backward substitution are the correct values $x_1 = 10.00$ and $x_2 = 1.000$. ■ ■ ■

The technique just described is called **maximal column pivoting**, or *partial pivoting*, and is implemented in the program GAUMCP62.

Although maximal column pivoting is sufficient for most linear systems, situations do arise when it is inadequate. For example, the linear system

$$E_1: \quad 30.00x_1 + 591400x_2 = 591700,$$
$$E_2: \quad 5.291x_1 - \quad 6.130x_2 = \quad 46.78,$$

is the same as that in Examples 1 and 2 except that all entries in the first equation have been multiplied by 10^4. Maximal column pivoting with four-digit arithmetic would lead to the same results as obtained in Example 1 since no row interchange would be performed. A technique known as **scaled-column pivoting** is needed for this system. The first step in this procedure is to define a scale factor s_k for each row:

$$s_k = \max_{j=1,2,\dots,n} |a_{kj}|.$$

The appropriate row interchange to place zeros in the first column is determined by choosing the first integer p with

$$\frac{|a_{p1}|}{s_p} = \max_{k=1,2,\ldots,n} \frac{|a_{k1}|}{s_k}$$

and performing $(E_1) \leftrightarrow (E_p)$. The effect of scaling is to ensure that the largest element in each row has a *relative* magnitude of 1 before the comparison for row interchange is performed. In the program GAUSCP63 the scaling is done only for comparison purposes, so the division by scaling factors produces no round-off error in the system.

EXAMPLE 3   Applying scaled-column pivoting to the system in Example 1 gives

$$s_1 = \max\{|30.00|, |591400|\} = 591400, \quad \text{and} \quad s_2 = \max\{|5.291|, |-6.130|\} = 6.130.$$

Consequently,

$$\frac{|a_{11}|}{s_1} = \frac{30.00}{591400} = 0.5073 \times 10^{-4} \quad \text{and} \quad \frac{|a_{21}|}{s_2} = \frac{5.291}{6.130} = 0.8631$$

and the interchange $(E_1) \to (E_2)$ is made. Applying Gaussian elimination to the new system produces the correct results: $x_1 = 10.00$ and $x_2 = 1.000$.   ∎ ∎ ∎

The scaled column pivoting procedure adds a total of

$$n(n-1) + \sum_{k=2}^{n}(k-1) = \frac{3}{2}n(n-1) \qquad \text{comparisons}$$

and

$$\sum_{k=2}^{n} k = \frac{n(n+1)}{2} - 1 \qquad \text{divisions}$$

to the Gaussian elimination procedure. The time required to perform a comparison is slightly more than that of an addition/subtraction. Since the total time to perform the basic Gaussian elimination procedure is $O(n^3/3)$ multiplications/divisions and $O(n^3/3)$ additions/subtractions, scaled-column pivoting does not add significantly to the computational time required to solve a system for large values of n.

If a system has computation difficulties that scaling pivoting cannot resolve, *maximal* (also called *total* or *full*) *pivoting* can be used. Maximal pivoting at the ith step searches all the entries a_{kj}, for $k = i, i+1, \ldots, n$ and $j = i, i+1, \ldots, n$, to find the entry with the largest magnitude. Both row and column interchanges are performed to bring this entry to the pivot position. The additional time required to incorporate maximal pivoting into Gaussian elimination is

$$\frac{n(n-1)(2n+5)}{6} \qquad \text{comparisons.}$$

This approximately doubles the amount of addition/subtraction time over ordinary Gaussian elimination.

E X E R C I S E   S E T   6.3

1. Use Gaussian elimination and three-digit chopping arithmetic to solve the following linear systems and compare the results to the actual solution.

   (a) $0.03x_1 + 58.9x_2 = 59.2$
   $5.31x_1 - 6.10x_2 = 47.0$
   Actual solution $(10, 1)$

   (b) $58.9x_1 + 0.03x_2 = 59.2$
   $-6.10x_1 + 5.31x_2 = 47.0$
   Actual solution $(1, 10)$

   (c) $3.03x_1 - 12.1x_2 + 14x_3 = -119$
   $-3.03x_1 + 12.1x_2 - 7x_3 = 120$
   $6.11x_1 - 14.2x_2 + 21x_3 = -139$
   Actual solution $\left(0, 10, \frac{1}{7}\right)$

   (d) $3.3330x_1 + 15920x_2 - 10.333x_3 = 7953$
   $2.2220x_1 + 16.710x_2 + 9.6120x_3 = 0.965$
   $-1.5611x_1 + 5.1792x_2 - 1.6855x_3 = 2.714$
   Actual solution $(1, 0.5, -1)$

   (e) $0.832x_1 + 0.448x_2 + 0.193x_3 = 1.00$

   $0.784x_1 - 0.421x_2 + 0.279x_3 = 0$

   $0.784x_1 + 0.421x_2 - 0.207x_3 = 0$
   Actual solution $(-0.11108022, 1.39628626, 2.41908030)$

   (f) $x_1 + \frac{1}{2}x_2 + \frac{1}{3}x_3 = 2$

   $\frac{1}{2}x_1 + \frac{1}{3}x_2 + \frac{1}{4}x_3 = -1$

   $\frac{1}{3}x_1 + \frac{1}{4}x_2 + \frac{1}{5}x_3 = 0$
   Actual solution $(54, -264, 240)$

   (g) $1.19x_1 + 2.11x_2 - 100x_3 + x_4 = 1.12$
   $14.2x_1 - 0.122x_2 + 12.2x_3 - x_4 = 3.44$
   $100x_2 - 99.9x_3 + x_4 = 2.15$
   $15.3x_1 + 0.110x_2 - 13.1x_3 - x_4 = 4.16$
   Actual solution $(0.17682530, 0.01269269, -0.02065405, -1.18260870)$

   (h) $\pi x_1 - ex_2 + \sqrt{2}x_3 - \sqrt{3}x_4 = \sqrt{11}$
   $\pi^2 x_1 + ex_2 - e^2 x_3 + \frac{3}{7}x_4 = 0$
   $\sqrt{5}x_1 - \sqrt{6}x_2 + x_3 - \sqrt{2}x_4 = \pi$
   $\pi^3 x_1 + e^2 x_2 - \sqrt{7}x_3 + \frac{1}{9}x_4 = \sqrt{2}$
   Actual solution $(0.78839378, -3.12541367, 0.16759660, 4.55700252)$.

2. Repeat Exercise 1 using three-digit rounding arithmetic.

3. Repeat Exercise 1 using Gaussian elimination with Maximal Column Pivoting.

4. Repeat Exercise 2 using Gaussian elimination with Maximal Column Pivoting.

5. Repeat Exercise 1 using Gaussian elimination with Scaled Column Pivoting.

6. Repeat Exercise 2 using Gaussian elimination with Scaled Column Pivoting.

7. Repeat Exercise 1 using Gaussian elimination and single-precision computer arithmetic.

8. Repeat Exercise 1 using Gaussian elimination with Maximal Column Pivoting and single-precision computer arithmetic.

9. Repeat Exercise 1 using Gaussian elimination with Scaled Column Pivoting and single-precision computer arithmetic.

6.4 Linear Algebra and Matrix Inversion

Early in this chapter we illustrated the convenience of matrix notation for the study of linear systems of equations, but there is a wealth of additional material in linear algebra that finds application in the study of approximation techniques. In this section we introduce some basic results and techniques that are needed for both theory and application. All the topics discussed here should be familiar to anyone who has studied matrix theory at the undergraduate level. This section could be omitted, but it is advisable to at least read the section to see the results from linear algebra that will be frequently called upon for service.

Two matrices A and B are **equal** if both are of the same size, say $n \times m$, and if $a_{ij} = b_{ij}$ for each $i = 1, 2, \ldots, n$ and $j = 1, 2, \ldots, m$.

This definition means, for example, that

$$\begin{bmatrix} 2 & -1 & 7 \\ 3 & 1 & 0 \end{bmatrix} \neq \begin{bmatrix} 2 & 3 \\ -1 & 1 \\ 7 & 0 \end{bmatrix},$$

since they differ in dimension.

If A and B are $n \times m$ matrices and λ is a real number, then the **sum** of A and B, denoted $A + B$, is the $n \times m$ matrix whose entries are $a_{ij} + b_{ij}$ and the **scalar product** of λ and A, denoted λA, is the $n \times m$ matrix whose entries are λa_{ij}.

If A is an $n \times m$ matrix and B an $m \times p$ matrix, the **matrix product** of A and B, denoted AB, is an $n \times p$ matrix C whose entries c_{ij} are given by

$$c_{ij} = \sum_{k=1}^{m} a_{ik} b_{kj} = a_{i1} b_{1j} + a_{i2} b_{2j} + \cdots + a_{im} b_{mj},$$

for each $i = 1, 2, \ldots, n$, and $j = 1, 2, \ldots, p$.

The computation of c_{ij} can be viewed as the multiplication of the entries of the ith row of A with corresponding entries in the jth column of B, followed by a summation. That is,

$$[a_{i1}, a_{i2}, \ldots, a_{im}] \begin{bmatrix} b_{1j} \\ b_{2j} \\ \vdots \\ b_{mj} \end{bmatrix} = [c_{ij}],$$

where

$$c_{ij} = a_{i1} b_{1j} + a_{i2} b_{2j} + \cdots + a_{im} b_{mj} = \sum_{k=1}^{m} a_{ik} b_{kj}.$$

This explains why the number of columns of A must equal the number of rows of B in order for the product AB to be defined.

EXAMPLE 1 Let

$$A = \begin{bmatrix} 2 & 1 & -1 \\ 3 & 1 & 2 \\ 0 & -2 & -3 \end{bmatrix}, \qquad B = \begin{bmatrix} 3 & 2 \\ -1 & 1 \\ 6 & 4 \end{bmatrix},$$

$$C = \begin{bmatrix} 2 & 1 & 0 \\ -1 & 3 & 2 \end{bmatrix}, \qquad \text{and} \qquad D = \begin{bmatrix} 1 & -1 & 1 \\ 2 & -1 & 2 \\ 3 & 0 & 3 \end{bmatrix}.$$

Then,

$$AD = \begin{bmatrix} 1 & -3 & 1 \\ 11 & -4 & 11 \\ -13 & 2 & -13 \end{bmatrix} \neq \begin{bmatrix} -1 & -2 & -6 \\ 1 & -3 & -10 \\ 6 & -3 & -12 \end{bmatrix} = DA.$$

Further,

$$BC = \begin{bmatrix} 4 & 9 & 4 \\ -3 & 2 & 2 \\ 8 & 18 & 8 \end{bmatrix} \qquad \text{and} \qquad CB = \begin{bmatrix} 5 & 5 \\ 6 & 9 \end{bmatrix}$$

are not even the same size. Finally,

$$AB = \begin{bmatrix} -1 & 1 \\ 20 & 15 \\ -16 & -14 \end{bmatrix},$$

but BA cannot be computed. ∎ ∎ ∎

A **square** matrix has the same number of rows as columns. A **diagonal** matrix is a square matrix $D = (d_{ij})$ with $d_{ij} = 0$ whenever $i \neq j$. The **identity** matrix of order n, $I_n = (\delta_{ij})$, is a diagonal matrix with entries

$$\delta_{ij} = \begin{cases} 1, & \text{if } i = j, \\ 0, & \text{if } i \neq j. \end{cases}$$

When the size of I_n is clear, this matrix is generally written simply as I. For example, the identity matrix of order three is

$$I = \begin{bmatrix} 1 & 0 & 0 \\ 0 & 1 & 0 \\ 0 & 0 & 1 \end{bmatrix}.$$

If A is any $n \times n$ matrix, then $AI = IA = A$.

An $n \times n$ **upper-triangular** matrix $U = (u_{ij})$ has, for each $j = 1, 2, \ldots, n$, the entries

$$u_{ij} = 0, \qquad \text{for each } i = j + 1, j + 2, \ldots, n;$$

and a **lower-triangular** matrix $L = (l_{ij})$ has for each $j = 1, 2, \ldots, n$, the entries

$$l_{ij} = 0, \qquad \text{for each } i = 1, 2, \ldots, j - 1.$$

(A diagonal matrix is both upper and lower triangular.)

In Example 1 we found that, in general, $AB \neq BA$, even when both products are defined. However, the other arithmetic properties associated with multiplication do hold. For example, when A, B, and C are matrices of the appropriate size and λ is a scalar, we have

$$A(BC) = (AB)C, \quad A(B + C) = AB + AC, \quad \text{and} \quad \lambda(AB) = (\lambda A)B = A(\lambda B).$$

An $n \times n$ matrix A is **nonsingular**, or *invertible*, if an $n \times n$ matrix A^{-1} exists with $AA^{-1} = A^{-1}A = I$. The matrix A^{-1} is the **inverse** of A. A matrix without an inverse is **singular**, or *noninvertible*.

EXAMPLE 2   Let

$$A = \begin{bmatrix} 1 & 2 & -1 \\ 2 & 1 & 0 \\ -1 & 1 & 2 \end{bmatrix} \quad \text{and} \quad B = \frac{1}{9} \begin{bmatrix} -2 & 5 & -1 \\ 4 & -1 & 2 \\ -3 & 3 & 3 \end{bmatrix}.$$

Since

$$AB = \begin{bmatrix} 1 & 2 & -1 \\ 2 & 1 & 0 \\ -1 & 1 & 2 \end{bmatrix} \cdot \frac{1}{9} \begin{bmatrix} -2 & 5 & -1 \\ 4 & -1 & 2 \\ -3 & 3 & 3 \end{bmatrix} = \begin{bmatrix} 1 & 0 & 0 \\ 0 & 1 & 0 \\ 0 & 0 & 1 \end{bmatrix}$$

and

$$BA = \frac{1}{9} \begin{bmatrix} -2 & 5 & -1 \\ 4 & -1 & 2 \\ -3 & 3 & 3 \end{bmatrix} \cdot \begin{bmatrix} 1 & 2 & -1 \\ 2 & 1 & 0 \\ -1 & 1 & 2 \end{bmatrix} = \begin{bmatrix} 1 & 0 & 0 \\ 0 & 1 & 0 \\ 0 & 0 & 1 \end{bmatrix},$$

A and B are nonsingular with $B = A^{-1}$ and $A = B^{-1}$.   ■ ■ ■

The reason for introducing this matrix operation at this time is that the linear system

$$a_{11}x_1 + a_{12}x_2 + \cdots + a_{1n}x_n = b_1,$$
$$a_{21}x_1 + a_{22}x_2 + \cdots + a_{2n}x_n = b_2,$$
$$\vdots$$
$$a_{n1}x_1 + a_{n2}x_2 + \cdots + a_{nn}x_n = b_n,$$

can be viewed as the matrix equation $A\mathbf{x} = \mathbf{b}$, where

$$A = \begin{bmatrix} a_{11} & a_{12} & \cdots & a_{1n} \\ a_{21} & a_{22} & \cdots & a_{2n} \\ \vdots & \vdots & & \vdots \\ a_{n1} & a_{n2} & \cdots & a_{nn} \end{bmatrix}, \quad \mathbf{x} = \begin{bmatrix} x_1 \\ x_2 \\ \vdots \\ x_n \end{bmatrix}, \quad \text{and} \quad \mathbf{b} = \begin{bmatrix} b_1 \\ b_2 \\ \vdots \\ b_n \end{bmatrix}.$$

If A is nonsingular matrix, then the solution $\mathbf{x}$ to the linear system $A\mathbf{x} = \mathbf{b}$ is given by $\mathbf{x} = A^{-1}(A\mathbf{x}) = A^{-1}\mathbf{b}$. In general, however, it is more difficult to determine A^{-1} than it is to solve the system $A\mathbf{x} = \mathbf{b}$, because the number of operations involved in determining A^{-1} is much larger. Even so, it is useful from a conceptual standpoint to describe a method for determining the inverse of a matrix.

EXAMPLE 3 To determine the inverse of the matrix

$$A = \begin{bmatrix} 1 & 2 & -1 \\ 2 & 1 & 0 \\ -1 & 1 & 2 \end{bmatrix},$$

let us first consider the product AB, where B is an arbitrary 3×3 matrix.

$$AB = \begin{bmatrix} 1 & 2 & -1 \\ 2 & 1 & 0 \\ -1 & 1 & 2 \end{bmatrix}\begin{bmatrix} b_{11} & b_{12} & b_{13} \\ b_{21} & b_{22} & b_{23} \\ b_{31} & b_{32} & b_{33} \end{bmatrix}$$

$$= \begin{bmatrix} b_{11} + 2b_{21} - b_{31} & b_{12} + 2b_{22} - b_{32} & b_{13} + 2b_{23} - b_{33} \\ 2b_{11} + b_{21} & 2b_{12} + b_{22} & 2b_{13} + b_{23} \\ -b_{11} + b_{21} + 2b_{31} & -b_{12} + b_{22} + 2b_{32} & -b_{13} + b_{23} + 2b_{33} \end{bmatrix}.$$

If $B = A^{-1}$, then $AB = I$, so we must have

$$b_{11} + 2b_{21} - b_{31} = 1, \qquad b_{12} + 2b_{22} - b_{32} = 0, \qquad b_{13} + 2b_{23} - b_{33} = 0$$
$$2b_{11} + b_{21} = 0, \qquad 2b_{12} + b_{22} = 1, \qquad 2b_{13} + b_{23} = 0$$
$$-b_{11} + b_{21} + 2b_{31} = 0, \qquad -b_{12} + b_{22} + 2b_{32} = 0, \qquad -b_{13} + b_{23} + 2b_{33} = 1.$$

Notice that the coefficients in each of the systems of equations are the same; the only change in the systems occurs on the right side of the equations. As a consequence, the computations can be performed on a larger augmented matrix, formed by combining the matrices for each of the systems

$$\begin{bmatrix} 1 & 2 & -1 & : & 1 & 0 & 0 \\ 2 & 1 & 0 & : & 0 & 1 & 0 \\ -1 & 1 & 2 & : & 0 & 0 & 1 \end{bmatrix}.$$

First, performing $(E_2 - 2E_1) \rightarrow (E_1)$ and $(E_3 + E_1) \rightarrow (E_3)$ gives

$$\begin{bmatrix} 1 & 2 & -1 & : & 1 & 0 & 0 \\ 0 & -3 & 2 & : & -2 & 1 & 0 \\ 0 & 3 & 1 & : & 1 & 0 & 1 \end{bmatrix}.$$

Next, performing $(E_3 + E_2) \rightarrow (E_3)$ produces

$$\begin{bmatrix} 1 & 2 & -1 & : & 1 & 0 & 0 \\ 0 & -3 & 2 & : & -2 & 1 & 0 \\ 0 & 0 & 3 & : & -1 & 1 & 1 \end{bmatrix}.$$

Backward substitution is performed on each of the three augmented matrices,

$$
\begin{bmatrix}
1 & 2 & -1 & \vdots & 1 \\
0 & -3 & 2 & \vdots & -2 \\
0 & 0 & 3 & \vdots & -1
\end{bmatrix},
\quad
\begin{bmatrix}
1 & 2 & -1 & \vdots & 0 \\
0 & -3 & 2 & \vdots & 1 \\
0 & 0 & 3 & \vdots & 1
\end{bmatrix},
\quad
\begin{bmatrix}
1 & 2 & -1 & \vdots & 0 \\
0 & -3 & 2 & \vdots & 0 \\
0 & 0 & 3 & \vdots & 1
\end{bmatrix},
$$

to eventually give

$$
\begin{array}{lll}
b_{11} = -\tfrac{2}{9}, & b_{12} = \tfrac{5}{9}, & b_{13} = -\tfrac{1}{9}, \\
b_{21} = \tfrac{4}{9}, & b_{22} = -\tfrac{1}{9}, \quad \text{and} & b_{23} = \tfrac{2}{9}, \\
b_{31} = -\tfrac{1}{3}, & b_{32} = \tfrac{1}{3}, & b_{33} = \tfrac{1}{3}.
\end{array}
$$

These are the entries of A^{-1}:

$$
A^{-1} =
\begin{bmatrix}
-\tfrac{2}{9} & \tfrac{5}{9} & -\tfrac{1}{9} \\
\tfrac{4}{9} & -\tfrac{1}{9} & \tfrac{2}{9} \\
-\tfrac{1}{3} & \tfrac{1}{3} & \tfrac{1}{3}
\end{bmatrix}
= \tfrac{1}{9}
\begin{bmatrix}
-2 & 5 & -1 \\
4 & -1 & 2 \\
-3 & 3 & 3
\end{bmatrix}.
$$

■ ■ ■

The **transpose** of an $n \times m$ matrix $A = (a_{ij})$ is the $m \times n$ matrix $A^t = (a_{ji})$. A square matrix A is **symmetric** if $A = A^t$.

EXAMPLE 4 The matrices

$$
A =
\begin{bmatrix}
7 & 2 & 0 \\
3 & 5 & -1 \\
0 & 5 & -6
\end{bmatrix},
\quad
B =
\begin{bmatrix}
2 & 4 & 7 \\
3 & -5 & -1
\end{bmatrix},
\quad
C =
\begin{bmatrix}
6 & 4 & -3 \\
4 & -2 & 0 \\
-3 & 0 & 1
\end{bmatrix}
$$

have transposes

$$
A^t =
\begin{bmatrix}
7 & 3 & 0 \\
2 & 5 & 5 \\
0 & -1 & -6
\end{bmatrix},
\quad
B^t =
\begin{bmatrix}
2 & 3 \\
4 & -5 \\
7 & -1
\end{bmatrix},
\quad
C^t =
\begin{bmatrix}
6 & 4 & -3 \\
4 & -2 & 0 \\
-3 & 0 & 1
\end{bmatrix}.
$$

The matrix C is symmetric, since $C^t = C$. The matrices A and B are not symmetric.

■ ■ ■

The following operations involving the transpose of a matrix hold whenever the operation is possible:

Transpose Facts

(a) $(A^t)^t = A$

(b) $(A + B)^t = A^t + B^t$.

(c) $(AB)^t = B^t A^t$

(d) If A^{-1} exists, $(A^{-1})^t = (A^t)^{-1}$.

The determinant of a square matrix is a number that can be useful in determining the existence and uniqueness of solutions to linear systems. We will denote the determinant of a matrix A by $\det A$, but it is also common to use the notation $|A|$.

Determinant of a Matrix

(a) If $A = [a]$ is a 1×1 matrix, then $\det A = a$.

(b) If A is an $n \times n$ matrix, the **minor** M_{ij} is the determinant of the $(n-1) \times (n-1)$ submatrix of A obtained by deleting the ith row and jth column of the matrix A. Then the determinant of A is given either by

$$\det A = \sum_{j=1}^{n}(-1)^{i+j} a_{ij} M_{ij} \qquad \text{for any } i = 1, 2, \ldots, n,$$

or by

$$\det A = \sum_{i=1}^{n}(-1)^{i+j} a_{ij} M_{ij} \qquad \text{for any } j = 1, 2, \ldots, n.$$

It can be shown that, if $n > 1$, to calculate the determinant of a general $n \times n$ matrix requires $O(n!)$ multiplications/divisions and additions/subtractions. Even for relatively small values of n, the number of calculations becomes unwieldy.

It appears that there are $2n$ different definitions of $\det A$, depending on which row or column is chosen. However, all definitions give the same numerical result. The flexibility in the definition is used in the following example. It is most convenient to compute $\det A$ across the row or down the column with the most zeros.

EXAMPLE 5 Let

$$A = \begin{bmatrix} 2 & -1 & 3 & 0 \\ 4 & -2 & 7 & 0 \\ -3 & -4 & 1 & 5 \\ 6 & -6 & 8 & 0 \end{bmatrix}.$$

To compute $\det A$, it is easiest to expand about the fourth column:

$$\det A = -a_{14}M_{14} + a_{24}M_{24} - a_{34}M_{34} + a_{44}M_{44} = -5M_{34}$$

$$= -5 \det \begin{bmatrix} 2 & -1 & 3 \\ 4 & -2 & 7 \\ 6 & -6 & 8 \end{bmatrix}$$

$$= -5 \left\{ 2 \det \begin{bmatrix} -2 & 7 \\ -6 & 8 \end{bmatrix} - (-1) \det \begin{bmatrix} 4 & 7 \\ 6 & 8 \end{bmatrix} + 3 \det \begin{bmatrix} 4 & -2 \\ 6 & -6 \end{bmatrix} \right\} = -30.$$

■ ■ ■

The following properties of determinants are useful in relating linear systems and Gaussian elimination to determinants:

Determinant Facts

Suppose A is an $n \times n$ matrix:

(a) If any row or column of A has only zero entries, then $\det A = 0$.

(b) If $\tilde{A}$ is obtained from A by the operation $(E_i) \leftrightarrow (E_k)$, with $i \neq k$, then $\det \tilde{A} = -\det A$.

(c) If A has two rows the same, then $\det A = 0$.

(d) If $\tilde{A}$ is obtained from A by the operation $(\lambda E_i) \rightarrow (E_i)$, then $\det \tilde{A} = \lambda \det A$.

(e) If $\tilde{A}$ is obtained from A by the operation $(E_i + \lambda E_k) \rightarrow (E_i)$ with $i \neq k$, then $\det \tilde{A} = \det A$.

(f) If B is also an $n \times n$ matrix, then $\det AB = \det A \det B$.

(g) $\det A^t = \det A$.

The key result relating nonsingularity, Gaussian elimination, linear systems, and determinants, is that the following statements are equivalent.

Equivalent Statements about an $n \times n$ Matrix A

(a) The equation $Ax = 0$ has the unique solution $\mathbf{x} = \mathbf{0}$.

(b) The system $Ax = \mathbf{b}$ has a unique solution for any n-dimensional column vector $\mathbf{b}$.

(c) The matrix A is nonsingular; that is, A^{-1} exists.

(d) $\det A \neq 0$.

(e) Gaussian elimination with row interchanges can be performed on the system $Ax = \mathbf{b}$ for any n-dimensional column vector $\mathbf{b}$.

EXERCISE SET 6.4

1. Determine which of the following matrices are nonsingular and compute the inverse of these matrices:

(a) $\begin{bmatrix} 4 & 2 & 6 \\ 3 & 0 & 7 \\ -2 & -1 & -3 \end{bmatrix}$
(b) $\begin{bmatrix} 1 & 2 & 0 \\ 2 & 1 & -1 \\ 3 & 1 & 1 \end{bmatrix}$
(c) $\begin{bmatrix} 2 & 0 & 0 \\ 0 & -3 & 0 \\ 0 & 0 & 1 \end{bmatrix}$

(d) $\begin{bmatrix} 4 & 0 & 0 \\ 0 & 0 & 0 \\ 0 & 0 & 3 \end{bmatrix}$
(e) $\begin{bmatrix} 1 & 1 & -1 & 1 \\ 1 & 2 & -4 & -2 \\ 2 & 1 & 1 & 5 \\ -1 & 0 & -2 & -4 \end{bmatrix}$
(f) $\begin{bmatrix} 2 & 3 & 1 & 2 \\ -2 & 4 & -1 & 5 \\ 3 & 7 & \frac{3}{2} & 1 \\ 6 & 9 & 3 & 7 \end{bmatrix}$

(g) $\begin{bmatrix} 4 & 0 & 0 & 0 \\ 6 & 7 & 0 & 0 \\ 9 & 11 & 1 & 0 \\ 5 & 4 & 1 & 1 \end{bmatrix}$ **(h)** $\begin{bmatrix} 2 & 0 & 1 & 2 \\ 1 & 1 & 0 & 2 \\ 2 & -1 & 3 & 1 \\ 3 & -1 & 4 & 3 \end{bmatrix}$ **(i)** $\begin{bmatrix} 1 & 0 & 0 & 0 \\ -2 & 1 & 0 & 0 \\ 5 & -4 & 1 & 0 \\ -5 & 3 & 0 & 1 \end{bmatrix}$

(j) $\begin{bmatrix} 1 & -1 & -1 & -1 \\ 0 & 1 & -1 & -1 \\ 0 & 0 & 1 & -1 \\ 0 & 0 & 0 & 1 \end{bmatrix}$

2. Compute the following matrix products:

(a) $\begin{bmatrix} 1 & 0 & 0 \\ -1 & 1 & 0 \\ 2 & 3 & 1 \end{bmatrix} \cdot \begin{bmatrix} 1 & 0 & 0 \\ 2 & 2 & 0 \\ 1 & -1 & 1 \end{bmatrix}$ **(b)** $\begin{bmatrix} 1 & 0 & 0 \\ 2 & 1 & 0 \\ -2 & -1 & 1 \end{bmatrix} \cdot \begin{bmatrix} 1 & -1 & 2 \\ 0 & 1 & 3 \\ 0 & 0 & 2 \end{bmatrix}$

(c) $\begin{bmatrix} 1 & 0 & 0 \\ 0 & 1 & 0 \\ 0 & -2 & 1 \end{bmatrix} \cdot \begin{bmatrix} 1 & 0 & 0 \\ 2 & 1 & 0 \\ -3 & 0 & 1 \end{bmatrix}$ **(d)** $\begin{bmatrix} 2 & -1 & 4 \\ 0 & -1 & 2 \\ 0 & 0 & 3 \end{bmatrix} \cdot \begin{bmatrix} 3 & -3 & 4 \\ 0 & 1 & 1 \\ 0 & 0 & 2 \end{bmatrix}$

3. Consider the four 3×3 linear systems with the same coefficient matrix:

$$\begin{aligned} 2x_1 - 3x_2 + x_3 &= 2, & 2x_1 - 3x_2 + x_3 &= 6, \\ x_1 + x_2 - x_3 &= -1, & x_1 + x_2 - x_3 &= 4, \\ -x_1 + x_2 - 3x_3 &= 0. & -x_1 + x_2 - 3x_3 &= 5. \end{aligned}$$

$$\begin{aligned} 2x_1 - 3x_2 + x_3 &= 0, & 2x_1 - 3x_2 + x_3 &= -1, \\ x_1 + x_2 - x_3 &= 1, & x_1 + x_2 - x_3 &= 0, \\ -x_1 + x_2 - 3x_3 &= -3. & -x_1 + x_2 - 3x_3 &= 0. \end{aligned}$$

(a) Solve the linear systems by applying Gaussian elimination to the augmented matrix

$$\begin{bmatrix} 2 & -3 & 1 & \vdots & 2 & 6 & 0 & -1 \\ 1 & 1 & -1 & \vdots & -1 & 4 & 1 & 0 \\ -1 & 1 & -3 & \vdots & 0 & 5 & -3 & 0 \end{bmatrix}.$$

(b) Solve the linear system by finding and multiplying by the inverse of

$$A = \begin{bmatrix} 2 & -3 & 1 \\ 1 & 1 & -1 \\ -1 & 1 & -3 \end{bmatrix}.$$

(c) Which method seems easier? Which method requires more operations?

4. Repeat Exercise 3 using the following linear systems:

$$\begin{aligned} x_1 - x_2 + 2x_3 - x_4 &= 6, & x_1 - x_2 + 2x_3 - x_4 &= 1, \\ x_1 \quad - x_3 + x_4 &= 4, & x_1 \quad - x_3 + x_4 &= 1, \\ 2x_1 + x_2 + 3x_3 - 4x_4 &= -2, & 2x_1 + x_2 + 3x_3 - 4x_4 &= 2, \\ - x_2 + x_3 - x_4 &= 5. & - x_2 + x_3 - x_4 &= -1. \end{aligned}$$

5. Compute the determinants of the following matrices:

(a) $\begin{bmatrix} 1 & 2 & 0 \\ 2 & 1 & -1 \\ 3 & 1 & 1 \end{bmatrix}$ **(b)** $\begin{bmatrix} 4 & 0 & 1 \\ 2 & 1 & 0 \\ 2 & 2 & 3 \end{bmatrix}$

(c) $\begin{bmatrix} 1 & 1 & -1 & 1 \\ 1 & 2 & -4 & -2 \\ 2 & 1 & 1 & 5 \\ -1 & 0 & -2 & -4 \end{bmatrix}$ **(d)** $\begin{bmatrix} 2 & 0 & 1 & 2 \\ 1 & 1 & 0 & 2 \\ 2 & -1 & 3 & 1 \\ 3 & -1 & 4 & 3 \end{bmatrix}$

6. Determine AB for the matrices listed below. Then compute $\det A$, $\det B$, and $\det AB$ and verify that $\det AB = \det A \det B$.

$$A = \begin{bmatrix} 4 & 6 & 1 & -1 \\ 2 & 1 & 0 & \frac{1}{2} \\ 3 & 0 & 0 & 1 \\ 1 & -1 & 1 & 1 \end{bmatrix} \quad \text{and} \quad B = \begin{bmatrix} 1 & 2 & 3 & 4 \\ 0 & 2 & -1 & 1 \\ 0 & 0 & 3 & 2 \\ 0 & 0 & 0 & -1 \end{bmatrix}.$$

7. Use the equivalent statements at the end of the section to prove that AB is nonsingular if and only if both A and B are nonsingular.

8. Prove the following statements or provide counterexamples to show they are not true.
   (a) The product of two symmetric matrices is symmetric.
   (b) The inverse of a nonsingular symmetric matrix is a nonsingular symmetric matrix.
   (c) If A and B are $n \times n$ matrices, then $(AB)^t = A^t B^t$.

9. (a) Show that the product of two $n \times n$ lower-triangular matrices is lower triangular.
   (b) Show that the product of two $n \times n$ upper-triangular matrices is upper triangular.
   (c) Show that the inverse of a nonsingular $n \times n$ lower-triangular matrix is lower triangular.

10. In a paper entitled "Population Waves," Bernadelli hypothesizes a type of simplified beetle, which has a natural life span of 3 years. The female of this species has a survival rate of 1/2 in the first year of life, has a survival rate of 1/3 from the second to third years, and gives birth to an average of six new females before expiring at the end of the third year. A matrix can be used to show the contribution an individual female beetle makes, in a probabilistic sense, to the female population of the species by letting a_{ij} in the matrix $A = (a_{ij})$ denote the contribution that a single female beetle of age j will make to the next year's female population of age i; that is,

$$A = \begin{bmatrix} 0 & 0 & 6 \\ \frac{1}{2} & 0 & 0 \\ 0 & \frac{1}{3} & 0 \end{bmatrix}.$$

   (a) The contribution that a female beetle will make to the population two years hence could be determined from the entries of A^2, of 3 years hence from A^3, and so on. Construct A^2 and A^3, and try to make a general statement about the contribution of a female beetle to the population in n years' time for any positive integral value of n.
   (b) Use your conclusions from part (a) to describe what will occur in future years to a population of these beetles that initially consists of 6000 female beetles in each of the three age groups.
   (c) Construct A^{-1} and describe its significance regarding the population of this species.

11. In Section 3.5 we found that the parametric form $(x(t), y(t))$ of the cubic Hermite polynomials through $(x(0), y(0)) = (x_0, y_0)$ and $(x(1), y(1)) = (x_1, y_1)$ with guide points $(x_0 + \alpha_0, y_0 + \beta_0)$ and $(x_1 - \alpha_1, y_1 - \beta_1)$, respectively, are given by

$$x(t) = [2(x_0 - x_1) + (\alpha_0 + \alpha_1)]t^3 + [3(x_1 - x_0) - \alpha_1 - 2\alpha_0]t^2 + \alpha_0 t + x_0$$

and

$$y(t) = [2(y_0 - y_1) + (\beta_0 + \beta_1)]t^3 + [3(y_1 - y_0) - \beta_1 - 2\beta_0]t^2 + \beta_0 t + y_0.$$

   The Bézier cubic polynomials have the form

$$\hat{x}(t) = [2(x_0 - x_1) + 3(\alpha_0 + \alpha_1)]t^3 + [3(x_1 - x_0) - 3(\alpha_1 + 2\alpha_0)]t^2 + 3\alpha_0 t + x_0$$

and

$$\hat{y}(t) = [2(y_0 - y_1) + 3(\beta_0 + \beta_1)]t^3 + [3(y_1 - y_0) - 3(\beta_1 + 2\beta_0)]t^2 + 3\beta_0 t + y_0.$$

(a) Show that the matrix

$$A = \begin{bmatrix} 7 & 4 & 4 & 0 \\ -6 & -3 & -6 & 0 \\ 0 & 0 & 3 & 0 \\ 0 & 0 & 0 & 1 \end{bmatrix}$$

maps the Hermite polynomial coefficients onto the Bézier polynomial coefficients.

(b) Determine a matrix B that maps the Bézier polynomial coefficients onto the Hermite polynomial coefficients.

6.5 Matrix Factorization

Gaussian elimination is the principal tool for the direct solution of linear systems of equations, so it should come as no surprise to learn that it appears in other guises. In this section we will see that the steps used to solve a system of the form $A\mathbf{x} = \mathbf{b}$ by Gaussian elimination can be used to factor the matrix A into a product of matrices that are easier to manipulate. The factorization is particularly useful when it has the form $A = LU$, where L is lower triangular and U is upper triangular. Although not all matrices have this type of representation, many do that occur frequently in the application of numerical techniques.

In Section 6.2 we found that Gaussian elimination applied to an arbitrary nonsingular system requires $O(n^3)$ operations to determine $\mathbf{x}$. If A has been factored into the triangular form $A = LU$, then we can solve for $\mathbf{x}$ more easily by using a two-step process. First we let $\mathbf{y} = U\mathbf{x}$ and solve the system $L\mathbf{y} = \mathbf{b}$ for $\mathbf{y}$. Since L is lower triangular, determining $\mathbf{y}$ from this equation requires only $O(n^2)$ operations. Once $\mathbf{y}$ is known, the upper triangular system $U\mathbf{x} = \mathbf{y}$ requires only an additional $O(n^2)$ operations to determine the solution $\mathbf{x}$. This means that the total number of operations needed to solve the system $A\mathbf{x} = \mathbf{b}$ is reduced from $O(n^3)$ to $O(n^2)$. In systems greater than 100 by 100, this can reduce the amount of calculation by more than 99%. Not surprisingly, the reductions from the factorization do not come free; determining the specific matrices L and U requires $O(n^3)$ operations. But once the factorization is determined, systems involving the matrix A can be solved in this simplified manner for any number of vectors $\mathbf{b}$.

EXAMPLE 1 The linear system

$$\begin{aligned} x_1 + x_2 \qquad + 3x_4 &= 4, \\ 2x_1 + x_2 - x_3 + x_4 &= 1, \\ 3x_1 - x_2 - x_3 + 2x_4 &= -3, \\ -x_1 + 2x_2 + 3x_3 - x_4 &= 4 \end{aligned}$$

was considered in Section 6.2. The sequence of operations $(E_2 - 2E_1) \rightarrow (E_2)$, $(E_3 - 3E_1) \rightarrow (E_3)$, $(E_4 - (-1)E_1) \rightarrow (E_4)$, $(E_3 - 4E_2) \rightarrow (E_3)$, $(E_4 - (-3)E_2) \rightarrow (E_4)$ converts the system to one that has the upper-triangular form:

$$\begin{aligned} x_1 + x_2 \qquad + 3x_4 &= 4, \\ - x_2 - x_3 - 5x_4 &= -7, \\ 3x_3 + 13x_4 &= 13, \\ - 13x_4 &= -13. \end{aligned}$$

Let U be the upper-triangular matrix with these coefficients as its entries and L be the lower-triangular matrix with 1s along the diagonal and the multipliers m_{kj} as entries below the diagonal. Then we have the factorization

$$A = \begin{bmatrix} 1 & 1 & 0 & 3 \\ 2 & 1 & -1 & 1 \\ 3 & -1 & -1 & 2 \\ -1 & 2 & 3 & -1 \end{bmatrix} = \begin{bmatrix} 1 & 0 & 0 & 0 \\ 2 & 1 & 0 & 0 \\ 3 & 4 & 1 & 0 \\ -1 & -3 & 0 & 1 \end{bmatrix} \begin{bmatrix} 1 & 1 & 0 & 3 \\ 0 & -1 & -1 & -5 \\ 0 & 0 & 3 & 13 \\ 0 & 0 & 0 & -13 \end{bmatrix} = LU.$$

This factorization permits us to easily solve any system involving the matrix A. For example, to solve

$$\begin{bmatrix} 1 & 0 & 0 & 0 \\ 2 & 1 & 0 & 0 \\ 3 & 4 & 1 & 0 \\ -1 & -3 & 0 & 1 \end{bmatrix} \begin{bmatrix} 1 & 1 & 0 & 3 \\ 0 & -1 & -1 & -5 \\ 0 & 0 & 3 & 13 \\ 0 & 0 & 0 & -13 \end{bmatrix} \begin{bmatrix} x_1 \\ x_2 \\ x_3 \\ x_4 \end{bmatrix} = \begin{bmatrix} 8 \\ 7 \\ 14 \\ -7 \end{bmatrix}$$

we first introduce the substitution $\mathbf{y} = U\mathbf{x}$. Then $L\mathbf{y} = \mathbf{b}$; that is,

$$\begin{bmatrix} 1 & 0 & 0 & 0 \\ 2 & 1 & 0 & 0 \\ 3 & 4 & 1 & 0 \\ -1 & -3 & 0 & 1 \end{bmatrix} \begin{bmatrix} y_1 \\ y_2 \\ y_3 \\ y_4 \end{bmatrix} = \begin{bmatrix} 8 \\ 7 \\ 14 \\ -7 \end{bmatrix}.$$

This system is solved for $\mathbf{y}$ by a simple forward-substitution process:

$$\begin{array}{llll} y_1 & = 8 \\ 2y_1 + y_2 & = 7, & \text{so} & y_2 = -9 \\ 3y_1 + 4y_2 + y_3 & = 14, & \text{so} & y_3 = 26 \\ -y_1 - 3y_2 + y_4 & = 7, & \text{so} & y_4 = -26. \end{array}$$

We then solve $U\mathbf{x} = \mathbf{y}$ for $\mathbf{x}$, the solution of the original system; that is,

$$\begin{bmatrix} 1 & 1 & 0 & 3 \\ 0 & -1 & -1 & -5 \\ 0 & 0 & 3 & 13 \\ 0 & 0 & 0 & -13 \end{bmatrix} \begin{bmatrix} x_1 \\ x_2 \\ x_3 \\ x_4 \end{bmatrix} = \begin{bmatrix} 8 \\ -9 \\ 26 \\ -26 \end{bmatrix}.$$

Using backward substitution we obtain $x_4 = 2$, $x_3 = 0$, $x_2 = -1$, $x_1 = 3$.   ■ ■ ■

In the factorization of $A = LU$, the following equations are used to generate the entries in L and U:

$$l_{11}u_{11} = a_{11};$$

for each $j = 2, 3, \ldots, n$,

$$u_{1j} = \frac{a_{1j}}{l_{11}} \quad \text{and} \quad l_{j1} = \frac{a_{j1}}{u_{11}};$$

for each $i = 2, 3, \ldots, n - 1$,

$$l_{ii}u_{ii} = a_{ii} - \sum_{k=1}^{i-1} l_{ik}u_{ki};$$

For each $j = i + 1, i + 2, \ldots, n$,

$$u_{ij} = \frac{1}{l_{ii}}\left[a_{ij} - \sum_{k=1}^{i-1} l_{ik}u_{kj}\right] \qquad \text{and} \qquad l_{ji} = \frac{1}{u_{ii}}\left[a_{ji} - \sum_{k=1}^{i-1} l_{jk}u_{ki}\right]$$

and

$$l_{nn}u_{nn} = a_{nn} - \sum_{k=1}^{n-1} l_{nk}u_{kn}.$$

A general procedure for factoring matrices into a product of triangular matrices is performed in the program DIFACT64. Although new matrices L and U are constructed, the values generated replace the corresponding entries of A that are no longer needed. Thus, the new matrix has entries $a_{ij} = l_{ij}$ for each $i = 2, 3, \ldots, n$ and $j = 1, 2, \ldots, i - 1$; and $a_{ij} = u_{ij}$ for each $i = 1, 2, \ldots, n$ and $j = i + 1, i + 2, \ldots, n$.

The factorization is particularly useful when a number of linear systems involving A must be solved, since the bulk of the operations need to be performed only once. To solve $LU\mathbf{x} = \mathbf{b}$, we first solve $L\mathbf{y} = \mathbf{b}$ for $\mathbf{y}$. Since L is lower triangular, we have

$$y_1 = \frac{b_1}{l_{11}}$$

and

$$y_i = \frac{1}{l_{ii}}\left[b_i - \sum_{j=1}^{i-1} l_{ij}y_j\right], \qquad \text{for each } i = 2, 3, \ldots, n.$$

Once $\mathbf{y}$ is calculated by this forward-substitution process, the upper-triangular system $U\mathbf{x} = \mathbf{y}$ is solved by backward substitution.

In the previous discussion we assumed that A is such that a linear system of the form $A\mathbf{x} = \mathbf{b}$ can be solved using Gaussian elimination that does not require row interchanges. From a practical standpoint, this factorization is useful only when row interchanges are not required to control the round-off error resulting from the use of finite-digit arithmetic. Although many systems we encounter when using approximation methods are of this type, factorization modifications must be made when row interchanges are required. We begin the discussion with the introduction of a class of matrices that are used to rearrange, or permute, rows of a given matrix.

An $n \times n$ **permutation matrix** P is a matrix with precisely one entry whose value is 1 in each column and each row and all of whose other entries are 0.

EXAMPLE 2   The matrix

$$P = \begin{bmatrix} 1 & 0 & 0 \\ 0 & 0 & 1 \\ 0 & 1 & 0 \end{bmatrix}$$

is a 3×3 permutation matrix. For any 3×3 matrix A, multiplying on the left by P has the effect of interchanging the second and third rows of A:

$$PA = \begin{bmatrix} 1 & 0 & 0 \\ 0 & 0 & 1 \\ 0 & 1 & 0 \end{bmatrix} \begin{bmatrix} a_{11} & a_{12} & a_{13} \\ a_{21} & a_{22} & a_{23} \\ a_{31} & a_{32} & a_{33} \end{bmatrix} = \begin{bmatrix} a_{11} & a_{12} & a_{13} \\ a_{31} & a_{32} & a_{33} \\ a_{21} & a_{22} & a_{23} \end{bmatrix}.$$

Similarly, multiplying on the right by P interchanges the second and third columns of A.

■ ■ ■

There are two useful properties of permutation matrices that relate to Gaussian elimination. The first of these is illustrated in the previous example and states that if $k_1, \ldots, k_n$ is a permutation of the integers $1, \ldots, n$ and the permutation matrix $P = (p_{ij})$ is defined by

$$p_{ij} = \begin{cases} 1, & \text{if } j = k_i, \\ 0, & \text{otherwise}, \end{cases}$$

then

$$PA = \begin{bmatrix} a_{k_1,1} & a_{k_1,2} & \cdots & a_{k_1,n} \\ a_{k_2,1} & a_{k_2,2} & \cdots & a_{k_2,n} \\ \vdots & \vdots & \vdots & \vdots \\ a_{k_n,1} & a_{k_n,2} & \cdots & a_{k_n,n} \end{bmatrix}.$$

The second is that if P is a permutation matrix, then P^{-1} exists and $P^{-1} = P^t$.

EXAMPLE 3   Since $a_{11} = 0$, the matrix

$$A = \begin{bmatrix} 0 & 1 & -1 & 1 \\ 1 & 1 & -1 & 2 \\ -1 & -1 & 1 & 0 \\ 1 & 2 & 0 & 2 \end{bmatrix}$$

does not have an LU factorization. However, using the row interchange $(E_1) \leftrightarrow (E_2)$, followed by $(E_3 + E_1) \rightarrow E_3$ and $(E_4 - E_1) \rightarrow E_4$, produces

$$\begin{bmatrix} 1 & 1 & -1 & 2 \\ 0 & 1 & -1 & 1 \\ 0 & 0 & 0 & 2 \\ 0 & 1 & 1 & 0 \end{bmatrix}.$$

Then the row interchange $(E_3) \leftrightarrow (E_4)$, followed by $(E_3 - E_2) \rightarrow E_3$ gives the matrix

$$U = \begin{bmatrix} 1 & 1 & -1 & 2 \\ 0 & 1 & -1 & 1 \\ 0 & 0 & 2 & -1 \\ 0 & 0 & 0 & 2 \end{bmatrix}.$$

The permutation matrix associated with the row interchanges $(E_i) \leftrightarrow (E_2)$ and $(E_3) \leftrightarrow (E_4)$ is

$$P = \begin{bmatrix} 0 & 1 & 0 & 0 \\ 1 & 0 & 0 & 0 \\ 0 & 0 & 0 & 1 \\ 0 & 0 & 1 & 0 \end{bmatrix}.$$

Gaussian elimination can be performed on PA without row interchanges to give the LU factorization

$$PA = \begin{bmatrix} 1 & 0 & 0 & 0 \\ 0 & 1 & 0 & 0 \\ 1 & 1 & 1 & 0 \\ -1 & 0 & 0 & 1 \end{bmatrix} \begin{bmatrix} 1 & 1 & -1 & -2 \\ 0 & 1 & -1 & 1 \\ 0 & 0 & 2 & -1 \\ 0 & 0 & 0 & 2 \end{bmatrix} = LU.$$

So

$$A = P^{-1}(LU) = P^t(LU) = (P^t L)U = \begin{bmatrix} 0 & 1 & 0 & 0 \\ 1 & 0 & 0 & 0 \\ -1 & 0 & 0 & 1 \\ 1 & 1 & 1 & 0 \end{bmatrix} \begin{bmatrix} 1 & 1 & -1 & 2 \\ 0 & 1 & -1 & 1 \\ 0 & 0 & 2 & -1 \\ 0 & 0 & 0 & 2 \end{bmatrix}.$$

■ ■ ■

EXERCISE SET 6.5

1. Factor the following matrices into the LU decomposition using direct factorization with $l_{ii} = 1$ for all i:

   (a) $\begin{bmatrix} 2 & -1 & 1 \\ 3 & 3 & 9 \\ 3 & 3 & 5 \end{bmatrix}$

   (b) $\begin{bmatrix} 2 & -1.5 & 3 \\ -1 & 0 & 2 \\ 4 & -4.5 & 5 \end{bmatrix}$

   (c) $\begin{bmatrix} 1.012 & -2.132 & 3.104 \\ -2.132 & 4.906 & -7.013 \\ 3.104 & -7.013 & 0.014 \end{bmatrix}$

   (d) $\begin{bmatrix} 3.107 & 2.101 & 0 \\ 0 & -1.213 & 2.101 \\ 0 & 0 & 2.179 \end{bmatrix}$

   (e) $\begin{bmatrix} 2 & 0 & 0 & 0 \\ 1 & 1.5 & 0 & 0 \\ 0 & -3 & 0.5 & 0 \\ 2 & -2 & 1 & 1 \end{bmatrix}$

   (f) $\begin{bmatrix} 2.1756 & 4.0231 & -2.1732 & 5.1967 \\ -4.0231 & 6.0000 & 0 & 1.1973 \\ -1.0000 & -5.2107 & 1.1111 & 0 \\ 6.0235 & 7.0000 & 0 & -4.1561 \end{bmatrix}$

2. Use the factorization determined in Exercise 1 to solve the following linear systems:

   (a) $2x_1 - x_2 + x_3 = -1$
   $3x_1 + 3x_2 + 9x_3 = 0$
   $3x_1 + 3x_2 + 5x_3 = 4$

   (b) $2x_1 = 3$
   $x_1 + 1.5x_2 = 4.5$
   $- 3x_2 + 0.5x_3 = -6.6$
   $2x_1 - 2x_2 + x_3 + x_4 = 0.8$

(c) $1.012x_1 - 2.132x_2 + 3.104x_3 = 1.984$
$-2.132x_1 + 4.096x_2 - 7.013x_3 = -5.049$
$3.104x_1 - 7.013x_2 + 0.014x_3 = -3.895$

(d) $3.107x_1 + 2.101x_2 \qquad\qquad = 1.001$
$- 1.213x_2 + 2.101x_3 = 0.000$
$2.179x_3 = 7.013$

(e) $2x_1 - 1.5x_2 + 3x_3 = 1$
$-x_1 \qquad + 2x_3 = 3$
$4x_1 - 4.5x_2 + 5x_3 = -1$

(f) $2.1756x_1 + 4.0231x_2 - 2.1732x_3 + 5.1967x_4 = 17.102$
$-4.0231x_1 + 6.0000x_2 \qquad\qquad + 1.1973x_4 = -6.1593$
$-1.0000x_1 - 5.2107x_2 + 1.1111x_3 \qquad\qquad = 3.0004$
$6.0235x_1 + 7.0000x_2 \qquad\qquad - 4.1561x_4 = 0.0000$

3. Obtain factorizations of the form $(P^t L)U$ for the following matrices:

(a)
$$A = \begin{bmatrix} 0 & 2 & 3 \\ 1 & 1 & -1 \\ 0 & -1 & 1 \end{bmatrix}$$

(b)
$$A = \begin{bmatrix} 1 & 2 & -1 \\ 1 & 2 & 3 \\ 2 & -1 & 4 \end{bmatrix}$$

(c)
$$A = \begin{bmatrix} 1 & 1 & 0 & 3 \\ 2 & 1 & -1 & 1 \\ -1 & 2 & 3 & -1 \\ 3 & -1 & -1 & 2 \end{bmatrix}$$

(d)
$$A = \begin{bmatrix} 1 & -2 & 3 & 0 \\ 1 & -2 & 3 & 1 \\ 1 & -2 & 2 & -2 \\ 2 & 1 & 3 & -1 \end{bmatrix}$$

6.6 Techniques for Special Matrices

Although this chapter has been concerned primarily with the effective application of Gaussian elimination for finding the solution to a linear system of equations, many of the results have wider application. It might be said that Gaussian elimination is the hub about which the chapter revolves, but the wheel itself is of equal interest and has application in many forms in the study of numerical methods. In this section we consider some matrices that are of special types, forms that will be used in other chapters of the book. In fact, one of these special types has already been observed. The matrices we used to find the solution to the cubic spline interpolations are called strictly diagonally dominant, a class that has some very convenient properties.

The $n \times n$ matrix A is said to be **strictly diagonally dominant** when

$$|a_{ii}| > \sum_{\substack{j=1, \\ j \neq i}}^{n} |a_{ij}|$$

holds for each $i = 1, 2, \ldots, n$.

EXAMPLE 1   Consider the matrices

$$A = \begin{bmatrix} 7 & 2 & 0 \\ 3 & 5 & -1 \\ 0 & 5 & -6 \end{bmatrix} \quad \text{and} \quad B = \begin{bmatrix} 6 & 4 & -3 \\ 4 & -2 & 0 \\ -3 & 0 & 1 \end{bmatrix}.$$

The nonsymmetric matrix A is strictly diagonally dominant, since $|7| > |2| + |0|, |5| > |3| + |-1|$, and $|-6| > |0| + |5|$. The symmetric matrix B is not strictly diagonally dominant, since for example, in the third row the absolute value of the diagonal element is $|6| < |4| + |-3| = 7$. It is interesting to note that A^t is not strictly diagonally dominant, nor, of course, is $B^t = B$. ■ ■ ■

The following result was needed in Section 3.6 to ensure that there are unique solutions to the linear systems needed to determine cubic spline interpolants.

Strictly Diagonally Dominant Matrices

A strictly diagonally dominant matrix A is nonsingular. Moreover, in this case, Gaussian elimination can be performed on any linear system of the form $Ax = b$ to obtain its unique solution without row or column interchanges, and the computations are stable with respect to the growth of round-off error.

A matrix A is **positive definite** if it is symmetric and if $x^t A x > 0$ for every n-dimensional column vector $x \neq 0$.

Using the definition to determine whether a matrix is positive definite can be difficult. Fortunately, there are more easily verified criteria for identifying members that are and are not of this important class.

Positive Definite Matrix Properties

If A is an $n \times n$ positive definite matrix, then

(a) A is nonsingular;

(b) $a_{ii} > 0$ for each $i = 1, 2, \ldots, n$;

(c) $\max_{1 \le k, j \le n} |a_{kj}| \le \max_{1 \le i \le n} |a_{ii}|$;

(d) $(a_{ij})^2 < a_{ii} a_{jj}$ for each $i \neq j$.

The next result parallels the strictly diagonally dominant results presented previously.

Positive Definite Matrix Equivalences

The following are equivalent for any $n \times n$ symmetric matrix A:

(a) A is positive definite.

(b) Gaussian elimination without row interchanges can be performed on the linear system $Ax = b$ with all pivot elements positive. (This ensures that the computations are stable with respect to the growth of round-off error.)

(c) A can be factored in the form LL^t, where L is lower-triangular with positive diagonal entries.

(d) A can be factored in the form LDL^t, where L is lower-triangular with ones on its diagonal and D is a diagonal matrix with positive diagonal entries.

The factorization in part (c) can be obtained by Choleski's factorization method. It is based on the following equations:

$$l_{11} = \sqrt{a_{11}};$$

for each $j = 2, 3, \ldots, n$,

$$l_{j1} = \frac{a_{j1}}{l_{11}};$$

for each $i = 2, 3, \ldots, n - 1$,

$$l_{ii} = \left[a_{ii} - \sum_{k=1}^{i-1} l_{ik}^2 \right]^{1/2};$$

for each $j = i + 1, i + 2, \ldots, n$,

$$l_{ji} = \frac{1}{l_{ii}} \left[a_{ji} - \sum_{k=1}^{i-1} l_{jk} l_{ik} \right]$$

and

$$l_{nn} = \left[a_{nn} - \sum_{k=1}^{n-1} l_{nk}^2 \right]^{1/2}.$$

These equations can be derived by writing out the system associated with $A = LL^t$. Choleski's method can be implemented using the program CHOLFC65.

In a similar manner to the general LU factorization, the factorization $A = LDL^t$ uses the equations

$$d_1 = a_{11};$$

for each $j = 2, 3, \ldots, n$,

$$l_{j1} = \frac{a_{j1}}{d_1};$$

for each $i = 2, 3, \ldots, n - 1$,

$$d_i = a_{ii} - \sum_{j=1}^{i-1} l_{ij}^2 d_j;$$

for each $j = i + 1, i + 2, \ldots, n$,

$$l_{ji} = \frac{1}{d_i} \left[a_{ji} - \sum_{k=1}^{i-1} l_{jk} l_{ik} d_k \right]$$

and

$$d_n = a_{nn} - \sum_{j=1}^{n-1} l_{nj}^2 d_j.$$

The LDL^t factorization can be accomplished with the program LDLFCT66.

EXAMPLE 2   The matrix

$$A = \begin{bmatrix} 4 & -1 & 1 \\ -1 & 4.25 & 2.75 \\ 1 & 2.75 & 3.5 \end{bmatrix}$$

is positive definite. The factorization LDL^t of A is

$$A = LDL^t = \begin{bmatrix} 1 & 0 & 0 \\ -0.25 & 1 & 0 \\ 0.25 & 0.75 & 1 \end{bmatrix} \begin{bmatrix} 4 & 0 & 0 \\ 0 & 4 & 0 \\ 0 & 0 & 1 \end{bmatrix} \begin{bmatrix} 1 & -0.25 & 0.25 \\ 0 & 1 & 0.75 \\ 0 & 0 & 1 \end{bmatrix},$$

and Choleski's method produces the factorization

$$A = LL^t = \begin{bmatrix} 2 & 0 & 0 \\ -0.5 & 2 & 0 \\ 0.5 & 1.5 & 1 \end{bmatrix} \begin{bmatrix} 2 & -0.5 & 0.5 \\ 0 & 2 & 1.5 \\ 0 & 0 & 1 \end{bmatrix}.$$

■ ■ ■

We can solve the linear system $Ax = b$ when A is positive definite by using Choleski's method to first factor A into the form LL^t. Let $y = L^t x$. The linear system $Ly = b$ is solved using forward substitution. Then the solution to the original system is obtained by using backward substitution to solve $L^t x = y$ with the equations

$$x_n = \frac{y_n}{l_{nn}}$$

and, for $i = n - 1, n - 2, \ldots, 1$,

$$x_i = \frac{1}{l_{ii}} \left[y_i - \sum_{j=i+1}^{n} l_{ji} x_j \right].$$

If $Ax = b$ is to be solved and the factorization $A = LDL^t$ is known, then we let $y = DL^t x$ and solve the system $Ly = b$ using forward substitution. The system $Dz = y$ is solved as

$$z_i = \frac{y_i}{d_i}, \qquad \text{for each } i = 1, 2, \ldots, n.$$

Then the system $L^t x = z$ is solved by backward substitution.

Any symmetric matrix A for which Gaussian elimination can be applied without row interchanges can be factored into the form LDL^t. In this general case, L is lower triangular with 1s on its diagonal and D is the diagonal matrix with the Gaussian elimination pivots on its diagonal. This result is widely applied, since symmetric matrices are common and easily recognized.

The last class of matrices considered are *band matrices*. In many applications the band matrices are also strictly diagonally dominant or positive definite. This combination of properties is very useful.

An $n \times n$ matrix is called a **band matrix** if integers p and q, with $1 < p, q < n$, exist with the property that $a_{ij} = 0$ whenever $i + p \le j$ or $j + q \le i$. The **bandwidth** is $w = p + q - 1$.

For example, the matrix

$$A = \begin{bmatrix} 7 & 2 & 0 \\ 3 & 5 & -1 \\ 0 & -5 & -6 \end{bmatrix}$$

is a band matrix with $p = q = 2$ and bandwidth 3.

Band matrices concentrate all their nonzero entries about the diagonal. Two special cases of band matrices that often occur in practice have $p = q = 2$ and $p = q = 4$.

The matrix of bandwidth 3, occurring when $p = q = 2$, has already been encountered in connection with the study of cubic spline approximations in Section 3.6. These matrices are called **tridiagonal** because they have the form

$$A = \begin{bmatrix} \alpha_1 & \gamma_1 & 0 & \cdots\cdots & 0 \\ \beta_2 & \alpha_2 & \gamma_2 & & \vdots \\ 0 & \beta_3 & \alpha_3 & & 0 \\ \vdots & & & & \gamma_{n-1} \\ 0 & \cdots\cdots & 0 & \beta_n & \alpha_n \end{bmatrix}.$$

Since the entries of tridiagonal matrices are predominantly zero, it is common to avoid the double subscript notation by relabeling the entries.

Tridiagonal matrices appear again in Chapter 11 in connection with the study of piecewise linear approximations to boundary-value problems. The case of $p = q = 4$ will also be used in that chapter for the solution of boundary-value problems, when the approximating functions assume the form of cubic splines.

The factorization methods can be simplified considerably in the case of band matrices because a large number of zeros appear in regular patterns. It is particularly interesting to observe the form the Crout (where $u_{ii} = 1$) and Doolittle (where $l_{ii} = 1$) methods assume in this case. To illustrate the situation, suppose a tridiagonal matrix

$$A = \begin{bmatrix} \alpha_1 & \gamma_1 & 0 & \cdots\cdots & 0 \\ \beta_2 & \alpha_2 & \gamma_2 & & \vdots \\ 0 & \beta_3 & \alpha_3 & & 0 \\ \vdots & & & & \gamma_{n-1} \\ 0 & \cdots\cdots & 0 & \beta_n & \alpha_n \end{bmatrix}.$$

can be factored into the triangular matrices L and U in the Crout form

$$L = \begin{bmatrix} l_1 & 0 & \cdots\cdots\cdots & 0 \\ \beta_2 & l_2 & & \vdots \\ 0 & \beta_3 & l_3 & \\ \vdots & & & 0 \\ 0 & \cdots\cdots & 0 & \beta_n & l_n \end{bmatrix} \quad \text{and} \quad U = \begin{bmatrix} 1 & u_1 & 0 & \cdots\cdots & 0 \\ 0 & 1 & u_2 & & \vdots \\ \vdots & & 1 & & 0 \\ & & & & u_{n-1} \\ 0 & \cdots\cdots\cdots & 0 & 1 \end{bmatrix}.$$

The entries are given by the following equations:

$$l_1 = \alpha_1 \quad \text{and} \quad u_1 = \frac{\gamma_1}{l_1};$$

for each $i = 2, 3, \ldots, n - 1$,

$$l_i = \alpha_i - \beta_i u_{i-1} \quad \text{and} \quad u_i = \frac{\gamma_i}{l_i};$$

and

$$l_n = \alpha_n - \beta_n u_{n-1}.$$

The linear system $A\mathbf{x} = LU\mathbf{x} = \mathbf{b}$ is solved using the equations

$$y_1 = \frac{b_1}{l_1}$$

and, for each $i = 2, 3, \ldots, n$,

$$y_i = \frac{1}{l_i}[b_i - \beta_i y_{i-1}],$$

which determines $\mathbf{y}$ in the linear system $L\mathbf{y} = \mathbf{b}$. The linear system $U\mathbf{x} = \mathbf{y}$ is solved using the equations

$$x_n = y_n$$

and, for each $i = n - 1, n - 2, \ldots, 1$,

$$x_i = y_i - u_i x_{i+1}.$$

The Crout factorization of a tridiagonal matrix can be performed with the program CRTRLS67.

EXAMPLE 3 To illustrate the procedure for tridiagonal matrices, consider the tridiagonal system of equations

$$
\begin{array}{rcrcrcrcl}
2x_1 & - & x_2 & & & & & = & 1, \\
-x_1 & + & 2x_2 & - & x_3 & & & = & 0, \\
& - & x_2 & + & 2x_3 & - & x_4 & = & 0, \\
& & & - & x_3 & + & 2x_4 & = & 1,
\end{array}
$$

whose augmented matrix is

$$
\left[
\begin{array}{cccc:c}
2 & -1 & 0 & 0 & 1 \\
-1 & 2 & -1 & 0 & 0 \\
0 & -1 & 2 & -1 & 0 \\
0 & 0 & -1 & 2 & 1
\end{array}
\right].
$$

The LU factorization is given by

$$
\begin{bmatrix}
2 & -1 & 0 & 0 \\
-1 & 2 & -1 & 0 \\
0 & -1 & 2 & -1 \\
0 & 0 & -1 & 2
\end{bmatrix}
=
\begin{bmatrix}
2 & 0 & 0 & 0 \\
-1 & \frac{3}{2} & 0 & 0 \\
0 & -1 & \frac{4}{5} & 0 \\
0 & 0 & -1 & \frac{5}{4}
\end{bmatrix}
\begin{bmatrix}
1 & -\frac{1}{2} & 0 & 0 \\
0 & 1 & -\frac{2}{3} & 0 \\
0 & 0 & 1 & -\frac{3}{4} \\
0 & 0 & 0 & 1
\end{bmatrix}
= LU.
$$

Solving the system $Ly = b$ gives $y = (\frac{1}{2}, \frac{1}{3}, \frac{1}{4}, 1)^t$, and the solution of $Ux = y$ is $x = (1, 1, 1, 1)^t$.   ■ ■ ■

The tridiagonal factorization can be applied whenever $l_i \neq 0$ for each $i = 1, 2, \ldots, n$. Two conditions, either of which ensure that this is true, are that the coefficient matrix of the system is positive definite or that it is strictly diagonally dominant. An additional condition that ensures this method can be applied is as follows:

Nonsingular Tridiagonal Matrices

Suppose that A is tridiagonal with $\beta_i \neq 0$ and $\gamma_i \neq 0$ for each $i = 2, 3, \ldots, n - 1$. If $|\alpha_1| > |\gamma_1|, |\alpha_i| \geq |\beta_i| + |\gamma_i|$ for each $i = 2, 3, n - 1$, and $|\alpha_n| > |\beta_n|$, then A is nonsingular and the values of l_i are nonzero for each $i = 1, 2, \ldots, n$.

EXERCISE SET 6.6

1. Determine which of the following matrices are (i) symmetric, (ii) singular, (iii) strictly diagonally dominant, (iv) positive definite:

(a) $\begin{bmatrix} 2 & 1 \\ 1 & 3 \end{bmatrix}$

(b) $\begin{bmatrix} -2 & 1 \\ 1 & -3 \end{bmatrix}$

(c) $\begin{bmatrix} 2 & 1 & 0 \\ 0 & 3 & 0 \\ 1 & 0 & 4 \end{bmatrix}$

(d) $\begin{bmatrix} 2 & 1 & 0 \\ 0 & 3 & 2 \\ 1 & 2 & 4 \end{bmatrix}$

(e) $\begin{bmatrix} 4 & 2 & 6 \\ 3 & 0 & 7 \\ -2 & -1 & -3 \end{bmatrix}$

(f) $\begin{bmatrix} 2 & -1 & 0 \\ -1 & 4 & 2 \\ 0 & 2 & 2 \end{bmatrix}$

(g) $\begin{bmatrix} 4 & 0 & 0 & 0 \\ 6 & 7 & 0 & 0 \\ 9 & 11 & 1 & 0 \\ 5 & 4 & 1 & 1 \end{bmatrix}$

(h) $\begin{bmatrix} 2 & 3 & 1 & 2 \\ -2 & 4 & -1 & 5 \\ 3 & 7 & 1.5 & 1 \\ 6 & -9 & 3 & 7 \end{bmatrix}$

2. Find a factorization of the form $A = LDL^t$ for the following matrices:

(a) $A = \begin{bmatrix} 2 & -1 & 0 \\ -1 & 2 & -1 \\ 0 & -1 & 2 \end{bmatrix}$

(b) $A = \begin{bmatrix} 4 & 1 & 1 & 1 \\ 1 & 3 & -1 & 1 \\ 1 & -1 & 2 & 0 \\ 1 & 1 & 0 & 2 \end{bmatrix}$

(c) $A = \begin{bmatrix} 4 & 1 & -1 & 0 \\ 1 & 3 & -1 & 0 \\ -1 & -1 & 5 & 2 \\ 0 & 0 & 2 & 4 \end{bmatrix}$

(d) $A = \begin{bmatrix} 6 & 2 & 1 & -1 \\ 2 & 4 & 1 & 0 \\ 1 & 1 & 4 & -1 \\ -1 & 0 & -1 & 3 \end{bmatrix}$

3. Find a factorization of the form $A = LL^t$ for the matrices in Exercise 2.

4. Use the factorizations in Exercise 2 to solve the following linear systems:

(a)
$$\begin{array}{rcrcrcr} 2x_1 & - & x_2 & & & = & 3 \\ -x_1 & + & 2x_2 & - & x_3 & = & -3 \\ & - & x_2 & + & 2x_3 & = & 1 \end{array}$$

(b) $4x_1 + x_2 + x_3 + x_4 = 0.65$
$x_1 + 3x_2 - x_3 + x_4 = 0.05$
$x_1 - x_2 + 2x_3 = 0$
$x_1 + x_2 + 2x_4 = 0.5$

(c) $4x_1 + x_2 - x_3 = 7$
$x_1 + 3x_2 - x_3 = 8$
$-x_1 - x_2 + 5x_3 + 2x_4 = -4$
$2x_3 + 4x_4 = 6$

(d) $6x_1 + 2x_2 + x_3 - x_4 = 0$
$2x_1 + 4x_2 + x_3 = 7$
$x_1 + x_2 + 4x_3 - x_4 = -1$
$x_1 - x_3 + 3x_4 = -2$

5. Use the factorizations in Exercise 3 to solve the linear systems in Exercise 4.

6. Solve the following linear systems using Crout factorization for tridiagonal systems:

   (a) $x_1 - x_2 = 0$
   $-2x_1 + 4x_2 - 2x_3 = -1$
   $- x_2 + 2x_3 = 1.5$

   (b) $3x_1 + x_2 = -1$
   $2x_1 + 4x_2 + x_3 = 7$
   $2x_2 + 5x_3 = 9$

   (c) $2x_1 - x_2 = 3$
   $-x_1 + 2x_2 - x_3 = -3$
   $- x_2 + 2x_3 = 1$

   (d) $0.5x_1 + 0.25x_2 = 0.35$
   $0.35x_1 + 0.8x_2 + 0.4x_3 = 0.77$
   $0.25x_2 + x_3 + 0.5x_4 = -0.5$
   $x_3 - 2x_4 = -2.25$

7. Let A be the 10×10 tridiagonal matrix given by $\alpha_i = 2, \gamma_i = \beta_i = -1$ for each $i = 2, \ldots, 9$ with $\alpha_1 = \alpha_{10} = 2, \gamma_1 = \beta_{10} = -1$. Let $\mathbf{b}$ be the ten-dimensional column vector given by $b_1 = b_{10} = 1$ and $b_i = 0$ for each $i = 2, 3, \ldots, 9$. Solve $A\mathbf{x} = \mathbf{b}$ using the Crout factorization for tridiagonal systems.

8. Find the LDL^t factorization for the following matrices if possible:

   (a)
   $$A = \begin{bmatrix} 3 & -3 & 6 \\ -3 & 2 & -7 \\ 6 & -7 & 13 \end{bmatrix}$$

   (b)
   $$A = \begin{bmatrix} 3 & -6 & 9 \\ -6 & 14 & -20 \\ 9 & -20 & 29 \end{bmatrix}$$

   (c)
   $$A = \begin{bmatrix} -1 & 2 & 0 & 1 \\ 2 & -3 & 2 & -1 \\ 0 & 2 & 5 & 6 \\ 1 & -1 & 6 & 12 \end{bmatrix}$$

   (d)
   $$A = \begin{bmatrix} 2 & -2 & 4 & -4 \\ -2 & 3 & -4 & 5 \\ 4 & -4 & 10 & -10 \\ -4 & 5 & -10 & 14 \end{bmatrix}$$

9. Which of the symmetric matrices in Exercise 8 are positive definite?

10. Suppose that A and B are strictly diagonally dominant $n \times n$ matrices.
    (a) Is $-A$ strictly diagonally dominant?
    (b) Is A^t strictly diagonally dominant?
    (c) Is $A + B$ strictly diagonally dominant?
    (d) Is A^2 strictly diagonally dominant?
    (e) Is $A - B$ strictly diagonally dominant?

11. Suppose that A and B are positive definite $n \times n$ matrices.
    (a) Is $-A$ positive definite?
    (b) Is A^t positive definite?
    (c) Is $A + B$ positive definite?
    (d) Is A^2 positive definite?
    (e) Is $A - B$ positive definite?

12. Suppose A and B commute; that is, $AB = BA$. Must A^t and B^t also commute?

13. Construct a 2×2 matrix A that is nonsymmetric but for which $\mathbf{x}^t A \mathbf{x} > 0$ for all $\mathbf{x} \neq \mathbf{0}$.

14. In a paper by Dorn and Burdick, it is reported that the average wing length that resulted from mating three mutant varieties of fruit flies (*Drosophila melanogaster*) can be expressed in the symmetric matrix form

$$A = \begin{bmatrix} 1.59 & 1.69 & 2.13 \\ 1.69 & 1.31 & 1.72 \\ 2.13 & 1.72 & 1.85 \end{bmatrix},$$

where a_{ij} denotes the average wing length of an offspring resulting from the mating of a male of type i with a female of type j.
    (a) What physical significance is associated with the symmetry of this matrix?
    (b) Is this matrix positive definite? If so, prove it; if not, find a nonzero vector $\mathbf{x}$ for which $\mathbf{x}^t A \mathbf{x} \leq 0$.

6.7   Survey of Methods and Software

In this chapter we have looked at direct methods for solving linear systems. A linear system consists of n equations in n unknowns expressed in matrix notation as $A\mathbf{x} = \mathbf{b}$. These techniques use a finite sequence of arithmetic operations to determine the exact solution of the system subject only to round-off error. We found that the linear system $A\mathbf{x} = \mathbf{b}$ has a unique solution if and only if A^{-1} exists, which is equivalent to $\det A \neq 0$. The solution of the linear system is the vector $\mathbf{x} = A^{-1}\mathbf{b}$.

Pivoting techniques were introduced to minimize the effects of round-off error, which can dominate the solution when using direct methods. We studied maximal column pivoting, scaled column pivoting, and total pivoting. We recommend the maximal column or scaled column pivoting methods for most problems, since these decrease the effects of round-off error without adding much extra computation. Total pivoting should be used if round-off error is suspected to be large. In Section 7.6 we will see some procedures for estimating this round-off error.

Gaussian elimination with minor modifications produces a factorization of the matrix A into LU, where L is lower triangular with 1s on the diagonal and U is upper triangular. Not all nonsingular matrices can be factored this way, but a permutation of the rows will always give a factorization of the form $PA = LU$, where P is the permutation matrix used to rearrange the rows of A. The advantage of the factorization is that the work is reduced when solving linear systems with the same coefficient matrix A and different right-hand sides $\mathbf{b}$.

When the matrix A is positive definite, factorizations take a simpler form. For example, the Choleski factorization has the form $A = LL^t$, where L is lower triangular. Positive definite matrices can also be factored in the form $A = LDL^t$, where L is lower triangular with 1s on the diagonal and D is diagonal. With these factorizations, manipulations involving A can be simplified. If A is tridiagonal, the LU factorization takes a particularly simple form, with L having 1s on the main diagonal and 0s elsewhere, except possibly on the diagonal immediately below the main diagonal. In addition, U has its only nonzero entries on the main diagonal and one diagonal above.

The direct methods are the methods of choice for most linear systems. For tridiagonal, banded, and positive definite matrices, the special methods are recommended. For the general case, Gaussian elimination or LU factorization methods, which allow pivoting, are recommended. In these cases, the effects of round-off error should be monitored. In Section 7.6 we discuss estimating errors in direct methods.

Large linear systems with primarily zero entries occurring in regular patterns can be solved efficiently by using a technique like one of those we discuss in Chapter 7. Systems of this type arise naturally, for example, when finite-difference techniques are used to solve boundary value problems, a common application in the numerical solution of partial-differential equations.

It can be very difficult to solve a large linear system that has primarily nonzero entries or one where the zero entries are not in a predictable pattern. The matrix associated with the system can be placed in secondary storage in partitioned form and portions read into main memory only as needed for calculation. Methods that require secondary storage can be either iterative or direct, but they generally require techniques from the fields of data structures and graph theory.

The software for matrix operations and the direct solution of linear systems implemented in IMSL and NAG is based on LINPACK, a subroutine package that is in the public domain. There is excellent documentation available with it and from the books written about it. We focus on several of the subroutines that are available in all three sources.

LINPACK consists of lower-level operations called Basic Linear Algebra Subprograms (BLAS) and higher-level subroutines for solving linear systems that call the lower-level routines. Level 1 of BLAS consists of vector operations with $O(n)$ input data and operation counts. Level 2 consists of the matrix-vector operations with $O(n^2)$ input data and operation counts. For example, in Level 1, the subroutine SCOPY overwrites a vector $\mathbf{y}$ with a vector $\mathbf{x}$, SSCAL computes a scalar a times a vector $\mathbf{x}$, SAXPY adds a scalar times a vector to a vector, and SDOT computes the inner product of two vectors. SNRM2 computes the Euclidean norm of a vector, a concept we will see in Chapter 7, and ISAMAX computes the index of the vector component that gives the maximum absolute value of all the components. In Level 2, MMULT computes the product of a matrix and a vector.

The subroutines in LINPACK for solving linear systems first factor the matrix A. The factorization depends on the type of matrix in the following way:

1. General matrix $PA = LU$;
2. Positive definite matrix $A = LL^t$;
3. Symmetric matrix $A = LDL^t$;
4. Tridiagonal matrix $A = LU$ (in banded form).

The subroutine STRSL solves a triangular linear system, in which the matrix can be either upper or lower triangular. This subroutine serves as a workhorse that is called by many of the other subroutines.

The subroutine SGEFA factors PA into LU as a preliminary operation to the subroutine SGESL, which then computes the solution to $Ax = \mathbf{b}$. The subroutine SGEDI is used to construct the inverse of a matrix A and to calculate the determinant of A once A has been factored via SGEFA.

The Choleski factorization of a positive definite matrix A is obtained with the subroutine SPOFA. The linear system $Ax = \mathbf{b}$ can then be solved using the subroutine SPOSL. Inverses and determinants of positive definite matrices, given Choleski factorization can be computed using SPODI. If A is symmetric, the LDL^t factorization is found using SSIFA. Linear systems can then be solved using SSISL. If inverses or determinants are desired, SSIDI can be used.

The IMSL Library includes counterparts to almost all the LINPACK subroutines and some extensions as well. The IMSL routines are named with regard to the tasks they perform as follows:

1. First three letters of the name indicate the task.
    (a) LSL solves a linear system.
    (b) LFT factors a coefficient matrix.
    (c) LFS solves a linear system given factors from LFT.
    (d) LFD calculates the determinants of given factors.
    (e) LIN computes the inverse of given factors.

2. Last two letters determine the type of matrix involved.
    (a) RG is general.
    (b) RT is triangular.
    (c) DS is positive definite.
    (d) SF is symmetric.
    (e) RB is banded.

For example, the routine LFTDS factors a real positive definite matrix. This is only a partial list of the routines and classes of matrices in the package.

The NAG Library has many subroutines for direct methods of solving linear systems similar to those in LINPACK and IMSL. For example, the subroutine FO4AEF solves linear systems using Crout factorization. The subroutine FO4ATF solves a single linear system using Crout factorization, as in FO4AEF. The subroutine FO4EAF solves a single linear system when the matrix is real and tridiagonal, and FO4ASF solves a system when the matrix is real and positive definite. Inverse matrices can be computed by FO1AAF, for an arbitrary real matrix, and by FO1ACF, if the matrix is positive definite. A determinant can be computed using FO3AAF. Factorizations can be obtained using FO1BTF for the LU factorization for a real matrix and using FO1LEF for a tridiagonal matrix. Linear systems can then be solved using FO4AYF. Choleski's factorization of a positive definite matrix can be obtained using FO1BXF, and a linear system can then be solved using FO4AZF. The NAG library also includes the lower-level matrix-vector manipulations.

MATLAB is an interactive program for matrix and vector operations. It provides an interface between the user at the terminal and the methods of matrix algebra contained in LINPACK and EISPACK. The user enters data into matrix or vector variables using the keyboard or existing files.

The representation of matrices and matrix operations is fundamental to MATLAB. A matrix A can be defined by the statement

$$A = [1\ 2\ -1;\ 2\ 1\ 0;\ -1\ 1\ 2]$$

which yields

$$A = \begin{bmatrix} 1 & 2 & -1 \\ 2 & 1 & 0 \\ -1 & 1 & 2 \end{bmatrix}.$$

If matrices A and B are defined, MATLAB allows matrix operations naturally as follows:

1. $C = A'$ assigns A^t to C.
2. $C = A + B$ assigns $A + B$ to C if A and B have the same dimension.
3. $C = A * B$ assigns AB to C if AB is defined.
4. $C = A \hat{} \ p$ assigns A^p to C if P is a positive integer and A is square.
5. $C = \text{inv}(A)$ assigns A^{-1} to C if A is nonsingular.
6. $c = \det(A)$ assigns the determinant of A to c if A is square.

The linear system $A\mathbf{x} = \mathbf{b}$ can be solved by the statement

$$\mathbf{x} = A \backslash \mathbf{b},$$

provided $\mathbf{b}$ is defined by $\mathbf{b} = [b_1; b_2; \ldots b_n]$, as a column vector. Matrix factorizations are also possible, as follows:

1. $[B, C] = lu(A)$ assigns $P^t L$ to B and U to C if $A = (P^t L)U$.
2. $B = chol(A)$ assigns L^t to B if A is positive definite and $A = LL^t$ is the Choleski factorization of A.

Iterative Methods for Solving Linear Systems

7.1 Introduction

The previous chapter considered the approximation of the solution of a linear system using direct methods, techniques that would produce the exact solution if all the calculations were performed using exact arithmetic. In this chapter we describe some popular *iterative* techniques. These methods would not return the exact solution even if all the calculations could be performed using exact arithmetic. In many instances, however, they are more effective than the direct methods, since they can require far less computational effort and round-off error is reduced. This is particularly true when the matrix is **sparse**; that is, when it has a high percentage of zero entries.

Some additional material from linear algebra is needed to describe the convergence of the iterative methods. Principally, we need to have a measure of how close two vectors are to one another, since the object of an iterative method is to determine an approximation that is within a certain tolerance of the exact solution. In Section 7.2, the notion of a norm is used to show how various forms of distance between vectors can be described. We will also see how this concept can be extended to describe the norm of—and, consequently, the distance between—matrices. In Section 7.3, matrix eigenvalues and eigenvectors are described and we consider the connection between these concepts and the convergence of an iterative method.

Section 7.4 describes the elementary Jacobi and Gauss–Seidel iterative methods. By analyzing the size of the largest eigenvalue of a matrix associated with an iterative method, we can determine conditions that predict the likelihood of convergence of the method. In Section 7.5 we introduce the SOR method. This is the most commonly applied iterative technique, since it tends to reduce the norm of approximation errors the fastest.

The final two sections in the chapter discuss some of the concerns that should be addressed when applying either an iterative or direct technique for approximating the solution to a linear system.

7.2 Convergence of Vectors

The distance between the real numbers x and y is the absolute value $|x - y|$. In Chapter 2 we saw that the stopping techniques for the iterative root-finding techniques used this measure to estimate the accuracy of the approximate solutions and to determine when the approximation was sufficiently accurate to accept the result. The iterative methods for solving systems of equations use similar logic, so the first step is to determine a measurement procedure for n-dimensional vectors, the form that is taken by the solution to a system of equations.

Let $\mathbb{R}^n$ denote the set of all n-dimensional column vectors with real number coefficients. It is a space-saving convenience to use the transpose notation presented in Section 6.4 when such a vector is represented in terms of its components. For example, the vector

$$\mathbf{x} = \begin{bmatrix} x_1 \\ x_2 \\ \vdots \\ x_n \end{bmatrix}$$

will generally be written $\mathbf{x} = (x_1, x_2, \ldots, x_n)^t$.

Vector Norm on $\mathbb{R}^n$

A vector norm on $\mathbb{R}^n$ is a function, $\|\cdot\|$, from $\mathbb{R}^n$ into $\mathbb{R}$ with the following properties:

   (i) $\|\mathbf{x}\| \geq 0$, for all $\mathbf{x} \in \mathbb{R}^n$.
   (ii) $\|\mathbf{x}\| = 0$ if and only if $\mathbf{x} = (0, 0, \ldots, 0)^t \equiv \mathbf{0}$.
   (iii) $\|\alpha \mathbf{x}\| = |\alpha|\, \|\mathbf{x}\|$, for all $\alpha \in \mathbb{R}$ and $\mathbf{x} \in \mathbb{R}^n$.
   (iv) $\|\mathbf{x} + \mathbf{y}\| \leq \|\mathbf{x}\| + \|\mathbf{y}\|$, for all $\mathbf{x}, \mathbf{y} \in \mathbb{R}^n$.

For our purposes, we need only two specific norms on $\mathbb{R}^n$. (A third is presented in Exercise 2.)

The l_2 and l_∞ norms for the vector $\mathbf{x} = (x_1, x_2, \ldots, x_n)^t$ are defined by

$$\|\mathbf{x}\|_2 = \left\{ \sum_{i=1}^{n} x_i^2 \right\}^{1/2} \qquad \text{and} \qquad \|\mathbf{x}\|_\infty = \max_{1 \leq i \leq n} |x_i|.$$

The l_2 norm is called the **Euclidean norm** of the vector $\mathbf{x}$, since it represents the usual notion of distance from the origin in case $\mathbf{x}$ is in $\mathbb{R}^1 \equiv \mathbb{R}$, $\mathbb{R}^2$, or $\mathbb{R}^3$. For example, the l_2 norm of the vector $\mathbf{x} = (x_1, x_2, x_3)^t$ denotes the length of the straight line joining the points $(0,0,0)$ and (x_1, x_2, x_3); that is, the length of the shortest path between those two points. Figure 7.1 shows the boundary of those vectors in $\mathbb{R}^2$ and $\mathbb{R}^3$ that have l_2 norm less than 1. Figure 7.2 is a similar illustration for the l_∞ norm.

Figure 7.1

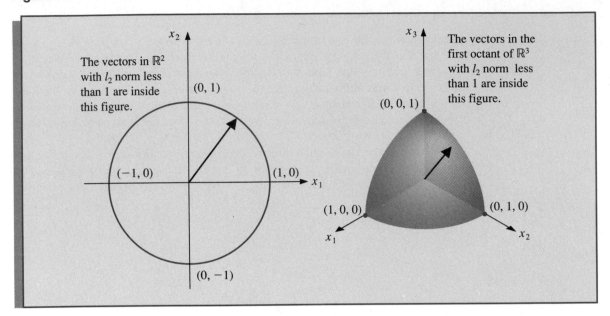

The vectors in $\mathbb{R}^2$ with l_2 norm less than 1 are inside this figure.

The vectors in the first octant of $\mathbb{R}^3$ with l_2 norm less than 1 are inside this figure.

Figure 7.2

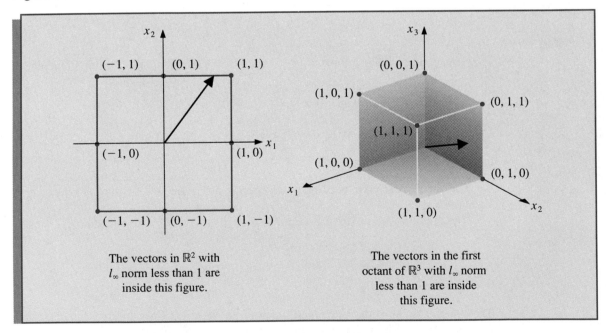

The vectors in $\mathbb{R}^2$ with l_∞ norm less than 1 are inside this figure.

The vectors in the first octant of $\mathbb{R}^3$ with l_∞ norm less than 1 are inside this figure.

EXAMPLE 1   The vector $\mathbf{x} = (-1, 1, -2)^t$ in $\mathbb{R}^3$ has norms

$$\|\mathbf{x}\|_2 = \sqrt{(-1)^2 + (1)^2 + (-2)^2} = \sqrt{6} \quad \text{and} \quad \|\mathbf{x}\|_\infty = \max\{|-1|, |1|, |-2|\} = 2.$$

■ ■ ■

Showing that $\|\mathbf{x}\|_\infty = \max_{1 \le i \le n} |x_i|$ satisfies the conditions necessary for a norm on $\mathbb{R}^n$ follows directly from the truth of similar statements concerning absolute values of real numbers. In the case of the l_2 norm, it is also easy to demonstrate the first three of the required properties, but the fourth,

$$\|\mathbf{x} + \mathbf{y}\|_2 \le \|\mathbf{x}\|_2 + \|\mathbf{y}\|_2,$$

is more difficult to show.

To demonstrate this inequality we need the Cauchy–Buniakowsky–Schwarz inequality, which states that for any $\mathbf{x} = (x_1, x_2, \ldots, x_n)^t$ and $\mathbf{y} = (y_1, y_2, \ldots, y_n)^t$,

$$\sum_{i=1}^{n} |x_i y_i| \le \left\{ \sum_{i=1}^{n} x_i^2 \right\}^{1/2} \left\{ \sum_{i=1}^{n} y_i^2 \right\}^{1/2}.$$

With this it follows that $\|\mathbf{x} + \mathbf{y}\|_2 \le \|\mathbf{x}\|_2 + \|\mathbf{y}\|_2$, since

$$\|\mathbf{x} + \mathbf{y}\|_2^2 = \sum_{i=1}^{n} x_i^2 + 2 \sum_{i=1}^{n} x_i y_i + \sum_{i=1}^{n} y_i^2 \le \sum_{i=1}^{n} x_i^2 + 2 \sum_{i=1}^{n} |x_i y_i| + \sum_{i=1}^{n} y_i^2$$

$$\le \sum_{i=1}^{n} x_i^2 + 2 \left\{ \sum_{i=1}^{n} x_i^2 \right\}^{1/2} \left\{ \sum_{i=1}^{n} y_i^2 \right\}^{1/2} + \sum_{i=1}^{n} y_i^2 = (\|\mathbf{x}\|_2 + \|\mathbf{y}\|_2)^2.$$

The norm of a vector gives a measure for the distance between the vector and the origin, so the distance between two vectors is the norm of the difference of the vectors.

Distance Between Two Vectors

If $\mathbf{x} = (x_1, x_2, \ldots, x_n)^t$ and $\mathbf{y} = (y_1, y_2, \ldots, y_n)^t$ are vectors in $\mathbb{R}^n$, the l_2 and l_∞ distances between $\mathbf{x}$ and $\mathbf{y}$ are defined by

$$\|\mathbf{x} - \mathbf{y}\|_2 = \left\{ \sum_{i=1}^{n} (x_i - y_i)^2 \right\}^{1/2} \qquad \text{and} \qquad \|\mathbf{x} - \mathbf{y}\|_\infty = \max_{1 \le i \le n} |x_i - y_i|.$$

EXAMPLE 2 The linear system

$$3.3330x_1 + 15920x_2 - 10.333x_3 = 15913,$$
$$2.2220x_1 + 16.710x_2 + 9.6120x_3 = 28.544,$$
$$1.5611x_1 + 5.1791x_2 + 1.6852x_3 = 8.4254,$$

has solution $(x_1, x_2, x_3)^t = (1.0000, 1.0000, 1.0000)^t$. If Gaussian elimination is performed in five-digit arithmetic using maximal column pivoting, the solution obtained is

$$\tilde{\mathbf{x}} = (\tilde{x}_1, \tilde{x}_2, \tilde{x}_3)^t = (1.2001, 0.99991, 0.92538)^t.$$

Measurements of $\mathbf{x} - \tilde{\mathbf{x}}$ are given by

$$\|\mathbf{x} - \tilde{\mathbf{x}}\|_\infty = \max\{|1.0000 - 1.2001|, |1.0000 - 0.99991|, |1.0000 - 0.92538|\}$$
$$= \max\{0.2001, 0.00009, 0.07462\} = 0.2001$$

and

$$\|\mathbf{x} - \tilde{\mathbf{x}}\|_2 = [(1.0000 - 1.2001)^2 + (1.0000 - 0.99991)^2 + (1.0000 - 0.92538)^2]^{\frac{1}{2}}$$
$$= [(0.2001)^2 + (0.00009)^2 + (0.07462)^2]^{\frac{1}{2}} = 0.21356.$$

Although the components $\tilde{x}_2$ and $\tilde{x}_3$ are good approximations to x_2 and x_3, the component $\tilde{x}_1$ is a poor approximation to x_1, and $|x_1 - \tilde{x}_1|$ dominates both norms.

■ ■ ■

The distance concept in $\mathbb{R}^n$ is used to define a limit of a sequence of vectors. A sequence $\{\mathbf{x}^{(k)}\}_{k=1}^\infty$ of vectors in $\mathbb{R}^n$ is said to **converge** to $\mathbf{x}$ with respect to the norm $\|\cdot\|$ if, given any $\epsilon > 0$, there exists an integer $N(\epsilon)$ such that

$$\|\mathbf{x}^{(k)} - \mathbf{x}\| < \epsilon, \qquad \text{for all } k \geq N(\epsilon).$$

EXAMPLE 3   Let $\mathbf{x}^{(k)} \in \mathbb{R}^4$ be defined by

$$\mathbf{x}^{(k)} = (x_1^{(k)}, x_2^{(k)}, x_3^{(k)}, x_4^{(k)})^t = \left(1, 2 + \frac{1}{k}, \frac{3}{k^2}, e^{-k} \sin k\right)^t.$$

Since $\lim_{k\to\infty} 1 = 1$, $\lim_{k\to\infty}(2 + 1/k) = 2$, $\lim_{k\to\infty} 3/k^2 = 0$, and $\lim_{k\to\infty} e^{-k} \sin k = 0$, and an integer $N(\epsilon)$ can be found so that we simultaneously have $|x_1 - 1|$, $|x_2^{(k)} - 2|$, $(x_3^{(k)} - 0)$, and $|x_4^{(k)} - 0|$ less than ϵ, for any given ϵ. This implies that the sequence $\{\mathbf{x}^{(k)}\}$ converges to $(1, 2, 0, 0)^t$ with respect to $\|\cdot\|_\infty$.   ■ ■ ■

To show directly that the sequence in Example 3 converges to $(1, 2, 0, 0)^t$ with respect to the l_2 norm is quite complicated. However, suppose that j is an index with the property that

$$\|\mathbf{x}\|_\infty = \max_{i=1,\ldots,n} |x_i| = |x_j|.$$

Then

$$\|\mathbf{x}\|_\infty^2 = |x_j|^2 = x_j^2 \leq \sum_{i=1}^n x_i^2 \leq \sum_{i=1}^n x_j^2 = n x_j^2 = n\|\mathbf{x}\|_\infty^2.$$

The inequality

$$\|\mathbf{x}\|_\infty \leq \|\mathbf{x}\|_2 \leq \sqrt{n}\|\mathbf{x}\|_\infty$$

implies that the sequence of vectors $\{\mathbf{x}^{(k)}\}$ converges to $\mathbf{x}$ in $\mathbb{R}^n$ with respect to $\|\cdot\|_2$ if and only if $\lim_{k\to\infty} x_i^{(k)} = x_i$ for each $i = 1, 2, \ldots, n$, since this is when the sequence converges in the l_∞ norm.

In fact, it can be shown that all norms on $\mathbb{R}^n$ are equivalent with respect to convergence; that is, if $\|\cdot\|$ and $\|\cdot\|'$ are any two norms on $\mathbb{R}^n$ and $\{\mathbf{x}^{(k)}\}_{k=1}^\infty$ has the limit $\mathbf{x}$ with respect to $\|\cdot\|$, then $\{\mathbf{x}^{(k)}\}_{k=1}^\infty$ has the limit $\mathbf{x}$ with respect to $\|\cdot\|'$. Since a vector sequence converges in the l_∞ norm precisely when each of its component sequences converge, we have the following:

Vector Sequence Convergence

The sequence of vectors $\{\mathbf{x}^{(k)}\}$ converges to $\mathbf{x}$ in $\mathbb{R}^n$ if and only if $\lim_{k \to \infty} x_i^{(k)} = x_i$ for each $i = 1, 2, \ldots, n$.

In the subsequent sections, we will need methods for determining the distance between $n \times n$ matrices. This again requires the use of a norm.

Matrix Norm

A matrix norm on the set of all $n \times n$ matrices is a real-valued function, $\| \cdot \|$, defined on this set, satisfying for all $n \times n$ matrices A and B and all real numbers α:

    (i) $\|A\| \geq 0,$
    (ii) $\|A\| = 0$ if and only if A is O, the matrix with all zero entries,
    (iii) $\|\alpha A\| = |\alpha| \, \|A\|,$
    (iv) $\|A + B\| \leq \|A\| + \|B\|,$
    (v) $\|AB\| \leq \|A\| \, \|B\|.$

A **distance between $n \times n$ matrices** A and B with respect to this matrix norm is $\|A - B\|$. Although matrix norms can be obtained in various ways, the only norms we consider are those that are natural consequences of a vector norm.

Natural Matrix Norm

If $\| \cdot \|$ is a vector norm on $\mathbb{R}^n$, the natural matrix norm on the set of $n \times n$ matrices given by $\| \cdot \|$ is defined by

$$\|A\| = \max_{\|\mathbf{x}\| = 1} \|A\mathbf{x}\|.$$

As a consequence, the matrix norms we will consider have the forms

$$\|A\|_\infty = \max_{\|\mathbf{x}\|_\infty = 1} \|A\mathbf{x}\|_\infty \qquad \text{(the } l_\infty \text{ norm)}$$

and

$$\|A\|_2 = \max_{\|\mathbf{x}\|_2 = 1} \|A\mathbf{x}\|_2, \qquad \text{(the } l_2 \text{ norm)}.$$

When $n = 2$ these norms have the geometric representations shown in Figures 7.3 and 7.4.

Figure 7.3

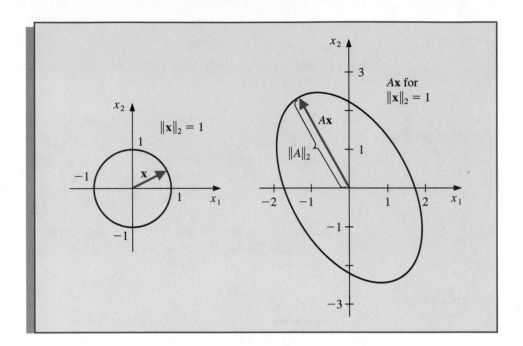

Figure 7.4

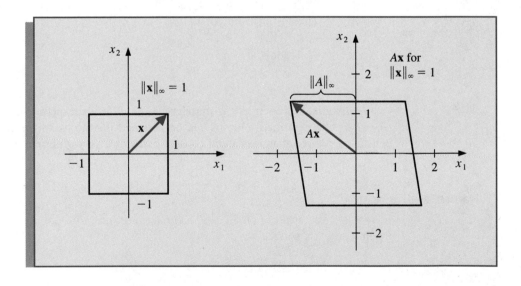

The l_∞ norm of a matrix has a representation with respect to the entries of the matrix that makes it particularly easy to compute:

$\|\cdot\|_\infty$ Norm of a Matrix

$$\|A\|_\infty = \max_{1 \le i \le n} \sum_{j=1}^{n} |a_{ij}|.$$

EXAMPLE 4 If

$$A = \begin{bmatrix} 1 & 2 & -1 \\ 0 & 3 & -1 \\ 5 & -1 & 1 \end{bmatrix},$$

then

$$\sum_{j=1}^{3} |a_{1j}| = |1| + |2| + |-1| = 4, \qquad \sum_{j=1}^{3} |a_{2j}| = |0| + |3| + |-1| = 4,$$

and

$$\sum_{j=1}^{3} |a_{3j}| = |5| + |-1| + |1| = 7.$$

So

$$\|A\|_\infty = \max\{4, 4, 7\} = 7.$$

■ ■ ■

The l_2 norm of a matrix is not as easily determined, but in the next section we will discover an alternative method for finding this norm.

EXERCISE SET 7.2

1. Find $\|\mathbf{x}\|_\infty$ and $\|\mathbf{x}\|_2$ for the following vectors:
   (a) $\mathbf{x} = (3, -4, 0, \frac{3}{2})^t$
   (b) $\mathbf{x} = (2, 1, -3, 4)^t$
   (c) $\mathbf{x} = (\sin k, \cos k, 2^k)^t$, for a fixed positive integer k
   (d) $\mathbf{x} = (4/(k+1), 2/k^2, k^2 e^{-k})^t$, for a fixed positive integer k

2. (a) Verify that the function $\|\cdot\|_1$, defined on $\mathbb{R}^n$ by

$$\|\mathbf{x}\|_1 = \sum_{i=1}^{n} |x_i|,$$

   is a norm on $\mathbb{R}^n$.
   (b) Find $\|\mathbf{x}\|_1$ for the vectors given in Exercise 1.

3. Find the limits of the following sequences of vectors:
   (a) $\mathbf{x}^{(k)} = (1/k, e^{1-k}, -2/k^2)^t$
   (b) $\mathbf{x}^{(k)} = (e^{-k} \cos k, k \sin(1/k), 3 + k^{-2})^t$
   (c) $\mathbf{x}^{(k)} = (ke^{-k^2}, (\cos k)/k, \sqrt{k^2 + k} - k)^t$
   (d) $\mathbf{x}^{(k)} = (e^{1/k}, (k^2 + 1)/(1 - k^2), (1/k^2)(1 + 3 + 5 + \cdots + (2k - 1)))^t$

4. Find $\|\cdot\|_\infty$ for the following matrices:
   (a) $\begin{bmatrix} 10 & 15 \\ 0 & 1 \end{bmatrix}$ (b) $\begin{bmatrix} 10 & 0 \\ 15 & 1 \end{bmatrix}$
   (c) $\begin{bmatrix} 2 & -1 & 0 \\ -1 & 2 & -1 \\ 0 & -1 & 2 \end{bmatrix}$ (d) $\begin{bmatrix} 4 & -1 & 7 \\ -1 & 4 & 0 \\ -7 & 0 & 4 \end{bmatrix}$

5. The following linear systems $Ax = b$ have x as the actual solution and $\tilde{x}$ as an approximate solution. Compute $\|x - \tilde{x}\|_\infty$ and $\|A\tilde{x} - b\|_\infty$.

(a)
$$\frac{1}{2}x_1 + \frac{1}{3}x_2 = \frac{1}{63}$$
$$\frac{1}{3}x_1 + \frac{1}{4}x_2 = \frac{1}{168}$$
$$x = \left(\frac{1}{7}, -\frac{1}{6}\right)^t$$
$$\tilde{x} = (0.142, -0.166)^t$$

(b)
$$x_1 + 2x_2 + 3x_3 = 1$$
$$2x_1 + 3x_2 + 4x_3 = -1$$
$$3x_1 + 4x_2 + 6x_3 = 2$$
$$x = (0, -7, 5)^t$$
$$\tilde{x} = (-0.33, -7.9, 5.8)^t$$

(c)
$$x_1 + 2x_2 + 3x_3 = 1$$
$$2x_1 + 3x_2 + 4x_3 = -1$$
$$3x_1 + 4x_2 + 6x_3 = 2$$
$$x = (0, -7, 5)^t$$
$$\tilde{x} = (-0.2, -7.5, 5.4)^t$$

(d)
$$0.04x_1 + 0.01x_2 - 0.01x_3 = 0.06$$
$$0.2x_1 + 0.5x_2 - 0.2x_3 = 0.3$$
$$x_1 + 2x_2 + 4x_3 = 11$$
$$x = (1.827586, 0.6551724, 1.965517)^t$$
$$\tilde{x} = (1.8, 0.64, 1.9)^t$$

6. The matrix norm $\|\cdot\|_1$, defined by $\|A\|_1 = \max_{\|x\|_1 = 1} \|Ax\|_1$, can be computed using the formula

$$\|A\|_1 = \max_{1 \le j \le n} \sum_{i=1}^{n} |a_{ij}|,$$

where the vector norm $\|\cdot\|_1$ is defined in Exercise 2. Find $\|\cdot\|_1$ for the matrices in Exercise 4.

7.3 Eigenvalues and Eigenvectors

An $n \times m$ matrix can be considered as a function that takes m-dimensional vectors into n-dimensional vectors. A square matrix takes the set of n dimensional vectors into itself. In this case certain nonzero vectors have x and Ax parallel, which means that a constant λ exists with $Ax = \lambda x$, or that $(A - \lambda I)x = 0$. There is a close connection between these numbers λ and the likelihood that an iterative method will converge. We consider this connection in this section.

For a square $n \times n$ matrix A, the **characteristic polynomial** of A is defined by

$$p(\lambda) = \det(A - \lambda I).$$

Because of the way the determinant of a matrix is defined, p is an nth-degree polynomial and, consequently, has at most n distinct zeros, some of which may be complex. These zeros of p are called the **eigenvalues** of the matrix A.

If λ is an eigenvalue, then $\det(A - \lambda I) = 0$, and the equivalence result at the end of Section 6.4 implies that $A - \lambda I$ is a singular matrix. As a consequence, the linear system defined by $(A - \lambda I)x = 0$ has a solution other than the zero vector. If $(A - \lambda I)x = 0$ and $x \ne 0$, then x is called an **eigenvector** of A corresponding to the eigenvalue λ.

If x is an eigenvector associated with the eigenvalue λ, then $Ax = \lambda x$, so the matrix A takes the vector x into a scalar multiple of itself. When λ is a real number and $\lambda > 1$, A has the effect of stretching x by a factor of λ. When $0 < \lambda < 1$, A shrinks x by a factor of λ. When $\lambda < 0$, the effects are similar, but the direction is reversed (see Figure 7.5).

Figure 7.5

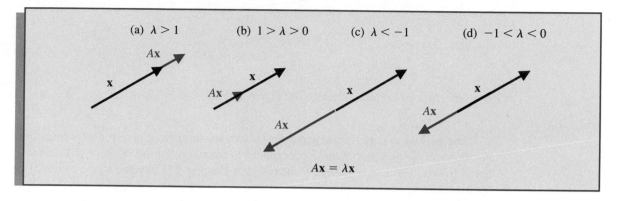

EXAMPLE 1   Let

$$
A = \begin{bmatrix} 1 & 0 & 2 \\ 0 & 1 & -1 \\ -1 & 1 & 1 \end{bmatrix}.
$$

To compute the eigenvalues of A, consider

$$
p(\lambda) = \det(A - \lambda I) = \det \begin{bmatrix} 1-\lambda & 0 & 2 \\ 0 & 1-\lambda & -1 \\ -1 & 1 & 1-\lambda \end{bmatrix} = (1-\lambda)(\lambda^2 - 2\lambda + 4).
$$

The eigenvalues of A are the solutions of $p(\lambda) = 0$: $\lambda_1 = 1$, $\lambda_2 = 1 + \sqrt{3}i$, and $\lambda_3 = 1 - \sqrt{3}i$.

An eigenvector $\mathbf{x}$ of A associated with λ_1 is a solution of the system $(A - \lambda_1 I)\mathbf{x} = \mathbf{0}$:

$$
\begin{bmatrix} 0 & 0 & 2 \\ 0 & 0 & -1 \\ -1 & 1 & 0 \end{bmatrix} \cdot \begin{bmatrix} x_1 \\ x_2 \\ x_3 \end{bmatrix} = \begin{bmatrix} 0 \\ 0 \\ 0 \end{bmatrix}.
$$

Thus

$$
2x_3 = 0, \qquad -x_3 = 0, \qquad \text{and} \qquad -x_1 + x_2 = 0,
$$

which implies that

$$
x_3 = 0, \qquad x_2 = x_1, \qquad \text{and} \qquad x_1 \text{ is arbitrary.}
$$

The choice $x_1 = 1$ produces the eigenvector $(1, 1, 0)^t$, corresponding to the eigenvalue $\lambda_1 = 1$. Since λ_2 and λ_3 are complex numbers, their corresponding eigenvectors are also complex. To find an eigenvector for λ_2, we solve the system

$$
\begin{bmatrix} 1-(1+\sqrt{3}i) & 0 & 2 \\ 0 & 1-(1+\sqrt{3}i) & -1 \\ -1 & 1 & 1-(1+\sqrt{3}i) \end{bmatrix} \begin{bmatrix} x_1 \\ x_2 \\ x_3 \end{bmatrix} = \begin{bmatrix} 0 \\ 0 \\ 0 \end{bmatrix}.
$$

One solution to this system is the vector

$$\left(-\tfrac{2\sqrt{3}}{3}i, \tfrac{\sqrt{3}}{3}i, 1\right)^t.$$

Similarly, the vector

$$\left(\tfrac{2\sqrt{3}}{3}i, -\tfrac{\sqrt{3}}{3}i, 1\right)^t$$

is an eigenvector corresponding to the eigenvalue $\lambda_3 = 1 - \sqrt{3}i$.   ■ ■ ■

The notions of eigenvalues and eigenvectors are introduced here for a specific computational convenience, but these concepts arise frequently in the study of physical systems. In fact, they are of sufficient interest that Chapter 9 is devoted to their numerical approximation.

The **spectral radius** $\rho(A)$ of a matrix A is defined by

$$\rho(A) = \max |\lambda|, \qquad \text{where } \lambda \text{ represents the eigenvalues of } A.$$

[*Note:* For complex $\lambda = \alpha + \beta i$, we have $|\lambda| = (\alpha^2 + \beta^2)^{\frac{1}{2}}$.]

EXAMPLE 2   For the matrix considered in Example 1,

$$\rho(A) = \max\{1, |1 + \sqrt{3}i|, |1 - \sqrt{3}i|\} = \max\{1, 2, 2\} = 2.$$ ■ ■ ■

The spectral radius is closely related to the norm of a matrix.

l_2 Matrix Norm Characterization

If A is an $n \times n$ matrix, then

  (a) $\|A\|_2 = [\rho(A^t A)]^{1/2}$,
  (b) $\rho(A) \le \|A\|$ for any natural norm.

The first part of this result is the computational method for determining the l_2 norm of matricies that we mentioned at the end of the previous section.

EXAMPLE 3   If

$$A = \begin{bmatrix} 1 & 1 & 0 \\ 1 & 2 & 1 \\ -1 & 1 & 2 \end{bmatrix},$$

then

$$A^t A = \begin{bmatrix} 1 & 1 & -1 \\ 1 & 2 & 1 \\ 0 & 1 & 2 \end{bmatrix}\begin{bmatrix} 1 & 1 & 0 \\ 1 & 2 & 1 \\ -1 & 1 & 2 \end{bmatrix} = \begin{bmatrix} 3 & 2 & -1 \\ 2 & 6 & 4 \\ -1 & 4 & 5 \end{bmatrix}.$$

To calculate $\rho(A^t A)$, we need the eigenvalues of $A^t A$:

$$0 = \det(A^t A - \lambda I) = \det \begin{bmatrix} 3 - \lambda & 2 & -1 \\ 2 & 6 - \lambda & 4 \\ -1 & 4 & 5 - \lambda \end{bmatrix}$$

$$= -\lambda^3 + 14\lambda^2 - 42\lambda$$

$$= -\lambda(\lambda^2 - 14\lambda + 42).$$

So

$$\lambda = 0 \quad \text{or} \quad \lambda = 7 \pm \sqrt{7},$$

and

$$\|A\|_2 = \sqrt{\rho(A^t A)} = \sqrt{\max\{0, 7 - \sqrt{7}, 7 + \sqrt{7}\}} = \sqrt{7 + \sqrt{7}} \approx 3.106.$$

■   ■   ■

In studying iterative matrix techniques, it is of particular importance to know when powers of a matrix become small (that is, when all of the entries approach zero). We call an $n \times n$ matrix A **convergent** if for each $i = 1, 2, \ldots, n$ and $j = 1, 2, \ldots, n$, we have

$$\lim_{k \to \infty} (A^k)_{ij} = 0.$$

EXAMPLE 4 Let

$$A = \begin{bmatrix} \frac{1}{2} & 0 \\ \frac{1}{4} & \frac{1}{2} \end{bmatrix}.$$

Computing powers of A, we obtain:

$$A^2 = \begin{bmatrix} \frac{1}{4} & 0 \\ \frac{1}{4} & \frac{1}{4} \end{bmatrix}, \quad A^3 = \begin{bmatrix} \frac{1}{8} & 0 \\ \frac{3}{16} & \frac{1}{8} \end{bmatrix}, \quad A^4 = \begin{bmatrix} \frac{1}{16} & 0 \\ \frac{1}{8} & \frac{1}{16} \end{bmatrix},$$

and, in general,

$$A^k = \begin{bmatrix} (\frac{1}{2})^k & 0 \\ \frac{k}{2^{k+1}} & (\frac{1}{2})^k \end{bmatrix}.$$

Since $\lim_{k \to \infty} \left(\frac{1}{2}\right)^k = 0$ and $\lim_{k \to \infty} \frac{k}{2^{k+1}} = 0$,

A is a convergent matrix. Note that $\rho(A) = \frac{1}{2}$, since $\frac{1}{2}$ is the only eigenvalue of A.

■   ■   ■

An important connection exists between the spectral radius of a matrix and the convergence of the matrix.

Convergent Matrix Equivalences

(a) A is a convergent matrix.

(b) $\lim\limits_{n\to\infty} \|A^n\| = 0$, for some natural norm.

(c) $\lim\limits_{n\to\infty} \|A^n\| = 0$, for all natural norms.

(d) $\rho(A) < 1$.

(e) $\lim\limits_{n\to\infty} A^n \mathbf{x} = \mathbf{0}$, for every $\mathbf{x}$.

EXERCISE SET 7.3

1. Compute the eigenvalues and associated eigenvectors of the following matrices:

   (a) $\begin{bmatrix} 2 & -1 \\ -1 & 2 \end{bmatrix}$

   (b) $\begin{bmatrix} 1 & 1 \\ 0 & 1 \end{bmatrix}$

   (c) $\begin{bmatrix} 0 & \frac{1}{2} \\ \frac{1}{2} & 0 \end{bmatrix}$

   (d) $\begin{bmatrix} 1 & 1 \\ -2 & -2 \end{bmatrix}$

   (e) $\begin{bmatrix} 2 & 1 & 0 \\ 1 & 2 & 0 \\ 0 & 0 & 3 \end{bmatrix}$

   (f) $\begin{bmatrix} -1 & 2 & 0 \\ 0 & 3 & 4 \\ 0 & 0 & 7 \end{bmatrix}$

   (g) $\begin{bmatrix} 2 & 1 & 1 \\ 2 & 3 & 2 \\ 1 & 1 & 2 \end{bmatrix}$

   (h) $\begin{bmatrix} 3 & 2 & -1 \\ 1 & -2 & 3 \\ 2 & 0 & 4 \end{bmatrix}$

2. Find the spectral radius for each matrix in Exercise 1.

3. Show that

$$A_1 = \begin{bmatrix} 1 & 0 \\ \frac{1}{4} & \frac{1}{2} \end{bmatrix}$$

   is not convergent, but

$$A_2 = \begin{bmatrix} \frac{1}{2} & 0 \\ 16 & \frac{1}{2} \end{bmatrix}$$

   is convergent.

4. Which of the matrices in Exercise 1 are convergent?

5. Compute the determinants of the matrices in Exercise 1, and show that $\det A = \prod\limits_{i=1}^{n} \lambda_i$, where $\lambda_1, \ldots, \lambda_n$ are the eigenvalues of A.

6. Find $\|\cdot\|_2$ for the matrices in Exercise 1.

7. In Exercise 10 of Section 6.4, we assumed that the contribution a female beetle of a certain type made to the future years' female beetle population could be expressed in terms of the

matrix

$$A = \begin{bmatrix} 0 & 0 & 6 \\ \frac{1}{2} & 0 & 0 \\ 0 & \frac{1}{3} & 0 \end{bmatrix},$$

where the entry in the ith row and jth column represents the probabilistic contribution of a beetle of age j to the next year's female population of age i.

(a) Does the matrix A have any real eigenvalues? If so, determine them and any associated eigenvectors.

(b) If a sample of this species was needed for laboratory test purposes that would have a constant proportion in each age group from year to year, what criteria could be imposed on the initial population to ensure that this would be satisfied?

7.4 The Jacobi and Gauss–Seidel Methods

In this section we describe the elementary Jacobi and Gauss–Seidel iterative methods. These are classic methods that date to the late eighteenth century, but they find current application in problems where the matrix is large and has mostly zero entries in predictable locations. Applications of this type are common, for example, in the study of large integrated circuits and in the numerical solution of boundary value problems and partial-differential equations.

An iterative technique for solving the $n \times n$ linear system $A\mathbf{x} = \mathbf{b}$ starts with an initial approximation $\mathbf{x}^{(0)}$ to the solution $\mathbf{x}$ and generates a sequence of vectors $\{\mathbf{x}^{(k)}\}_{k=1}^{\infty}$ that converges to $\mathbf{x}$. These iterative techniques involve a process that converts the system $A\mathbf{x} = \mathbf{b}$ into an equivalent system of the form $\mathbf{x} = T\mathbf{x} + \mathbf{c}$ for some $n \times n$ matrix T and vector $\mathbf{c}$.

After the initial vector $\mathbf{x}^{(0)}$ is selected, the sequence of approximate solution vectors is generated by computing

$$\mathbf{x}^{(k)} = T\mathbf{x}^{(k-1)} + \mathbf{c}$$

for each $k = 1, 2, 3, \ldots$.

The following result provides an important connection between the eigenvalues of the matrix T and the expectation that the iterative method will converge:

Convergence and the Spectral Radius

For any $\mathbf{x}^{(0)}$ in $\mathbb{R}^n$, the sequence

$$\mathbf{x}^{(k)} = T\mathbf{x}^{(k-1)} + \mathbf{c}$$

converges to the unique solution of $\mathbf{x} = T\mathbf{x} + \mathbf{c}$ precisely when $\rho(T) < 1$.

EXAMPLE 1   The linear system $A\mathbf{x} = \mathbf{b}$ given by

$$
\begin{aligned}
E_1 : 10x_1 - x_2 + 2x_3 &= 6, \\
E_2 : -x_1 + 11x_2 - x_3 + 3x_4 &= 25, \\
E_3 : 2x_1 - x_2 + 10x_3 - x_4 &= -11, \\
E_4 : 3x_2 - x_3 + 8x_4 &= 15
\end{aligned}
$$

has solution $\mathbf{x} = (1, 2, -1, 1)^t$. To convert $A\mathbf{x} = \mathbf{b}$ to the form $\mathbf{x} = T\mathbf{x} + \mathbf{c}$, solve equation E_i for x_i, obtaining

$$
\begin{aligned}
x_1 &= \phantom{-\tfrac{1}{5}x_1 +} \tfrac{1}{10}x_2 - \tfrac{1}{5}x_3 \phantom{+ \tfrac{1}{10}x_4} + \tfrac{3}{5}, \\
x_2 &= \tfrac{1}{11}x_1 \phantom{+ \tfrac{1}{10}x_2} + \tfrac{1}{11}x_3 - \tfrac{3}{11}x_4 + \tfrac{25}{11}, \\
x_3 &= -\tfrac{1}{5}x_1 + \tfrac{1}{10}x_2 \phantom{+ \tfrac{1}{11}x_3} + \tfrac{1}{10}x_4 - \tfrac{11}{10}, \\
x_4 &= \phantom{-\tfrac{1}{5}x_1} - \tfrac{3}{8}x_2 + \tfrac{1}{8}x_3 \phantom{+ \tfrac{1}{10}x_4} + \tfrac{15}{8}.
\end{aligned}
$$

Then $A\mathbf{x} = \mathbf{b}$ will have the form $\mathbf{x} = T\mathbf{x} + \mathbf{c}$ with

$$
T = \begin{bmatrix}
0 & \tfrac{1}{10} & -\tfrac{1}{5} & 0 \\
\tfrac{1}{11} & 0 & \tfrac{1}{11} & -\tfrac{3}{11} \\
-\tfrac{1}{5} & \tfrac{1}{10} & 0 & \tfrac{1}{10} \\
0 & -\tfrac{3}{8} & \tfrac{1}{8} & 0
\end{bmatrix}
\quad \text{and} \quad
\mathbf{c} = \begin{bmatrix}
\tfrac{3}{5} \\
\tfrac{25}{11} \\
-\tfrac{11}{10} \\
\tfrac{15}{8}
\end{bmatrix}.
$$

For an initial approximation, suppose $\mathbf{x}^{(0)} = (0, 0, 0, 0)^t$. Then $\mathbf{x}^{(1)}$ is given by:

$$
\begin{aligned}
x_1^{(1)} &= \phantom{-\tfrac{1}{5}x_1^{(0)} +} \tfrac{1}{10}x_2^{(0)} - \tfrac{1}{5}x_3^{(0)} \phantom{+ \tfrac{1}{10}x_4^{(0)}} + \tfrac{3}{5} = 0.6000, \\
x_2^{(1)} &= \tfrac{1}{11}x_1^{(0)} \phantom{+ \tfrac{1}{10}x_2^{(0)}} + \tfrac{1}{11}x_3^{(0)} - \tfrac{3}{11}x_4^{(0)} + \tfrac{25}{11} = 2.2727, \\
x_3^{(1)} &= -\tfrac{1}{5}x_1^{(0)} + \tfrac{1}{10}x_2^{(0)} \phantom{+ \tfrac{1}{11}x_3^{(0)}} + \tfrac{1}{10}x_4^{(0)} - \tfrac{11}{10} = -1.1000, \\
x_4^{(1)} &= \phantom{-\tfrac{1}{5}x_1^{(0)}} - \tfrac{3}{8}x_2^{(0)} + \tfrac{1}{8}x_3^{(0)} \phantom{+ \tfrac{1}{10}x_4^{(0)}} + \tfrac{15}{8} = 1.8750.
\end{aligned}
$$

Additional iterates, $\mathbf{x}^{(k)} = (x_1^{(k)}, x_2^{(k)}, x_3^{(k)}, x_4^{(k)})^t$, are generated in a similar manner and are presented in Table 7.1 on page 257. The decision to stop after ten iterations was based on the criterion

$$
\frac{\|\mathbf{x}^{(10)} - \mathbf{x}^{(9)}\|_\infty}{\|\mathbf{x}^{(10)}\|_\infty} = \frac{8.0 \times 10^{-4}}{1.9998} < 10^{-3}.
$$

In fact, $\|\mathbf{x}^{(10)} - \mathbf{x}\|_\infty \approx 0.0002$.   ■ ■ ■

The method of Example 1 is called the **Jacobi** iterative method. It consists of solving the ith equation in $A\mathbf{x} = \mathbf{b}$ for x_i to obtain, provided $a_{ii} \neq 0$,

$$
x_i = \sum_{\substack{j=1 \\ j \neq i}}^{n} \left(-\frac{a_{ij}x_j}{a_{ii}} \right) + \frac{b_i}{a_{ii}}, \qquad \text{for } i = 1, 2, \ldots, n,
$$

Table 7.1

k	$x_1^{(k)}$	$x_2^{(k)}$	$x_3^{(k)}$	$x_4^{(k)}$
0	0.0000	0.0000	0.0000	0.0000
1	0.6000	2.2727	−1.1000	1.8750
2	1.0473	1.7159	−0.8052	0.8852
3	0.9326	2.0533	−1.0493	1.1309
4	1.0152	1.9537	−0.9681	0.9739
5	0.9890	2.0114	−1.0103	1.0214
6	1.0032	1.9922	−0.9945	0.9944
7	0.9981	2.0023	−1.0020	1.0036
8	1.0006	1.9987	−0.9990	0.9989
9	0.9997	2.0004	−1.0004	1.0006
10	1.0001	1.9998	−0.9998	0.9998

and generating each $x_i^{(k)}$ from components of $\mathbf{x}^{(k-1)}$ for $k \geq 1$ by

$$(7.1) \qquad x_i^{(k)} = \frac{\sum_{\substack{j=1 \\ j \neq i}}^{n} \left(-a_{ij} x_j^{(k-1)}\right) + b_i}{a_{ii}}, \qquad \text{for } i = 1, 2, \ldots, n.$$

The method is written in the form $\mathbf{x}^{(k)} = T\mathbf{x}^{(k-1)} + \mathbf{c}$ by splitting A into its diagonal and off-diagonal parts. To see this, let D be the diagonal matrix whose diagonal is the same as A, $-L$ be the strictly lower-triangular part of A, and $-U$ be the strictly upper-triangular part of A. With this notation, A is split into

$$A = \begin{bmatrix} a_{11} & a_{12} & \cdots & a_{1n} \\ a_{21} & a_{22} & \cdots & a_{2n} \\ \vdots & \vdots & & \vdots \\ a_{n1} & a_{n2} & \cdots & a_{nn} \end{bmatrix} = \begin{bmatrix} a_{11} & 0 & \cdots & 0 \\ 0 & a_{22} & & \vdots \\ \vdots & & \ddots & 0 \\ 0 & \cdots & 0 & a_{nn} \end{bmatrix}$$

$$- \begin{bmatrix} 0 & \cdots & \cdots & 0 \\ -a_{21} & \ddots & & \vdots \\ \vdots & \ddots & \ddots & \vdots \\ -a_{n1} & \cdots & -a_{n,n-1} & 0 \end{bmatrix} - \begin{bmatrix} 0 & -a_{12} & \cdots & -a_{1n} \\ \vdots & \ddots & \ddots & \vdots \\ & & \ddots & -a_{n-1,n} \\ 0 & \cdots & \cdots & 0 \end{bmatrix}$$

$$= D - L - U.$$

The equation $A\mathbf{x} = \mathbf{b}$ or $(D - L - U)\mathbf{x} = \mathbf{b}$ is then transformed into

$$D\mathbf{x} = (L + U)\mathbf{x} + \mathbf{b},$$

and, if D^{-1} exists; that is, if $a_{ii} \neq 0$ for each i, then

$$\mathbf{x} = D^{-1}(L + U)\mathbf{x} + D^{-1}\mathbf{b}.$$

This results in the matrix form of the Jacobi iterative technique:

$$\mathbf{x}^{(k)} = T_j \mathbf{x}^{(k-1)} + \mathbf{c}_j,$$

where $T_j = D^{-1}(L + U)$ and $\mathbf{c}_j = D^{-1}\mathbf{b}$.

The program JACITR71 implements the Jacobi method. If $a_{ii} \neq 0$, for some i, and the system is nonsingular, a reordering of the equations is performed so that no $a_{ii} = 0$. To speed convergence, the equations should be arranged so that a_{ii} is as large as possible.

A possible improvement on the Jacobi method can be seen by reconsidering Eq. (7.1). In this equation the components of $\mathbf{x}^{(k-1)}$ are used to compute $x_i^{(k)}$. Since $x_1^{(k)}, \ldots, x_{i-1}^{(k)}$ have already been computed and are probably better approximations to the actual solutions $x_1, \ldots, x_{i-1}$ than $x_1^{(k-1)}, \ldots, x_{i-1}^{(k-1)}$, we can compute $x_i^{(k)}$ using these most recently calculated values; that is, we can use

(7.2)
$$x_i^{(k)} = \frac{-\sum_{j=1}^{i-1}(a_{ij}x_j^{(k)}) - \sum_{j=i+1}^{n}(a_{ij}x_j^{(k-1)}) + b_i}{a_{ii}},$$

for each $i = 1, 2, \ldots, n$. This modification is called the **Gauss–Seidel** iterative technique and is illustrated in the following example.

EXAMPLE 2   The linear system given by

$$
\begin{array}{rcrcrcrcr}
10x_1 & - & x_2 & + & 2x_3 & & & = & 6, \\
-x_1 & + & 11x_2 & - & x_3 & + & 3x_4 & = & 25, \\
2x_1 & - & x_2 & + & 10x_3 & - & x_4 & = & -11, \\
& & 3x_2 & - & x_3 & + & 8x_4 & = & 15
\end{array}
$$

was solved in Example 1 by the Jacobi iterative method. Using Eq. (7.2) gives the equations

$$
\begin{array}{rclcccc}
x_1^{(k)} &=& & \frac{1}{10}x_2^{(k-1)} & - \frac{1}{5}x_3^{(k-1)} & & + \frac{3}{5}, \\[4pt]
x_2^{(k)} &=& \frac{1}{11}x_1^{(k)} & & + \frac{1}{11}x_3^{(k-1)} & - \frac{3}{11}x_4^{(k-1)} & + \frac{25}{11}, \\[4pt]
x_3^{(k)} &=& -\frac{1}{5}x_1^{(k)} & + \frac{1}{10}x_2^{(k)} & & + \frac{1}{10}x_4^{(k-1)} & - \frac{11}{10}, \\[4pt]
x_4^{(k)} &=& & -\frac{3}{8}x_2^{(k)} & + \frac{1}{8}x_3^{(k)} & & + \frac{15}{8}.
\end{array}
$$

Letting $\mathbf{x}^{(0)} = (0, 0, 0, 0)^t$, we generate the iterates in Table 7.2. Since

$$\frac{\|\mathbf{x}^{(5)} - \mathbf{x}^{(4)}\|_\infty}{\|\mathbf{x}^{(5)}\|_\infty} = \frac{0.0008}{2.000} = 4 \times 10^{-4},$$

$\mathbf{x}^{(5)}$ is accepted as a reasonable approximation to the solution. Note that Jacobi's method in Example 1 required twice the iterations for the same accuracy.   ■ ■ ■

Table 7.2

k	0	1	2	3	4	5
$x_1^{(k)}$	0.0000	0.6000	1.030	1.0065	1.0009	1.0001
$x_2^{(k)}$	0.0000	2.3272	2.037	2.0036	2.0003	2.0000
$x_3^{(k)}$	0.0000	−0.9873	−1.014	−1.0025	−1.0003	−1.0000
$x_4^{(k)}$	0.0000	0.8789	0.9844	0.9983	0.9999	1.0000

To write the Gauss–Seidel method in matrix form, multiply both sides of Eq. (7.2) by a_{ii} and collect all kth iterate terms, to give

$$a_{ii}x_1^{(k)} + a_{i2}x_2^{(k)} + \cdots + a_{ii}x_i^{(k)} = -a_{i,i+1}x_{i+1}^{(k-1)} - \cdots - a_{in}x_n^{(k-1)} + b_i,$$

for each $i = 1, 2, \ldots, n$. Writing all n equations gives

$$a_{11}x_1^{(k)} = -a_{12}x_2^{(k-1)} - a_{13}x_3^{(k-1)} - \cdots - a_{1n}x_n^{(k-1)} + b_1,$$
$$a_{21}x_1^{(k)} + a_{22}x_2^{(k)} = -a_{23}x_3^{(k-1)} - \cdots - a_{2n}x_n^{(k-1)} + b_2,$$
$$\vdots$$

$$a_{n1}x_1^{(k)} + a_{n2}x_2^{(k)} + \cdots + a_{nn}x_n^{(k)} = b_n.$$

With the definitions of D, L, and U that we used previously, we have the Gauss–Seidel method represented by

$$(D - L)\mathbf{x}^{(k)} = U\mathbf{x}^{(k-1)} + \mathbf{b}$$

or, if $(D - L)^{-1}$ exists, by

$$\mathbf{x}^{(k)} = T_g \mathbf{x}^{(k-1)} + \mathbf{c}_g, \qquad \text{for each } k = 1, 2, \ldots,$$

where $T_g = (D-L)^{-1}U$ and $c_g = (D-L)^{-1}\mathbf{b}$. Since $\det(D-L) = \prod_{i=1}^{n} a_{ii}$, the lower-triangular matrix $D - L$ is nonsingular precisely when $a_{ii} \neq 0$ for each $i = 1, 2, \ldots, n$. The Gauss–Seidel method is performed by the program GSEITR72.

The preceding discussion and the results of Examples 1 and 2 seem to imply that the Gauss–Seidel method is superior to the Jacobi method. This is generally, but not always, true. There are linear systems for which the Jacobi method converges and the Gauss–Seidel method does not and others for which the Gauss–Seidel method converges and the Jacobi method does not. However, if A is strictly diagonally dominant, then for any $\mathbf{b}$ and any choice of $\mathbf{x}^{(0)}$, the Jacobi and Gauss–Seidel methods will both converge to the unique solution of $A\mathbf{x} = \mathbf{b}$.

EXERCISE SET 7.4

1. Find the first two iterations of the Jacobi method for the following linear systems, using $\mathbf{x}^{(0)} = \mathbf{0}$:

(a) $3x_1 - x_2 + x_3 = 1$
$3x_1 + 6x_2 + 2x_3 = 0$
$3x_1 + 3x_2 + 7x_3 = 4$

(b) $10x_1 - x_2 = 9$
$-x_1 + 10x_2 - 2x_3 = 7$
$- 2x_2 + 10x_3 = 6$

(c) $2x_1 - 2x_2 + x_3 + x_4 = 0.8$
$- 3x_2 + 0.5x_3 + x_4 = -6.6$
$5x_3 - x_4 = 4.5$
$2x_4 = 3$

(d) $10x_1 + x_2 - 2x_3 = 6$
$x_1 + 10x_2 - x_3 + 3x_4 = 25$
$-2x_1 - x_2 + 8x_3 - x_4 = -11$
$3x_2 - x_3 + 5x_4 = -11$

(e) $2x_1 + x_2 - x_3 + x_4 = 4$
$x_1 - 2x_2 + x_3 - 2x_4 = 5$
$2x_1 + 2x_2 + 5x_3 - x_4 = 6$
$x_1 - x_2 + x_3 + 4x_4 = 7$

(f) $4x_1 + x_2 + x_3 + x_5 = 6$
$-x_1 - 3x_2 + x_3 + x_4 = 6$
$2x_1 + x_2 + 5x_3 - x_4 - x_5 = 6$
$-x_1 - x_2 - x_3 + 4x_4 = 6$
$2x_2 - x_3 + x_4 + 4x_5 = 6$

(g) $4x_1 + x_2 - x_3 + x_4 = -2$
$x_1 + 4x_2 - x_3 - x_4 = -1$
$-x_1 - x_2 + 5x_3 + x_4 = 0$
$x_1 - x_2 + x_3 + 3x_4 = 1$

(h) $4x_1 - x_2 - x_4 = 0$
$-x_1 + 4x_2 - x_3 - x_5 = 5$
$- x_2 + 4x_3 - x_6 = 0$
$-x_1 + 4x_4 - x_5 = 6$
$- x_2 - x_4 + 4x_5 - x_6 = -2$
$- x_3 - x_5 + 4x_6 = 6$

2. Repeat Exercise 1 using the Gauss–Seidel method.

3. Apply the Jacobi method to solve the linear systems in Exercise 1, if possible. Use a tolerance of 10^{-3} and let the maximum number of iterations be 25.

4. Repeat Exercise 3 using the Gauss–Seidel method.

7.5   The SOR Method

The SOR method is similar to the Jacobi and Gauss–Seidel methods, but it uses a scaling factor to more rapidly reduce the approximation error. In contrast to the classic methods discussed in the previous section, the SOR technique is a more recent innovation, being used extensively only since 1962, when it was described to the general scientific community by Richard Varga in his book *Matrix Iterative Analysis*.

The SOR technique is one of a class of *relaxation* methods that compute approximations $\mathbf{x}^{(k)}$ by the formula

$$x_i^{(k)} = (1 - \omega)x_i^{(k-1)} + \frac{\omega}{a_{ii}}\left[b_i - \sum_{j=1}^{i-1} a_{ij}x_j^{(k)} - \sum_{j=i+1}^{n} a_{ij}x_j^{(k-1)}\right],$$

where ω is the scaling factor.

When ω is chosen in $(0, 1)$, the procedures are called **under-relaxation methods** and can be used to obtain convergence of some systems that are not convergent by the Gauss–Seidel method. When $1 < \omega$, the procedures are called **over-relaxation methods**, which are used to accelerate the convergence for systems that are convergent by the Gauss–Seidel technique. These methods are abbreviated **SOR** for **successive over-relaxation**, and are particularly useful for solving the linear systems that occur in the numerical solution of certain partial-differential equations.

To determine the matrix form of the SOR method, we rewrite the preceeding equation as

$$a_{ii}x_i^{(k)} + \omega \sum_{j=1}^{i-1} a_{ij}x_j^{(k)} = (1 - \omega)a_{ii}x_i^{(k-1)} - \omega \sum_{j=i+1}^{n} a_{ij}x_j^{(k-1)} + \omega b_i,$$

so that in vector form we have

$$(D - \omega L)\mathbf{x}^{(k)} = [(1 - \omega)D + \omega U]\mathbf{x}^{(k-1)} + \omega \mathbf{b}.$$

If $(D - \omega L)^{-1}$ exists, then

$$\mathbf{x}^{(k)} = T_\omega \mathbf{x}^{(k-1)} + \mathbf{c}_\omega,$$

where $T_\omega = (D - \omega L)^{-1}[(1 - \omega)D + \omega U]$ and $c_\omega = \omega(D - \omega L)^{-1}\mathbf{b}$.

The SOR technique can be applied using the program SORITR73.

EXAMPLE 1 The linear system $A\mathbf{x} = \mathbf{b}$ given by

$$\begin{aligned} 4x_1 + 3x_2 \quad\quad &= \quad 24, \\ 3x_1 + 4x_2 - x_3 &= \quad 30, \\ - x_2 + 4x_3 &= -24 \end{aligned}$$

has the solution $(3, 4, -5)^t$. The Gauss–Seidel method and the SOR method with $\omega = 1.25$ will be used to solve this system, using $\mathbf{x}^{(0)} = (1, 1, 1)^t$ for both methods. For each $k = 1, 2, \ldots$, the equations for the Gauss–Seidel method are

$$\begin{aligned} x_1^{(k)} &= -0.75x_2^{(k-1)} + 6, \\ x_2^{(k)} &= -0.75x_1^{(k)} + 0.25x_3^{(k-1)} + 7.5, \\ x_3^{(k)} &= 0.25x_2^{(k)} - 6, \end{aligned}$$

and the equations for the SOR method with $\omega = 1.25$ are

$$\begin{aligned} x_1^{(k)} &= -0.25x_1^{(k-1)} - 0.9375x_2^{(k-1)} + 7.5, \\ x_2^{(k)} &= -0.9375x_1^{(k)} - 0.25x_2^{(k-1)} + 0.3125x_3^{(k-1)} + 9.375, \\ x_3^{(k)} &= 0.3125x_2^{(k)} - 0.25x_3^{(k-1)} - 7.5. \end{aligned}$$

The first seven iterates for each method are listed in Tables 7.3 and 7.4. To be accurate to seven decimal places, the Gauss–Seidel method required 34 iterations as opposed to 14 iterations for the SOR method with $\omega = 1.25$. ■ ■ ■

Table 7.3   Gauss–Seidel

k	0	1	2	3	4	5	6	7
$x_1^{(k)}$	1	5.250000	3.1406250	3.0878906	3.0549316	3.0343323	3.0214577	3.0134110
$x_2^{(k)}$	1	3.812500	3.8828125	3.9267578	3.9542236	3.9713898	3.9821186	3.9888241
$x_3^{(k)}$	1	-5.046875	-5.0292969	-5.0183105	-5.0114441	-5.0071526	-5.0044703	-5.0027940

Table 7.4   SOR with $\omega = 1.25$

k	0	1	2	3	4	5	6	7
$x_1^{(k)}$	1	6.312500	2.6223145	3.1333027	2.9570512	3.0037211	2.9963276	3.0000498
$x_2^{(k)}$	1	3.5195313	3.9585266	4.0102646	4.0074838	4.0029250	4.0009262	4.0002586
$x_3^{(k)}$	1	-6.6501465	-4.6004238	-5.0966863	-4.9734897	-5.0057135	-4.9982822	-5.0003486

The obvious question to ask is how the appropriate value of ω is chosen. Although no complete answer to this question is known for the general $n \times n$ linear system, the following result can be used in certain situations.

SOR Method Convergence

If A is a positive definite matrix and $0 < \omega < 2$, then the SOR method converges for any choice of initial approximate solution vector $\mathbf{x}^{(0)}$.

If, in addition, A is tridiagonal, then $\rho(T_g) = [\rho(T_j)]^2 < 1$, and the optimal choice of ω for the SOR method is

$$\omega = \frac{2}{1 + \sqrt{1 - \rho(T_g)}} = \frac{2}{1 + \sqrt{1 - [\rho(T_j)]^2}}.$$

With this choice of ω, $\rho(T_\omega) = \omega - 1$.

EXAMPLE 2   The matrix

$$A = \begin{bmatrix} 4 & 3 & 0 \\ 3 & 4 & -1 \\ 0 & -1 & 4 \end{bmatrix}$$

given in Example 1 is positive definite and tridiagonal. Since

$$T_j = D^{-1}(L + U) = \begin{bmatrix} \frac{1}{4} & 0 & 0 \\ 0 & \frac{1}{4} & 0 \\ 0 & 0 & \frac{1}{4} \end{bmatrix} \begin{bmatrix} 0 & -3 & 0 \\ -3 & 0 & 1 \\ 0 & 1 & 0 \end{bmatrix} = \begin{bmatrix} 0 & -0.75 & 0 \\ -0.75 & 0 & 0.25 \\ 0 & 0.25 & 0 \end{bmatrix},$$

we have

$$\det(T_j - \lambda I) = -\lambda(\lambda^2 - 0.625) \qquad \text{and} \qquad \rho(T_j) = \sqrt{0.625}.$$

Hence, the optimal choice of ω is

$$\omega = \frac{2}{1 + \sqrt{1 - \rho(T_g)}} = \frac{2}{1 + \sqrt{1 - [\rho(T_j)]^2}} = \frac{2}{1 + \sqrt{1 - 0.625}} \approx 1.24.$$

This explains the rapid convergence obtained in Example 1 by using $\omega = 1.25$.

■ ■ ■

EXERCISE SET 7.5

1. Find the first two iterations of the SOR method with $\omega = 1.1$ for the following linear systems, using $\mathbf{x}^{(0)} = \mathbf{0}$:

(a)
$$\begin{aligned}
3x_1 - x_2 + x_3 &= 1 \\
3x_1 + 6x_2 + 2x_3 &= 0 \\
3x_1 + 3x_2 + 7x_3 &= 4
\end{aligned}$$

(b)
$$\begin{aligned}
10x_1 - x_2 &= 9 \\
-x_1 + 10x_2 - 2x_3 &= 7 \\
- 2x_2 + 10x_3 &= 6
\end{aligned}$$

(c)
$$\begin{aligned}
2x_1 - 2x_2 + x_3 + x_4 &= 0.8 \\
- 3x_2 + 0.5x_3 + x_4 &= -6.6 \\
5x_3 - x_4 &= 4.5 \\
2x_4 &= 3
\end{aligned}$$

(d)
$$\begin{aligned}
10x_1 + x_2 - 2x_3 &= 6 \\
x_1 + 10x_2 - x_3 + 3x_4 &= 25 \\
-2x_1 - x_2 + 8x_3 - x_4 &= -11 \\
3x_2 - x_3 + 5x_4 &= -11
\end{aligned}$$

(e)
$$\begin{aligned}
2x_1 + x_2 - x_3 + x_4 &= 4 \\
x_1 - 2x_2 + x_3 - 2x_4 &= 5 \\
2x_1 + 2x_2 + 5x_3 - x_4 &= 6 \\
x_1 - x_2 + x_3 + 4x_4 &= 7
\end{aligned}$$

(f)
$$\begin{aligned}
4x_1 + x_2 + x_3 + x_5 &= 6 \\
-x_1 - 3x_2 + x_3 + x_4 &= 6 \\
2x_1 + x_2 + 5x_3 - x_4 - x_5 &= 6 \\
-x_1 - x_2 - x_3 + 4x_4 &= 6 \\
2x_2 - x_3 + x_4 + 4x_5 &= 6
\end{aligned}$$

(g)
$$\begin{aligned}
4x_1 + x_2 - x_3 + x_4 &= -2 \\
x_1 + 4x_2 - x_3 - x_4 &= -1 \\
-x_1 - x_2 + 5x_3 + x_4 &= 0 \\
x_1 - x_2 + x_3 + 3x_4 &= 1
\end{aligned}$$

(h)
$$\begin{aligned}
4x_1 - x_2 - x_4 &= 0 \\
-x_1 + 4x_2 - x_3 - x_5 &= 5 \\
- x_2 + 4x_3 - x_6 &= 0 \\
-x_1 + 4x_4 - x_5 &= 6 \\
- x_2 - x_4 + 4x_5 - x_6 &= -2 \\
- x_3 - x_5 + 4x_6 &= 6
\end{aligned}$$

2. Repeat Exercise 1 using $\omega = 1.3$.

3. Apply the SOR method to solve the linear systems in Exercise 1, if possible. Use $\omega = 1.2$, a tolerance of 10^{-3}, and let the maximum number of iterations be 25.

4. Determine which matrices in Exercise 1 are tridiagonal and positive definite. Repeat Exercise 3 for these matrices using the optimal choice of ω.

7.6 Stability of Matrix Techniques

This section considers convergence concerns that should be addressed when applying either an iterative or direct method to a linear system. There is no universally superior technique for approximating the solution to these systems, but some methods will commonly give better results than others, particularly when certain conditions are known about the matrix and the likely range of the solution.

It seems intuitively reasonable that if $\tilde{\mathbf{x}}$ is an approximation to the solution $\mathbf{x}$ of $A\mathbf{x} = \mathbf{b}$, and the **residual** vector, defined by $\mathbf{b} - A\tilde{\mathbf{x}}$, has the property that $\|\mathbf{b} - A\tilde{\mathbf{x}}\|$ is small, then $\|\mathbf{x} - \tilde{\mathbf{x}}\|$ should be small as well. This is often the case, but certain systems, which occur quite often in practice, fail to have this property.

EXAMPLE 1   The linear system $A\mathbf{x} = \mathbf{b}$ given by

$$\begin{bmatrix} 1 & 1 \\ 1.0001 & 2 \end{bmatrix}\begin{bmatrix} x_1 \\ x_2 \end{bmatrix} = \begin{bmatrix} 3 \\ 3.0001 \end{bmatrix}$$

has the unique solution $\mathbf{x} = (1, 1)^t$. The poor approximation $\tilde{\mathbf{x}} = (3, 0)^t$ has the residual vector

$$\mathbf{b} - A\tilde{\mathbf{x}} = \begin{bmatrix} 3 \\ 3.0001 \end{bmatrix} - \begin{bmatrix} 1 & 2 \\ 1.0001 & 2 \end{bmatrix}\begin{bmatrix} 3 \\ 0 \end{bmatrix} = \begin{bmatrix} 0 \\ 0.0002 \end{bmatrix},$$

so $\|\mathbf{b} - A\tilde{\mathbf{x}}\|_\infty = 0.0002$. Although the norm of the residual vector is small, the approximation $\tilde{\mathbf{x}} = (3, 0)^t$ is obviously quite poor; in fact, $\|\mathbf{x} - \tilde{\mathbf{x}}\|_\infty = 2$.   ■ ■ ■

The difficulty in Example 1 is explained quite simply by noting that the solution to the system represents the intersection of the lines

$$l_1: \qquad x_1 + 2x_2 = 3 \qquad \text{and} \qquad l_2: \quad 1.0001x_1 + 2x_2 = 3.0001.$$

The point $(3, 0)$ lies on l_1, and the lines are nearly parallel. This implies that $(3, 0)$ also lies close to l_2, even though it differs significantly from the solution of the system, which is the intersection point $(1, 1)$. (See Figure 7.6.)

Figure 7.6

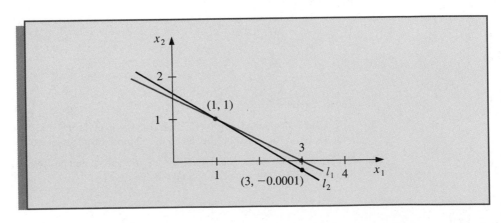

Example 1 was clearly constructed to show the difficulties that might—and in fact do—arise. Had the lines not been nearly coincident, we would expect a small residual vector to imply an accurate approximation. In the general situation, we cannot rely on the geometry of the system to give an indication of when problems might occur. We can, however, obtain this information by considering the norms of the matrix and its inverse.

Residual Vector Error Bounds

If $\tilde{\mathbf{x}}$ is an approximation to the solution of $A\mathbf{x} = \mathbf{b}$ and A is a nonsingular matrix, then for any natural norm,

$$\|\mathbf{x} - \tilde{\mathbf{x}}\| \leq \|\mathbf{b} - A\tilde{\mathbf{x}}\| \cdot \|A^{-1}\|$$

and

$$\frac{\|\mathbf{x} - \tilde{\mathbf{x}}\|}{\|\mathbf{x}\|} \leq \|A\| \cdot \|A^{-1}\| \frac{\|\mathbf{b} - A\tilde{\mathbf{x}}\|}{\|\mathbf{b}\|}, \qquad \text{provided } \mathbf{x} \neq \mathbf{0} \text{ and } \mathbf{b} \neq \mathbf{0}.$$

This implies that $\|A^{-1}\|$ and $\|A\| \cdot \|A^{-1}\|$ provide an indication of the connection between the residual vector and the accuracy of the approximation. In general, the relative error $\|\mathbf{x} - \tilde{\mathbf{x}}\| / \|\mathbf{x}\|$ is of most interest, and this error is bounded by the product of $\|A\| \cdot \|A^{-1}\|$ with the relative residual for this approximation, $\|\mathbf{b} - A\tilde{\mathbf{x}}\| / \|\mathbf{b}\|$. Any convenient norm can be used for this approximation; the only requirement is that it be used consistently throughout.

The **condition number** $K(A)$ of the nonsingular matrix A relative to a norm $\| \cdot \|$ is

$$K(A) = \|A\| \cdot \|A^{-1}\|.$$

With this notation, we can reexpress the inequalities in the previous result as

$$\|\mathbf{x} - \tilde{\mathbf{x}}\| \leq K(A)\frac{\|\mathbf{b} - A\tilde{\mathbf{x}}\|}{\|A\|} \qquad \text{and} \qquad \frac{\|\mathbf{x} - \tilde{\mathbf{x}}\|}{\|\mathbf{x}\|} \leq K(A)\frac{\|\mathbf{b} - A\tilde{\mathbf{x}}\|}{\|\mathbf{b}\|}.$$

For any nonsingular matrix A and natural norm $\| \cdot \|$,

$$1 = \|I\| = \|A \cdot A^{-1}\| \leq \|A\| \cdot \|A^{-1}\| = K(A).$$

A matrix A is well-behaved (called **well-conditioned**) if $K(A)$ is close to 1 and is not well-behaved (called **ill-conditioned**) when $K(A)$ is significantly greater than 1. Behavior in this instance refers to the relative security that a small residual vector implies a correspondingly accurate approximate solution.

EXAMPLE 2 The matrix for the system considered in Example 1 was

$$A = \begin{bmatrix} 1 & 2 \\ 1.0001 & 2 \end{bmatrix},$$

which has $\|A\|_\infty = 3.0001$. This norm would not be considered large. However,

$$A^{-1} = \begin{bmatrix} -10000 & 10000 \\ 5000.5 & -5000 \end{bmatrix}, \qquad \text{so} \qquad \|A^{-1}\|_\infty = 20000,$$

and, for the infinity norm, $K(A) = (20000)(3.0001) = 60002$. The size of the condition number for this example should certainly keep us from making hasty accuracy decisions based on the residual of an approximation. ■ ■ ■

EXERCISE SET 7.6

1. Compute the condition numbers of the following matrices relative to $\|\cdot\|_\infty$:

   (a) $\begin{bmatrix} \frac{1}{2} & \frac{1}{3} \\ \frac{1}{3} & \frac{1}{4} \end{bmatrix}$

   (b) $\begin{bmatrix} 3.9 & 1.6 \\ 6.8 & 2.9 \end{bmatrix}$

   (c) $\begin{bmatrix} 4.56 & 2.18 \\ 2.79 & 1.38 \end{bmatrix}$

   (d) $\begin{bmatrix} 1 & 2 \\ 1.00001 & 2 \end{bmatrix}$

   (e) $\begin{bmatrix} 1.003 & 58.09 \\ 5.550 & 321.8 \end{bmatrix}$

   (f) $\begin{bmatrix} 1 & -1 & -1 \\ 0 & 1 & -1 \\ 0 & 0 & -1 \end{bmatrix}$

   (g) $\begin{bmatrix} 1 & 2 & 3 \\ 2 & 3 & 4 \\ 3 & 4 & 6 \end{bmatrix}$

   (h) $\begin{bmatrix} 0.04 & 0.01 & -0.01 \\ 0.2 & 0.5 & -0.2 \\ 1 & 2 & 4 \end{bmatrix}$

2. The following linear systems $A\mathbf{x} = \mathbf{b}$ have $\mathbf{x}$ as the actual solution and $\tilde{\mathbf{x}}$ as an approximate solution. Using the results of Exercise 1, compute $\|\mathbf{x} - \tilde{\mathbf{x}}\|_\infty$ and $\dfrac{\|\mathbf{b} - A\tilde{\mathbf{x}}\|_\infty}{\|A\|_\infty} K_\infty(A)$.

   (a) $\frac{1}{2}x_1 + \frac{1}{3}x_2 = \frac{1}{63}$
   $\frac{1}{3}x_1 + \frac{1}{4}x_2 = \frac{1}{168}$

   $\mathbf{x} = \left(\frac{1}{7}, -\frac{1}{6}\right)^t$
   $\tilde{\mathbf{x}} = (0.142, -0.166)^t$

   (b) $3.9x_1 + 1.6x_2 = 5.5$
   $6.8x_1 + 2.9x_2 = 9.7$

   $\mathbf{x} = (1, 1)^t$
   $\tilde{\mathbf{x}} = (0.98, 1.1)^t$

   (c) $4.56x_1 + 2.18x_2 = 6.74$
   $2.79x_1 + 1.38x_2 = 4.13$

   $\mathbf{x} = (1.4140551, 0.1339031)^t$
   $\tilde{\mathbf{x}} = (1.4, 0.14)^t$

   (d) $x_1 + 2x_2 = 3$
   $1.0001x_1 + 2x_2 = 3.0001$

   $\mathbf{x} = (1, 1)^t$
   $\tilde{\mathbf{x}} = (0.96, 1.02)^t$

   (e) $1.003x_1 + 58.09x_2 = 68.12$
   $5.550x_1 + 321.8x_2 = 377.3$

   $\mathbf{x} = (10, 1)^t$
   $\tilde{\mathbf{x}} = (-10, 1)^t$

   (f) $x_1 - x_2 - x_3 = 2\pi$
   $x_2 - x_3 = 0$
   $-x_3 = \pi$

   $\mathbf{x} = (0, -\pi, -\pi)^t$
   $\tilde{\mathbf{x}} = (-0.1, -3.15, -3.14)^t$

   (g) $x_1 + 2x_2 + 3x_3 = 1$
   $2x_1 + 3x_2 + 4x_3 = -1$
   $3x_1 + 4x_2 + 6x_3 = 2$

   $\mathbf{x} = (0, -7, 5)^t$
   $\tilde{\mathbf{x}} = (-0.2, -7.5, 5.4)^t$

   (h) $0.04x_1 + 0.01x_2 - 0.01x_3 = 0.06$
   $0.2x_1 + 0.5x_2 - 0.2x_3 = 0.3$
   $x_1 + 2x_2 + 4x_3 = 11$

   $\mathbf{x} = (1.827586, 0.6551724, 1.965517)^t$
   $\tilde{\mathbf{x}} = (1.8, 0.64, 1.9)^t$

3. The linear system $A\mathbf{x} = \mathbf{b}$ given by

$$\begin{bmatrix} 1 & 2 \\ 1.0001 & 2 \end{bmatrix} \begin{bmatrix} x_1 \\ x_2 \end{bmatrix} = \begin{bmatrix} 3 \\ 3.0001 \end{bmatrix}$$

has solution $(1, 1)^t$. Change A slightly to

$$\begin{bmatrix} 1 & 2 \\ 0.9999 & 2 \end{bmatrix}$$

and consider the linear system

$$\begin{bmatrix} 1 & 2 \\ 0.9999 & 2 \end{bmatrix} \begin{bmatrix} x_1 \\ x_2 \end{bmatrix} = \begin{bmatrix} 3 \\ 3.0001 \end{bmatrix}.$$

Compute the new solution using five-digit rounding arithmetic, and compare the change in A to the change in $\mathbf{x}$.

4. The linear system $A\mathbf{x} = \mathbf{b}$ given by

$$\begin{bmatrix} 1 & 2 \\ 1.00001 & 2 \end{bmatrix} \begin{bmatrix} x_1 \\ x_2 \end{bmatrix} = \begin{bmatrix} 3 \\ 3.00001 \end{bmatrix}$$

has solution $(1, 1)^t$. Use seven-digit rounding arithmetic to find the solution of the perturbed system

$$\begin{bmatrix} 1 & 2 \\ 1.000011 & 2 \end{bmatrix} \begin{bmatrix} x_1 \\ x_2 \end{bmatrix} = \begin{bmatrix} 3.00001 \\ 3.00003 \end{bmatrix},$$

and compare the change in A and $\mathbf{b}$ to the change in $\mathbf{x}$.

5. (a) Use single precision on a computer to solve the following linear system using the Gaussian elimination method.

$$\begin{aligned}
\tfrac{1}{3}x_1 - \tfrac{1}{3}x_2 - \tfrac{1}{3}x_3 - \tfrac{1}{3}x_4 - \tfrac{1}{3}x_5 &= 1 \\
\tfrac{1}{3}x_2 - \tfrac{1}{3}x_3 - \tfrac{1}{3}x_4 - \tfrac{1}{3}x_5 &= 0 \\
\tfrac{1}{3}x_3 - \tfrac{1}{3}x_4 - \tfrac{1}{3}x_5 &= -1 \\
\tfrac{1}{3}x_4 - \tfrac{1}{3}x_5 &= 2 \\
\tfrac{1}{3}x_5 &= 7
\end{aligned}$$

(b) Compute the condition number of the matrix for the system relative to $\| \cdot \|_\infty$.

(c) Find the exact solution to the linear system.

6. The $n \times n$ **Hilbert** matrix $H^{(n)}$ defined by

$$H_{ij}^{(n)} = \tfrac{1}{i+j-1}, \qquad 1 \le i, j \le n$$

is an ill-conditioned matrix that arises in solving the normal equations for the coefficients of the least squares polynomial (see Section 8.3).

(a) Show that

$$[H^{(4)}]^{-1} = \begin{bmatrix} 16 & -120 & 240 & -140 \\ -120 & 1200 & -2700 & 1680 \\ 240 & -2700 & 6480 & -4200 \\ -140 & 1680 & -4200 & 2800 \end{bmatrix},$$

and compute $K(H^{(4)})$ relative to $\|\cdot\|_\infty$.

(b) Show that

$$[H^{(5)}]^{-1} = \begin{bmatrix} 52 & -300 & 1050 & -1400 & 630 \\ -300 & 4800 & -18900 & 26880 & -12600 \\ 1050 & -18900 & 79380 & -117600 & 56700 \\ -1400 & 26880 & -117600 & 179200 & -88200 \\ 630 & -12600 & 56700 & -88200 & 44100 \end{bmatrix},$$

and compute $K(H^{(5)})$ relative to $\|\cdot\|_\infty$.

(c) Solve the linear system

$$H^{(4)} \begin{bmatrix} x_1 \\ x_2 \\ x_3 \\ x_4 \end{bmatrix} = \begin{bmatrix} 1 \\ 0 \\ 0 \\ 1 \end{bmatrix}$$

using three-digit rounding arithmetic, and compare the actual error to the residual vector error bound.

7.7 Survey of Methods and Software

In this chapter we have studied iterative techniques to approximate the solution of linear systems. We began with the Jacobi method and the Gauss–Seidel method to introduce the iterative methods. Both methods require an arbitrary initial approximation $\mathbf{x}^{(0)}$ and generate a sequence of vectors $\mathbf{x}^{(i+1)}$ using an equation of the form

$$\mathbf{x}^{(i+1)} = T\mathbf{x}^{(i)} + \mathbf{c}.$$

It was noted that the method will converge if and only if the spectral radius of the iteration matrix $\rho(T) < 1$ and, the smaller the spectral radius, the faster is the convergence. Analysis of the Gauss–Seidel technique leads to the SOR iterative method, which involves a parameter ω to speed convergence.

These iterative methods and modifications are used extensively in the solution of linear systems that arise in the numerical solution of boundary value problems and partial-differential equations (see Chapters 11 and 12). These systems are often very large, on the order of 10,000 equations in 10,000 unknowns, and are sparse with their nonzero entries in predictable positions. The iterative methods are also useful for other large sparse systems and are easily adapted for efficient use on parallel computers.

The packages LINPACK and LAPACK contain only direct methods for the solution of linear systems. Neither the IMSL library nor the NAG library contains subroutines for the iterative solution of linear systems, since neither library focuses on methods for boundary value problems or partial differentials which require the solution of large sparse systems. However, the public domain packages ITPACK, SLAP, and SPARSPAK contain iterative methods.

The concepts of condition number and poorly conditioned matrices were introduced in the last section of the chapter. Many of the subroutines for solving a linear system or for factoring a matrix into its LU factorization include checks for ill-conditioned matrices and also give an estimate of the condition number.

The subroutine SGECO in LINPACK factors the real matrix A into its LU factorization and gives the row ordering for the permutation matrix P, where $PA = LU$. The subroutine also gives the condition number of A. LINPACK has other subroutines for special matrices; for example, SPOCO performs the Choleski factorization of a positive definite matrix A and estimates its condition number.

The IMSL library has subroutines that estimate the condition number. For example, the subroutine LFCRG computes the LU factorization $PA = LU$ of the matrix A and also gives an estimate of the condition number. The NAG library has similar subroutines.

LAPACK, LINPACK, the IMSL library, and the NAG library have subroutines that improve on a solution to a linear system that is poorly conditioned. The subroutines test the condition number and then use iterative refinement to obtain the most accurate solution possible given the precision of the computer.

MATLAB can be used to obtain condition numbers of matrices using the functions COND and RCOND and to compute norms of matrices and vectors using the function NORM.

CHAPTER 8

Approximation Theory

8.1 Introduction

Approximation theory involves two types of problems. One arises when a function is given explicitly, but we wish to find a "simpler" type of function, such as a polynomial, for representation. The other problem concerns fitting functions to given data and finding the "best" function in a certain class that can be used to represent the data. We will begin the chapter with this problem.

8.2 Discrete Least Squares Approximation

Consider the problem of estimating the values of a function at nontabulated points, given the experimental data in Table 8.1.

Table 8.1

x_i	y_i
1	1.3
2	3.5
3	4.2
4	5.0
5	7.0
6	8.8
7	10.1
8	12.5
9	13.0
10	15.6

Interpolation requires a function that assumes the value of y_i at x_i for each $i = 1, 2, \ldots, 10$. Figure 8.1 shows a graph of the values in Table 8.1. From this graph, it appears that the actual relationship between x and y is linear. It is likely that no line

Figure 8.1

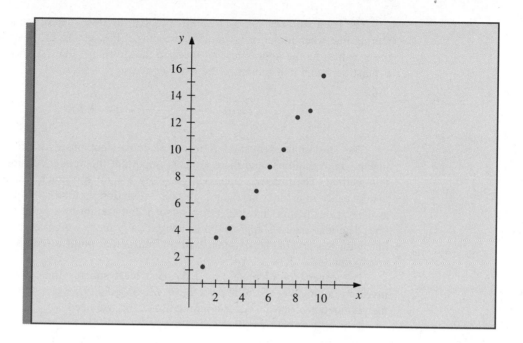

precisely fits the data because of errors in the data-collection procedure. In this case, it is unreasonable to require that the approximating function agree exactly with the given data; in fact, such a function would introduce oscillations that should not be present. A better approach for a problem of this type would be to find the "best" (in some sense) approximating line, even though it might not agree precisely with the data at any point.

Let $ax_i + b$ denote the ith value on the approximating line and y_i the ith given y-value. The problem of finding the equation of the best linear approximation in the absolute sense requires that values of a and b be found to minimize

$$E_\infty(a, b) = \max_{i = 1, 2, \dots, 10}\{|y_i - (ax_i + b)|\}.$$

This is commonly called a **minimax** problem and cannot be handled by elementary techniques. Another approach to determining the best linear approximation involves finding values of a and b to minimize

$$E_1(a, b) = \sum_{i=1}^{10} |y_i - (ax_i + b)|.$$

This quantity is called the **absolute deviation.** To minimize a function of two variables, we need to set its partial derivatives to zero and simultaneously solve the resulting equation. In the case of the absolute deviation, we would need to find a and b, with

$$0 = \frac{\partial}{\partial a} \sum_{i=1}^{10} |y_i - (ax_i + b)| \qquad \text{and} \qquad 0 = \frac{\partial}{\partial b} \sum_{i=1}^{10} |y_i - (ax_i + b)|.$$

The difficulty with this procedure is that the absolute-value function is not differentiable at zero, and solutions to this pair of equations cannot necessarily be obtained.

The **least squares** approach to this problem involves determining the best approximating line when the error involved is the sum of the squares of the differences between the y-values on the approximating line and the given y-values. Hence, constants a and b must be found that minimize the *total least squares error:*

$$E_2(a, b) = \sum_{i=1}^{10} [y_i - (ax_i + b)]^2.$$

The least squares method is the most convenient procedure for determining best linear approximations, and there are also important theoretical considerations that favor this method. The minimax approach generally assigns too much weight to a bit of data that is badly in error, whereas the absolute deviation method simply averages the error at the various points and does not give sufficient weight to a point that is badly out of line. The least squares approach puts substantially more weight on a point that is out of line with the rest of the data but will not allow that point to completely dominate the approximation.

The general problem of fitting the best least squares line to a collection of data involves minimizing the total error $E_2(a, b) = \sum_{i=1}^{m}[y_i - (ax_i + b)]^2$ with respect to the parameters a and b. For a minimum to occur, we need

$$0 = \frac{\partial}{\partial a} \sum_{i=1}^{m} [y_i - (ax_i + b)]^2 = 2\sum_{i=1}^{m} (y_i - ax_i - b)(-x_i)$$

and

$$0 = \frac{\partial}{\partial b} \sum_{i=1}^{m} (y_i - ax_i - b)^2 = 2\sum_{i=1}^{m} (y_i - ax_i - b)(-1).$$

These equations simplify to the **normal equations**

$$a\sum_{i=1}^{m} x_i^2 + b\sum_{i=1}^{m} x_i = \sum_{i=1}^{n} x_i y_i \quad \text{and} \quad a\sum_{i=1}^{m} x_i + b \cdot m = \sum_{i=1}^{m} y_i.$$

The solution to this system is as follows:

Linear Least Squares

The linear least squares solution for a given collection (x_i, y_i), for $i = 1, 2, \ldots, m$, has the form $y = ax + b$ when

$$a = \frac{m(\sum_{i=1}^{m} x_i y_i) - (\sum_{i=1}^{m} x_i)(\sum_{i=1}^{m} y_i)}{m(\sum_{i=1}^{m} x_i^2) - (\sum_{i=1}^{m} x_i)^2}$$

and

$$b = \frac{(\sum_{i=1}^{m} x_i^2)(\sum_{i=1}^{m} y_i) - (\sum_{i=1}^{m} x_i y_i)(\sum_{i=1}^{m} x_i)}{m(\sum_{i=1}^{m} x_i^2) - (\sum_{i=1}^{m} x_i)^2}.$$

EXAMPLE 1 Consider the data presented in Table 8.1. To find the least squares line approximating this data, extend the table as shown in the third and fourth columns of Table 8.2, and sum the columns.

Table 8.2

x_i	y_i	x_i^2	$x_i y_i$	$P(x_i) = 1.538x_i - 0.360$
1	1.3	1	1.3	1.18
2	3.5	4	7.0	2.72
3	4.2	9	12.6	4.25
4	5.0	16	20.0	5.79
5	7.0	25	35.0	7.33
6	8.8	36	52.8	8.87
7	10.1	49	70.7	10.41
8	12.5	64	100.0	11.94
9	13.0	81	117.0	13.48
10	15.6	100	156.0	15.02
55	81.0	385	572.4	$E = \sum_{i=1}^{10}(y_i - P(x_i))^2 \approx 2.34$

Solving the normal equations produces

$$a = \frac{10(572.4) - 55(81)}{10(385) - (55)^2} = 1.538 \quad \text{and} \quad b = \frac{385(81) - 55(572.4)}{10(385) - (55)^2} = -0.360.$$

The graph of this line and the data points are shown in Figure 8.2. The approximate values given by the least squares technique at the data points are in the final column in Table 8.2. ∎ ∎ ∎

Figure 8.2

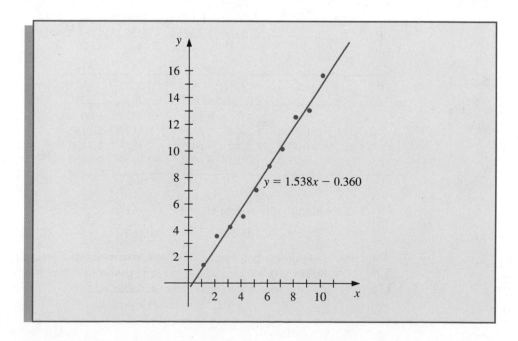

The problem of approximating a set of data $\{(x_i, y_i)|i = 1, 2, \ldots, m\}$ with an algebraic polynomial $P_n(x) = \sum_{k=0}^{n} a_k x^k$ of degree $n < m - 1$ using least squares is handled in a similar manner. It requires choosing the constants $a_0, a_1, \ldots, a_n$ to minimize the *total least squares error:*

$$E = \sum_{i=1}^{m} (y_i - P_n(x_i))^2.$$

For E to be minimized, it is necessary that $\partial E/\partial a_j = 0$ for each $j = 0, 1, \ldots, n$. This gives $n + 1$ **normal equations** in the $n + 1$ unknowns, a_j,

$$a_0 \sum_{i=1}^{m} x_i^0 + a_1 \sum_{i=1}^{m} x_i^1 + a_2 \sum_{i=1}^{m} x_i^2 + \cdots + a_n \sum_{i=1}^{m} x_i^n = \sum_{i=1}^{m} y_i x_i^0,$$

$$a_0 \sum_{i=1}^{m} x_i^1 + a_1 \sum_{i=1}^{m} x_i^2 + a_2 \sum_{i=1}^{m} x_i^3 + \cdots + a_n \sum_{i=1}^{m} x_i^{n+1} = \sum_{i=1}^{m} y_i x_i^1,$$

$$\vdots$$

$$a_0 \sum_{i=1}^{m} x_i^n + a_1 \sum_{i=1}^{m} x_i^{n+1} + a_2 \sum_{i=1}^{m} x_i^{n+2} + \cdots + a_n \sum_{i=1}^{m} x_i^{2n} = \sum_{i=1}^{m} y_i x_i^n.$$

The normal equations will have a unique solution, provided that the x_i are distinct.

EXAMPLE 2   Fit the data in the first two rows of Table 8.3 with the discrete least squares polynomial of degree two. For this problem, $n = 2, m = 5$, and the three normal equations are

$$5a_0 + 2.5a_1 + 1.875a_2 = 8.7680,$$
$$2.5a_0 + 1.875a_1 + 1.5625a_2 = 5.4514,$$
$$1.875a_0 + 1.5625a_1 + 1.3828a_2 = 4.4015.$$

Table 8.3

i	1	2	3	4	5
x_i	0	0.25	0.50	0.75	1.00
y_i	1.0000	1.2840	1.6487	2.1170	2.7183
$P(x_i)$	1.0052	1.2740	1.6482	2.1279	2.7130
$y_i - P(x_i)$	-0.0052	0.0100	0.0005	-0.0109	0.0053

The solution to this system is

$$a_0 = 1.0052, \qquad a_1 = 0.8641, \qquad a_2 = 0.8437.$$

Thus, the least squares polynomial of degree two fitting the preceding data is $P_2(x) = 1.0052 + 0.8641x + 0.8437x^2$, whose graph is shown in Figure 8.3. At the given values of x_i, we have the approximations shown in Table 8.3.

Figure 8.3

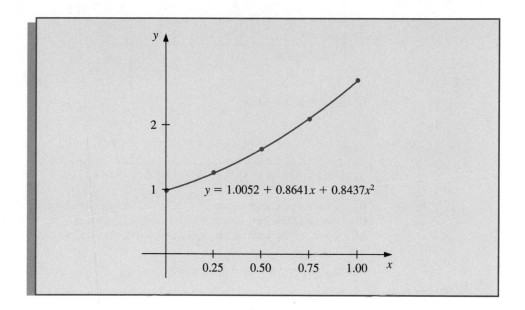

The total error

$$\sum_{i=1}^{5}(y_i - P(x_i))^2 = 2.76 \times 10^{-4}$$

is the least that can be obtained by using a quadratic polynomial of degree at most two.

■ ■ ■

EXERCISE SET 8.2

1. Compute the linear least squares polynomial for the data of Example 2.

2. Compute the least squares polynomial of degree two for the data of Example 1 and compare the total error for the two polynomials.

3. Find the least squares polynomials of degree one, two, and three for the data in the following table. Compute the total error in each case. Graph the data and the polynomials.

x_i	1.0	1.1	1.3	1.5	1.9	2.1
y_i	1.84	1.96	2.21	2.45	2.94	3.18

4. Find the least squares polynomials of degree one, two, and three for the data in the table on page 276 and compute the total error in each case. Graph the data and the polynomials.

x_i	0	0.15	0.31	0.5	0.6	0.75
y_i	1.0	1.004	1.031	1.117	1.223	1.422

5. Given the following data:

x_i	4.0	4.2	4.5	4.7	5.1	5.5	5.9	6.3	6.8	7.1
y_i	102.56	113.18	130.11	142.05	167.53	195.14	224.87	256.73	299.50	326.72

(a) Construct the least squares approximation of degree one and compute the total error.
(b) Construct the least squares approximation of degree two and compute the total error.
(c) Construct the least squares approximation of degree three and compute the total error.

6. Repeat Exercise 5 for the following data:

x_i	0.2	0.3	0.6	0.9	1.1	1.3	1.4	1.6
y_i	0.050446	0.098426	0.33277	0.72660	1.0972	1.5697	1.8487	2.5015

7. The following table lists the college grade-point averages of 20 mathematics and computer science majors, together with the scores that these students received on the mathematics portion of the ACT (American College Testing Program) test while in high school. Plot the data, and find the equation of the least squares line for these data.

ACT Score	Grade-Point Average	ACT Score	Grade-Point Average
28	3.84	29	3.75
25	3.21	28	3.65
28	3.23	27	3.87
27	3.63	29	3.75
28	3.75	21	1.66
33	3.20	28	3.12
28	3.41	28	2.96
29	3.38	26	2.92
23	3.53	30	3.10
27	2.03	24	2.81

8. To determine a functional relationship between the attenuation coefficient and the thickness of a sample of taconite, V. P. Singh fit a collection of data by using a linear least squares polynomial. The collection of data on page 277 is taken from a graph in that paper. Find the best linear least squares approximation for the data.

Thickness (cm)	Attenuation Coefficient (db/cm)
0.040	26.5
0.041	28.1
0.055	25.2
0.056	26.0
0.062	24.0
0.071	25.0
0.071	26.4
0.078	27.2
0.082	25.6
0.090	25.0
0.092	26.8
0.100	24.8
0.105	27.0
0.120	25.0
0.123	27.3
0.130	26.9
0.140	26.2

8.3 Continuous Least Squares Approximation

Suppose $f \in C[a, b]$ and we would like to determine P_n, a polynomial of degree at most n, to minimize the error

$$\int_a^b (f(x) - P_n(x))^2 \, dx.$$

As in the previous section we will determine a least squares approximating polynomial — that is, find numbers $a_0, a_1, \ldots, a_n$ to minimize the expression for the *total error*

$$E(a_0, a_1, \ldots, a_n) = \int_a^b \left(f(x) - \sum_{k=0}^n a_k x^k \right)^2 dx,$$

where

$$P_n(x) = a_n x^n + a_{n-1} x^{n-1} + \cdots + a_1 x + a_0 = \sum_{k=0}^n a_k x^k.$$

A necessary condition for the numbers $a_0, a_1, \ldots, a_n$ to minimize the total error E is that

$$\frac{\partial E}{\partial a_j} = 0 \qquad \text{for each } j = 0, 1, \ldots, n.$$

Since

$$E = \int_a^b [f(x)]^2\, dx - 2\sum_{k=0}^n a_k \int_a^b x^k f(x)\, dx + \int_a^b \left(\sum_{k=0}^n a_k x^k\right)^2 dx,$$

$$\frac{\partial E}{\partial a_j} = -2\int_a^b x^j f(x)\, dx + 2\sum_{k=0}^n a_k \int_a^b x^{j+k}\, dx.$$

Setting these to zero and rearranging we obtain the $(n+1)$ linear **normal equations**

$$\sum_{k=0}^n a_k \int_a^b x^{j+k}\, dx = \int_a^b x^j f(x)\, dx, \qquad \text{for each } j = 0, 1, \ldots, n,$$

which must be solved for the $(n+1)$ unknowns a_j. The normal equations have a unique solution provided that $f \in C[a, b]$.

EXAMPLE 1   Find the least squares approximating polynomial of degree two for the function $f(x) = \sin \pi x$ on the interval $[0, 1]$. The normal equations for $P_2(x) = a_2 x^2 + a_1 x + a_0$ are

$$a_0 \int_0^1 1\, dx + a_1 \int_0^1 x\, dx + a_2 \int_0^1 x^2\, dx = \int_0^1 \sin \pi x\, dx,$$

$$a_0 \int_0^1 x\, dx + a_1 \int_0^1 x^2\, dx + a_2 \int_0^1 x^3\, dx = \int_0^1 x \sin \pi x\, dx,$$

$$a_0 \int_0^1 x^2\, dx + a_1 \int_0^1 x^3\, dx + a_2 \int_0^1 x^4\, dx = \int_0^1 x^2 \sin \pi x\, dx.$$

Performing the integration yields

$$a_0 + \frac{1}{2}a_1 + \frac{1}{3}a_2 = \frac{2}{\pi}, \qquad \frac{1}{2}a_0 + \frac{1}{3}a_1 + \frac{1}{4}a_2 = \frac{1}{\pi}, \qquad \frac{1}{3}a_0 + \frac{1}{4}a_1 + \frac{1}{5}a_2 = \frac{\pi^2 - 4}{\pi^3}.$$

These three equations in three unknowns can be solved to obtain

$$a_0 = \frac{12\pi^2 - 120}{\pi^3} \approx -0.050465 \quad \text{and} \quad a_1 = -a_2 = \frac{720 - 60\pi^2}{\pi^3} \approx 4.12251.$$

Consequently, the least squares polynomial approximation of degree two for $f(x) = \sin \pi x$ on $[0, 1]$ is $P_2(x) = -4.12251x^2 + 4.12251x - 0.050465$. (See Figure 8.4.)   ■ ■ ■

Example 1 illustrates the difficulty in obtaining a least squares polynomial approximation. An $(n+1) \times (n+1)$ linear system for the unknowns $a_0, \ldots, a_n$ must be solved, and the coefficients in the linear system are of the form

$$\int_a^b x^{j+k}\, dx = \frac{b^{j+k+1} - a^{j+k+1}}{j + k + 1}.$$

Figure 8.4

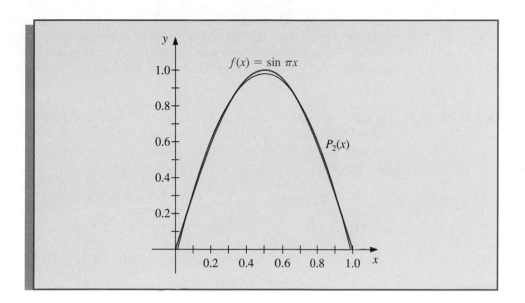

The matrix in the linear system is known as a *Hilbert matrix*. It is a classic example for demonstrating round-off error difficulties; no pivoting technique can be satisfactorily used.

Another disadvantage to the technique used in Example 1 is similar to the situation that occurred when the Lagrange polynomials were first introduced in Section 3.3. The calculations that were performed in obtaining the best nth-degree polynomial do not lessen the amount of work required to obtain the best polynomials of higher degree. Both of these problems are resolved by resorting to a technique that reduces the $n + 1$ equations in $n + 1$ unknowns to $n + 1$ equations, each of which contains only one unknown. This simplifies the problem to one that can be easily solved, but the technique requires some new concepts.

The set of functions $\{\phi_0, \phi_1, \ldots, \phi_n\}$ is said to be **linearly independent** on $[a, b]$ if, whenever

$$c_0\phi_0(x) + c_1\phi_1(x) + \cdots + c_n\phi_n(x) = 0 \qquad \text{for all } x \in [a, b],$$

then $c_0 = c_1 = \cdots = c_n = 0$. Otherwise the set of functions is said to be **linearly dependent.**

Linearly independent sets of functions are basic to our discussion and, since the functions we are using for approximations are polynomials, the following result is fundamental:

Linearly Independent Sets of Polynomials

If ϕ_j is a polynomial of degree j for each $j = 0, 1, \ldots, n$, then $\{\phi_0, \ldots, \phi_n\}$ is linearly independent on any interval $[a, b]$.

The situation illustrated in the following example demonstrates a fact that holds in a much more general setting. Let $\prod_n$ be the **set of all polynomials of degree at most n**. If $\{\phi_0, \phi_1, \ldots, \phi_n\}$ is any collection of linearly independent polynomials in $\prod_n$, then each polynomial in $\prod_n$ can be written uniquely as a linear combination of $\{\phi_0, \phi_1, \ldots, \phi_n\}$.

EXAMPLE 2 Let $\phi_0(x) = 2$, $\phi_1(x) = x - 3$, and $\phi_2(x) = x^2 + 2x + 7$. Then $\{\phi_0, \phi_1, \phi_2\}$ is linearly independent on any interval $[a, b]$. Suppose $Q(x) = a_0 + a_1 x + a_2 x^2$. We will show that there exist constants c_0, c_1, and c_2 such that $Q(x) = c_0 \phi_0(x) + c_1 \phi_1(x) + c_2 \phi_2(x)$. Note first that

$$1 = \tfrac{1}{2}\phi_0(x), \qquad x = \phi_1(x) + 3 = \phi_1(x) + \tfrac{3}{2}\phi_0(x),$$

and

$$x^2 = \phi_2(x) - 2x - 7 = \phi_2(x) - 2[\phi_1(x) + \tfrac{3}{2}\phi_0(x)] - 7[\tfrac{1}{2}\phi_0(x)]$$
$$= \phi_2(x) - 2\phi_1(x) - \tfrac{13}{2}\phi_0(x).$$

Hence,

$$Q(x) = a_0\left[\tfrac{1}{2}\phi_0(x)\right] + a_1\left[\phi_1(x) + \tfrac{3}{2}\phi_0(x)\right] + a_2\left[\phi_2(x) - 2\phi_1(x) - \tfrac{13}{2}\phi_0(x)\right]$$
$$= \left[\tfrac{1}{2}a_0 + \tfrac{3}{2}a_1 - \tfrac{13}{2}a_2\right]\phi_0(x) + [a_1 - 2a_2]\phi_1(x) + a_2\phi_2(x),$$

so any quadratic polynomial can be expressed as a linear combination of the functions ϕ_0, ϕ_1, and ϕ_2. ■ ■ ■

To discuss general function approximation requires the introduction of the notions of weight functions and orthogonality.

An integrable function w is called a **weight function** on the interval I if $w(x) \geq 0$ for all x in I, and if w is not identically zero on any subinterval of I.

The purpose of a weight function is to assign varying degrees of importance to approximations on certain portions of the interval. For example, the weight function

$$w(x) = \frac{1}{\sqrt{1 - x^2}}$$

places less emphasis near the center of the interval $(-1, 1)$ and more emphasis when $|x|$ is near 1 (see Figure 8.5). This weight function is used in the next section.

Suppose $\{\phi_0, \phi_1, \ldots, \phi_n\}$ is a set of linearly independent functions on $[a, b]$, w is a weight function for $[a, b]$, and, for $f \in C[a, b]$, a linear combination

$$P(x) = \sum_{k=0}^{n} a_k \phi_k(x)$$

Figure 8.5

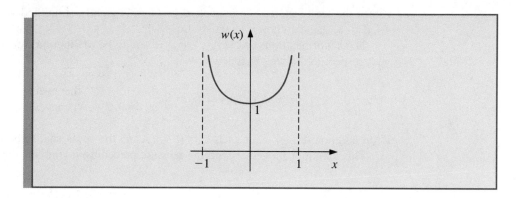

is sought to minimize the error

$$E(a_0, \ldots, a_n) = \int_a^b w(x) \left[f(x) - \sum_{k=0}^n a_k \phi_k(x) \right]^2 dx.$$

This problem reduces to the situation considered at the beginning of this section in the special case when $w(x) \equiv 1$ and $\phi_k(x) = x^k$ for each $k = 0, 1, \ldots, n$. The normal equations associated with this problem are derived from the fact that for each $j = 0, 1, \ldots, n$,

$$0 = \frac{\partial E}{\partial a_j} = 2 \int_a^b w(x) \left[f(x) - \sum_{k=0}^n a_k \phi_k(x) \right] \phi_j(x) \, dx.$$

The system of normal equations can be written

$$\int_a^b w(x) f(x) \phi_j(x) \, dx = \sum_{k=0}^n a_k \int_a^b w(x) \phi_k(x) \phi_j(x) \, dx, \qquad \text{for each } j = 0, 1, \ldots, n.$$

If the functions $\phi_0, \phi_1, \ldots, \phi_n$ can be chosen so that

$$(8.1) \qquad \int_a^b w(x) \phi_k(x) \phi_j(x) \, dx = \begin{cases} 0, & \text{when } j \neq k, \\ \alpha_k > 0, & \text{when } j = k, \end{cases}$$

for some numbers α_k, then the normal equations reduce to

$$\int_a^b w(x) f(x) \phi_j(x) \, dx = a_j \int_a^b w(x) [\phi_j(x)]^2 \, dx = a_j \alpha_j,$$

for each $j = 0, 1, \ldots, n$, and are easily solved:

$$a_j = \frac{1}{\alpha_j} \int_a^b w(x) f(x) \phi_j(x) \, dx.$$

Hence the least squares approximation problem is greatly simplified when the functions $\phi_0, \phi_1, \ldots, \phi_n$ are chosen to satisfy Eq. (8.1).

The set of functions $\{\phi_0, \phi_1, \ldots, \phi_n\}$ is said to be **orthogonal** for the interval $[a, b]$ with respect to the weight function w if

$$\int_a^b w(x)\phi_j(x)\phi_k(x)\,dx = \begin{cases} 0, & \text{whenever } j \neq k, \\ \alpha_k > 0, & \text{whenever } j = k. \end{cases}$$

If, in addition, $\alpha_k = 1$ for each $k = 0, 1, \ldots, n$, the set is said to be **orthonormal**.

This definition, together with the remarks preceding it, implies the following:

Least Squares for Orthogonal Functions

If $\{\phi_0, \phi_1, \ldots, \phi_n\}$ is an orthogonal set of functions on an interval $[a, b]$ with respect to the weight function w, then the least squares approximation to f on $[a, b]$ with respect to w is

$$P(x) = \sum_{k=0}^n a_k \phi_k(x),$$

where

$$a_k = \frac{\int_a^b w(x)\phi_k(x)f(x)\,dx}{\int_a^b w(x)[\phi_k(x)]^2\,dx} = \frac{1}{\alpha_k}\int_a^b w(x)\phi_k(x)f(x)\,dx.$$

The next result, which is based on the *Gram–Schmidt* process, describes a recursive procedure for constructing orthogonal polynomials on $[a, b]$ with respect to a weight function w.

Recursive Generation of Orthogonal Polynomials

The set of polynomials $\{\phi_0, \phi_1, \ldots, \phi_n\}$ defined in the following way is linearly independent and orthogonal on $[a, b]$ with respect to the weight function w.

$$\phi_0(x) \equiv 1, \qquad \phi_1(x) = x - B_1,$$

where

$$B_1 = \frac{\int_a^b xw(x)[\phi_0(x)]^2\,dx}{\int_a^b w(x)[\phi_0(x)]^2\,dx},$$

and when $k \geq 2$,

$$\phi_k(x) = (x - B_k)\phi_{k-1}(x) - C_k\phi_{k-2}(x),$$

where

$$B_k = \frac{\int_a^b x w(x)[\phi_{k-1}(x)]^2\, dx}{\int_a^b w(x)[\phi_{k-1}(x)]^2\, dx} \quad \text{and} \quad C_k = \frac{\int_a^b x w(x)\phi_{k-1}(x)\phi_{k-2}(x)\, dx}{\int_a^b w(x)[\phi_{k-2}(x)]^2\, dx}.$$

Moreover, for any polynomial Q_k of degree $k < n$,

$$\int_a^b w(x)\phi_n(x)Q_k(x)\, dx = 0.$$

EXAMPLE 3 The set of **Legendre polynomials**, $\{P_n\}$, is orthogonal on $[-1, 1]$ with respect to the weight function $w(x) \equiv 1$. The classical definition of the Legendre polynomials requires that $P_n(1) = 1$ for each n, and a recursive relation can be used to generate the polynomials when $n \geq 2$. This normalization will not be needed in our discussion, and the least squares approximating polynomials generated in either case will be essentially the same. Using the recursive procedure, $P_0(x) \equiv 1$, so

$$B_1 = \frac{\int_{-1}^1 x\, dx}{\int_{-1}^1 dx} = 0 \quad \text{and} \quad P_1(x) = (x - B_1)P_0(x) = x.$$

Also,

$$B_2 = \frac{\int_{-1}^1 x^3\, dx}{\int_{-1}^1 x^2\, dx} = 0 \quad \text{and} \quad C_2 = \frac{\int_{-1}^1 x^2\, dx}{\int_{-1}^1 1\, dx} = \frac{1}{3},$$

so

$$P_2(x) = (x - B_2)P_1(x) - C_2 P_0(x) = (x - 0)x - \tfrac{1}{3} \cdot 1 = x^2 - \tfrac{1}{3}.$$

Higher-degree Legendre polynomials are derived in the same manner. The next three are $P_3(x) = x^3 - (\tfrac{3}{5})x$, $P_4(x) = x^4 - (\tfrac{6}{7})x^2 + \tfrac{3}{35}$, and $P_5(x) = \tfrac{10x^3}{9} + \tfrac{5x}{21}$. Figure 8.6 (page 284) shows the graphs of these polynomials. ■ ■ ■

Figure 8.6

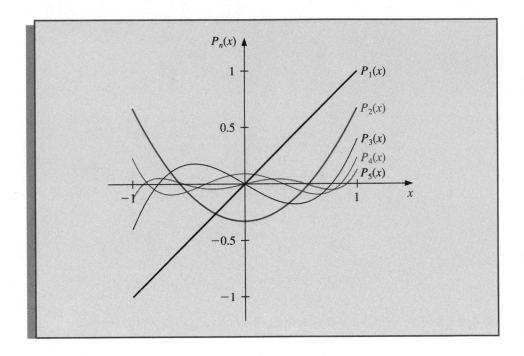

EXERCISE SET 8.3

1. Find the linear least squares polynomial approximation to $f(x)$ on the indicated interval.
    (a) $f(x) = x^2 + 3x + 2$, $[0, 1]$ **(b)** $f(x) = x^3$, $[0, 2]$
    (c) $f(x) = \frac{1}{x}$, $[1, 3]$ **(d)** $f(x) = e^x$, $[0, 2]$
    (e) $f(x) = \frac{1}{2}\cos x + \frac{1}{3}\sin 2x$, $[0, 1]$ **(f)** $f(x) = x \ln x$, $[1, 3]$

2. Find the least squares polynomial approximation of degree two to the functions and intervals in Exercise 1.

3. Find the linear least squares polynomial approximation on the interval $[-1, 1]$ for the following functions:
    (a) $f(x) = x^2 - 2x + 3$ **(b)** $f(x) = x^3$
    (c) $f(x) = \frac{1}{x+2}$ **(d)** $f(x) = e^x$
    (e) $f(x) = \frac{1}{2}\cos x + \frac{1}{3}\sin 2x$ **(f)** $f(x) = \ln(x + 2)$

4. Find the least squares polynomial approximation of degree two on the interval $[-1, 1]$ for the functions in Exercise 3.

5. Compute the total error for the approximations in Exercise 3.

6. Compute the total error for the approximations in Exercise 4.

7. Use the Gram–Schmidt process to construct $\phi_0(x)$, $\phi_1(x)$, $\phi_2(x)$, and $\phi_3(x)$ for the following intervals:
    (a) $[0, 1]$ **(b)** $[0, 2]$ **(c)** $[1, 3]$

8. Repeat Exercise 1 using the results of Exercise 7.

9. Repeat Exercise 2 using the results of Exercise 7.

10. Use the results of Exercise 7 to obtain the least squares approximation polynomial of degree three for the functions in Exercise 1.

11. Use the Gram–Schmidt procedure to calculate L_1, L_2, and L_3, where $\{L_0, L_1, L_2, L_3\}$ is an orthogonal set of polynomials on $(0, \infty)$ with respect to the weight functions $w(x) = e^{-x}$ and $L_0(x) \equiv 1$. The polynomials obtained from this procedure are called the **Laguerre polynomials.**

12. Use the Laguerre polynomials calculated in Exercise 11 to compute the least squares polynomials of degree one, two, and three on the interval $(0, \infty)$ with respect to the weight function $w(x) = e^{-x}$ for the following functions:

(a) $f(x) = x^2$ **(b)** $f(x) = e^{-x}$

(c) $f(x) = x^3$ **(d)** $f(x) = e^{-2x}$

8.4 Chebyshev Polynomials

The **Chebyshev polynomials** $\{T_n\}$ are orthogonal on $(-1, 1)$ with respect to the weight function $w(x) = (1 - x^2)^{-1/2}$. Although they can be derived by the method in the previous section, it is easier to give their definition and then show that they satisfy the required orthogonality properties.

For $x \in [-1, 1]$, define

$$T_n(x) = \cos(n \arccos x) \qquad \text{for each } n \geq 0.$$

First note that

$$T_0(x) = \cos 0 = 1 \qquad \text{and} \qquad T_1(x) = \cos(\arccos x) = x.$$

For $n \geq 1$ we introduce the substitution $\theta = \arccos x$ to change this equation to

$$T_n(\theta(x)) \equiv T_n(\theta) = \cos(n\theta), \qquad \text{where } \theta \in [0, \pi].$$

A recurrence relation is derived by noting that

$$T_{n+1}(\theta) = \cos(n\theta + \theta) = \cos(n\theta)\cos\theta - \sin(n\theta)\sin\theta$$

and

$$T_{n-1}(\theta) = \cos(n\theta - \theta) = \cos(n\theta)\cos\theta + \sin(n\theta)\sin\theta.$$

Adding these equations gives

$$T_{n+1}(\theta) = 2\cos(n\theta)\cos\theta - T_{n-1}(\theta).$$

Returning to the variable x we have the following:

Chebyshev Polynomials

$$T_0(x) = 1, \qquad T_1(x) = x,$$

and, for $n \geq 1$,

$$T_{n+1}(x) = 2xT_n(x) - T_{n-1}(x).$$

The recurrence relation implies that $T_n(x)$ is a polynomial of degree n with leading coefficient 2^{n-1}, when $n \geq 1$. The next three Chebyshev polynomials are

$$T_2(x) = 2xT_1(x) - T_0(x) = 2x^2 - 1,$$
$$T_3(x) = 2xT_2(x) - T_1(x) = 4x^3 - 3x,$$

and

$$T_4(x) = 2xT_3(x) - T_2(x) = 8x^4 - 8x^2 + 1.$$

The graphs of T_1, T_2, T_3, and T_4 are shown in Figure 8.7.

Figure 8.7

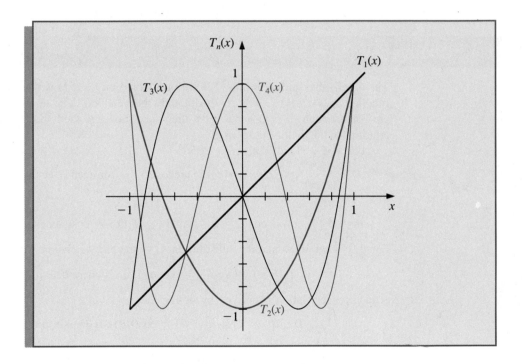

To show the orthogonality of the Chebyshev polynomials, consider

$$\int_{-1}^{1} \frac{T_n(x)T_m(x)}{\sqrt{1-x^2}} \, dx = \int_{-1}^{1} \frac{\cos(n \arccos x) \cos(m \arccos x)}{\sqrt{1-x^2}} \, dx.$$

Reintroducing the substitution $\theta = \arccos x$ gives

$$d\theta = -\frac{1}{\sqrt{1-x^2}} \, dx$$

and

$$\int_{-1}^{1} \frac{T_n(x)T_m(x)}{\sqrt{1-x^2}} \, dx = -\int_{\pi}^{0} \cos(n\theta) \cos(m\theta) \, d\theta = \int_{0}^{\pi} \cos(n\theta) \cos(m\theta) \, d\theta.$$

Suppose $n \neq m$. Since

$$\cos(n\theta)\cos(m\theta) = \tfrac{1}{2}[\cos(n+m)\theta + \cos(n-m)\theta],$$

we have

$$\int_{-1}^{1} \frac{T_n(x)T_m(x)}{\sqrt{1-x^2}}\, dx = \frac{1}{2}\int_0^\pi \cos((n+m)\theta)\, d\theta + \frac{1}{2}\int_0^\pi \cos((n-m)\theta)\, d\theta$$

$$= \left[\frac{1}{2(n+m)}\sin((n+m)\theta) + \frac{1}{2(n-m)}\sin((n-m)\theta\right]_0^\pi = 0.$$

By a similar technique, it can also be shown that

$$\int_{-1}^{1} \frac{[T_n(x)]^2}{\sqrt{1-x^2}}\, dx = \frac{\pi}{2} \qquad \text{for each } n \geq 1.$$

One of the important results about the Chebyshev polynomials concerns their zeros and extrema.

Zeros and Extrema of Chebyshev Polynomials

The Chebyshev polynomial T_n of degree $n \geq 1$ has n simple zeros in $[-1, 1]$ at

$$\bar{x}_k = \cos\left(\frac{2k-1}{2n}\pi\right) \qquad \text{for each } k = 1, 2, \ldots, n.$$

Moreover, T_n assumes its absolute extrema on $[-1, 1]$ at

$$\bar{x}_k' = \cos\left(\frac{k\pi}{n}\right) \qquad \text{with} \qquad T_n(\bar{x}_k') = (-1)^k \qquad \text{for each } k = 0, 1, \ldots, n.$$

The *monic Chebyshev polynomial* (polynomial with leading coefficient 1) $\tilde{T}_n$ is derived from the Chebyshev polynomial T_n by dividing by the leading coefficient 2^{n-1}, when $n \geq 1$. So

$$\tilde{T}_0(x) = 1, \qquad \text{and} \qquad \tilde{T}_n(x) = 2^{1-n}T_n(x), \qquad \text{for each } n \geq 1.$$

These polynomials satisfy the recurrence relation

$$\tilde{T}_2(x) = x\tilde{T}_1(x) - \frac{1}{2}\tilde{T}_0(x); \qquad \tilde{T}_{n+1}(x) = x\tilde{T}_n(x) - \frac{1}{4}\tilde{T}_{n-1}(x),$$

for each $n \geq 2$. Because of the linear relationship between $\tilde{T}_n$ and T_n, the zeros of $\tilde{T}_n$ also occur at

$$\bar{x}_k = \cos\left(\frac{2k-1}{2n}\pi\right), \qquad \text{for each } k = 1, 2, \ldots, n,$$

and the extreme values of $\tilde{T}_n$ occur at

$$\bar{x}_k' = \cos\left(\frac{k\pi}{n}\right) \quad \text{with} \quad \tilde{T}_n(\bar{x}_k') = \frac{(-1)^k}{2^{n-1}}, \quad \text{for each } k = 0, 1, 2, \ldots, n.$$

Let $\tilde{\Pi}_n$ denote **the set of all monic polynomials of degree n**. We have an important minimization property that distinguishes the polynomials $\tilde{T}_n$ from the other members of $\tilde{\Pi}_n$.

Minimum Property of Monic Chebyshev Polynomials

The polynomials $\tilde{T}_n$, when $n \geq 1$, have the property that

$$\frac{1}{2^{n-1}} = \max_{x \in [-1,1]} |\tilde{T}_n(x)| \leq \max_{x \in [-1,1]} |P_n(x)| \quad \text{for all } P_n \in \tilde{\Pi}_n.$$

Moreover, equality can occur only if $P_n = \tilde{T}_n$.

This result is used to answer the question of where to place interpolating nodes to minimize the error in Lagrange interpolation. The error form for the Lagrange polynomial applied to the interval $[-1, 1]$ states that if $x_0, \ldots, x_n$ are distinct numbers in the interval $[-1, 1]$ and $f \in C^{n+1}[-1, 1]$, then, for each $x \in [-1, 1]$, a number $\xi(x)$ exists in $(-1, 1)$ with

$$f(x) - P(x) = \frac{f^{(n+1)}(\xi(x))}{(n+1)!}(x - x_0)\cdots(x - x_n),$$

where P denotes the Lagrange interpolating polynomial. There is no control over $\xi(x)$, so to minimize the error by shrewd placement of the nodes $x_0, \ldots, x_n$ is equivalent to finding the $x_0, \ldots, x_n$ to minimize the quantity

$$|(x - x_0)(x - x_1)\cdots(x - x_n)|$$

throughout the interval $[-1, 1]$. Since $(x - x_0)(x - x_1)\cdots(x - x_n)$ is a monic polynomial of degree $(n + 1)$, the minimum is obtained when

$$(x - x_0)(x - x_1)\cdots(x - x_n) = \tilde{T}_{n+1}(x).$$

When x_k is chosen to be the $(k + 1)$st zero of $\tilde{T}_{n+1}$, for each $k = 0, 1, \ldots, n$; that is, when x_k is

$$\bar{x}_{k+1} = \cos\frac{2k + 1}{2(n + 1)}\pi,$$

the maximum value of $|(x - x_0)(x - x_1)\cdots(x - x_n)|$ is minimized. Since

$$\max_{x \in [-1,1]} |\tilde{T}_{n+1}(x)| = \frac{1}{2^n},$$

this also implies that

$$\frac{1}{2^n} = \max_{x\in[-1,1]} \left|(x - \bar{x}_1)(x - \bar{x}_2)\cdots(x - \bar{x}_{n+1})\right| \le \max_{x\in[-1,1]} \left|(x - x_0)(x - x_1)\cdots(x - x_n)\right|,$$

for any choice of $x_0, x_1, \ldots, x_n$ in the interval $[-1, 1]$.

Minimizing Lagrange Interpolation Error

If P is the interpolating polynomial of degree at most n with nodes at the roots of $T_{n+1}(x)$, then, for any $f \in C^{n+1}[-1, 1]$,

$$\max_{x\in[-1,1]} |f(x) - P(x)| \le \frac{1}{2^n(n+1)!} \max_{x\in[-1,1]} \left|f^{(n+1)}(x)\right|.$$

The technique for choosing points to minimize the interpolating error can be easily extended to a general closed interval $[a, b]$ by using the change of variable

$$\tilde{x} = \tfrac{1}{2}[(b - a)x + a + b]$$

to transform the numbers $\bar{x}_k$ in the interval $[-1, 1]$ into the corresponding numbers in the interval $[a, b]$.

EXAMPLE 1 Let $f(x) = xe^x$ on $[0, 1.5]$. Two interpolation polynomials of degree at most three will be constructed. First, the equally spaced nodes $x_0 = 0, x_1 = 0.5, x_2 = 1$, and $x_3 = 1.5$ are used. This interpolating polynomial is

$$P_3(x) = 1.3875x^3 + 0.057570x^2 + 1.2730x.$$

For the second interpolating polynomial, shift the zeros $\bar{x}_k = \cos((2k + 1)/8)\pi$, for $k = 0, 1, 2, 3$, of $\tilde{T}_4$ from $[-1, 1]$ to $[0, 1.5]$, using the linear transformation

$$\tilde{x}_k = \tfrac{1}{2}[(1.5 - 0)\bar{x}_k + (1.5 + 0)] = 0.75 + 0.75\bar{x}_k$$

to obtain

$$\tilde{x}_0 = 1.44291, \qquad \tilde{x}_1 = 1.03701, \qquad \tilde{x}_2 = 0.46299, \qquad \text{and} \qquad \tilde{x}_3 = 0.05709.$$

For these nodes, the interpolation polynomial of degree at most three is

$$\tilde{P}_3(x) = 1.3811x^3 + 0.044445x^2 + 1.3030x - 0.014357.$$

For comparison, Table 8.4 lists various values of x, together with the values of $f(x), P_3(x)$, and $\tilde{P}_3(x)$. Although the error using P_3 is less than using $\tilde{P}_3$ near the middle of the table, the maximum error involved with using $\tilde{P}_3$ is considerably less. (See Figure 8.8.) ∎ ∎ ∎

Table 8.4

x	$f(x) = xe^x$	$P_3(x)$	$\|xe^x - P_3(x)\|$	$\tilde{P}_3(x)$	$\|xe^x - \tilde{P}_3(x)\|$
0.15	0.1743	0.1969	0.0226	0.1868	0.0125
0.25	0.3210	0.3435	0.0225	0.3358	0.0148
0.35	0.4967	0.5121	0.0154	0.5064	0.0097
0.65	1.245	1.233	0.0120	1.231	0.0140
0.75	1.588	1.572	0.0160	1.571	0.0170
0.85	1.989	1.976	0.0130	0.973	0.0160
1.15	3.632	3.650	0.0180	3.643	0.0110
1.25	4.363	4.391	0.0280	4.381	0.0180
1.35	5.208	5.237	0.0290	5.224	0.0160

Figure 8.8

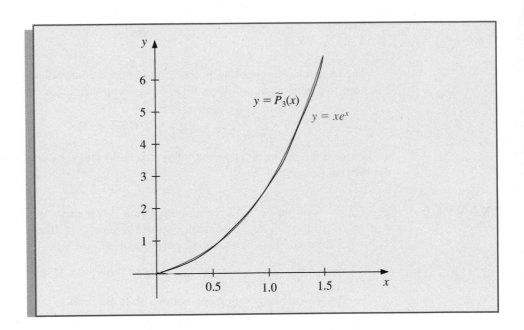

EXERCISE SET 8.4

1. Use the zeros of $\tilde{\mathcal{T}}_3$ to construct an interpolating polynomial of degree two for the following functions on the interval $[-1, 1]$:
   (a) $f(x) = e^x$
   (b) $f(x) = \sin x$
   (c) $f(x) = \ln(x + 2)$
   (d) $f(x) = x^4$

2. Find a bound for the maximum error of the approximation in Exercise 1 on the interval $[-1, 1]$.

3. Use the zeros of $\tilde{T}_4$ to construct an interpolating polynomial of degree three for the functions in Exercise 1.

4. Repeat Exercise 2 for the approximations computed in Exercise 3.

5. Use the zeros of $\tilde{T}_3$ and transformations of the given interval to construct an interpolating polynomial of degree two for the following functions:
    (a) $f(x) = \frac{1}{x}$, $[1, 3]$
    (b) $f(x) = e^{-x}$, $[0, 2]$
    (c) $f(x) = \frac{1}{2}\cos x + \frac{1}{3}\sin 2x$, $[0, 1]$
    (d) $f(x) = x \ln x$, $[1, 3]$

6. Use the zeros of $\tilde{T}_4$ to construct an interpolating polynomial of degree three for the functions in Exercise 5.

7. Show that for any positive integers i and j with $i > j$,

    $$T_i(x)T_j(x) = \tfrac{1}{2}[T_{i+j}(x) + T_{i-j}(x)].$$

8. Show that for each Chebyshev polynomial $T_n(x)$,

    $$\int_{-1}^{1} \frac{[T_n(x)]^2}{\sqrt{1-x^2}}\, dx = \frac{\pi}{2}.$$

8.5 Rational Function Approximation

The class of algebraic polynomials has some distinct advantages for use in approximation. There is a sufficient number of polynomials to approximate any continuous function on a closed interval to within an arbitrary tolerance, polynomials are easily evaluated at arbitrary values, and the derivatives and integrals of polynomials exist and are easily determined. The disadvantages of using polynomials for approximation is their tendency for oscillation. This often causes error bounds in polynomial approximation to significantly exceed the average approximation error, since error bounds are determined by the maximum approximation error. We now consider methods that spread the approximation error more evenly over the approximation interval.

A **rational function** r of degree N has the form

$$r(x) = \frac{p(x)}{q(x)},$$

where p and q are polynomials whose degrees sum to N.

Since every polynomial is a rational function (simply let $q(x) \equiv 1$), approximation by rational functions gives results with no greater error bounds than approximation by polynomials. However, rational functions whose numerator and denominator have the same or nearly the same degree generally produce approximation results superior to polynomial methods for the same amount of computational effort. Rational functions have the added advantage of permitting efficient approximation of functions that have infinite discontinuities near the interval of approximation. Polynomial approximation is generally unacceptable in this situation.

Suppose r is a rational function of degree $N = n + m$ of the form

$$r(x) = \frac{p(x)}{q(x)} = \frac{p_0 + p_1 x + \cdots + p_n x^n}{q_0 + q_1 x + \cdots + q_m x^m}$$

that is used to approximate a function f on a closed interval I containing zero. For r to be defined at zero requires that $q_0 \neq 0$. We can assume that $q_0 = 1$, for if this is not the case we simply replace $p(x)$ by $p(x)/q_0$ and $q(x)/q_0$. Consequently, there are $N + 1$ parameters $q_1, q_2, \ldots, q_m, p_0, p_1, \ldots, p_n$ available for the approximation of f by r. The **Padé approximation technique** chooses the $N + 1$ parameters so that $f^{(k)}(0) = r^{(k)}(0)$ for each $k = 0, 1, \ldots, N$. Padé approximation is the extension of Taylor polynomial approximation to rational functions. In fact, when $n = N$ and $m = 0$, the Padé approximation is the Nth Taylor polynomial expanded about zero—that is, the Nth Maclaurin polynomial.

Consider the difference

$$f(x) - r(x) = f(x) - \frac{p(x)}{q(x)} = \frac{f(x)q(x) - p(x)}{q(x)} = \frac{f(x)\sum_{i=0}^{m} q_i x^i - \sum_{i=0}^{n} p_i x^i}{q(x)}$$

and suppose f has the Maclaurin series expansion $f(x) = \sum_{i=0}^{\infty} a_i x^i$. Then

$$\textbf{(8.2)} \qquad f(x) - r(x) = \frac{\sum_{i=0}^{\infty} a_i x^i \sum_{i=0}^{m} q_i x^i - \sum_{i=0}^{n} p_i x^i}{q(x)}.$$

The object is to choose the constants $q_1, q_2, \ldots, q_m$ and $p_0, p_1, \ldots, p_n$ so that

$$f^{(k)}(0) - r^{(k)}(0) = 0 \qquad \text{for each } k = 0, 1, \ldots, N.$$

This is equivalent to $f - r$ having a root of multiplicity $N + 1$ at zero. As a consequence, we choose $q_1, q_2, \ldots, q_m$ and $p_0, p_1, \ldots, p_n$ so that the numerator on the right side of Eq. (8.2),

$$(a_0 + a_1 x + \cdots)(1 + q_1 x + \cdots + q_m x^m) - (p_0 + p_1 x + \cdots + p_n x^n),$$

has no terms of degree less than or equal to N. To simplify notation, we define $p_{n+1} = p_{n+2} = \cdots = p_N = 0$ and $q_{m+1} = q_{m+2} = \cdots = q_N = 0$. The coefficient of x^k is then

$$\left(\sum_{i=0}^{k} a_i q_{k-i} \right) - p_k,$$

and the rational function for Padé approximation results from the solution of the $N + 1$ linear equations

$$\sum_{i=0}^{k} a_i q_{k-i} = p_k, \qquad k = 0, 1, \ldots, N$$

in the $N + 1$ unknowns $q_1, q_2, \ldots, q_m, p_0, p_1, \ldots, p_n$.

The Padé technique can be implemented using the program PADEMD81.

EXAMPLE 1 The Maclaurin series expansion for e^{-x} is

$$\sum_{i=0}^{\infty} \frac{(-1)^i}{i!} x^i.$$

To find the Padé approximation to e^{-x} of degree five with $n = 3$ and $m = 2$ requires choosing $p_0, p_1, p_2, p_3, q_1,$ and q_2 so that the coefficients of x^k for $k = 0, 1, \ldots, 5$ are zero in the expression

$$\left(1 - x + \frac{x^2}{2} - \frac{x^3}{6} + \cdots\right)(1 + q_1 x + q_2 x^2) - (p_0 + p_1 x + p_2 x^2 + p_3 x^3).$$

Expanding and collecting terms produces

$$x^5: \quad -\tfrac{1}{120} + \tfrac{1}{24} q_1 - \tfrac{1}{6} q_2 = 0; \qquad x^2: \quad \tfrac{1}{2} - q_1 + q_2 = p_2;$$
$$x^4: \quad \tfrac{1}{24} - \tfrac{1}{6} q_1 + \tfrac{1}{2} q_2 = 0; \qquad x^1: \quad -1 + q_1 = p_1;$$
$$x^3: \quad -\tfrac{1}{6} + \tfrac{1}{2} q_1 - q_2 = p_3; \qquad x^0: \quad 1 = p_0.$$

The solution to this system is

$$p_0 = 1, \quad p_1 = -\frac{3}{5}, \quad p_2 = \frac{3}{20}, \quad p_3 = -\frac{1}{60}, \quad q_1 = \frac{2}{5}, \quad \text{and} \quad q_2 = \frac{1}{20},$$

so the Padé approximation is

$$r(x) = \frac{1 - \frac{3}{5} x + \frac{3}{20} x^2 - \frac{1}{60} x^3}{1 + \frac{2}{5} x + \frac{1}{20} x^2}.$$

Table 8.5 lists values of $r(x)$ and $P_5(x)$, the fifth Maclaurin polynomial. The Padé approximation is clearly superior in this example. ■ ■ ■

Table 8.5

x	e^{-x}	$P_5(x)$	$|e^{-x} - P_5(x)|$	$r(x)$	$|e^{-x} - r(x)|$
0.2	0.81873075	0.81873067	8.64×10^{-8}	0.81873075	7.55×10^{-9}
0.4	0.67032005	0.67031467	5.38×10^{-6}	0.67031963	4.11×10^{-7}
0.6	0.54881164	0.54875200	5.96×10^{-5}	0.54880763	4.00×10^{-6}
0.8	0.44932896	0.44900267	3.26×10^{-4}	0.44930966	1.93×10^{-5}
1.0	0.36787944	0.36666667	1.21×10^{-3}	0.36781609	6.33×10^{-5}

It is interesting to compare the number of arithmetic operations required for calculations of $P_5(x)$ and $r(x)$ in Example 1. Using nested multiplication, $P_5(x)$ can be expressed as

$$P_5(x) = 1 - x\left(1 - x\left(\tfrac{1}{2} - x\left(\tfrac{1}{6} - x\left(\tfrac{1}{24} - \tfrac{1}{120} x\right)\right)\right)\right).$$

Assuming that the coefficients of $1, x, x^2, x^3, x^4$, and x^5 are represented as decimals, a single calculation of $P_5(x)$ in nested form requires five multiplications and five additions/subtractions. Using nested multiplication, $r(x)$ is expressed as

$$r(x) = \frac{1 - x(\frac{3}{5} - x(\frac{3}{20} - \frac{1}{60}x))}{1 + x(\frac{2}{5} + \frac{1}{20}x)},$$

so a single calculation of $r(x)$ requires five multiplications, five additions/subtractions, and one division. Hence, computational effort appears to favor the polynomial approximation. However, by reexpressing $r(x)$ by continued division, we can write

$$r(x) = \frac{1 - \frac{3}{5}x + \frac{3}{20}x^2 - \frac{1}{60}x^3}{1 + \frac{2}{5}x + \frac{1}{20}x^2}$$

$$= \frac{-\frac{1}{3}x^3 + 3x^2 - 12x + 20}{x^2 + 8x + 20}$$

$$= -\frac{1}{3}x + \frac{17}{3} + \frac{(-\frac{152}{3}x - \frac{280}{3})}{x^2 + 8x + 20)}$$

$$= -\frac{1}{3}x + \frac{17}{3} + \frac{-\frac{152}{3}}{\left(\dfrac{x^2 + 8x + 20}{x + \frac{35}{19}}\right)}$$

or

$$r(x) = -\frac{1}{3}x + \frac{17}{3} + \frac{-\frac{152}{3}}{\left(x + \frac{117}{19} + \dfrac{\frac{3125}{361}}{x + \frac{35}{19}}\right)}.$$

Written in this form, a single calculation of $r(x)$ requires one multiplication, five additions/subtractions, and two divisions. If the amount of computation required for division is approximately the same as for multiplication, the computational effort required for an evaluation of $P_5(x)$ significantly exceeds that required for an evaluation of $r(x)$. Expressing a rational function approximation in this form is called **continued-fraction** approximation. This is a classical approximation technique of current interest because of the computational efficiency of this representation. It is, however, a specialized technique, and one we will not discuss further.

Although the rational function approximation in Example 1 gave results superior to the polynomial approximation of the same degree, the approximation has a wide variation in accuracy; the approximation at 0.2 is accurate to within 8×10^{-9}, whereas, at 1.0, the approximation and the function agree only to within 7×10^{-5}. This accuracy variation is expected, because the Padé approximation is based on a Taylor polynomial representation of e^{-x}, and the Taylor representation has a wide variation of accuracy in $[0.2, 1.0]$.

To obtain more uniformly accurate rational function approximations, we use the set of Chebyshev polynomials, a class that exhibits more uniform behavior. The general Chebyshev rational function approximation method proceeds in the same manner as Padé approximation, except that each x^k term in the Padé approximation is replaced by the kth-degree Chebyshev polynomial $T_k(x)$.

Suppose we want to approximate the function f by an Nth-degree rational function written in the form

$$r(x) = \frac{\sum_{k=0}^{n} p_k T_k(x)}{\sum_{k=0}^{m} q_k T_k(x)}, \qquad \text{where } N = n + m \text{ and } q_0 = 1.$$

Writing $f(x)$ in a series involving Chebyshev polynomials gives

$$f(x) - r(x) = \sum_{k=0}^{\infty} a_k T_k(x) - \frac{\sum_{k=0}^{n} p_k T_k(x)}{\sum_{k=0}^{m} q_k T_k(x)}$$

or

$$f(x) - r(x) = \frac{\sum_{k=0}^{\infty} a_k T_k(x) \sum_{k=0}^{m} q_k T_k(x) - \sum_{k=0}^{n} p_k T_k(x)}{\sum_{k=0}^{m} q_k T_k(x)}.$$

The coefficients $q_1, q_2, \ldots, q_m$ and $p_0, p_1, \ldots, p_n$ are chosen so that the numerator on the right-hand side of this equation has zero coefficients for $T_k(x)$ when $k = 0, 1, \ldots, N$. This implies that

$$(a_0 T_0(x) + a_1 T_1(x) + \cdots)(T_0(x) + q_1 T_1(x) + \cdots + q_m T_m(x))$$
$$-(p_0 T_0(x) + p_1 T_1(x) + \cdots + p_n T_n(x))$$

has no terms of degree less than or equal to N.

Two problems arise with the Chebyshev procedure that make it more difficult to implement than the Padé method. One occurs because the product of the polynomial $q(x)$ and the series for $f(x)$ involves products of Chebyshev polynomials. This problem is resolved by making use of the relationship

$$T_i(x)T_j(x) = \tfrac{1}{2}[T_{i+j}(x) + T_{|i-j|}(x)].$$

The other problem is more difficult to resolve and involves the computation of the Chebyshev series for $f(x)$. In theory, this is not difficult, for if

$$f(x) = \sum_{k=0}^{\infty} a_k T_k(x),$$

then the orthogonality of the Chebyshev polynomials implies that

$$a_0 = \frac{1}{\pi} \int_{-1}^{1} \frac{f(x)}{\sqrt{1-x^2}}\, dx \qquad \text{and} \qquad a_k = \frac{2}{\pi} \int_{-1}^{1} \frac{f(x)T_k(x)}{\sqrt{1-x^2}}\, dx, \qquad \text{where } k \geq 1.$$

Practically, however, these integrals can seldom be evaluated in closed form, and a numerical integration technique is required for each evaluation.

This Chebyshev rational approximation can be generated using the program CHEBYM82.

EXAMPLE 2   The first five terms of the Chebyshev expansion for e^{-x} are

$$\tilde{P}_5(x) = 1.266066T_0(x) - 1.130318T_1(x) + 0.271495T_2(x) - 0.044337T_3(x)$$
$$+ 0.005474T_4(x) - 0.000543T_5(x).$$

To determine the Chebyshev rational approximation of degree five with $n = 3$ and $m = 2$ requires choosing p_0, p_1, p_2, p_3, q_1, and q_2 so that for $k = 0, 1, 2, 3, 4$, and 5, the coefficients of $T_k(x)$ are zero in the expansion

$$\tilde{P}_5(x)[T_0(x) + q_1T_1(x) + q_2T_2(x)] - [p_0T_0(x) + p_1T_1(x) + p_2T_2(x) + p_3T_3(x)].$$

Using the product relation for the Chebyshev polynomials and collecting terms gives the equations

$$
\begin{aligned}
T_0: & \quad 1.266066 - 0.565159q_1 + 0.1357485q_2 = p_0, \\
T_1: & \quad -1.130318 + 1.401814q_1 - 0.583275q_2 = p_1, \\
T_2: & \quad 0.271495 - 0.587328q_1 + 1.268803q_2 = p_2, \\
T_3: & \quad -0.044337 + 0.138485q_1 - 0.565431q_2 = p_3, \\
T_4: & \quad 0.005474 - 0.022440q_1 + 0.135748q_2 = 0, \\
T_5: & \quad -0.000543 + 0.002737q_1 - 0.022169q_2 = 0.
\end{aligned}
$$

The solution to this system produces the rational function

$$r_T(x) = \frac{1.055265T_0(x) - 0.613016T_1(x) + 0.077478T_2(x) - 0.004506T_3(x)}{T_0(x) + 0.378331T_1(x) + 0.022216T_2(x)}.$$

Converting the Chebyshev polynomials to powers of x gives

$$r_T(x) = \frac{0.977787 - 0.599499x + 0.154956x^2 - 0.018022x^3}{0.977784 + 0.378331x + 0.044432x^2}.$$

Table 8.6 lists values of $r_T(x)$ and, for comparison purposes, the values of $r(x)$ obtained in Example 1. Note that the approximation given by $r(x)$ is superior to that of $r_T(x)$ for $x = 0.2$ and 0.4, but that the maximum error for $r(x)$ is 6.33×10^{-5} compared to 9.13×10^{-6} for $r_T(x)$.   ■ ■ ■

Table 8.6

x	e^{-x}	$r(x)$	$\|e^{-x} - r(x)\|$	$r_T(x)$	$\|e^{-x} - r_T(x)\|$
0.2	0.81873075	0.81873075	7.55×10^{-9}	0.81872510	5.66×10^{-6}
0.4	0.67032005	0.67031963	4.11×10^{-7}	0.67031310	6.95×10^{-6}
0.6	0.54881164	0.54880763	4.00×10^{-6}	0.54881292	1.28×10^{-6}
0.8	0.44932896	0.44930966	1.93×10^{-5}	0.44933809	9.13×10^{-6}
1.0	0.36787944	0.36781609	6.33×10^{-5}	0.36787155	7.89×10^{-6}

The Chebyshev method does not produce the best rational function approximation in the sense of the approximation whose maximum approximation error is minimal. The method can, however, be used as a starting point for an iterative method known as the second Remes' algorithm that converges to the best approximation.

EXERCISE SET 8.5

1. Determine all Padé approximations for $f(x) = e^{2x}$ of degree two. Compare the results at $x_i = 0.2i$, for $i = 1, 2, 3, 4, 5$, with the actual values $f(x_i)$.

2. Determine all Padé approximations for $f(x) = x \ln(x + 1)$ of degree three. Compare the results at $x_i = 0.2i$, for $i = 1, 2, 3, 4, 5$, with the actual values $f(x_i)$.

3. Determine the Padé approximation of degree five with $n = 2$ and $m = 3$ for $f(x) = e^x$. Compare the results at $x_i = 0.2i$, for $i = 1, 2, 3, 4, 5$, with those from the fifth Maclaurin polynomial.

4. Repeat Exercise 3 using instead the Padé approximation of degree five with $n = 3$ and $m = 2$.

5. Determine the Padé approximation of degree six with $n = m = 3$ for $f(x) = \sin x$. Compare the results at $x_i = 0.1i$, for $i = 0, 1, 2, 3, 4, 5$, with the exact results and with the results of the sixth Maclaurin polynomial.

6. Determine the Padé approximations of degree six with (a) $n = 2, m = 4$ and (b) $n = 4, m = 2$ for $f(x) = \sin x$. Compare the results at each x_i to those obtained in Exercise 5.

7. Table 8.9 lists results of the Padé approximation of degree five with $n = 3$ and $m = 2$, the fifth Maclaurin polynomial, and the exact values of $f(x) = e^{-x}$ when $x_i = 0.2i$, for $i = 1, 2, 3, 4, 5$. Compare these results with those produced from the other Padé approximations of degree five.
   - (a) $n = 0, m = 5$
   - (b) $n = 1, m = 4$
   - (c) $n = 3, m = 2$
   - (d) $n = 4, m = 1$

8. Express the following rational functions in continued-fraction form:
   - (a) $\dfrac{x^2 + 3x + 2}{x^2 - x + 1}$
   - (b) $\dfrac{4x^2 + 3x - 7}{2x^3 + x^2 - x + 5}$
   - (c) $\dfrac{2x^3 - 3x^2 + 4x - 5}{x^2 + 2x + 4}$
   - (d) $\dfrac{2x^3 + x^2 - x + 3}{3x^3 + 2x^2 - x + 1}$

9. Use three-digit rounding arithmetic to compute the values of the rational fractions in Exercise 8 for $x = 1.23$
   - (a) Precisely as listed in the exercise,
   - (b) Using nested multiplication in the numerator and denominator,
   - (c) In continued-fraction form.

   Compare the approximations to the exact result.

10. Determine the Chebyshev expansion for each case.
    - (a) $f(x) = e^{-x}$ (first two terms only)
    - (b) $f(x) = \cos x$ (first three terms only)
    - (c) $f(x) = \sin x$ (first four terms only)
    - (d) $f(x) = e^x$ (first five terms only)

11. Find all the Chebyshev rational approximations of degree two for $f(x) = e^{-x}$. Which give the best approximations to $f(x) = e^{-x}$ at $x = 0.25, 0.5$, and 1?

12. Find all the Chebyshev rational approximations of degree three for $f(x) = \cos x$. Which give the best approximations to $f(x) = \cos x$ at $x = \frac{\pi}{4}$ and $\frac{\pi}{3}$?

13. Find the Chebyshev rational approximation of degree four with $n = m = 2$ for $f(x) = \sin x$. Compare the results at $x_i = 0.1i$, for $i = 0, 1, 2, 3, 4, 5$, from this approximation with those obtained in Exercise 5 using a sixth-degree Padé approximation.

14. Find all Chebyshev rational approximations of degree five for $f(x) = e^x$. Compare the results at $x_i = 0.2i$, for $i = 1, 2, 3, 4, 5$, from this approximation with those obtained in Exercises 3 and 4.

8.6 Trigonometric Polynomial Approximation

Trigonometric functions are used to approximate functions that have periodic behavior, functions with the property that for some constant T, $f(x + T) = f(x)$ for all x. We can translate the general problem so that $T = 2\pi$ and restrict the approximation to the interval $[-\pi, \pi]$.

For each positive integer n, the set $\mathcal{T}_n$ of **trigonometric polynomials** of degree less than or equal to n is the set of all linear combinations of $\{\phi_0, \phi_1, \ldots, \phi_{2n-1}\}$, where

$$\phi_0(x) = \frac{1}{\sqrt{2\pi}},$$

$$\phi_k(x) = \frac{1}{\sqrt{\pi}} \cos kx, \qquad \text{for each } k = 1, 2, \ldots, n,$$

and

$$\phi_{n+k}(x) = \frac{1}{\sqrt{\pi}} \sin kx, \qquad \text{for each } k = 1, 2, \ldots, n - 1.$$

(Some sources include an additional function in the set, $\phi_{2n}(x) = (1/\sqrt{\pi}) \sin nx$. We will not follow this convention.)

The set $\{\phi_0, \phi_1, \ldots, \phi_{2n-1}\}$ is orthonormal on $[-\pi, \pi]$ with respect to the weight function $w(x) \equiv 1$. This follows from a demonstration similar to that which shows that the Chebyshev polynomials are orthogonal on $[-1, 1]$. For example, if $k \neq j$ and $j \neq 0$,

$$\int_{-\pi}^{\pi} \phi_{n+k}(x)\phi_j(x) \, dx = \int_{-\pi}^{\pi} \frac{1}{\sqrt{\pi}} \sin kx \frac{1}{\sqrt{\pi}} \cos jx \, dx = \frac{1}{\pi} \int_{-\pi}^{\pi} \sin kx \cos jx \, dx.$$

The trigonometric identity

$$\sin kx \cos jx = \tfrac{1}{2} \sin(k + j)x + \tfrac{1}{2} \sin(k - j)x$$

can now be used to give

$$\int_{-\pi}^{\pi} \phi_{n+k}(x)\phi_j(x) \, dx = \frac{1}{2\pi} \int_{-\pi}^{\pi} [\sin(k + j)x + \sin(k - j)x] \, dx$$

$$= \frac{1}{2\pi} \left[\frac{-\cos(k + j)x}{k + j} - \frac{\cos(k - j)x}{k - j} \right]_{-\pi}^{\pi} = 0,$$

since $\cos(k + j)\pi = \cos(k + j)(-\pi)$ and $\cos(k - j)\pi = \cos(k - j)(-\pi)$. The result also holds when $k = j$, for in this case $\sin(k - j)x = \sin 0 = 0$.

Showing orthogonality for the other possibilities from $\{\phi_0, \phi_1, \ldots, \phi_{2n-1}\}$ is similar and uses the appropriate trigonometric identities from the collection

$$\sin j \cos k = \tfrac{1}{2}[\sin(j + k) + \sin(j - k)],$$
$$\sin j \sin k = \tfrac{1}{2}[\cos(j - k) - \cos(j + k)],$$
$$\cos j \cos k = \tfrac{1}{2}[\cos(j + k) + \cos(j - k)]$$

to convert the products into sums.

Given $f \in C[-\pi, \pi]$, the least squares approximation by functions in $\mathcal{T}_n$ is defined by

$$S_n(x) = \sum_{k=0}^{2n-1} a_k \phi_k(x),$$

where

$$a_k = \int_{-\pi}^{\pi} f(x)\phi_k(x)\,dx \qquad \text{for each } k = 0, 1, \ldots, 2n - 1.$$

The limit of S_n as $n \to \infty$ is called the **Fourier series** of f. Fourier series are used to describe the solution of various ordinary and partial-differential equations that occur in physical situations.

EXAMPLE 1 To determine the trigonometric polynomial from $\mathcal{T}_n$ that approximates

$$f(x) = |x| \qquad \text{for } -\pi < x < \pi$$

requires finding

$$a_0 = \int_{-\pi}^{\pi} |x| \frac{1}{\sqrt{2\pi}}\,dx = -\frac{1}{\sqrt{2\pi}} \int_{-\pi}^{0} x\,dx + \frac{1}{\sqrt{2\pi}} \int_{0}^{\pi} x\,dx$$

$$= \frac{2}{\sqrt{2\pi}} \int_{0}^{\pi} x\,dx = \frac{\sqrt{2\pi^2}}{2\sqrt{\pi}},$$

$$a_k = \frac{1}{\sqrt{\pi}} \int_{-\pi}^{\pi} |x| \cos kx\,dx = \frac{2}{\sqrt{\pi}} \int_{0}^{\pi} x \cos kx\,dx$$

$$= \frac{2}{\sqrt{\pi}k^2}[(-1)^k - 1], \qquad \text{for each } k = 1, 2, \ldots, n,$$

and the coefficients a_{n+k}. The coefficients a_{n+k} in the Fourier expansion are commonly denoted b_k; that is, $b_k = a_{n+k}$ for $k = 1, 2, \ldots, n - 1$. In our example, we have

$$b_k = \frac{1}{\sqrt{\pi}} \int_{-\pi}^{\pi} |x| \sin kx\,dx = 0, \qquad \text{for each } k = 1, 2, \ldots, n - 1,$$

since the integrand is an odd function. The trigonometric polynomial from $\mathcal{T}_n$ approximating f is, therefore,

$$S_n(x) = \frac{\pi}{2} + \frac{2}{\pi} \sum_{k=0}^{n} \frac{(-1)^k - 1}{k^2} \cos kx.$$

The first few trigonometric polynomials for $f(x) = |x|$ are shown in Figure 8.9.

Figure 8.9

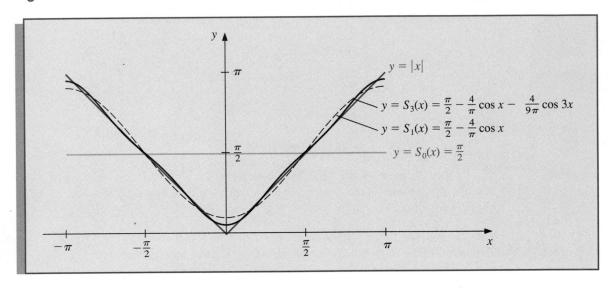

The Fourier series for f is

$$S(x) = \lim_{n \to \infty} S_n(x) = \frac{\pi}{2} + \frac{2}{\pi} \sum_{k=1}^{\infty} \frac{(-1)^k - 1}{k^2} \cos kx.$$

Since $|\cos kx| \le 1$ for every k and x, the series converges and $S(x)$ exists for all real numbers x.   ■ ■ ■

There is a discrete analog to Fourier series that is useful for the least squares approximation and interpolation of large amounts of data when the data is given at equally spaced points. Suppose that a collection of $2m$ paired data points $\{(x_j, y_j)\}_{j=0}^{2m-1}$ is given, with the first elements in the pairs equally partitioning a closed interval. For convenience, we assume that the interval is $[-\pi, \pi]$ and that, as shown in Figure 8.10,

$$x_j = -\pi + \left(\frac{j}{m}\right)\pi \qquad \text{for each } j = 0, 1, \ldots, 2m - 1.$$

Figure 8.10

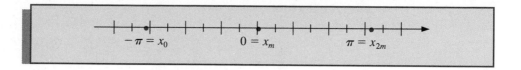

If this were not the case, a simple linear transformation could be used to translate the data into this form.

For a fixed $n < m$, consider the set of functions $\hat{\mathcal{T}}_n$ consisting of all the linear combinations of

$$\hat{\phi}_0(x) = \tfrac{1}{2},$$

$$\hat{\phi}_k(x) = \cos kx, \qquad \text{for each } k = 1, 2, \ldots, n,$$

and

$$\hat{\phi}_{n+k}(x) = \sin kx, \qquad \text{for each } k = 1, 2, \ldots, n-1.$$

The goal is to determine the linear combination of these functions for which

$$E(S_n) = \sum_{j=0}^{2m-1} \{y_j - S_n(x_j)\}^2$$

is minimized. That is, we wish to minimize the total error

$$E(S_n) = \sum_{j=0}^{2m-1} \left\{ y_j - \left[\frac{a_0}{2} + a_n \cos nx_j + \sum_{k=1}^{n-1} (a_k \cos kx_j + a_{n+k} \sin kx_j) \right] \right\}^2.$$

The determination of the constants is simplified by the fact that the set is orthogonal with respect to summation over the equally spaced points $\{x_j\}_{j=0}^{2m-1}$ in $[-\pi, \pi]$. By this we mean that, for each $k \neq l$,

$$\sum_{j=0}^{2m-1} \hat{\phi}_k(x_j)\hat{\phi}_l(x_j) = 0.$$

The orthogonality follows from the fact that if r and m are positive integers with $r < 2m$, we have

$$\sum_{j=0}^{2m-1} \cos rx_j = 0 \qquad \text{and} \qquad \sum_{j=0}^{2m-1} \sin rx_j = 0.$$

To obtain the constants a_k for $k = 0, 1, \ldots, n$ and $b_k \equiv a_{n+k}$ for $k = 1, 2, \ldots,$ $n-1$ in the summation

$$S_n(x) = \frac{a_0}{2} + a_n \cos nx + \sum_{k=1}^{n-1} (a_k \cos kx + b_k \sin kx),$$

we minimize the least squares sum

$$E(a_0, \ldots, a_n, b_1, \ldots, b_{n-1}) = \sum_{j=0}^{2m-1} [y_j - S_n(x_j)]^2$$

by setting to zero the partial derivatives of E with respect to the a_k's and the b_k's. This implies that

$$a_k = \frac{1}{m} \sum_{j=0}^{2m-1} y_j \cos kx_j, \qquad \text{for each } k = 0, 1, \ldots, n,$$

and

$$b_k = \frac{1}{m} \sum_{j=0}^{2m-1} y_j \sin kx_j, \qquad \text{for each } k = 1, 2, \ldots, n-1.$$

EXAMPLE 2 Let $f(x) = x^4 - 3x^3 + 2x^2 - \tan x(x-2)$. To find the least squares approximation S_3 for the data $\{(x_j, y_j)\}_{j=0}^{9}$, where $x_j = j/5$ and $y_j = f(x_j)$, requires that the data points $\{(x_j, y_j)\}_{j=0}^{9}$ first be translated from $[0, 2]$ to $[-\pi, \pi]$. The required linear transformation is

$$z_j = \pi(x_j - 1),$$

and the translated data is of the form

$$\left\{ \left(z_j, f\left(1 + \frac{z_j}{\pi}\right) \right) \right\}_{j=0}^{9}.$$

Consequently, the least squares trigonometric polynomial is

$$S_3(z) = \frac{a_0}{2} + a_3 \cos 3z + \sum_{k=1}^{2} (a_k \cos kz + b_k \sin kz),$$

where

$$a_k = \frac{1}{5} \sum_{j=0}^{9} f\left(1 + \frac{z_j}{\pi}\right) \cos kz_j, \qquad \text{for } k = 0, 1, 2, 3,$$

and

$$b_k = \frac{1}{5} \sum_{j=0}^{9} f\left(1 + \frac{z_j}{\pi}\right) \sin kz_j, \qquad \text{for } k = 1, 2.$$

Evaluating these sums produces the approximation

$$S_3(z) = 0.76201 + 0.77177 \cos z + 0.017423 \cos 2z + 0.0065673 \cos 3z$$
$$- 0.38676 \sin z + 0.047806 \sin 2z.$$

Converting back to the variable x gives

$$S_3(x) = 0.76201 + 0.77177 \cos \pi(x - 1) + 0.017423 \cos 2\pi(x - 1)$$
$$+ 0.0065673 \cos 3\pi(x - 1) - 0.38676 \sin \pi(x - 1) + 0.047806 \sin 2\pi(x - 1).$$

Table 8.7 lists values of $f(x)$ and $S_3(x)$. ■ ■ ■

Table 8.7

x	$f(x)$	$S_3(x)$	$|f(x) - S_3(x)|$
0.125	0.26440	0.24060	2.38×10^{-2}
0.375	0.84081	0.85154	1.07×10^{-2}
0.625	1.36150	1.36248	9.74×10^{-4}
0.875	1.61282	1.60406	8.75×10^{-3}
1.125	1.36672	1.37566	8.94×10^{-3}
1.375	0.71697	0.71545	1.52×10^{-3}
1.625	0.07909	0.06929	9.80×10^{-3}
1.875	-0.14576	-0.12302	2.27×10^{-2}

EXERCISE SET 8.6

1. Find the trigonometric least squares approximation polynomial S_2 in $\mathcal{T}_2$ for $f(x) = x$ on $[-\pi, \pi]$.

2. Find the general trigonometric least squares approximation polynomial S_n in $\mathcal{T}_n$ for $f(x) = x$ on $[-\pi, \pi]$.

3. Find the trigonometric least squares approximation polynomial S_2 in $\mathcal{T}_2$ for $f(x) = e^x$ on $[-\pi, \pi]$.

4. Find the general least squares approximation polynomial S_n in $\mathcal{T}_n$ for

$$f(x) = \begin{cases} -1, & \text{if } -\pi < x \leq 0, \\ 1, & \text{if } 0 < x < \pi. \end{cases}$$

5. Determine the trigonometric least squares polynomial S_n in $\tilde{\mathcal{T}}_n$ on the interval $[-\pi, \pi]$ for the following functions, using the given values of m and n:
   (a) $f(x) = \cos 2x$, $\quad m = 4, n = 2$
   (b) $f(x) = \cos 3x$, $\quad m = 4, n = 2$
   (c) $f(x) = \sin \frac{1}{2}x + 2 \cos \frac{1}{3}x$, $\quad m = 6, n = 3$
   (d) $f(x) = x^2 \cos x$, $\quad m = 6, n = 3$

6. Compute the error $E(S_n)$ for each of the functions in Exercise 5.

7. Determine the trigonometric least squares polynomial S_3 in $\tilde{\mathcal{T}}_3$ using $m = 4$ for $f(x) = e^x \cos 2x$ on the interval $[-\pi, \pi]$. Compute the error $E(S_n)$.

8. Repeat Exercise 7 using $m = 8$. Compare the values of the approximating polynomials with the values of f at the points $\xi_j = -\pi + 0.2j\pi, 0 \leq j \leq 10$. Which approximation is better?

9. Let $f(x) = 2 \tan x - \sec 2x, 2 \le x \le 4$. Determine the trigonometric polynomials S_n in $\hat{\mathcal{I}}_n$, using the following values of n and m. Compute the error in each case.
   (a) $n = 3, m = 6$
   (b) $n = 4, m = 6$

10. (a) Determine the least squares trigonometric polynomial S_4 in $\hat{\mathcal{I}}_4$ for $f(x) = x^2 \sin x$ on the interval $[0, 1]$, using $m = 16$.
   (b) Compute $\int_0^1 S_4(x)\, dx$.
   (c) Compare the integral in part (b) to $\int_0^1 x^2 \sin x\, dx$.

8.7 Fast Fourier Transforms

The *interpolatory* trigonometric polynomial on the $2m$ data points $\{(x_j, y_j)\}_{j=0}^{2m-1}$ is the least squares polynomial from $\hat{\mathcal{I}}_m$ for this collection of points. This least squares trigonometric polynomial was found in Section 8.6 to have the form

$$S_n(x) = \frac{a_0}{2} + a_n \cos nx + \sum_{k=1}^{n-1}(a_k \cos kx + b_k \sin kx),$$

where

$$a_k = \frac{1}{m} \sum_{j=0}^{2m-1} y_j \cos kx_j \quad \text{for each } k = 0, 1, \ldots, n$$

and

$$b_k = \frac{1}{m} \sum_{j=0}^{2m-1} y_j \sin kx_j \quad \text{for each } k = 1, 2, \ldots, n-1.$$

We will use this form with $n = m$ for interpolation with only a minor modification. For interpolation, we balance the system by replacing the term a_m with $a_m/2$. The interpolatory polynomial then has the form

$$S_m(x) = \frac{a_0 + a_m \cos mx}{2} + \sum_{k=1}^{m-1}[a_k \cos kx + b_k \sin kx],$$

where the coefficients a_k and b_k are given at the beginning of the section.

The interpolation of large amounts of equally spaced data by trigonometric polynomials can produce very accurate results. It is the appropriate approximation technique in areas involving digital filters, antenna field patterns, quantum mechanics, optics, and certain simulation problems. Until the middle of the 1960s, the method had not been extensively applied due to the number of arithmetic calculations required for the determination of the constants in the approximation. The interpolation of $2m$ data points requires approximately $(2m)^2$ multiplications and $(2m)^2$ additions by the direct calculation technique. The approximation of many thousands of data points is not unusual in areas requiring trigonometric interpolation, so the direct methods for evaluating the constants require multiplication and addition operations numbering in the millions. The computation time for this many calculations is prohibitive, and the round-off error would generally dominate the approximation.

In 1965 a paper by J. W. Cooley and J. W. Tukey described an alternative method of calculating the constants in the interpolating trigonometric polynomial. This method requires only $O(m \log_2 m)$ multiplications and $O(m \log_2 m)$ additions, provided m is chosen in an appropriate manner. For a problem with thousands of data points, this reduces the number of calculations from millions to thousands. The method had actually been discovered a number of years before the Cooley–Tukey paper appeared but had gone unnoticed by most researchers until that time.

The method described by Cooley and Tukey is generally known as the **fast Fourier Transform (FFT) algorithm** and has led to a revolution in the use of interpolatory trigonometric polynomials. The method consists of organizing the problem so that the number of data points being used can be easily factored, particularly into powers of two.

The relationship between the number of data points $2m$ and the degree of the trigonometric polynomial used in the fast Fourier transform procedure allows some notational simplification. The nodes are given by

$$x_j = -\pi + \left(\frac{j}{m}\right)\pi,$$

for each $j = 0, 1, \ldots, 2m - 1$, and the coefficients as

$$a_k = \frac{1}{m} \sum_{j=0}^{2m-1} y_j \cos k x_j$$

and

$$b_k = \frac{1}{m} \sum_{j=0}^{2m-1} y_j \sin k x_j,$$

for each $k = 0, 1, \ldots, m$. For notational convenience, b_0 and b_m have been added to the collection, but both are zero and do not contribute to the sum.

Instead of directly evaluating the constants a_k and b_k, the fast Fourier transform procedure computes the complex coefficients c_k in the formula

$$F(x) = \frac{1}{m} \sum_{k=0}^{2m-1} c_k e^{ikx},$$

where

$$c_k = \sum_{j=0}^{2m-1} y_j e^{\frac{\pi i j k}{m}}, \qquad \text{for each } k = 0, 1, \ldots, 2m - 1.$$

Once the constants c_k have been determined, a_k and b_k can be recovered. To do this we need *Euler's formula*, which states that for all real numbers z we have

$$e^{iz} = \cos z + i \sin z,$$

where the complex number i satisfies $i^2 = -1$. Then, for each $k = 0, 1, \ldots, m$,

$$\frac{1}{m} c_k e^{-i\pi k} = \frac{1}{m} \sum_{j=0}^{2m-1} y_j e^{\frac{\pi i j k}{m}} e^{-i\pi k} = \frac{1}{m} \sum_{j=0}^{2m-1} y_j e^{ik[-\pi+(j/m)\pi]}$$

$$= \frac{1}{m} \sum_{j=0}^{2m-1} y_j (\cos k x_j + i \sin k x_j),$$

so

$$\frac{1}{m} c_k e^{-i\pi k} = a_k + i b_k.$$

The operation reduction feature of the fast Fourier transform results from calculating the coefficients c_k in clusters. The following example illustrates the technique.

EXAMPLE 1   To illustrate the reduced computational effort that results from the fast Fourier transform procedure, consider the technique applied to $8 = 2^3$ data points $\{(x_j, y_j)\}_{j=0}^{7}$, where $x_j = -\pi + j\pi/4$.

Then

$$S_4(x) = \frac{a_0 + a_4 \cos 4x}{2} + \sum_{k=1}^{3} (a_k \cos kx + b_k \sin kx),$$

where

$$a_k = \frac{1}{4} \sum_{j=0}^{7} y_j \cos k x_j \quad \text{and} \quad b_k = \frac{1}{4} \sum_{j=0}^{7} y_j \sin k x_j, \quad \text{for each } k = 0, 1, 2, 3, 4.$$

Define

$$F(x) = \frac{1}{4} \sum_{j=0}^{7} c_k e^{ikx},$$

where

$$c_k = \sum_{j=0}^{7} y_j e^{ijk\pi/4}, \quad \text{for } k = 0, 1, \ldots, 7.$$

By direct calculation, the complex constants c_k are given by

$$c_0 = y_0 + y_1 + y_2 + y_3 + y_4 + y_5 + y_6 + y_7;$$

$$c_1 = y_0 + \frac{i+1}{\sqrt{2}} y_1 + i y_2 + \frac{i-1}{\sqrt{2}} y_3 - y_4 - \frac{i+1}{\sqrt{2}} y_5 - i y_6 - \frac{i-1}{\sqrt{2}} y_7;$$

$$c_2 = y_0 + i y_1 - y_2 - i y_3 + y_4 + i y_5 - y_6 - i y_7;$$

$$c_3 = y_0 + \frac{i-1}{\sqrt{2}}y_1 - iy_2 + \frac{i+1}{\sqrt{2}}y_3 - y_4 - \frac{i-1}{\sqrt{2}}y_5 + iy_6 - \frac{i+1}{\sqrt{2}}y_7;$$

$$c_4 = y_0 - y_1 + y_2 - y_3 + y_4 - y_5 + y_6 - y_7;$$

$$c_5 = y_0 - \frac{i+1}{\sqrt{2}}y_1 + iy_2 - \frac{i-1}{\sqrt{2}}y_3 - y_4 + \frac{i+1}{\sqrt{2}}y_5 - iy_6 + \frac{i-1}{\sqrt{2}}y_7;$$

$$c_6 = y_0 - iy_1 - y_2 + iy_3 + y_4 - iy_5 - y_6 + iy_7;$$

and

$$c_7 = y_0 - \frac{i-1}{\sqrt{2}}y_1 - iy_2 - \frac{i+1}{\sqrt{2}}y_3 - y_4 + \frac{i-1}{\sqrt{2}}y_5 + iy_6 + \frac{i+1}{\sqrt{2}}y_7.$$

Because of the small size of the collection of data points, many of the coefficients of the y_j in these equations are 1 or -1. This frequency will decrease in a larger application, so to count the computational operations accurately, multiplication by 1 or -1 will be included, even though it would not be necessary in this example. With this understanding, 64 multiplications/divisions and 56 additions/subtractions are required for the direct computation of $c_0, c_1, \ldots, c_7$.

To apply the fast Fourier transform procedure, we first define

$$d_0 = \tfrac{1}{2}(c_0 + c_4) = y_0 + y_2 + y_4 + y_6;$$

$$d_1 = \tfrac{1}{2}(c_0 - c_4) = y_1 + y_3 + y_5 + y_7;$$

$$d_2 = \tfrac{1}{2}(c_1 + c_5) = y_0 + iy_2 - y_4 - iy_6;$$

$$d_3 = \tfrac{1}{2}(c_1 - c_5) = \frac{i+1}{\sqrt{2}}(y_1 + iy_3 - y_5 - iy_7);$$

$$d_4 = \tfrac{1}{2}(c_2 - c_6) = y_0 - y_2 + y_4 - y_6;$$

$$d_5 = \tfrac{1}{2}(c_2 - c_6) = i(y_1, y_3 + y_5 - y_7);$$

$$d_6 = \tfrac{1}{2}(c_3 + c_7) = y_0 - iy_2 - y_4 + iy_6;$$

and

$$d_7 = \tfrac{1}{2}(c_3 - c_7) = \frac{i-1}{\sqrt{2}}(y_1 - iy_3 - y_5 + iy_7).$$

We then define

$$e_0 = \tfrac{1}{2}(d_0 + d_4) = y_0 + y_4;$$

$$e_1 = \tfrac{1}{2}(d_0 - d_4) = y_2 + y_6;$$

$$e_2 = \tfrac{1}{2}(id_1 + d_5) = i(y_1 + y_5);$$

$$e_3 = \tfrac{1}{2}(id_1 - d_5) = i(y_3 + y_7);$$

$$e_4 = \tfrac{1}{2}(d_2 + d_6) = y_0 - y_4;$$
$$e_5 = \tfrac{1}{2}(d_2 - d_6) = i(y_2 - y_6);$$
$$e_6 = \tfrac{1}{2}(id_3 + d_7) = \frac{i-1}{\sqrt{2}}(y_1 - y_5);$$

and

$$e_7 = \tfrac{1}{2}(id_3 - d_7) = \frac{-i-1}{\sqrt{2}}(y_3 - y_7).$$

Finally, we define

$$f_0 = \tfrac{1}{2}(e_0 + e_4) = y_0;$$
$$f_1 = \tfrac{1}{2}(e_0 - e_4) = y_4;$$
$$f_2 = \tfrac{1}{2}(ie_1 + e_5) = iy_2;$$
$$f_3 = \tfrac{1}{2}(ie_1 - e_5) = iy_6;$$

$$f_4 = \frac{1}{2}\left(\frac{i+1}{\sqrt{2}}e_2 + e_6\right) = \frac{i-1}{\sqrt{2}}y_1;$$

$$f_5 = \frac{1}{2}\left(\frac{i+1}{\sqrt{2}}e_2 - e_6\right) = \frac{i-1}{\sqrt{2}}y_5;$$

$$f_6 = \frac{1}{2}\left(\frac{i+1}{\sqrt{2}}e_3 + e_7\right) = -\frac{i+1}{\sqrt{2}}y_3;$$

and

$$f_7 = \frac{1}{2}\left(\frac{i+1}{\sqrt{2}}e_3 - e_7\right) = -\frac{i+1}{\sqrt{2}}y_7.$$

The $c_0, \ldots, c_7$; $d_0, \ldots, d_7$; $e_0, \ldots, e_7$; and $f_0, \ldots, f_7$ are independent of the particular data points; they depend only on the fact that $m = 4$. For each m there is a unique set of constants $\{c_k\}_{k=0}^{2m-1}$; $\{d_k\}_{k=0}^{2m-1}$; $\{e_k\}_{k=0}^{2m-1}$; and $\{f_k\}_{k=0}^{2m-1}$. This portion of the work is not needed for a particular application. Only the calculations that follow are required:

1. $f_0 = y_0;$ $f_1 = y_4;$ $f_2 = iy_2;$ $f_3 = iy_6;$

$$f_4 = \frac{i-1}{\sqrt{2}}y_1; \qquad f_5 = \frac{i-1}{\sqrt{2}}y_5; \qquad f_6 = -\frac{i+1}{\sqrt{2}}y_3;$$

$$f_7 = -\frac{i+1}{\sqrt{2}}y_7.$$

2. $e_0 = f_0 + f_1;$ $e_1 = -i(f_2 + f_3);$ $e_2 = \frac{-i+1}{\sqrt{2}}(f_4 + f_5);$

$$e_3 = \frac{-i+1}{\sqrt{2}}(f_6 + f_7); \qquad e_4 = f_0 - f_1; \qquad e_5 = f_2 - f_3;$$

$$e_6 = f_4 - f_5; \qquad e_7 = f_6 - f_7.$$

3. $d_0 = e_0 + e_1;$ $\quad d_1 = -i(e_2 + e_3);$ $\quad d_2 = e_4 + e_5;$ $\quad d_3 = -i(e_6 + e_7);$
$\quad d_4 = e_0 - e_1;$ $\quad d_5 = e_2 - e_3;$ $\quad d_6 = e_4 - e_5;$ $\quad d_7 = e_6 - e_7.$
4. $c_0 = d_0 + d_1;$ $\quad c_1 = d_2 + d_3;$ $\quad c_2 = d_4 + d_5;$ $\quad c_3 = d_6 + d_7;$
$\quad c_4 = d_0 - d_1;$ $\quad c_5 = d_2 - d_3;$ $\quad c_6 = d_4 - d_5;$ $\quad c_7 = d_6 - d_7.$

Computing the constants $c_0, c_1, \ldots, c_7$ in this manner requires the number of operations shown in Table 8.8. Note again that multiplication by 1 or -1 has been included in the count, even though this does not require computational effort.

Table 8.8

Step	Multiplications/Divisions	Additions/Subtractions
1	8	0
2	8	9
3	8	8
4	0	8
Total	24	24

The lack of multiplications/divisions in step 4 reflects the fact that for any m, the coefficients $\{c_k\}_{k=0}^{2m-1}$ are computed from $\{d_k\}_{k=0}^{2m-1}$ in the same manner:

$$c_k = d_{2k} + d_{2k+1}, \quad \text{and} \quad c_{k+m} = d_{2k} - d_{2k+1}, \quad \text{for } k = 0, 1, \ldots, m-1,$$

so no complex multiplication is used.

In summary, the direct computation of the coefficients $c_0, c_1, \ldots, c_7$ requires 64 multiplications/divisions and 56 additions/subtractions. The fast Fourier transform technique reduces the computations to 24 multiplications/divisions and 24 additions/subtractions.

∎ ∎ ∎

The program FFTRNS83 performs the fast Fourier transform when $m = 2^p$ for some positive integer p. Modifications of the technique can be made when m takes other forms.

EXAMPLE 2 Let $f(x) = x^4 - 3x^3 + 2x^2 - \tan x(x-2)$. To determine the trigonometric interpolating polynomial of degree four for the data $\{(x_j, y_j)\}_{j=0}^7$, where $x_j = j/4$ and $y_j = f(x_j)$, requires a transformation of the interval $[0, 2]$ to $[-\pi, \pi]$. The translation is given by

$$z_j = \pi(x_j - 1)$$

so that the input data to the fast Fourier transform method is

$$\left\{z_j, f\left(1 + \frac{z_j}{\pi}\right)\right\}_{j=0}^7 .$$

The interpolating polynomial in z is

$$S_4(z) = 0.761979 + 0.771841 \cos z + 0.0173037 \cos 2z + 0.00686304 \cos 3z$$
$$- 0.000578545 \cos 4z - 0.386374 \sin z + 0.0468750 \sin 2z$$
$$- 0.0113738 \sin 3z .$$

The trigonometric polynomial $S_4(x)$ on $[0,2]$ is obtained by substituting $z = \pi(x - 1)$ into $S_4(z)$. The graphs of $y = f(x)$ and $y = S_4(x)$ are shown in Figure 8.11. Values of $f(x)$ and $S_4(x)$ are given in Table 8.9.   ■ ■ ■

Table 8.9

x	$f(x)$	$S_4(x)$	$\lvert f(x) - S_4(x) \rvert$
0.125	0.26440	0.25001	1.44×10^{-2}
0.375	0.84081	0.84647	5.66×10^{-3}
0.625	1.36150	1.35824	3.27×10^{-3}
0.875	1.61282	1.61515	2.33×10^{-3}
1.125	1.36672	1.36471	2.02×10^{-3}
1.375	0.71697	0.71931	2.33×10^{-3}
1.625	0.07909	0.07496	4.14×10^{-3}
1.875	-0.14576	-0.13301	1.27×10^{-2}

Figure 8.11

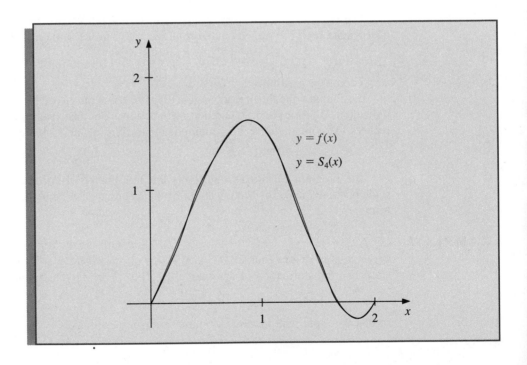

EXERCISE SET 8.7

1. Determine the trigonometric interpolatory polynomial $S_2(x)$ of degree two on $[-\pi, \pi]$ for the following functions, and graph $f(x) - S_2(x)$:
   (a) $f(x) = \pi(x - \pi)$
   (b) $f(x) = x(\pi - x)$

(c) $f(x) = |x|$

(d) $f(x) = \begin{cases} -1, & -\pi \le x \le 0 \\ 1, & 0 < x \le \pi \end{cases}$

2. Determine the trigonometric interpolating polynomial of degree four for $f(x) = x(\pi - x)$ on the interval $[-\pi, \pi]$.

3. Use the fast Fourier transform method to compute the trigonometric interpolatory polynomial of degree four on $[-\pi, \pi]$ for the following functions:
   (a) $f(x) = \pi(x - \pi)$
   (b) $f(x) = |x|$
   (c) $f(x) = \cos \pi x - 2 \sin \pi x$
   (d) $f(x) = x \cos x^2 + e^x \cos e^x$

4. (a) Determine the trigonometric interpolating polynomial $S_4(x)$ of degree four for $f(x) = x^2 \sin x$ on the interval $[0, 1]$.
   (b) Compute $\int_0^1 S_4(x) \, dx$.
   (c) Compare the integral in part (b) to $\int_0^1 x^2 \sin x \, dx$.

5. Use the approximations obtained in Exercise 3 to approximate the following integrals. Compare the approximation to the actual value.
   (a) $\int_{-\pi}^{\pi} \pi(x - \pi) \, dx$
   (b) $\int_{-\pi}^{\pi} |x| \, dx$
   (c) $\int_{-\pi}^{\pi} (\cos \pi x - 2 \sin \pi x) \, dx$
   (d) $\int_{-\pi}^{\pi} (x \cos x^2 + e^x \cos e^x) \, dx$

6. Use the fast Fourier transform method to determine the trigonometric interpolating polynomial of degree sixteen for $f(x) = x^2 \cos x$ on $[-\pi, \pi]$.

8.8 Survey of Methods and Software

In this chapter we have considered the approximation of functions with simpler functions. The simpler functions used were polynomials, rational functions, and trigonometric polynomials. We considered two types of approximations, discrete and continuous. Discrete approximations arise when approximating a finite set of data with a simpler function. Continuous approximations are used when the function to be approximated is known.

Discrete least squares techniques are recommended when the function is specified by giving a set of data which may not have exact function values. Least squares fit of data can take the form of a linear or other polynomial approximation or even an exponential form. These approximations are computed by solving sets of normal equations, as given in Section 8.2. If the data are periodic, the trigonometric least squares fits may be appropriate. Because of the orthonormality of the trigonometric basis functions, the least squares trigonometric approximation does not require the solution of a linear system. For large amounts of periodic data, interpolation by trigonometric polynomials is also recommended. An efficient method of computing the trigonometric interpolating polynomial is given by the fast Fourier transform.

When the function to be approximated can be evaluated at any required argument, the approximations often seek to minimize an integral instead of a sum. Continuous least squares polynomial approximations were considered in Section 8.3. Efficient computation of least squares polynomials led to orthonormal sets of polynomials, such as the Legendre and Chebyshev polynomials. Approximation by rational functions was studied in Section 8.5, where Padé approximation as a generalization of the Maclaurin polynomial was presented, and the extension to Chebyshev rational approximation was introduced. Both methods allow a more uniform method of approximation than polynomials. Continuous least squares approximation by trigonometric functions was discussed in Section 8.6, especially as it relates to Fourier series.

The IMSL library provides a number of routines for approximation. The subroutine RLINE gives the least squares line for a set of data points and returns statistics such as means and variances. Polynomial least squares approximations can be obtained with the subroutine RCURV. The subroutine FNLSQ computes the discrete least squares approximation for a choice of basis functions specified by the user. The subroutine BSLSQ computes a least squares cubic spline approximation and RATCH computes the rational weighted Chebyshev approximation to a continuous function f on an interval $[a, b]$. The subroutine FFTCB computes the fast Fourier transform for a given set of data.

The NAG library has many subroutines for function approximation. Least squares polynomial approximation is given in the subroutine E02ADF. This subroutine is quite versatile in that it computes least squares polynomials for varying degrees and also gives the total least squares error for each. It uses Chebyshev polynomials to minimize round-off error and enhance accuracy.

The routine E02AEF can be used to evaluate the approximation obtained by E02ADF. NAG also supplies the routine E02BAF to compute least squares cubic spline fits, E02GAF to compute the best 1-norm linear fit, and E02GCF to compute the best ∞-norm fit. The routine E02RAF computes the Padé approximation. The NAG library also includes many routines for fast Fourier transforms, one of which is C06ECF.

MATLAB can be used to compute the discrete least squares approximating polynomial. Given data $\{(x_i, y_i) | i = 1, 2, \ldots, m\}$, the least squares polynomial of degree n can be obtained by

$$p = polyfit(x, y, n).$$

MATLAB will also calculate the fast Fourier transform of given data with the function fft. MATLAB has many features for the signal-processing application.

Approximating Eigenvalues

9.1 Introduction

Eigenvalues and eigenvectors were introduced in Chapter 7 in connection with the convergence of iterative methods for approximating the solution to a linear system. To determine the eigenvalues of an $n \times n$ matrix A, we construct the characteristic polynomial

$$p(\lambda) = \det(A - \lambda I)$$

and then determine its zeros. Finding the determinant of an n by n matrix is computationally expensive, and finding good approximations to the roots of $p(\lambda)$ is also difficult. In this chapter we will explore other means for approximating the eigenvalues of a matrix.

9.2 Isolating Eigenvalues

In Chapter 7 we found that an iterative technique for solving a linear system will converge if all the eigenvalues associated with the problem have magnitude less than 1, but cannot be expected to converge if there are eigenvalues with magnitude greater than 1. The exact values of the eigenvalues in this case are not of primary importance—only the region of the complex plane in which they lie. Even when we need to know the eigenvalues, the fact that many of the techniques for their approximation are iterative implies that determining regions in which they lie is a first step in the direction of determining the approximation, since that step provides us with the initial approximation that iterative methods need.

Before proceeding with the techniques for approximating eigenvalues, we need some more results from linear algebra to help us isolate them. We give here all the additional linear algebra that is used in this chapter so that it is in one place for ease of reference. The first definitions and results parallel those in Section 8.2 for sets of polynomials.

The set of nonzero vectors $\{\mathbf{v}^{(1)}, \mathbf{v}^{(2)}, \mathbf{v}^{(3)}, \ldots, \mathbf{v}^{(k)}\}$ is said to be **linearly independent** if, whenever

$$\mathbf{0} = \alpha_1 \mathbf{v}^{(1)} + \alpha_2 \mathbf{v}^{(2)} + \alpha_3 \mathbf{v}^{(3)} + \cdots + \alpha_k \mathbf{v}^{(k)},$$

we have $\alpha_1 = \alpha_2 = \alpha_3 = \cdots = \alpha_k = 0$. A set of vectors that is not linearly independent is called **linearly dependent**.

Unique Representation of Vectors in $\mathbb{R}^n$

If $\{\mathbf{v}^{(1)}, \mathbf{v}^{(2)}, \mathbf{v}^{(3)}, \ldots, \mathbf{v}^{(n)}\}$ is a set of n linearly independent vectors in $\mathbb{R}^n$, then any vector $\mathbf{x} \in \mathbb{R}^n$ can be written uniquely as

$$\mathbf{x} = \beta_1 \mathbf{v}^{(1)} + \beta_2 \mathbf{v}^{(2)} + \beta_3 \mathbf{v}^{(3)} + \cdots + \beta_n \mathbf{v}^{(n)}$$

for some collection of constants $\beta_1, \beta_2, \ldots, \beta_n$.

Any collection of n linearly independent vectors in $\mathbb{R}^n$ is called a **basis** for $\mathbb{R}^n$.

EXAMPLE 1 Let $\mathbf{v}^{(1)} = (1, 0, 0)^t$, $\mathbf{v}^{(2)} = (-1, 1, 1)^t$, and $\mathbf{v}^{(3)} = (0, 4, 2)^t$. If α_1, α_2, and α_3 are numbers with

$$\mathbf{0} = \alpha_1 \mathbf{v}^{(1)} + \alpha_2 \mathbf{v}^{(2)} + \alpha_3 \mathbf{v}^{(3)},$$

then

$$\begin{aligned}
(0, 0, 0)^t &= \alpha_1 (1, 0, 0)^t + \alpha_2 (-1, 1, 1)^t + \alpha_3 (0, 4, 2)^t \\
&= (\alpha_1 - \alpha_2, \alpha_2 + 4\alpha_3, \alpha_2 + 2\alpha_3)^t,
\end{aligned}$$

so

$$\alpha_1 - \alpha_2 = 0, \qquad \alpha_2 + 4\alpha_3 = 0, \qquad \text{and} \qquad \alpha_2 + 2\alpha_3 = 0.$$

Since the only solution to this system is $\alpha_1 = \alpha_2 = \alpha_3 = 0$, the set $\{\mathbf{v}^{(1)}, \mathbf{v}^{(2)}, \mathbf{v}^{(3)}\}$ is linearly independent in $\mathbb{R}^3$ and is a basis for $\mathbb{R}^3$. A vector $\mathbf{x} = (x_1, x_2, x_3)^t$ in $\mathbb{R}^3$ can be written as

$$\mathbf{x} = \beta_1 \mathbf{v}^{(1)} + \beta_2 \mathbf{v}^{(2)} + \beta_3 \mathbf{v}^{(3)}$$

by choosing

$$\beta_1 = x_1 - x_2 + 2x_3, \qquad \beta_2 = 2x_3 - x_2, \qquad \text{and} \qquad \beta_3 = \tfrac{1}{2}(x_2 - x_3).$$

■ ■ ■

The next result is used in the following section to develop the Power method for approximating eigenvalues.

Linear Independence of Eigenvectors

If A is a matrix and $\lambda_1, \ldots, \lambda_k$ are distinct eigenvalues of A with associated eigenvectors $\mathbf{x}^{(1)}, \mathbf{x}^{(2)}, \ldots, \mathbf{x}^{(k)}$, then $\{\mathbf{x}^{(1)}, \mathbf{x}^{(2)}, \ldots, \mathbf{x}^{(k)}\}$ is linearly independent.

A set of vectors $\{\mathbf{v}^{(1)}, \mathbf{v}^{(2)}, \ldots, \mathbf{v}^{(n)}\}$ is **orthogonal** if $(\mathbf{v}^{(i)})^t \mathbf{v}^{(j)} = 0$ for all $i \neq j$. If, in addition, $(\mathbf{v}^{(i)})^t \mathbf{v}^{(i)} = 1$ for all $i = 1, 2, \ldots, n$, then the set is **orthonormal**.

Since $\mathbf{x}^t \mathbf{x} = \|\mathbf{x}\|_2^2$, a set of orthogonal vectors $\{\mathbf{v}^{(1)}, \mathbf{v}^{(2)}, \ldots, \mathbf{v}^{(n)}\}$ is orthonormal if and only if

$$\|\mathbf{v}^{(i)}\|_2 = 1 \qquad \text{for each } i = 1, 2, \ldots, n.$$

Any orthogonal set of vectors that does not contain the zero vector is linearly independent.

EXAMPLE 2 The vectors $\mathbf{v}^{(1)} = (0, 4, 2)^t$, $\mathbf{v}^{(2)} = (-1, -\frac{1}{5}, \frac{2}{5})^t$, and $\mathbf{v}^{(3)} = (\frac{1}{6}, -\frac{1}{6}, \frac{1}{3})^t$ form an orthogonal set. For these vectors we have

$$\|\mathbf{v}^{(1)}\|_2 = 2\sqrt{5}, \qquad \|\mathbf{v}^{(2)}\|_2 = \frac{\sqrt{30}}{5}, \qquad \text{and } \|\mathbf{v}^{(3)}\|_2 = \frac{\sqrt{6}}{6}.$$

The vectors

$$\mathbf{u}^{(1)} = \frac{\mathbf{v}^{(1)}}{\|\mathbf{v}^{(1)}\|_2} = \left(0, \frac{2\sqrt{5}}{5}, \frac{\sqrt{5}}{5}\right)^t, \qquad \mathbf{u}^{(2)} = \frac{\mathbf{v}^{(2)}}{\|\mathbf{v}^{(2)}\|_2} = \left(-\frac{\sqrt{30}}{6}, -\frac{\sqrt{30}}{30}, \frac{\sqrt{30}}{15}\right)^t,$$

and

$$\mathbf{u}^{(3)} = \frac{\mathbf{v}^{(3)}}{\|\mathbf{v}^{(3)}\|_2} = \left(\frac{\sqrt{6}}{6}, -\frac{\sqrt{6}}{6}, \frac{\sqrt{6}}{3}\right)^t$$

form an orthonormal set, since they inherit orthogonality from $\mathbf{v}^{(1)}$, $\mathbf{v}^{(2)}$, and $\mathbf{v}^{(3)}$, and, in addition,

$$\|\mathbf{u}^{(1)}\|_2 = \|\mathbf{u}^{(2)}\|_2 = \|\mathbf{u}^{(3)}\|_2 = 1. \qquad \blacksquare \quad \blacksquare \quad \blacksquare$$

An $n \times n$ matrix P is **orthogonal** if $P^{-1} = P^t$. This terminology follows from the fact that the columns of an orthogonal matrix form an orthogonal—in fact orthonormal— set of vectors.

EXAMPLE 3 The orthogonal matrix P formed from the orthonormal set of vectors found in Example 2 is

$$P = [\mathbf{u}^{(1)}, \mathbf{u}^{(2)}, \mathbf{u}^{(3)}] = \begin{bmatrix} 0 & -\frac{\sqrt{30}}{6} & \frac{\sqrt{6}}{6} \\ \frac{2\sqrt{5}}{5} & -\frac{\sqrt{30}}{30} & -\frac{\sqrt{6}}{6} \\ \frac{\sqrt{5}}{5} & \frac{\sqrt{30}}{15} & \frac{\sqrt{6}}{6} \end{bmatrix}.$$

Note that

$$PP^t = \begin{bmatrix} 0 & -\frac{\sqrt{30}}{6} & \frac{\sqrt{6}}{6} \\ \frac{2\sqrt{5}}{5} & -\frac{\sqrt{30}}{30} & -\frac{\sqrt{6}}{6} \\ \frac{\sqrt{5}}{5} & \frac{\sqrt{30}}{15} & \frac{\sqrt{6}}{6} \end{bmatrix} \cdot \begin{bmatrix} 0 & \frac{2\sqrt{5}}{5} & \frac{\sqrt{5}}{5} \\ -\frac{\sqrt{30}}{6} & -\frac{\sqrt{30}}{30} & \frac{\sqrt{30}}{15} \\ \frac{\sqrt{6}}{6} & -\frac{\sqrt{6}}{6} & \frac{\sqrt{6}}{6} \end{bmatrix} = \begin{bmatrix} 1 & 0 & 0 \\ 0 & 1 & 0 \\ 0 & 0 & 1 \end{bmatrix}.$$

It is also true that $P^t P = I$, so $P^t = P^{-1}$. $\qquad \blacksquare \quad \blacksquare \quad \blacksquare$

Two $n \times n$ matrices A and B are **similar** if a matrix S exists with $A = S^{-1}BS$. The important feature of similar matrices is that they share eigenvalues. The next result follows from observing that if $\lambda\mathbf{x} = A\mathbf{x} = S^{-1}BS\mathbf{x}$, then $BS\mathbf{x} = \lambda S\mathbf{x}$. Also, if $\mathbf{x} \neq \mathbf{0}$ and S is nonsingular, then $S\mathbf{x} \neq \mathbf{0}$ and $S\mathbf{x}$ is an eigenvector of B corresponding to its eigenvalue λ.

Eigenvalues and Eigenvectors of Similar Matrices

Suppose A and B are similar $n \times n$ matrices and λ is an eigenvalue of A with associated eigenvector $\mathbf{x}$. Then λ is also an eigenvalue of B and if $A = S^{-1}BS$, then $S\mathbf{x}$ is an eigenvector associated with λ and the matrix B.

The determination of eigenvalues is easy for a triangular matrix A, for in this case λ is a solution to the equation

$$0 = \det(A - \lambda I) = \prod_{i=1}^{n}(a_{ii} - \lambda)$$

precisely when $\lambda = a_{ii}$ for some i. The next result provides a relationship, called a **similarity transformation**, between arbitrary matrices and triangular matrices.

Schur's Theorem

Let A be an arbitrary $n \times n$ matrix. A nonsingular matrix U exists with the property that

$$T = U^{-1}AU,$$

where T is an upper-triangular matrix whose diagonal entries consist of the eigenvalues of A.

Schur's Theorem is an existence theorem that ensures that the triangular matrix T exists, but the theorom does not provide a constructive means for determining T. In most instances, the similarity transformation is difficult to determine. The restriction to symmetric matrices reduces the complication, since in this case the transformation matrix is orthogonal.

Eigenvalues of Symmetric Matrices

Suppose that A is an $n \times n$ symmetric matrix.

1. If D is the diagonal matrix whose diagonal entries are the eigenvalues of A, then there exists an orthogonal matrix P such that $D = P^{-1}AP = P^tAP$.
2. There exist n eigenvectors of A that form an orthonormal set and are the columns of the matrix P described in (1).
3. The eigenvalues are all real numbers.
4. A is positive definite if and only if all the eigenvalues of A are positive.

The final result of the section concerns bounds for the approximation of eigenvalues.

Gerschgorin Circle Theorem

Let A be an $n \times n$ matrix and R_i denote the circle in the complex plane with center a_{ii} and radius $\sum_{\substack{j=1 \\ j \neq i}}^{n} |a_{ij}|$; that is,

$$R_i = \left\{ z \in C \,\middle|\, |z - a_{ii}| \leq \sum_{\substack{j=1 \\ j \neq i}}^{n} |a_{ij}| \right\},$$

where C denotes the complex plane. The eigenvalues of A are contained within $R = \cup_{i=1}^{n} R_i$. Moreover, the union of any k of these circles that do not intersect the remaining $(n - k)$ contain precisely k (counting multiplicities) of the eigenvalues.

EXAMPLE 4   For the matrix

$$A = \begin{bmatrix} 4 & 1 & 1 \\ 0 & 2 & 1 \\ -2 & 0 & 9 \end{bmatrix},$$

the circles in the Gerschgorin Theorem are (see Figure 9.1):

$$R_1 = \{ z \in C \,\big|\, |z - 4| \leq 2 \}, \qquad R_2 = \{ z \in C \,\big|\, |z - 2| \leq 1 \},$$

and
$$R_3 = \{ z \in C \,\big|\, |z - 9| \leq 2 \}.$$

Since R_1 and R_2 are disjoint from R_3, there are two eigenvalues within $R_1 \cup R_2$ and one within R_3. Moreover, since $\rho(A) = \max_{1 \leq i \leq 3} |\lambda_i|$, we have $7 \leq \rho(A) \leq 11$. ∎ ∎ ∎

Figure 9.1

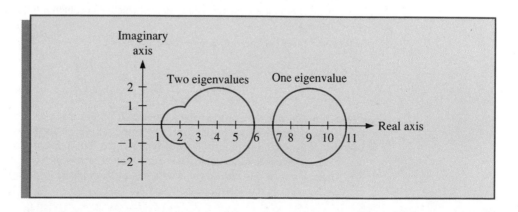

EXERCISE SET 9.2

1.   Find eigenvalues and associated eigenvectors for the following 3×3 matrices. Is there a set of 3 linearly independent eigenvectors?

(a)
$$A = \begin{bmatrix} 2 & -3 & 6 \\ 0 & 3 & -4 \\ 0 & 2 & -3 \end{bmatrix}$$

(b)
$$A = \begin{bmatrix} 2 & 1 & -1 \\ 0 & 2 & 1 \\ 0 & 0 & 3 \end{bmatrix}$$

(c)
$$A = \begin{bmatrix} 2 & 0 & 1 \\ 0 & 2 & 0 \\ 0 & 0 & 2 \end{bmatrix}$$

(d)
$$A = \begin{bmatrix} 1 & 0 & 0 \\ -1 & 0 & 1 \\ -1 & -1 & 2 \end{bmatrix}$$

(e)
$$A = \begin{bmatrix} 2 & 0 & 1 \\ 0 & 2 & 0 \\ 1 & 0 & 2 \end{bmatrix}$$

(f)
$$A = \begin{bmatrix} 2 & -1 & -1 \\ -1 & 2 & -1 \\ -1 & -1 & 2 \end{bmatrix}$$

(g)
$$A = \begin{bmatrix} 1 & 1 & 1 \\ 1 & 1 & 0 \\ 1 & 0 & 1 \end{bmatrix}$$

(h)
$$A = \begin{bmatrix} 2 & 1 & 1 \\ 1 & 2 & 1 \\ 1 & 1 & 2 \end{bmatrix}$$

2. The matrices in Exercises 1(e), (f), (g), and (h) are symmetric.
   (a) Are any of these positive definite?
   (b) Use the eigenvectors found in Exercise 1 to construct an orthogonal matrix P for which $P^t A P = D$, a diagonal matrix.

3. Use the Gerschgorin Circle Theorem to determine bounds for the eigenvalues of the following matrices:

(a)
$$\begin{bmatrix} 1 & 0 & 0 \\ -1 & 0 & 1 \\ -1 & -1 & 2 \end{bmatrix}$$

(b)
$$\begin{bmatrix} 4 & -1 & 0 \\ -1 & 4 & -1 \\ -1 & -1 & 4 \end{bmatrix}$$

(c)
$$\begin{bmatrix} 3 & 2 & 1 \\ 2 & 3 & 0 \\ 1 & 0 & 3 \end{bmatrix}$$

(d)
$$\begin{bmatrix} 4.75 & 2.25 & -0.25 \\ 2.25 & 4.75 & 1.25 \\ -0.25 & 1.25 & 4.75 \end{bmatrix}$$

(e)
$$\begin{bmatrix} -4 & 0 & 1 & 3 \\ 0 & -4 & 2 & 1 \\ 1 & 2 & -2 & 0 \\ 3 & 1 & 0 & -4 \end{bmatrix}$$

(f)
$$\begin{bmatrix} 1 & 0 & -1 & 1 \\ 2 & 2 & -1 & 1 \\ 0 & 1 & 3 & -2 \\ 1 & 0 & 1 & 4 \end{bmatrix}$$

4. Show that any four vectors in $\mathbb{R}^3$ must be linearly dependent.

5. Let $\{v_1, \ldots, v_n\}$ be a set of orthogonal nonzero vectors. Show that $\{v_1, \ldots, v_n\}$ is a linearly independent set.

6. Let $\{v_1, \ldots, v_n\}$ be a set of orthonormal nonzero vectors in $\mathbb{R}^n$ and let x be in $\mathbb{R}^n$. Show that $x = \sum_{k=1}^{n} c_k v_k$, where $c_k = v_k^t x$.

7. Show that if A is an $n \times n$ matrix with n distinct eigenvalues, then A has n linearly independent eigenvectors.

8. In Exercise 14 of Section 6.6, a symmetric matrix

$$A = \begin{bmatrix} 1.59 & 1.69 & 2.13 \\ 1.69 & 1.31 & 1.72 \\ 2.13 & 1.72 & 1.85 \end{bmatrix}$$

was used to describe the average wing lengths of fruit flies that were offspring resulting from the mating of three mutants of the flies. The entry a_{ij} represents the average wing length of a fly that is the offspring of a male fly of type i and a female fly of type j.
   (a) Find the eigenvalues and associated eigenvectors of this matrix.
   (b) Use a result in this section to answer the question posed in part (b) of Exercise 14 in Section 6.6. Is this matrix positive definite?

9.3 The Power Method

The Power method is an iterative technique used to determine the dominant eigenvalue of a matrix; that is, the eigenvalue with the largest magnitude. By modifying the method slightly it can also be used to determine other eigenvalues. One useful feature of the

Power method is that it produces not only an eigenvalue, but an associated eigenvector as well. In fact, the Power method is often applied to find an eigenvector for an eigenvalue that is determined by some other means.

To apply the Power method, we must assume that the $n \times n$ matrix A has n eigenvalues $\lambda_1, \lambda_2, \ldots, \lambda_n$ with an associated collection of linearly independent eigenvectors $\{\mathbf{v}^{(1)}, \mathbf{v}^{(2)}, \mathbf{v}^{(3)}, \ldots, \mathbf{v}^{(n)}\}$. Moreover, we assume A has precisely one eigenvalue that is largest in magnitude.

Let $\lambda_1, \lambda_2, \ldots, \lambda_n$ denote the eigenvalues of A with $|\lambda_1| > |\lambda_2| \geq |\lambda_3| \geq \cdots \geq |\lambda_n| \geq 0$. If $\mathbf{x}$ is any vector in $\mathbb{R}^n$, the fact that $\{\mathbf{v}^{(1)}, \mathbf{v}^{(2)}, \mathbf{v}^{(3)}, \ldots, \mathbf{v}^{(n)}\}$ is linearly independent implies that constants $\alpha_1, \alpha_2, \ldots, \alpha_n$ exist with

$$\mathbf{x} = \sum_{j=1}^{n} \alpha_j \mathbf{v}^{(j)}.$$

Multiplying both sides of this equation successively by $A, A^2, \ldots, A^k$, we obtain:

$$A\mathbf{x} = \sum_{j=1}^{n} \alpha_j A \mathbf{v}^{(j)} = \sum_{j=1}^{n} \alpha_j \lambda_j \mathbf{v}^{(j)},$$

$$A^2\mathbf{x} = \sum_{j=1}^{n} \alpha_j \lambda_j A \mathbf{v}^{(j)} = \sum_{j=1}^{n} \alpha_j \lambda_j^2 \mathbf{v}^{(j)},$$

$$\vdots$$

$$A^k\mathbf{x} = \sum_{j=1}^{n} \alpha_j \lambda_j^k \mathbf{v}^{(j)}.$$

If λ_1^k is factored from each term on the right side of the last equation, then

$$A^k\mathbf{x} = \lambda_1^k \sum_{j=1}^{n} \alpha_j \left(\frac{\lambda_j}{\lambda_1} \right)^k \mathbf{v}^{(j)}.$$

Since $|\lambda_1| > |\lambda_j|$ for all $j = 2, 3, \ldots, n$, we have $\lim_{k \to \infty} (\lambda_j / \lambda_1)^k = 0$, and

$$\lim_{k \to \infty} A^k\mathbf{x} = \lim_{k \to \infty} \lambda_1^k \alpha_1 \mathbf{v}^{(1)}.$$

This sequence converges to zero if $|\lambda_1| < 1$ and diverges if $|\lambda_1| \geq 1$, provided, of course, that $\alpha_1 \neq 0$. Advantage can be made of this relationship by scaling the powers of $A^k\mathbf{x}$ in an appropriate manner to ensure that the limit is finite and nonzero. The scaling begins by choosing $\mathbf{x}$ to be a unit vector $\mathbf{x}^{(0)}$ relative to $\| \cdot \|_\infty$ and a component $x_{p_0}^{(0)}$ of $\mathbf{x}^{(0)}$ with

$$x_{p_0}^{(0)} = 1 = \|\mathbf{x}^{(0)}\|_\infty.$$

Let $\mathbf{y}^{(1)} = A\mathbf{x}^{(0)}$, and define $\mu^{(1)} = y_{p_0}^{(1)}$. With this notation,

$$\mu^{(1)} = y_{p_0}^{(1)} = \frac{y_{p_0}^{(1)}}{x_{p_0}^{(0)}} = \frac{\alpha_1 \lambda_1 v_{p_0}^{(1)} + \sum_{j=2}^{n} \alpha_j \lambda_j v_{p_0}^{(j)}}{\alpha_1 v_{p_0}^{(1)} + \sum_{j=2}^{n} \alpha_j v_{p_0}^{(j)}} = \lambda_1 \left[\frac{\alpha_1 v_{p_0}^{(1)} + \sum_{j=2}^{n} \alpha_j (\lambda_j / \lambda_1) v_{p_0}^{(j)}}{\alpha_1 v_{p_0}^{(1)} + \sum_{j=2}^{n} \alpha_j v_{p_0}^{(j)}} \right].$$

Then let p_1 be the least integer such that

$$|y_{p_1}^{(1)}| = \|\mathbf{y}^{(1)}\|_\infty$$

and define

$$\mathbf{x}^{(1)} = \frac{1}{y_{p_1}}\mathbf{y}^{(1)} = \frac{1}{y_{p_1}}A\mathbf{x}^{(0)}.$$

This continues with each iteration.

The Power method has the disadvantage that it is unknown at the outset whether the matrix has a single dominant eigenvalue. Nor is it known how $\mathbf{x}^{(0)}$ should be chosen to ensure that its representation in terms of the eigenvectors of the matrix will contain a nonzero contribution from the eigenvector associated with the dominant eigenvalue, should it exist.

Choosing the least integer, p_m, for which $|y_{p_m}^{(m)}| = \|\mathbf{y}^{(m)}\|_\infty$ will generally ensure that this index eventually becomes invariant. The rate at which $\{\mu^{(m)}\}_{m=1}^\infty$ converges to λ_1 is determined by the ratios $|\lambda_j/\lambda_1|^m$, for $j = 2, 3, \ldots, n$, and, in particular, by $|\lambda_2/\lambda_1|^m$; that is, the convergence is of order $O(|\lambda_2/\lambda_1|^m)$. Further, there is a constant k such that for large m

$$|\mu^{(m)} - \lambda_1| \approx k\left|\frac{\lambda_2}{\lambda_1}\right|^m.$$

This implies that

$$\lim_{m\to\infty} \frac{|\mu^{(m+1)} - \lambda_1|}{|\mu^{(m)} - \lambda_1|} \approx \left|\frac{\lambda_2}{\lambda_1}\right| < 1.$$

The sequence $\{\mu^{(m)}\}$ converges linearly to λ_1, so Aitken's Δ^2 procedure can be used to speed the convergence.

In actuality, it is not necessary for the matrix to have n distinct eigenvalues for the Power method to converge. If the matrix has a unique dominant eigenvalue, λ_1, with multiplicity r greater than 1 and $\mathbf{v}^{(1)}, \mathbf{v}^{(2)}, \ldots, \mathbf{v}^{(r)}$ are linearly independent eigenvectors associated with λ_1, the procedure will still converge to λ_1. The sequence of vectors $\{\mathbf{x}^{(m)}\}_{m=0}^\infty$ will in this case converge to an eigenvector of λ_1 of norm one that is a linear combination of $\mathbf{v}^{(1)}, \mathbf{v}^{(2)}, \ldots, \mathbf{v}^{(r)}$ and depends on the choice of the initial vector $\mathbf{x}^{(0)}$.

The program POWERM91 implements the Power method.

EXAMPLE 1 The matrix

$$A = \begin{bmatrix} -4 & 14 & 0 \\ -5 & 13 & 0 \\ -1 & 0 & 2 \end{bmatrix}$$

has eigenvalues $\lambda_1 = 6$, $\lambda_2 = 3$, and $\lambda_3 = 2$, so the Power method will converge. Let $\mathbf{x}^{(0)} = (1, 1, 1)^t$. Then

$$\mathbf{y}^{(1)} = A\mathbf{x}^{(0)} = (10, 8, 1)^t;$$

so

$$\|\mathbf{y}^{(1)}\|_\infty = 10, \qquad \mu^{(1)} = y_1^{(1)} = 10, \qquad \text{and} \qquad \mathbf{x}^{(1)} = \frac{\mathbf{y}^{(1)}}{10} = (1, 0.8, 0.1)^t.$$

Continuing in this manner leads to the values in Table 9.1, where $\tilde{\mu}^{(m)}$ represents the sequence generated by the Aitken's $\triangle^2$ procedure.

Table 9.1

m	$\mathbf{x}^{(m)}$	$\mu^{(m)}$	$\tilde{\mu}^{(m)}$
0	$(1,1,1)^t$		
1	$(1,0.8,0.1)^t$	10	6.266667
2	$(1,0.75,-0.111)^t$	7.2	6.062473
3	$(1,0.730769,-0.188034)^t$	6.5	6.015054
4	$(1,0.722200,-0.220850)^t$	6.230769	6.004202
5	$(1,0.718182,-0.235915)^t$	6.111000	6.000855
6	$(1,0.716216,-0.243095)^t$	6.054546	6.000240
7	$(1,0.715247,-0.246588)^t$	6.027027	6.000058
8	$(1,0.714765,-0.248306)^t$	6.013453	6.000017
9	$(1,0.714525,-0.249157)^t$	6.006711	6.000003
10	$(1,0.714405,-0.249579)^t$	6.003352	6.000000
11	$(1,0.714346,-0.249790)^t$	6.001675	6.000000
12	$(1,0.714316,-0.249895)^t$	6.000837	

An approximation to the dominant eigenvalue, 6, at this stage is $\tilde{\mu}^{(10)} = 6.000000$ with approximate unit eigenvector $(1, 0.714316, 0.249895)^t$. ■ ■ ■

When A is symmetric, a variation in the choice of the vectors $\mathbf{x}^{(m)}$, $\mathbf{y}^{(m)}$, and scalars $\mu^{(m)}$ can be made to significantly improve the rate of convergence of the sequence $\{\mu^{(m)}\}_{m=1}^\infty$ to the dominant eigenvalue λ_1. First select $\mathbf{x}^{(0)}$ with $\|\mathbf{x}^{(0)}\|_2 = 1$. For each $m = 1, 2, \ldots$, define

$$\mu^{(m)} = \left(\mathbf{x}^{(m-1)}\right)^t A\mathbf{x}^{(m-1)} \qquad \text{and} \qquad \mathbf{x}^{(m)} = \frac{1}{\|A\mathbf{x}^{(m-1)}\|_2} A\mathbf{x}^{(m-1)}.$$

The rate of convergence of the Power method is $O((\lambda_2/\lambda_1)^m)$, but with this modification the rate of convergence for symmetric matrices is $O((\lambda_2/\lambda_1)^{2m})$. The program SYMPWR92 implements the Symmetric Power method in this way.

EXAMPLE 2 The matrix

$$A = \begin{bmatrix} 4 & -1 & 1 \\ -1 & 3 & -2 \\ 1 & -2 & 3 \end{bmatrix}$$

is symmetric with eigenvalues $\lambda_1 = 6$, $\lambda_2 = 3$, and $\lambda_3 = 1$. Tables 9.2 and 9.3 list, respectively, the first ten iterations of the Power method and the Symmetric Power method assuming, in each case, that $\mathbf{y}^{(0)} = \mathbf{x}^{(0)} = (1, 0, 0)^t$.

Table 9.2

m	$\mathbf{y}^{(m)}$	$\mu^{(m)}$	$\mathbf{x}^{(m)}$
0			$(1,0,0)$
1	$(4,-1,1)^t$	4	$(1,-0.25,0.25)^t$
2	$(4.5,-2.25,2.25)^t$	4.5	$(1,-0.5,0.5)^t$
3	$(5,-3.5,3.5)^t$	5	$(1,-0.7,0.7)^t$
4	$(5.4,-4.5,4.5)^t$	5.4	$(1,-8.33\bar{3},0.833\bar{3})^t$
5	$(5.66\bar{6},-5.166\bar{6},5.166\bar{6})^t$	$5.66\bar{6}$	$(1,-0.911765,0.911765)^t$
6	$(5.823529,-5.558824,5.558824)^t$	5.823529	$(1,-0.954545,0.954545)^t$
7	$(5.909091,-5.772727,5.772727)^t$	5.909091	$(1,-0.976923,0.976923)^t$
8	$(5.953846,-5.884615,5.884615)^t$	5.953846	$(1,-0.988372,0.988372)^t$
9	$(5.976744,-5.941861,5.941861)^t$	5.976744	$(1,-0.994163,0.994163)^t$
10	$(5.988327,-5.970817,5.970817)^t$	5.988327	$(1,-0.997076,0.997076)^t$

Notice the significant improvement that the Symmetric Power method provides. The approximations to the eigenvectors produced in the Power method are converging to $(1, -1, 0)^t$, a vector with $\|(1, -1, 0)^t\|_\infty = 1$. In the Symmetric Power method the convergence is to the parallel vector $(\frac{\sqrt{2}}{2}, -\frac{\sqrt{2}}{2}, 0)^t$, since $\|(\frac{\sqrt{2}}{2}, -\frac{\sqrt{2}}{2}, 0)^t\|_2 = 1$.

■ ■ ■

Table 9.3

m	$\mathbf{y}^{(m)}$	$\mu^{(m)}$	$\mathbf{x}^{(m)}$
0	$(1,0,0)^t$		$(1,0,0)^t$
1	$(4,-1,1)^t$	4	$(0.942809,-0.235702,0.235702)^t$
2	$(4.242641,-2.121320,2.121320)^t$	5	$(0.816497,-0.408248,0.408248)^t$
3	$(4.082483,-2.857738,2.857738)^t$	5.666667	$(0.710669,-0.497468,0.497468)^t$
4	$(3.837613,-3.198011,3.198011)^t$	5.909091	$(0.646997,-0.539164,0.539164)^t$
5	$(3.666314,-3.342816,3.342816)^t$	5.976744	$(0.612836,-0.558763,0.558763)^t$
6	$(3.568871,-3.406650,3.406650)^t$	5.994152	$(0.595247,-0.568190,0.568190)^t$
7	$(3.517370,-3.436200,3.436200)^t$	5.998536	$(0.586336,-0.572805,0.572805)^t$
8	$(3.490952,-3.450359,3.450359)^t$	5.999634	$(0.581852,-0.575086,0.575086)^t$
9	$(3.477580,-3.457283,3.457283)^t$	5.999908	$(0.579603,-0.576220,0.576220)^t$
10	$(3.470854,-3.460706,3.460706)^t$	5.999977	$(0.578477,-0.576786,0.576786)^t$

The **Inverse Power method** is a modification of the Power method and is used to determine the eigenvalue of A that is closest to a specified number q.

Suppose that the matrix A has eigenvalues $\lambda_1, \ldots, \lambda_n$ with linearly independent eigenvectors $\mathbf{v}^{(1)}, \mathbf{v}^{(2)}, \ldots, \mathbf{v}^{(n)}$. The eigenvalues of $(A - qI)^{-1}$, where $q \neq \lambda_i$ for each $i = 1, 2, \ldots, n$, are

$$\frac{1}{\lambda_1 - q}, \frac{1}{\lambda_2 - q}, \ldots, \frac{1}{\lambda_n - q}$$

with eigenvectors $\mathbf{v}^{(1)}, \mathbf{v}^{(2)}, \ldots, \mathbf{v}^{(n)}$. Applying the Power method to $(A - qI)^{-1}$ gives

$$\mathbf{y}^{(m)} = (A - qI)^{-1}\mathbf{x}^{(m-1)},$$

$$\mu^{(m)} = y^{(m)}_{p_{m-1}} = \frac{y^{(m)}_{p_{m-1}}}{x^{(m-1)}_{p_{m-1}}} = \frac{\sum_{j=1}^{n} \alpha_j \dfrac{1}{(\lambda_j - q)^m} v^{(j)}_{p_{m-1}}}{\sum_{j=1}^{n} \alpha_j \dfrac{1}{(\lambda_j - q)^{m-1}} v^{(j)}_{p_{m-1}}},$$

and

$$\mathbf{x}^{(m)} = \frac{\mathbf{y}^{(m)}}{y^{(m)}_{p_m}},$$

where, at each step, p_m represents an integer for which $\left|y^{(m)}_{p_m}\right| = \|\mathbf{y}^{(m)}\|_\infty$. The sequence $\{\mu^{(m)}\}$ converges to $1/(\lambda_k - q)$, where

$$\frac{1}{|\lambda_k - q|} = \max_{1 \le i \le n} \frac{1}{|\lambda_i - q|},$$

and λ_k is the eigenvalue of A closest to q.

For a fixed k, we can write

$$\mu^{(m)} = \frac{1}{\lambda_k - q}\left[\frac{\alpha_k v^{(k)}_{p_{m-1}} + \sum_{\substack{j=1 \\ j \ne k}}^{n} \alpha_j \left[\dfrac{\lambda_k - q}{\lambda_j - q}\right]^m v^{(j)}_{p_{m-1}}}{\alpha_k v^{(k)}_{p_{m-1}} + \sum_{\substack{j=1 \\ j \ne k}}^{n} \alpha_j \left[\dfrac{\lambda_k - q}{\lambda_j - q}\right]^{m-1} (j)_{p_{m-1}}}\right].$$

The choice of q determines the convergence, provided that $1/(\lambda_k - q)$ is a unique dominant eigenvalue of $(A - qI)^{-1}$ (although it may be a multiple eigenvalue). The closer q is to an eigenvalue λ_k, the faster the convergence, since the convergence is of order

$$O\left(\left|\frac{(\lambda_k - q)^{-1}}{(\lambda - q)^{-1}}\right|^m\right) = O\left(\left|\frac{(\lambda_k - q)}{(\lambda - q)}\right|^m\right),$$

where λ represents the eigenvalue of A that is second closest to q.

The vector $\mathbf{y}^{(m)}$ is obtained from the equation

$$(A - qI)\mathbf{y}^{(m)} = \mathbf{x}^{(m-1)}.$$

It is recommended that Gaussian elimination with pivoting be used to solve this system.

Although the Inverse Power method requires the solution of an $n \times n$ system at each step, the multipliers can be saved to reduce the computation. The selection of q can be based on the Gerschgorin Theorem or on any other means of localizing an eigenvalue.

The value of q comes from the initial approximation to the eigenvector $\mathbf{x}^{(0)}$:

$$q = \frac{\mathbf{x}^{(0)t} A \mathbf{x}^{(0)}}{\mathbf{x}^{(0)t} \mathbf{x}^{(0)}}.$$

This choice of q results from the observation that if $\mathbf{x}$ is an eigenvector of A with respect to the eigenvalue λ, then $A\mathbf{x} = \lambda\mathbf{x}$. So $\mathbf{x}^t A\mathbf{x} = \lambda\mathbf{x}^t\mathbf{x}$ and

$$\lambda = \frac{\mathbf{x}^t A\mathbf{x}}{\mathbf{x}^t\mathbf{x}}.$$

If q is close to an eigenvalue, the convergence will be quite rapid.

Since the convergence of the Inverse Power method is linear, Aitken's Δ^2 procedure can be used to speed convergence. The following example illustrates the fast convergence of the Inverse Power method if q is close to an eigenvalue.

EXAMPLE 3   The matrix

$$A = \begin{bmatrix} -4 & 14 & 0 \\ -5 & 13 & 0 \\ -1 & 0 & 2 \end{bmatrix}$$

was considered in Example 1. The Power method gave the approximation $\mu^{(12)} = 6.000837$ using $\mathbf{x}^{(0)} = (1, 1, 1)^t$. With this same initial vector, we have

$$q = \frac{\mathbf{x}^{(0)t}A\mathbf{x}^{(0)}}{\mathbf{x}^{(0)t}\mathbf{x}^{(0)}} = \frac{19}{3} = 6.333333.$$

The results of applying the program INVPOW93 for the Inverse Power method with this value q are listed in Table 9.4   ■ ■ ■

Table 9.4

m	$\mathbf{x}^{(m)}$	$\mu^{(m)}$
0	$(1,1,1)^t$	
1	$(1,0.720727,-0.194042)^t$	6.183183
2	$(1,0.715518,-0.245052)^t$	6.017244
3	$(1,0.714409,-0.249522)^t$	6.001719
4	$(1,0.714298,-0.249953)^t$	6.000175
5	$(1,0.714287,-0.250000)^t$	6.000021
6	$(1,0.714286,-0.249999)^t$	6.000005

Numerous techniques are available for obtaining approximations to the other eigenvalues of a matrix once an approximation to the dominant eigenvalue has been computed. We will restrict our presentation to **deflation techniques**. These techniques involve forming a new matrix B whose eigenvalues are the same as those of A, except that the dominant eigenvalue of A is replaced by the eigenvalue 0 in B. The following result justifies the procedure:

Eigenvalues and Eigenvectors of Deflated Matrices

Suppose $\lambda_1, \lambda_2, \ldots, \lambda_n$ are eigenvalues of A with associated eigenvectors $\mathbf{v}^{(1)}, \mathbf{v}^{(2)}, \ldots,$ $\mathbf{v}^{(n)}$ and that λ_1 has multiplicity one. If $\mathbf{x}$ is any vector with the property that $\mathbf{x}^t \mathbf{v}^{(1)} = 1$, then the matrix

$$B = A - \lambda_1 \mathbf{v}^{(1)} \mathbf{x}^t$$

has eigenvalues $0, \lambda_2, \lambda_3, \ldots, \lambda_n$ with associated eigenvectors $\mathbf{v}^{(1)}, \mathbf{w}^{(2)}, \mathbf{w}^{(3)}, \ldots, \mathbf{w}^{(n)}$, where $\mathbf{v}^{(i)}$ and $\mathbf{w}^{(i)}$ are related by the equation

$$\mathbf{v}^{(i)} = (\lambda_i - \lambda_1)\mathbf{w}^{(i)} + \lambda_1(\mathbf{x}^t \mathbf{w}^{(i)})\mathbf{v}^{(1)},$$

for each $i = 2, 3, \ldots, n$.

Wielandt's deflation results from defining

$$\mathbf{x} = \frac{1}{\lambda_1 v_i^{(1)}} \begin{bmatrix} a_{i1} \\ a_{i2} \\ \vdots \\ a_{in} \end{bmatrix},$$

where $v_i^{(1)}$ is a coordinate of $\mathbf{v}^{(1)}$ that is nonzero and the values $a_{i1}, a_{i2}, \ldots, a_{in}$ are the entries in the ith row of A. With this definition,

$$\mathbf{x}^t \mathbf{v}^{(1)} = \frac{1}{\lambda_1 v_i^{(1)}} [a_{i1}, a_{i2}, \ldots, a_{in}](v_1^{(1)}, v_2^{(1)}, \ldots, v_n^{(1)})^t = \frac{1}{\lambda_1 v_i^{(1)}} \sum_{j=1}^{n} a_{ij} v_j^{(1)},$$

where the sum is the ith coordinate of the product $A\mathbf{v}^{(1)}$. Since $A\mathbf{v}^{(1)} = \lambda_1 \mathbf{v}^{(1)}$, this implies that

$$\mathbf{x}^t \mathbf{v}^{(1)} = \frac{1}{\lambda_1 v_i^{(1)}} (\lambda_1 v_i^{(1)}) = 1.$$

So $\mathbf{x}$ satisfies the hypotheses of the result concerning the eigenvalues of deflated matrices. Moreover, the ith row of $B = A - \lambda_1 \mathbf{v}^{(1)} \mathbf{x}^t$ consists entirely of zero entries.

If $\lambda \neq 0$ is an eigenvalue with associated eigenvector $\mathbf{w}$, the relation $B\mathbf{w} = \lambda\mathbf{w}$ implies that the ith coordinate of $\mathbf{w}$ must also be zero. Consequently, the ith column of the matrix B makes no contribution to the product $B\mathbf{w} = \lambda\mathbf{w}$. Thus, the matrix B can be replaced by an $(n-1) \times (n-1)$ matrix B', obtained by deleting the ith row and column from B. The matrix B' has eigenvalues $\lambda_2, \lambda_3, \ldots, \lambda_n$. If $|\lambda_2| > |\lambda_3|$, the Power method is reapplied to the matrix B' to determine this new dominant eigenvalue and an eigenvector, $\mathbf{w}^{(2)'}$, associated with λ_2, with respect to the matrix B'. To find the associated eigenvector $\mathbf{w}^{(2)}$ for the matrix B, insert a zero coordinate between the coordinates $w_{i-1}^{(2)'}$ and $w_i^{(2)'}$ of the $(n-1)$-dimensional vector $\mathbf{w}^{(2)'}$ and then calculate $\mathbf{v}^{(2)}$ by using the result for deflated matrices. This deflation technique can be performed using the program WIEDEF94.

EXAMPLE 4   From Example 2, we know that the matrix

$$A = \begin{bmatrix} 4 & -1 & 1 \\ -1 & 3 & -2 \\ 1 & -2 & 3 \end{bmatrix}$$

has eigenvalues $\lambda_1 = 6$, $\lambda_2 = 3$, and $\lambda_3 = 1$. Assuming that the dominant eigenvalue $\lambda_1 = 6$ and associated unit eigenvector $\mathbf{v}^{(1)} = (1, -1, 1)^t$ have been calculated, the procedure just outlined for obtaining λ_2 proceeds as follows:

$$\mathbf{x} = \frac{1}{6}\begin{bmatrix} 4 \\ -1 \\ 1 \end{bmatrix} = \left(\tfrac{2}{3}, -\tfrac{1}{6}, \tfrac{1}{6}\right)^t,$$

$$\mathbf{v}^{(1)}\mathbf{x}^t = \begin{bmatrix} 1 \\ -1 \\ 1 \end{bmatrix}\left[\tfrac{2}{3}, -\tfrac{1}{6}, \tfrac{1}{6}\right] = \begin{bmatrix} \tfrac{2}{3} & -\tfrac{1}{6} & \tfrac{1}{6} \\ -\tfrac{2}{3} & \tfrac{1}{6} & -\tfrac{1}{6} \\ \tfrac{2}{3} & -\tfrac{1}{6} & \tfrac{1}{6} \end{bmatrix};$$

and

$$B = A - \lambda_1\mathbf{v}^{(1)}\mathbf{x}^t = \begin{bmatrix} 4 & -1 & 1 \\ -1 & 3 & -2 \\ 1 & -2 & 3 \end{bmatrix} - 6\begin{bmatrix} \tfrac{2}{3} & -\tfrac{1}{6} & \tfrac{1}{6} \\ -\tfrac{2}{3} & \tfrac{1}{6} & -\tfrac{1}{6} \\ \tfrac{2}{3} & -\tfrac{1}{6} & \tfrac{1}{6} \end{bmatrix}$$

$$= \begin{bmatrix} 0 & 0 & 0 \\ 3 & 2 & -1 \\ -3 & -1 & 2 \end{bmatrix}.$$

Deleting the first row and column gives

$$B' = \begin{bmatrix} 2 & -1 \\ -1 & 2 \end{bmatrix},$$

which has eigenvalues $\lambda_2 = 3$ and $\lambda_3 = 1$. For $\lambda_2 = 3$ the eigenvector $\mathbf{w}^{(2)'}$ can be obtained by solving the second-order linear system

$$(B' - 3I)\mathbf{w}^{(2)'} = \mathbf{0},$$

resulting in

$$\mathbf{w}^{(2)'} = (1, -1)^t.$$

Thus, $\mathbf{w}^{(2)} = (0, 1, -1)^t$ and we have:

$$\mathbf{v}^{(2)} = (3 - 6)(0, 1, -1)^t + 6\left[\left[\tfrac{2}{3}, -\tfrac{1}{6}, \tfrac{1}{6}\right](0, 1, -1)^t\right](1, -1, 1)^t = (-2, -1, 1)^t.$$

■ ■ ■

Although deflation can be used to find approximations to all the eigenvalues and eigenvectors of a matrix, the process is susceptible to round-off error. Techniques based on similarity transformations are presented in the next two sections. These methods are generally preferable when approximations to all the eigenvalues are needed.

EXERCISE SET 9.3

1. Find the first three iterations obtained by the Power method applied to the following matrices:

(a) $\begin{bmatrix} 2 & 1 & 1 \\ 1 & 2 & 1 \\ 1 & 1 & 2 \end{bmatrix}$; use $\mathbf{x}^{(0)} = (1, -1, 2)^t$.

(b) $\begin{bmatrix} 1 & 1 & 1 \\ 1 & 1 & 0 \\ 1 & 0 & 1 \end{bmatrix}$; use $\mathbf{x}^{(0)} = (-1, 0, 1)^t$.

(c) $\begin{bmatrix} 1 & -1 & 0 \\ -2 & 4 & -2 \\ 0 & -1 & 1 \end{bmatrix}$; use $\mathbf{x}^{(0)} = (1, -1, 1)^t$.

(d) $\begin{bmatrix} 1 & -1 & 0 \\ -2 & 4 & -2 \\ 0 & -1 & 2 \end{bmatrix}$; use $\mathbf{x}^{(0)} = (-1, 2, 1)^t$.

(e) $\begin{bmatrix} 4 & 1 & 1 & 1 \\ 1 & 3 & -1 & 1 \\ 1 & -1 & 2 & 0 \\ 1 & 1 & 0 & 2 \end{bmatrix}$; use $\mathbf{x}^{(0)} = (1, -2, 0, 3)^t$.

(f) $\begin{bmatrix} 5 & -2 & -\frac{1}{2} & \frac{3}{2} \\ -2 & 5 & \frac{3}{2} & -\frac{1}{2} \\ -\frac{1}{2} & \frac{3}{2} & 5 & -2 \\ \frac{3}{2} & -\frac{1}{2} & -2 & 5 \end{bmatrix}$; use $\mathbf{x}^{(0)} = (1, 1, 0, -3)^t$.

(g) $\begin{bmatrix} 3 & 1 & -2 & 1 \\ 1 & 8 & -1 & 0 \\ -2 & -1 & 3 & -1 \\ 1 & 0 & -1 & 8 \end{bmatrix}$; use $\mathbf{x}^{(0)} = (2, 1, 0, -1)^t$.

(h) $\begin{bmatrix} -4 & 0 & \frac{1}{2} & \frac{1}{2} \\ \frac{1}{2} & -2 & 0 & \frac{1}{2} \\ \frac{1}{2} & \frac{1}{2} & 0 & 0 \\ 0 & 1 & 1 & 4 \end{bmatrix}$; use $\mathbf{x}^{(0)} = (1, -1, -1, 1)^t$.

2. Repeat Exercise 1 using the Inverse Power method.

3. Find the first three iterations obtained by the Symmetric Power method applied to the following matrices:

(a) $\begin{bmatrix} 2 & 1 & 1 \\ 1 & 2 & 1 \\ 1 & 1 & 2 \end{bmatrix}$; use $\mathbf{x}^{(0)} = (1, -1, 2)^t$.

(b) $\begin{bmatrix} 1 & 1 & 1 \\ 1 & 1 & 0 \\ 1 & 0 & 1 \end{bmatrix}$; use $\mathbf{x}^{(0)} = (-1, 0, 1)^t$.

(c) $\begin{bmatrix} 2 & -1 & 0 \\ -1 & 2 & -1 \\ 0 & -1 & 2 \end{bmatrix}$; use $\mathbf{x}^{(0)} = (1, 0, 0)^t$.

(d) $\begin{bmatrix} 4.75 & 2.25 & -0.25 \\ 2.25 & 4.75 & 1.25 \\ -0.25 & 1.25 & 4.75 \end{bmatrix}$; use $\mathbf{x}^{(0)} = (0, 1, 0)^t$.

(e) $\begin{bmatrix} 4 & 1 & -1 & 0 \\ 1 & 3 & -1 & 0 \\ -1 & -1 & 5 & 2 \\ 0 & 0 & 2 & 4 \end{bmatrix}$; use $\mathbf{x}^{(0)} = (0, 1, 0, 0)^t$.

(f) $\begin{bmatrix} 4 & 1 & 1 & 1 \\ 1 & 3 & -1 & 1 \\ 1 & -1 & 2 & 0 \\ 1 & 1 & 0 & 2 \end{bmatrix}$; use $\mathbf{x}^{(0)} = (1, 0, 0, 0)^t$.

(g) $\begin{bmatrix} 3 & 1 & -2 & 1 \\ 1 & 8 & -1 & 0 \\ -2 & -1 & 3 & -1 \\ 1 & 0 & -1 & 8 \end{bmatrix}$; use $\mathbf{x}^{(0)} = (2, 1, 0, -1)^t$.

(h) $\begin{bmatrix} 5 & -2 & -\frac{1}{2} & \frac{3}{2} \\ -2 & 5 & \frac{3}{2} & -\frac{1}{2} \\ -\frac{1}{2} & \frac{3}{2} & 5 & -2 \\ \frac{3}{2} & -\frac{1}{2} & -2 & 5 \end{bmatrix}$; use $\mathbf{x}^{(0)} = (1, 1, 0, -3)^t$.

4. Use the Power method and Wielandt deflation to approximate the two most dominant eigenvalues for the matrices in Exercise 1. Iterate until a tolerance of 10^{-4} is achieved or until the number of iterations exceeds 25.

5. Repeat Exercise 4 using Aitken's Δ^2 technique and the Power method to approximate the first eigenvalue.

6. Use the Symmetric Power method to compute the largest eigenvalue (in absolute value) of the matrices given in Exercise 3. Iterate until a tolerance of 10^{-4} is achieved or until the number of iterations exceeds 25.

7. Repeat Exercise 5 using the Inverse Power method.

8. Repeat Exercise 6 using the Inverse Power method.

9. Following along the line of Exercise 10 in Section 6.4 and Exercise 7 in Section 7.3, suppose that a species of beetle has a life span of 4 years, that a female in the first year has a survival rate of $\frac{1}{2}$, in the second year a survival rate of $\frac{1}{4}$, and in the third year a survival rate of $\frac{1}{8}$. Suppose, additionally, that a female gives birth, on the average, to two new females in the third year and to four new females in the fourth year. The matrix describing a single female's contribution in one year to the female population in the succeeding year is

$$A = \begin{bmatrix} 0 & 0 & 2 & 4 \\ \frac{1}{2} & 0 & 0 & 0 \\ 0 & \frac{1}{4} & 0 & 0 \\ 0 & 0 & \frac{1}{8} & 0 \end{bmatrix},$$

where again the entry in the ith row and jth column denotes the probabilistic contribution that a female of age j makes on the next year's female population of age i.

(a) Use the Gerschgorin Circle Theorem to determine a region in the complex plane containing all the eigenvalues of A.

(b) Use the Power method to determine the dominant eigenvalue of the matrix and its associated eigenvector.

(c) Use Wielandt deflation to determine any remaining eigenvalues and eigenvectors of A.

(d) Find the eigenvalues of A by using the characteristic polynomial of A and Newton's method.

(e) What is your long-range prediction for the population of these beetles?

9.4 Householder's Method

In the next section we use the QR method to reduce a symmetric tridiagonal matrix to a nearly diagonal matrix to which it is similar. The diagonal entries of the reduced matrix are approximations to the eigenvalues of both matrices. In this section, we consider the associated problem of reducing an arbitrary symmetric matrix to a similar tridiagonal matrix using a method devised by Alton Householder. Although there is a clear connection between the problems we are solving in these two sections, Householder's method has wide application in areas other than eigenvalue approximation.

Householder's method is used to find a symmetric tridiagonal matrix B that is similar to a given symmetric matrix A. Schur's Theorem ensures that a symmetric matrix A is similar to a diagonal matrix D, since an orthogonal matrix Q exists with the property that $D = Q^{-1}AQ = Q^t AQ$. However, the matrix Q (and consequently D) is generally difficult to compute, so Householder's method offers a compromise.

Let $\mathbf{w} \in \mathbb{R}^n$ with $\mathbf{w}^t \mathbf{w} = 1$. The $n \times n$ matrix

$$P = I - 2\mathbf{w}\mathbf{w}^t$$

is called a **Householder transformation**.

Householder transformations are used to selectively zero out blocks of entries in vectors or in columns of matrices in a manner that is extremely stable with respect to round-off error. An important property of Householder transformations follows.

Householder Transformations

If $P = I - 2\mathbf{w}\mathbf{w}^t$ is a Householder transformation, then P is symmetric and orthogonal, so $P^{-1} = P$.

Householder's method begins by determining a transformation $P^{(1)}$ with the property that $A^{(2)} = P^{(1)}AP^{(1)}$ will have

$$a_{j1}^{(2)} = 0 \qquad \text{for each } j = 3, 4, \ldots, n.$$

By symmetry, this also implies that $a_{1j}^{(2)} = 0$ for each $j = 3, 4, \ldots, n$.

The vector $\mathbf{w} = (w_1, w_2, \ldots, w_n)^t$ is chosen so that $\mathbf{w}^t\mathbf{w} = 1$, $a_{j1}^{(2)} = 0$, for each $j = 3, 4, \ldots, n$, and $a_{11}^{(2)} = a_{11}$. This imposes n conditions on the n unknowns $w_1, w_2, \ldots, w_n$. Setting $w_1 = 0$ ensures that $a_{11}^{(2)} = a_{11}$. We want

$$P^{(1)} = I - 2\mathbf{w}\mathbf{w}^t$$

to satisfy

(9.1) $$P^{(1)}(a_{11}, a_{21}, a_{31}, \ldots, a_{n1})^t = (a_{11}, \alpha, 0, \ldots, 0)^t,$$

where α will be chosen later. To simplify notation, let

$$\hat{\mathbf{w}} = (w_2, w_3, \ldots, w_n)^t \in \mathbb{R}^{n-1}, \qquad \hat{\mathbf{y}} = (a_{21}, a_{31}, \ldots, a_{n1})^t \in \mathbb{R}^{n-1},$$

and $\hat{P}$ be the $(n-1) \times (n-1)$ Householder transformation

$$\hat{P} = I_{n-1} - 2\hat{\mathbf{w}}\hat{\mathbf{w}}^t.$$

Eq. (9.1) then becomes

$$P^{(1)}\begin{bmatrix} a_{11} \\ a_{21} \\ a_{31} \\ \vdots \\ a_{n1} \end{bmatrix} = \begin{bmatrix} 1 & 0 & \cdots & 0 \\ 0 & & & \\ \vdots & & \hat{P} & \\ \vdots & & & \\ 0 & & & \end{bmatrix} \cdot \begin{bmatrix} a_{11} \\ \\ \hat{\mathbf{y}} \\ \\ \end{bmatrix} = \begin{bmatrix} a_{11} \\ \\ \hat{P}\hat{\mathbf{y}} \\ \\ \end{bmatrix} = \begin{bmatrix} a_{11} \\ \alpha \\ 0 \\ \vdots \\ 0 \end{bmatrix}$$

with

(9.2) $$\hat{P}\hat{\mathbf{y}} = (I_{n-1} - 2\hat{\mathbf{w}}\hat{\mathbf{w}}^t)\hat{\mathbf{y}} = \hat{\mathbf{y}} - 2\hat{\mathbf{w}}^t\hat{\mathbf{y}}\hat{\mathbf{w}} = (\alpha, 0, \ldots, 0)^t.$$

Let $r = \mathbf{w}^t\mathbf{y}$; then

$$(\alpha, 0, \ldots, 0)^t = (a_{21} - 2rw_2, a_{31} - 2rw_3, \ldots, a_{n1} - 2rw_n)^t.$$

Equating components gives

$$\alpha = a_{21} - 2rw_2 \qquad 0 = a_{j1} - 2rw_j, \quad \text{for each } j = 3, \ldots, n.$$

Thus,

(9.3) $2rw_2 = a_{21} - \alpha$ and $2rw_j = a_{j1}$,   for each $j = 3, \ldots, n$.

Squaring both sides of each of the equations and summing gives

$$4r^2 \sum_{j=2}^{n} w_j^2 = (a_{21} - \alpha)^2 + \sum_{j=3}^{n} a_{j1}^2.$$

Since $\mathbf{w}^t \mathbf{w} = 1$ and $w_1 = 0$, we have $\sum_{j=2}^{n} w_j^2 = 1$ and

(9.4)
$$4r^2 = \sum_{j=2}^{n} a_{j1}^2 - 2\alpha a_{21} + \alpha^2.$$

Equation (9.2) and the fact that P is orthogonal imply that

$$\alpha^2 = (\alpha, 0, \ldots, 0)(\alpha, 0, \ldots, 0)^t = (\hat{P}\hat{\mathbf{y}})^t \hat{P}\hat{\mathbf{y}} = \hat{\mathbf{y}}^t \hat{P}^t \hat{P}\hat{\mathbf{y}} = \hat{\mathbf{y}}^t \hat{\mathbf{y}}.$$

Thus,

$$\alpha^2 = \sum_{j=2}^{n} a_{j1}^2,$$

which, when substituted into Eq. (9.4), gives

$$2r^2 = \sum_{j=2}^{n} a_{j1}^2 - \alpha a_{21}.$$

To ensure that $2r^2 = 0$ only if $a_{21} = a_{31} = \cdots = a_{n1} = 0$, we choose

$$\alpha = -(\text{sign } a_{21})\left(\sum_{j=2}^{n} a_{j1}^2\right)^{1/2}$$

which implies that

$$2r^2 = \sum_{j=2}^{n} a_{j1}^2 + |a_{21}|\left(\sum_{j=2}^{n} a_{j1}^2\right)^{1/2}.$$

With this choice of α and $2r^2$, we solve the equations in (9.3) to obtain

$$w_2 = \frac{a_{21} - \alpha}{2r} \quad \text{and} \quad w_j = \frac{a_{j1}}{2r}, \quad \text{for each } j = 3, \ldots, n.$$

To summarize the choice of $P^{(1)}$, we have

$$\alpha = -(\text{sign } a_{21})\left(\sum_{j=2}^{n} a_{j1}^2\right)^{1/2},$$

$$r = \left(\tfrac{1}{2}\alpha^2 - \tfrac{1}{2}a_{21}\alpha\right)^{1/2},$$

$$w_1 = 0,$$

$$w_2 = \frac{a_{21} - \alpha}{2r},$$

and

$$w_j = \frac{a_{j1}}{2r}, \qquad \text{for each } j = 3, \ldots, n.$$

With this choice,

$$A^{(2)} = P^{(1)}AP^{(1)} = \begin{bmatrix} a_{11}^{(2)} & a_{12}^{(2)} & 0 & \cdots & 0 \\ a_{21}^{(2)} & a_{22}^{(2)} & a_{23}^{(2)} & \cdots & a_{2n}^{(2)} \\ 0 & a_{32}^{(2)} & a_{33}^{(2)} & \cdots & a_{3n}^{(2)} \\ \vdots & \vdots & \vdots & & \vdots \\ 0 & a_{n2}^{(2)} & a_{n3}^{(2)} & \cdots & a_{nn}^{(2)} \end{bmatrix}.$$

Having found $P^{(1)}$ and computed $A^{(2)}$, the process is repeated for $k = 2, 3, \ldots, n-2$ as follows:

$$\alpha = -(\text{sign } a_{k+1,k}^{(k)})\left(\sum_{j=k+1}^{n} (a_{jk}^{(k)})^2 \right)^{1/2},$$

$$r = \left(\tfrac{1}{2}\alpha^2 - \tfrac{1}{2}\alpha a_{k+1,k}^{(k)} \right)^{1/2},$$

$$w_1^{(k)} = w_2^{(k)} = \cdots = w_k^{(k)} = 0,$$

$$w_{k+1}^{(k)} = \frac{a_{k+1,k}^{(k)} - \alpha}{2r}$$

$$w_j^{(k)} = \frac{a_{jk}^{(k)}}{2r}, \qquad \text{for each } j = k+2, k+3, \ldots, n,$$

$$P^{(k)} = I - 2\mathbf{w}^{(k)} \cdot \left(\mathbf{w}^{(k)}\right)^t,$$

and

$$A^{(k+1)} = P^{(k)}A^{(k)}P^{(k)},$$

where

$$A^{(k+1)} = \begin{bmatrix} a_{11}^{(k+1)} & a_{12}^{(k+1)} & 0 & \cdots\cdots\cdots\cdots\cdots\cdots\cdots & 0 \\ a_{21}^{(k+1)} & & & & \vdots \\ 0 & & & 0 \cdots\cdots\cdots 0 \\ \vdots & & a_{k+1,k}^{(k+1)} & a_{k+1,k+1}^{(k+1)} & a_{k+1,k+2}^{(k+1)} & \cdots & a_{k+1,n}^{(k+1)} \\ \vdots & & & 0 & \vdots & & \vdots \\ 0 \cdots\cdots\cdots 0 & a_{n,k+1}^{(k+1)} & \cdots\cdots\cdots & a_{nn}^{(k+1)} \end{bmatrix}.$$

Continuing in this manner, the tridiagonal and symmetric matrix $A^{(n-1)}$ is formed, where

$$A^{(n-1)} = P^{(n-2)}P^{(n-3)} \cdots P^{(1)}AP^{(1)} \cdots P^{(n-3)}P^{(n-2)}.$$

The program HSEHLD95 performs Householder's method in this manner on a symmetric matrix.

EXAMPLE 1 The 4×4 matrix

$$A = \begin{bmatrix} 4 & 1 & -2 & 2 \\ 1 & 2 & 0 & 1 \\ -2 & 0 & 3 & -2 \\ 2 & 1 & -2 & -1 \end{bmatrix}$$

is symmetric. For the first application of a Householder transformation:

$$\alpha = -(1)\left(\sum_{j=2}^{4} a_{j1}^2\right)^{1/2} = -3, \quad r = \left(\frac{1}{2}(-3)^2 - \frac{1}{2}(1)(-3)\right)^{1/2} = \sqrt{6},$$

$$\mathbf{w} = \left(0, \frac{\sqrt{6}}{3}, -\frac{\sqrt{6}}{6}, \frac{\sqrt{6}}{6}\right)^t, \quad P^{(1)} = \begin{bmatrix} 1 & 0 & 0 & 0 \\ 0 & 1 & 0 & 0 \\ 0 & 0 & 1 & 0 \\ 0 & 0 & 0 & 1 \end{bmatrix} - 2\left(\frac{\sqrt{6}}{6}\right)^2 \begin{bmatrix} 0 \\ 2 \\ -1 \\ 1 \end{bmatrix} \cdot (0, 2, -1, 1)^t$$

$$= \begin{bmatrix} 1 & 0 & 0 & 0 \\ 0 & -\frac{1}{3} & \frac{2}{3} & -\frac{2}{3} \\ 0 & \frac{2}{3} & \frac{2}{3} & \frac{1}{3} \\ 0 & -\frac{2}{3} & \frac{1}{3} & \frac{2}{3} \end{bmatrix},$$

and

$$A^{(2)} = \begin{bmatrix} 4 & -3 & 0 & 0 \\ -3 & \frac{10}{3} & 1 & \frac{4}{3} \\ 0 & 1 & \frac{5}{3} & -\frac{4}{3} \\ 0 & \frac{4}{3} & -\frac{4}{3} & -1 \end{bmatrix}.$$

Continuing to the second iteration,

$$\alpha = -\frac{5}{3}, \quad r = \frac{2\sqrt{5}}{3}, \quad \mathbf{w} = \left(0, 0, 2\sqrt{5}, \frac{\sqrt{5}}{5}\right)^t,$$

and

$$P^{(2)} = \begin{bmatrix} 1 & 0 & 0 & 0 \\ 0 & 1 & 0 & 0 \\ 0 & 0 & -\frac{3}{5} & -\frac{4}{5} \\ 0 & 0 & -\frac{4}{5} & \frac{3}{5} \end{bmatrix},$$

and the symmetric tridiagonal matrix is

$$A^{(3)} = \begin{bmatrix} 4 & -3 & 0 & 0 \\ -3 & \frac{10}{3} & -\frac{5}{3} & 0 \\ 0 & -\frac{5}{3} & -\frac{33}{25} & \frac{68}{75} \\ 0 & 0 & \frac{68}{75} & \frac{149}{75} \end{bmatrix}.$$ ■ ■ ■

In the next section, we will examine how the QR method can then be applied to $A^{(n-1)}$ to determine the eigenvalues of $A^{(n-1)}$, which are the same as those of the original matrix A.

EXERCISE SET 9.4

1   Use Householder's method to place the following matrices in tridiagonal form:

(a) $\begin{bmatrix} 12 & 10 & 4 \\ 10 & 8 & -5 \\ 4 & -5 & 3 \end{bmatrix}$ (b) $\begin{bmatrix} 2 & -1 & -1 \\ -1 & 2 & -1 \\ -1 & -1 & 2 \end{bmatrix}$ (c) $\begin{bmatrix} 2 & 1 & 1 \\ 1 & 2 & 1 \\ 1 & 1 & 2 \end{bmatrix}$

(d) $\begin{bmatrix} 1 & 1 & 1 \\ 1 & 1 & 0 \\ 1 & 0 & 1 \end{bmatrix}$ (e) $\begin{bmatrix} 2 & 0 & 1 \\ 0 & 3 & -2 \\ 1 & -2 & -1 \end{bmatrix}$ (f) $\begin{bmatrix} 4.75 & 2.25 & -0.25 \\ 2.25 & 4.75 & 1.25 \\ -0.25 & 1.25 & 4.75 \end{bmatrix}$

2.   Use Householder's method to place the following matrices in tridiagonal form:

(a) $\begin{bmatrix} 4 & -1 & -1 & 0 \\ -1 & 4 & 0 & -1 \\ -1 & 0 & 4 & -1 \\ 0 & -1 & -1 & 4 \end{bmatrix}$ (b) $\begin{bmatrix} 5 & -2 & -0.5 & 1.5 \\ -2 & 5 & 1.5 & -0.5 \\ -0.5 & 1.5 & 5 & -2 \\ 1.5 & -0.5 & -2 & 5 \end{bmatrix}$

(c) $\begin{bmatrix} -4 & 0 & 1 & 3 \\ 0 & -4 & 2 & 1 \\ 1 & 2 & -2 & 0 \\ 3 & 1 & 0 & -4 \end{bmatrix}$ (d) $\begin{bmatrix} 4 & 1 & 1 & 1 \\ 1 & 3 & -1 & 1 \\ 1 & -1 & 2 & 0 \\ 1 & 1 & 0 & 2 \end{bmatrix}$

(e) $\begin{bmatrix} 8 & 0.25 & 0.5 & 2 & -1 \\ 0.25 & -4 & 0 & 1 & 2 \\ 0.5 & 0 & 5 & 0.75 & -1 \\ 2 & 1 & 0.75 & 5 & -0.5 \\ -1 & 2 & -1 & -0.5 & 6 \end{bmatrix}$ (f) $\begin{bmatrix} 2 & -1 & -1 & 0 & 0 \\ -1 & 3 & 0 & -2 & 0 \\ -1 & 0 & 4 & 2 & 1 \\ 0 & -2 & 2 & 8 & 3 \\ 0 & 0 & 1 & 3 & 9 \end{bmatrix}$

9.5   The QR Method

To apply the *QR* method, we begin with a symmetric matrix in tridiagonal form; that is, the only nonzero entries in the matrix lie either on the diagonal or on the subdiagonals directly above or below the diagonal. If this is not the form of the symmetric matrix, the first step is to apply Householder's method to compute a symmetric, tridiagonal matrix similar to the given matrix.

If A is a symmetric tridiagonal matrix, we can simplify the notation somewhat by labeling the entries of A as follows:

$$A = \begin{bmatrix} a_1 & b_2 & 0 & \cdots & 0 \\ b_2 & a_2 & b_3 & \ddots & \vdots \\ 0 & b_3 & a_3 & \ddots & 0 \\ \vdots & \ddots & \ddots & \ddots & b_n \\ 0 & \cdots & 0 & b_n & a_n \end{bmatrix}.$$

First observe that when $b_j = 0$, for some $2 < j < n$, the problem can be reduced to considering, instead of A, the smaller matrices

$$\begin{bmatrix} a_1 & b_2 & 0 & \cdots & 0 \\ b_2 & a_2 & b_3 & \ddots & \vdots \\ 0 & b_3 & a_3 & \ddots & 0 \\ \vdots & \ddots & \ddots & \ddots & b_{j-1} \\ 0 & \cdots & 0 & b_{j-1} & a_{j-1} \end{bmatrix} \quad \text{and} \quad \begin{bmatrix} a_j & b_{j+1} & 0 & \cdots & 0 \\ b_{j+1} & a_{j+1} & b_{j+2} & \ddots & \vdots \\ 0 & b_{j+2} & a_{j+3} & \ddots & 0 \\ \vdots & \ddots & \ddots & \ddots & b_n \\ 0 & \cdots & 0 & b_n & a_n \end{bmatrix}.$$

If $b_2 = 0$ or $b_n = 0$, then the 1×1 matrix $[a_1]$ or $[a_n]$ immediately produces an eigenvalue a_1 or a_n of A.

If none of the b_j are zero, the QR method proceeds by forming a sequence of matrices $A = A^{(1)}, A^{(2)}, A^{(3)}, \ldots$, as follows:

1. $A^{(1)} = A$ is factored as a product $A^{(1)} = Q^{(1)}R^{(1)}$, where $Q^{(1)}$ is orthogonal and $R^{(1)}$ is upper triangular.
2. $A^{(2)}$ is defined as $A^{(2)} = R^{(1)}Q^{(1)}$ and factored as a product $A^{(2)} = Q^{(2)}R^{(2)}$, where $Q^{(2)}$ is orthogonal and $R^{(2)}$ is upper triangular.

In general, $A^{(i)}$ is factored as a product $A^{(i)} = Q^{(i)}R^{(i)}$ of an orthogonal matrix $Q^{(i)}$ and an upper triangular matrix $R^{(i)}$. Then $A^{(i+1)}$ is defined by the product of $R^{(i)}$ and $Q^{(i)}$ in the reverse direction $A^{(i+1)} = R^{(i)}Q^{(i)}$. Since $Q^{(i)}$ is orthogonal,

$$A^{(i+1)} = R^{(i)}Q^{(i)} = (Q^{(i)^t}A^{(i)})Q^{(i)} = Q^{(i)^t}A^{(i)}Q^{(i)},$$

and $A^{(i+1)}$ is symmetric and tridiagonal with the same eigenvalues as $A^{(i)}$. Continuing by induction, $A^{(i+1)}$ has the same eigenvalues as the original matrix A. The success of the procedure is a result of the fact that $A^{(i+1)}$ tends to a diagonal matrix with the eigenvalues of A along the diagonal.

To describe the construction of the factoring matrices $Q^{(i)}$ and $R^{(i)}$, we first introduce the notion of a *rotation matrix*.

A **rotation matrix** P is an orthogonal matrix that differs from the identity matrix in at most four elements. These four elements are of the form

$$p_{ii} = p_{jj} = \cos\theta \quad \text{and} \quad p_{ij} = -p_{ji} = \sin\theta$$

for some θ and some $i \neq j$.

For any rotation matrix P, the matrix AP differs from A only in the ith and jth columns and the matrix PA differs from A only in the ith and jth rows. For any $i \neq j$, the angle θ can be chosen so that the product PA has a zero entry for $(PA)_{ij}$.

The factorization of $A^{(1)}$ into $A^{(1)} = Q^{(1)}R^{(1)}$ uses a product of $n-1$ rotation matrices of this type to construct

$$R^{(1)} = P_n P_{n-1} \ldots P_2 A^{(1)}.$$

We first choose the rotation matrix P_2 so that the matrix

$$A_2^{(1)} = P_2 A^{(1)}$$

has a zero in the $(2, 1)$ position; that is, in the second row and first column. Since the multiplication $P_2 A^{(1)}$ affects both rows 1 and 2 of $A^{(1)}$, the new matrix does not necessarily retain zero entries in positions $(1, 3)$, $(1, 4)$, ..., and $(1, n)$. However, $A^{(1)}$ is tridiagonal, so the $(1, 4)$, ..., $(1, n)$ entries of $A_2^{(1)}$ are zero. Only the $(1, 3)$-entry, the one in the first row and third column, can become nonzero.

In general, the matrix P_k is chosen so that the $(k, k-1)$-entry in $A_k^{(1)} = P_k A_{k-1}^{(1)}$ is zero, which results in the $(k-1, k+1)$-entry becoming nonzero. The matrix $A_k^{(1)}$ has the form

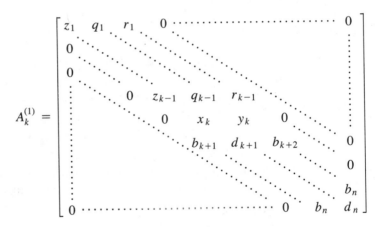

and P_{k+1} has the form

$$P_{k+1} = \left[\begin{array}{c|cc|c} I_{k-1} & & O & O \\ \hline & c_{k+1} & s_{k+1} & \\ O & & & O \\ & -s_{k+1} & c_{k+1} & \\ \hline O & & O & I_{n-k-1} \end{array} \right], \quad \leftarrow \text{row } k$$

$$\underset{\text{column } k}{\uparrow}$$

where O denotes the appropriately dimensioned matrix with all zero entries.

The constants $c_{k+1} = \cos\theta_{k+1}$ and $s_{k+1} = \sin\theta_{k+1}$ in P_{k+1} are chosen so that the $(k+1, k)$-entry in $A_{k+1}^{(1)}$ is zero; that is,

$$s_{k+1} x_k - c_{k+1} b_{k+1} = 0.$$

Since $c_{k+1}^2 + s_{k+1}^2 = 1$, the solution to this equation is

$$s_{k+1} = \frac{b_{k+1}}{\sqrt{b_{k+1}^2 + x_k^2}} \qquad \text{and} \qquad c_{k+1} = \frac{x_k}{\sqrt{b_{k+1}^2 + x_k^2}},$$

and $A^{(1)}_{k+1}$ has the form

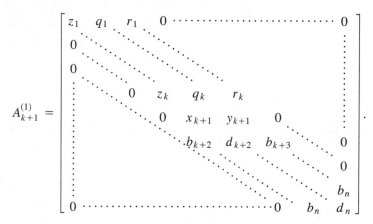

$$A^{(1)}_{k+1} = \begin{bmatrix} z_1 & q_1 & r_1 & 0 & \cdots & & & & & 0 \\ 0 & & & & & & & & & \\ 0 & & & & & & & & & \vdots \\ & & 0 & z_k & q_k & r_k & & & & \\ & & & 0 & x_{k+1} & y_{k+1} & 0 & & & \\ & & & & b_{k+2} & d_{k+2} & b_{k+3} & & 0 & \\ & & & & & & & & & 0 \\ & & & & & & & & & b_n \\ 0 & \cdots & & & & & 0 & & b_n & d_n \end{bmatrix}.$$

Proceeding with this construction in the sequence $P_2, \ldots, P_n$ produces the upper-triangular matrix

$$R^{(1)} \equiv A^{(1)}_n = \begin{bmatrix} z_1 & q_1 & r_1 & 0 & \cdots & 0 \\ 0 & & & & & \vdots \\ & & & & & 0 \\ \vdots & & & & & \\ & & & & & r_{n-2} \\ & & & & z_{n-1} & q_{n-1} \\ 0 & \cdots & & & 0 & x_n \end{bmatrix}.$$

The orthogonal matrix $Q^{(1)}$ is defined as $Q^{(1)} = P^t_2 P^t_3 \ldots P^t_n$, so

$$A^{(2)} = R^{(1)} Q^{(1)} = R^{(1)} P^t_2 P^t_3 \ldots P^t_{n-1} = P_n P_{n-1} \ldots P_2 A P^t_2 P^t_3 \ldots P^t_n.$$

The matrix $A^{(2)}$ is tridiagonal, and, in general, the entries off the diagonal will be smaller in magnitude than the corresponding entries in $A^{(1)}$. The process is repeated to construct $A^{(3)}$, $A^{(4)}$, and so on.

If the eigenvalues of A have the form $|\lambda_1| > |\lambda_2| > \cdots > |\lambda_n|$, then the QR method converges; that is, the matrices tend to a diagonal matrix with the eigenvalues on the diagonal. The rate of convergence depends on the ratio $\left| \dfrac{\lambda_{j+1}}{\lambda_j} \right|$, so the convergence might be slow if this ratio is close to 1.

If the eigenvalues of A have distinct moduli, the rate of convergence of the entry $b^{(i+1)}_{j+1}$ to zero in the matrix $A^{(i+1)}$ is

$$O\left(\left| \frac{\lambda_{i+1}}{\lambda_j} \right|^{i+1} \right),$$

where the eigenvalues are ordered by $|\lambda_1| > |\lambda_2| > \cdots > |\lambda_n|$. The rate of convergence of $b^{(i+1)}_{j+1}$ to zero determines the rate at which the entry $a^{(i+1)}_j$ converges to the jth eigenvalue λ_j.

To accelerate the convergence, a shifting technique is employed similar to that used with the Inverse Power method. A constant s is selected close to an eigenvalue of A. This modifies the factorization to a question of choosing $Q^{(i)}$ and $R^{(i)}$ so that

$$A^{(i)} - sI = Q^{(i)}R^{(i)};$$

correspondingly, the matrix $A^{(i+1)}$ is defined to be

$$A^{(i+1)} = R^{(i)}Q^{(i)} + sI.$$

With this modification, the rate of convergence of $b_{j+1}^{(i+1)}$ to zero becomes

$$0\left(\left|\frac{\lambda_{j+1} - s}{\lambda_j - s}\right|^{i+1}\right),$$

which can result in a significant improvement over the original rate of convergence of $a_j^{(i+1)}$ to λ_j if s is close to λ_{j+1} but not close to λ_j.

The shift s is computed at each step as the eigenvalue closest to $a_n^{(i)}$ of the 2 by 2 matrix

$$\begin{bmatrix} a_{n-1}^{(i)} & b_n^{(i)} \\ b_n^{(i)} & a_n^{(i)} \end{bmatrix}.$$

With this shift, $b_n^{(i+1)}$ converges to zero and $a_n^{(i+1)}$ converges to the eigenvalue λ_n. If $b_j^{(i+1)}$ converges to zero for some $j \neq n$, the splitting technique is employed. The shifts are accumulated and added to the approximation after convergence is obtained. The program QRSYMT96 performs the QR method for symmetric tridiagonal matrices in this way.

EXAMPLE 1   Let

$$A = \begin{bmatrix} 3 & 1 & 0 \\ 1 & 3 & 1 \\ 0 & 1 & 3 \end{bmatrix} = \begin{bmatrix} a_1^{(1)} & b_2^{(1)} & 0 \\ b_2^{(1)} & a_2^{(1)} & b_3^{(1)} \\ 0 & b_3^{(1)} & a_3^{(1)} \end{bmatrix}.$$

To find the acceleration parameter for shifting requires finding the eigenvalues of

$$\begin{bmatrix} a_2^{(1)} & b_3^{(1)} \\ b_3^{(1)} & a_3^{(1)} \end{bmatrix} = \begin{bmatrix} 3 & 1 \\ 1 & 3 \end{bmatrix},$$

which are $\mu_1 = 4$ and $\mu_2 = 2$. The choice of eigenvalue closest to $a_3^{(1)} = 3$ is arbitrary, and we choose $\mu_2 = 2$ and shift by this amount. Then $s_1 = 2$ and

$$\begin{bmatrix} d_1 & b_2^{(1)} & 0 \\ b_2^{(1)} & d_2 & b_3^{(1)} \\ 0 & b_3^{(1)} & d_3 \end{bmatrix} = \begin{bmatrix} 1 & 1 & 0 \\ 1 & 1 & 1 \\ 0 & 1 & 1 \end{bmatrix}.$$

Continuing the computation gives

$$x_1 = 1, \qquad y_1 = 1, \qquad z_1 = \sqrt{2}, \qquad c_2 = \frac{\sqrt{2}}{2}, \qquad s_2 = \frac{\sqrt{2}}{2},$$

$$q_1 = \sqrt{2}, \qquad x_2 = 0, \qquad r_1 = \frac{\sqrt{2}}{2}, \qquad \text{and} \qquad y_2 = \frac{\sqrt{2}}{2},$$

so

$$A_2^{(1)} = \begin{bmatrix} \sqrt{2} & \sqrt{2} & \frac{\sqrt{2}}{2} \\ 0 & 0 & \sqrt{2} \\ 0 & 1 & 1 \end{bmatrix}.$$

Further,

$$z_2 = 1, \quad c_3 = 0, \quad s_3 = 1, \quad q_2 = 1, \quad \text{and} \quad x_3 = -\frac{\sqrt{2}}{2},$$

so

$$R^{(1)} = A_3^{(1)} = \begin{bmatrix} \sqrt{2} & \sqrt{2} & \frac{\sqrt{2}}{2} \\ 0 & 1 & 1 \\ 0 & 0 & -\frac{\sqrt{2}}{2} \end{bmatrix}.$$

To compute $A^{(2)}$, we have

$$z_3 = -\frac{\sqrt{2}}{2}, \quad a_1^{(2)} = 2, \quad b_2^{(2)} = \frac{\sqrt{2}}{2}, \quad a_2^{(2)} = 1, \quad b_3^{(2)} = -\frac{\sqrt{2}}{2}, \quad \text{and} \quad a_3^{(2)} = 0,$$

so

$$A^{(2)} = R^{(1)}Q^{(1)} = \begin{bmatrix} 2 & \frac{\sqrt{2}}{2} & 0 \\ \frac{\sqrt{2}}{2} & 1 & -\frac{\sqrt{2}}{2} \\ 0 & -\frac{\sqrt{2}}{2} & 0 \end{bmatrix}.$$

One iteration of the QR method is complete. Since neither $b_2^{(2)} = \sqrt{2}/2$ nor $b_3^{(2)} = -\sqrt{2}/2$ is small, another iteration of the QR method will be performed. This iteration provides a shift of $s_2 = (1 - \sqrt{3})/2$ and the matrix

$$A^{(3)} = \begin{bmatrix} 2.6720277 & 0.37597448 & 0 \\ 0.37597448 & 1.4736080 & 0.030396964 \\ 0 & 0.030396964 & -0.047559530 \end{bmatrix}.$$

If $b_3^{(3)} = 0.030396964$ is sufficiently small, then the approximation to the eigenvalue λ_3 is 1.5864151, the sum of $a_3^{(3)}$ and the shifts $s_1 + s_2 = 2 + \frac{1}{2}(1 - \sqrt{3})$. Deleting the third row and column gives

$$A^{(3)} = \begin{bmatrix} 2.6720277 & 0.37597448 \\ 0.37597448 & 1.4736080 \end{bmatrix},$$

which has eigenvalues $\mu_1 = 2.7802140$ and $\mu_2 = 1.3654218$. Adding the shifts gives the approximations

$$\lambda_1 \approx 4.4141886 \quad \text{and} \quad \lambda_2 \approx 2.9993964.$$

Since the actual eigenvalues of the matrix A are 4.41420, 3.00000, and 1.58579, the QR method gave four significant digits of accuracy in only two iterations.

EXERCISE SET 9.5

1. Apply two iterations of the QR method to the following matrices:

(a) $\begin{bmatrix} 2 & -1 & 0 \\ -1 & 2 & -1 \\ 0 & -1 & 2 \end{bmatrix}$

(b) $\begin{bmatrix} 3 & 1 & 0 \\ 1 & 4 & 2 \\ 0 & 2 & 1 \end{bmatrix}$

(c) $\begin{bmatrix} 2 & 1 & 0 \\ 1 & 2 & 1 \\ 0 & 1 & 2 \end{bmatrix}$

(d) $\begin{bmatrix} 4 & -1 & 0 \\ -1 & 3 & -1 \\ 0 & -1 & 2 \end{bmatrix}$

(e) $\begin{bmatrix} 1 & 1 & 0 & 0 \\ 1 & 2 & -1 & 0 \\ 0 & -1 & 3 & 1 \\ 0 & 0 & 1 & 4 \end{bmatrix}$

(f) $\begin{bmatrix} -2 & 1 & 0 & 0 \\ 1 & -3 & -1 & 0 \\ 0 & -1 & 1 & 1 \\ 0 & 0 & 1 & 3 \end{bmatrix}$

(g) $\begin{bmatrix} 0.5 & 0.25 & 0 & 0 \\ 0.25 & 0.8 & 0.4 & 0 \\ 0 & 0.4 & 0.6 & 0.1 \\ 0 & 0 & 0.1 & 1 \end{bmatrix}$

(h) $\begin{bmatrix} 4 & -3 & 0 & 0 \\ -3 & \frac{10}{3} & -\frac{5}{3} & 0 \\ 0 & -\frac{5}{3} & -\frac{99}{75} & \frac{68}{75} \\ 0 & 0 & \frac{68}{75} & \frac{149}{75} \end{bmatrix}$

2. Use the QR method to determine, to within 10^{-5}, all the eigenvalues of the following matrices:

(a) $\begin{bmatrix} 2 & -1 & 0 \\ -1 & -1 & -2 \\ 0 & -2 & 3 \end{bmatrix}$

(b) $\begin{bmatrix} 3 & 1 & 0 \\ 1 & 4 & 2 \\ 0 & 2 & 3 \end{bmatrix}$

(c) $\begin{bmatrix} 4 & 2 & 0 & 0 & 0 \\ 2 & 4 & 2 & 0 & 0 \\ 0 & 2 & 4 & 2 & 0 \\ 0 & 0 & 2 & 4 & 2 \\ 0 & 0 & 0 & 2 & 4 \end{bmatrix}$

(d) $\begin{bmatrix} 5 & -1 & 0 & 0 & 0 \\ -1 & 4.5 & 0.2 & 0 & 0 \\ 0 & 0.2 & 1 & -0.4 & 0 \\ 0 & 0 & -0.4 & 3 & 1 \\ 0 & 0 & 0 & 1 & 3 \end{bmatrix}$

3. Use the QR method to determine, to within 10^{-5}, all the eigenvalues for the matrices given in Exercise 1.

4. Use the Inverse Power method to determine, to within 10^{-5}, the eigenvectors of the matrices in Exercise 1.

9.6 Survey of Methods and Software

This chapter discussed the approximation of eigenvalues and eigenvectors. The Gerschgorin circles give a rough approximation to the location of the eigenvalues of a matrix. The Power method can be used to find the dominant eigenvalue and an associated eigenvector for an arbitrary matrix A. If A is symmetric, the Symmetric Power method gives faster convergence to the dominant eigenvalue and an associated eigenvector. The Inverse Power method finds the eigenvalue closest to a given value and an associated eigenvector. This method is often used to refine an approximate eigenvalue and to compute an eigenvector once an eigenvalue has been found by some other method.

Deflation methods, such as Wielandt deflation, obtain other eigenvalues once the dominant eigenvalue is known. These methods are used if only a few eigenvalues are required and should be applied with caution, since they can be unstable with respect to round-off error.

Methods based on similarity transformations, such as Householder's method, are used to convert a symmetric matrix into a similar tridiagonal matrix. Techniques such as the QR method can then be applied to a tridiagonal matrix to obtain approximations to all the eigenvalues. The associated eigenvectors can be found by using an iterative method, such as the Inverse Power method, or by applying the transformations to eigenvectors obtained directly from the QR method.

The subroutines in the IMSL and NAG libraries are based on the subroutines contained in the EISPACK and LAPACK packages, which were discussed in Section 1.5. In general, the subroutines transform a matrix into the appropriate form for the QR method or one of its modifications, such as the QL method. The subroutines approximate all the eigenvalues and can approximate an associated eigenvector for each eigenvalue. There are special routines that find all the eigenvalues within an interval or region or find only the largest or smallest eigenvalue. Subroutines are also available to approximate the accuracy of the eigenvalue and the sensitivity of the process to round-off error.

The IMSL subroutine EVLRG produces all eigenvalues of A in increasing order of magnitude. This subroutine first balances the matrix A using a version of the EISPACK routine BALANC, so that the sums of the magnitudes of the entries in each row and in each column are approximately the same. This leads to greater stability in the ensuing computations. EVLRG next performs orthogonal similarity transformations, such as in Householder's method, to reduce A to a similar upper Hessenberg matrix. An upper Hessenberg matrix has zeros below the lower subdiagonal—that is, in positions (i, j) for $j + 2 \le i$. It is symmetric and tridiagonal precisely when the original matrix is symmetric. This portion is similar to the EISPACK subroutine ORTHES. Finally, the shifted QR method is performed to obtain all the eigenvalues. This part is similar to the subroutine HQR in EISPACK. The IMSL subroutine EVCRG is the same as EVRLG except that corresponding eigenvectors are computed.

The subroutine EVLSF computes the eigenvalues of the real symmetric matrix A. The matrix A is first reduced to tridiagonal form using a modification of the EISPACK routine TRED2. Then the eigenvalues are computed using a modification of the EISPACK routine IMTQL2, which is a variation of the QR method called the implicit QL method. The subroutine EVCSF is the same as EVLSF, except that the eigenvectors are also calculated. Finally, EVLRH and EVCRH compute all eigenvalues of the upper Hessenberg matrix A and, additionally, EVCRH computes the eigenvectors. These subroutines are based on the subroutines HQR and HQR2, respectively, in EISPACK.

The NAG library has similar subroutines based on the EISPACK routines. The subroutine F02AFF computes the eigenvalues of the real matrix A. The subroutine F02AGF is the same as F02AFF except the eigenvectors are also calculated. The matrix A is first balanced and then reduced to upper Hessenberg form for the QR method. The subroutine F02AAF is used when the matrix is real and symmetric. The eigenvalues, which are real, are computed in increasing order. If the eigenvectors are also needed, the subroutine F02ABF can be applied. In either case, the matrix A is reduced to tridiagonal form using Householder's method and the eigenvalues are then computed using the QL method. The subroutine F01AGF implements Householder's method directly for symmetric matrices to produce a similar tridiagonal symmetric matrix. Routines are also available in the NAG library for directly balancing real matrices, recovering eigenvectors if a matrix was first balanced, and performing other operations on special types of matrices.

Since MATLAB incorporates the subroutines of EISPACK, MATLAB can calculate eigenvalues. For example, if

$$A = [3 \ 1 \ 0; \ 1 \ 4 \ 2; \ 0 \ 2 \ 3],$$

then $eig(A)$ uses the QR method to compute the eigenvalues of A as 3.000000, 1.208712, and 5.791288.

Solutions of Systems
of Nonlinear Equations

10.1 Introduction

A large part of the material in this book has involved the solution of systems of equations. Even so, to this point, the methods have been appropriate only for systems of *linear* equations, equations in the variables $x_1, x_2, \ldots, x_n$ of the form

$$a_{i1}x_1 + a_{i2}x_2 + \cdots + a_{in}x_n = b_i,$$

for $i = 1, 2, \ldots, n$. If you have wondered why we have not considered more general systems of equations, the reason is simple. It is much harder to approximate the solutions to a system of general, or *nonlinear*, equations.

Solving a system of nonlinear equations is a problem that is avoided when possible, customarily by approximating the nonlinear system by a system of linear equations. When this is unsatisfactory, the problem must be tackled directly. The most straightforward method of approach is to adapt the methods from Chapter 2 that approximate the solutions of a single nonlinear equation in one variable to apply when the single-variable problem is replaced by a vector problem that incorporates all the variables.

The principle tool in Chapter 2 was Newton's method, a technique that is generally quadratically convergent. This is the first technique we modify to solve systems of nonlinear equations. Newton's method as modified for systems of equations is quite costly to apply, so in Section 10.3 we describe how a modified Secant method can be used to obtain approximations more easily, although with a loss of the extremely rapid convergence that Newton's method provides.

Section 10.4 describes the method of Steepest Descent. This technique is only linearly convergent, but it does not require the precise starting approximations needed for more rapidly converging techniques. It is often used to find a good initial approximation for Newton's method or one of its modifications, just as the Bisection technique is applied for this purpose in the case of a single nonlinear equation.

A system of nonlinear equations has the form

$$f_1(x_1, x_2, \ldots, x_n) = 0,$$
$$f_2(x_1, x_2, \ldots, x_n) = 0,$$
$$\vdots$$
$$f_n(x_1, x_2, \ldots, x_n) = 0,$$

where each function f_i can be thought of as mapping a vector $\mathbf{x} = (x_1, x_2, \ldots, x_n)^t$ of the n-dimensional space $\mathbb{R}^n$ into the real line $\mathbb{R}$. A geometric representation of a nonlinear system when $n = 2$ is given in Figure 10.1.

Figure 10.1

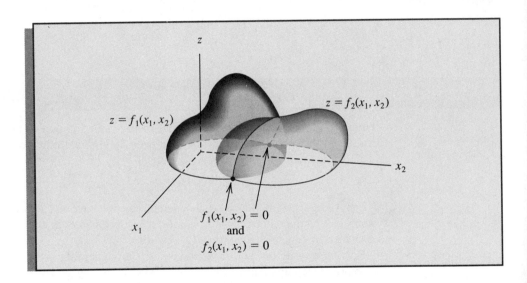

A general system of n nonlinear equations in n unknowns can alternatively be represented by defining a function $\mathbf{F}$, mapping $\mathbb{R}^n$ into $\mathbb{R}^n$, by

$$\mathbf{F}(x_1, x_2, \ldots, x_n) = (f_1(x_1, x_2, \ldots, x_n), f_2(x_1, x_2, \ldots, x_n), \ldots, f_n(x_1, x_2, \ldots, x_n))^t.$$

If vector notation is used to represent the variables $x_1, x_2, \ldots, x_n$, the nonlinear system assumes the form

$$\mathbf{F}(\mathbf{x}) = \mathbf{0}.$$

The functions $f_1, f_2, \ldots, f_n$ are the **coordinate functions** of $\mathbf{F}$.

EXAMPLE 1   The three by three nonlinear system

$$2x_1 - \cos(x_2 x_3) - \tfrac{1}{2} = 0,$$
$$x_1^2 - 81(x_2 + 0.1)^2 + \sin x_3 + 1.06 = 0,$$
$$e^{-x_1 x_2} + 20x_3 + \frac{10\pi - 3}{3} = 0$$

can be placed in the form $\mathbf{F}(\mathbf{x}) = \mathbf{0}$ by first defining the three coordinate functions $f_1, f_2,$ and f_3 from $\mathbb{R}^3$ to $\mathbb{R}$ as

$$f_1(x_1, x_2, x_3) = 3x_1 - \cos(x_2 x_3) - \tfrac{1}{2},$$

$$f_2(x_1, x_2, x_3) = x_1^2 - 81(x_2 + 0.1)^2 + \sin x_3 + 1.06,$$

$$f_3(x_1, x_2, x_3) = e^{-x_1 x_2} + 20 x_3 + \frac{10\pi - 3}{3},$$

and then defining $\mathbf{F}$ from $\mathbb{R}^3 \to \mathbb{R}^3$ by

$$\mathbf{F}(\mathbf{x}) = \mathbf{F}(x_1, x_2, x_3)$$
$$= (f_1(x_1, x_2, x_3), f_2(x_1, x_2, x_3), f_3(x_1, x_2, x_3))^t$$
$$= \left(3x_1 - \cos(x_1 x_2) - \tfrac{1}{2}, x_1^2 - 81(x_2 + 0.1)^2 + \sin x_3 + 1.06,\right.$$
$$\left. e^{-x_1 x_2} + 20 x_3 + \frac{10\pi - 3}{3}\right)^t.$$

■ ■ ■

Before discussing the solution of a system of nonlinear equations, we need some results concerning continuity and differentiability of functions from $\mathbb{R}^n$ into $\mathbb{R}$ and $\mathbb{R}^n$ into $\mathbb{R}^n$. These results parallel those given in Section 1.1 for a function from $\mathbb{R}$ into $\mathbb{R}$.

Let f be a function defined on a set $D \subset \mathbb{R}^n$ and mapping $\mathbb{R}^n$ into $\mathbb{R}$. The function f has the **limit** L at $\mathbf{x}_0$, written

$$\lim_{\mathbf{x} \to \mathbf{x}_0} f(\mathbf{x}) = L,$$

if, given any number $\epsilon > 0$, a number $\delta > 0$ exists with the property that

$$|f(\mathbf{x}) - L| < \epsilon \qquad \text{whenever} \qquad \mathbf{x} \in D \qquad \text{and} \qquad 0 < \|\mathbf{x} - \mathbf{x}_0\| < \delta.$$

Any convenient norm can be used to satisfy the condition in this definition. The specific value of δ depends on the norm chosen, but the existence of δ is independent of the norm.

The function f is **continuous** at $\mathbf{x}_0 \in D$ provided $\lim_{\mathbf{x} \to \mathbf{x}_0} f(\mathbf{x})$ exists and is $f(\mathbf{x}_0)$. In addition, f is **continuous on a set** D provided f is continuous at every point of D. This is expressed by writing $f \in C(D)$.

We define the limit and continuity concepts for functions from $\mathbb{R}^n$ into $\mathbb{R}^n$ by considering the coordinate functions from $\mathbb{R}^n$ into $\mathbb{R}$.

Let $\mathbf{F}$ be a function from $D \subset \mathbb{R}^n$ into $\mathbb{R}^n$ and suppose $\mathbf{F}$ has the representation

$$\mathbf{F}(\mathbf{x}) = (f_1(\mathbf{x}), f_2(\mathbf{x}), \ldots, f_n(\mathbf{x}))^t.$$

We define

$$\lim_{\mathbf{x} \to \mathbf{x}_0} \mathbf{F}(\mathbf{x}) = \mathbf{L} = (L_1, L_2, \ldots, L_n)^t$$

precisely when $\lim_{\mathbf{x} \to \mathbf{x}_0} f_i(\mathbf{x}) = L_i$ for each $i = 1, 2, \ldots, n$.

The function $\mathbf{F}$ is **continuous** at $\mathbf{x}_0 \in D$ provided $\lim_{\mathbf{x} \to \mathbf{x}_0} \mathbf{F}(\mathbf{x})$ exists and is $\mathbf{F}(\mathbf{x}_0)$. In addition, $\mathbf{F}$ is **continuous on the set** D if $\mathbf{F}$ is continuous at each $\mathbf{x}$ in D.

10.2 Newton's Method for Systems

Newton's method for approximating the solution p to the single nonlinear equation

$$f(x) = 0$$

requires an initial approximation p_0 to p and generates a sequence defined by

$$p_k = p_{k-1} - \frac{f(p_{k-1})}{f'(p_{k-1})}, \qquad \text{for } k \geq 1.$$

To modify this technique to find the vector solution **p** to the vector equation

$$\mathbf{F(x)} = \mathbf{0},$$

we first need to determine an initial approximation vector $\mathbf{p}^{(0)}$. We must then decide how to modify the single-variable Newton's method to a vector function method that will have the same convergence properties but not require division, since this operation is not defined for vectors. We must also replace the derivative of f in the single-variable version of Newton's method with something that is appropriate for the vector function **F**.

The derivative f' of the single-variable function f describes how the values of the function change relative to a change in its independent variable. The vector function **F** has n different variables $x_1, x_2, \ldots, x_n$ and n different component functions $f_1, f_2, \ldots, f_n$, each of which can change as any one of the variables changes. The appropriate derivative modification from the single-variable Newton's method to the vector form must involve all these n^2 possible changes. The natural way to represent n^2 items is by an $n \times n$ matrix. Since each change in a component function f_i with respect to the change in the variable x_j is described by the partial derivative

$$\frac{\partial f_i}{\partial x_j},$$

the $n \times n$ matrix that replaces the derivative that occurs in the single-variable case is

$$J(\mathbf{x}) = \begin{bmatrix} \dfrac{\partial f_1(\mathbf{x})}{\partial x_1} & \dfrac{\partial f_1(\mathbf{x})}{\partial x_2} & \cdots & \dfrac{\partial f_1(\mathbf{x})}{\partial x_n} \\[2mm] \dfrac{\partial f_2(\mathbf{x})}{\partial x_1} & \dfrac{\partial f_2(\mathbf{x})}{\partial x_2} & \cdots & \dfrac{\partial f_2(\mathbf{x})}{\partial x_2} \\[2mm] \vdots & \vdots & & \vdots \\[2mm] \dfrac{\partial f_n(\mathbf{x})}{\partial x_1} & \dfrac{\partial f_n(\mathbf{x})}{\partial x_2} & \cdots & \dfrac{\partial f_n(\mathbf{x})}{\partial x_n} \end{bmatrix}.$$

The matrix $J(\mathbf{x})$ is called the **Jacobian** matrix and has a number of applications in analysis. It might, in particular, be familiar due to its application in the multiple integration of a function of several variables over a region which requires a change of variables to be performed.

Newton's method for systems replaces the derivative in the single variable case with the $n \times n$ Jacobian matrix in the vector situation and substitutes the division by the derivative with the inversion of the Jacobian matrix. As a consequence, Newton's method for finding the solution $\mathbf{p}$ to the nonlinear system of equations represented by the vector equation $\mathbf{F(x)} = \mathbf{0}$ has the form

$$\mathbf{p}^{(k)} = \mathbf{p}^{(k-1)} - [J(\mathbf{p}^{(k-1)})]^{-1}\mathbf{F}(\mathbf{p}^{(k-1)}), \qquad \text{for } k \geq 1,$$

given the initial approximation $\mathbf{p}^{(0)}$ to the solution $\mathbf{p}$.

A weakness in Newton's method for systems arises from the necessity of inverting the matrix $J(\mathbf{p}^{(k-1)})$ at each iteration. In practice, explicit computation of the inverse of $J(\mathbf{p}^{(k-1)})$ is avoided by performing the operation in a two-step manner. First, a vector $\mathbf{y}^{(k-1)}$ is found that will satisfy $J(\mathbf{p}^{(k-1)})\mathbf{y}^{(k-1)} = -\mathbf{F}(\mathbf{p}^{(k-1)})$. After this has been accomplished, the new approximation, $\mathbf{p}^{(k)}$, is obtained by adding $\mathbf{y}^{(k-1)}$ to $\mathbf{p}^{(k-1)}$. The general implementation of Newton's method for systems of nonlinear equations can be performed using the program NWTSY101.

EXAMPLE 1 The nonlinear system

$$3x_1 - \cos(x_2 x_3) - \tfrac{1}{2} = 0,$$

$$x_1^2 - 81(x_2 + 0.1)^2 + \sin x_3 + 1.06 = 0,$$

$$e^{-x_1 x_2} + 20x_3 + \frac{10\pi - 3}{3} = 0$$

has an approximate solution at $(0.5, 0, -0.52359877)^t$. Newton's method will be used to obtain this approximation when the initial approximation is $\mathbf{p}^{(0)} = (0.1, 0.1, -0.1)^t$.

The Jacobian matrix $J(\mathbf{x})$ for this system is

$$J(x_1, x_2, x_3) = \begin{bmatrix} 3 & x_3 \sin x_2 x_3 & x_2 \sin x_2 x_3 \\ 2x_1 & -162(x_2 + 0.1) & \cos x_3 \\ -x_2 e^{-x_1 x_2} & -x_1 e^{-x_1 x_2} & 20 \end{bmatrix}$$

and

$$\begin{bmatrix} p_1^{(k)} \\ p_2^{(k)} \\ p_3^{(k)} \end{bmatrix} = \begin{bmatrix} p_1^{(k-1)} \\ p_2^{(k-1)} \\ p_3^{(k-1)} \end{bmatrix} + \begin{bmatrix} y_1^{(k-1)} \\ y_2^{(k-1)} \\ y_3^{(k-1)} \end{bmatrix},$$

where

$$\begin{bmatrix} y_1^{(k-1)} \\ y_2^{(k-1)} \\ y_3^{(k-1)} \end{bmatrix} = -(J(p_1^{(k-1)}, p_2^{(k-1)}, p_3^{(k-1)}))^{-1}\mathbf{F}(p_1^{(k-1)}, p_2^{(k-1)}, p_3^{(k-1)}).$$

Thus, at the kth step, the linear system $J(\mathbf{p}^{(k-1)})\mathbf{y}^{(k-1)} = -\mathbf{F}(\mathbf{p}^{(k-1)})$ must be solved, where

$$J(\mathbf{p}^{(k-1)}) = \begin{bmatrix} 3 & p_3^{(k-1)} \sin p_2^{(k-1)} p_3^{(k-1)} & p_2^{(k-1)} \sin p_2^{(k-1)} p_3^{(k-1)} \\ 2p_1^{(k-1)} & -162(p_2^{(k-1)} + 0.1) & \cos p_3^{(k-1)} \\ -p_2^{(k-1)} e^{-p_1^{(k-1)} p_2^{(k-1)}} & -p_1^{(k-1)} e^{-p_1^{(k-1)} p_2^{(k-1)}} & 20 \end{bmatrix}$$

and

$$\mathbf{F}(\mathbf{p}^{(k-1)}) = \begin{bmatrix} 3p_1^{(k-1)} - \cos p_2^{(k-1)} p_3^{(k-1)} - \dfrac{1}{2} \\ (p_1^{(k-1)})^2 - 81(p_2^{(k-1)} + 0.1)^2 + \sin p_3^{(k-1)} + 1.06 \\ e^{-p_1^{(k-1)} p_2^{(k-1)}} + 20 p_3^{(k-1)} + \dfrac{10\pi - 3}{3} \end{bmatrix}.$$

The results using this iterative procedure are shown in Table 10.1.   ■ ■ ■

Table 10.1

k	$p_1^{(k)}$	$p_2^{(k)}$	$p_3^{(k)}$	$\|\mathbf{p}^{(k)} - \mathbf{p}^{(k-1)}\|_\infty$
0	0.10000000	0.10000000	-0.10000000	
1	0.50003702	0.01946686	-0.52152047	0.422
2	0.50004593	0.00158859	-0.52355711	1.79×10^{-2}
3	0.50000034	0.00001244	-0.52359845	1.58×10^{-3}
4	0.50000000	0.00000000	-0.52359877	1.24×10^{-5}
5	0.50000000	0.00000000	-0.52359877	0

The previous example illustrates that Newton's method can converge very rapidly once an approximation is obtained that is near the true solution. However, it is not always easy to determine starting values that will lead to a solution, and the method is computationally expensive. In the next section, we consider a method for overcoming the latter weakness. Good starting values can usually be found by the method discussed in Section 10.4.

EXERCISE SET 10.2

1. The nonlinear system

$$x_1(1 - x_1) + 4x_2 = 12$$
$$(x_1 - 2)^2 + (2x_2 - 3)^2 = 25$$

has two solutions.
(a) Approximate the solutions graphically.
(b) Use the approximations from part (a) as initial approximations for Newton's method and calculate solutions to within 10^{-5} in the l_∞ norm.

2. The nonlinear system

$$5x_1^2 - x_2^2 = 0$$
$$x_2 - 0.25(\sin x_1 + \cos x_2) = 0$$

has a solution near $(\frac{1}{4}, \frac{1}{4})^t$. Use Newton's method to find a solution that is accurate to within 10^{-5} in the l_∞ norm.

3. Use Newton's method to find a solution to the following nonlinear systems. Iterate until successive approximations agree to within 10^{-6} in the l_∞ norm.

(a) $4x_1^2 - 20x_1 + \frac{1}{4}x_2^2 + 8 = 0$
$\frac{1}{2}x_1x_2^2 + 2x_1 - 5x_2 + 8 = 0$

(b) $3x_1^2 - x_2^2 = 0$
$3x_1x_2^2 - x_1^3 - 1 = 0$

(c) $\ln(x_1^2 + x_2^2) - \sin(x_1x_2) = \ln 2 + \ln \pi$
$e^{x_1 - x_2} + \cos(x_1x_2) = 0$

(d) $\sin(4\pi x_1 x_2) - 2x_2 - x_1 = 0$
$\left(\frac{4\pi - 1}{4\pi}\right)(e^{2x_1} - e) + 4ex_2^2 - 2ex_1 = 0$

4. Use Newton's method to find a solution to the following nonlinear systems in the given domain. Iterate until successive approximations agree to within 10^{-6} in the l_∞ norm.

(a) $15x_1 + x_2^2 - 4x_3 = 13$
$x_1^2 + 10x_2 - x_3 = 11$
$x_2^3 - 25x_3 = -22$

$D = \{(x_1, x_2, x_3)^t \mid 0 \le x_1 \le 2, 0 \le x_2 \le 2, 0 \le x_3 \le 2\}$

(b) $x_1 + \cos(x_1x_2x_3) - 1 = 0$
$(1 - x_1)^{\frac{1}{4}} + x_2 + 0.05x_3^2 - 0.15x_3 - 1 = 0$
$-x_1^2 - 0.1x_2^2 + 0.01x_2 + x_3 - 1 = 0$

$D = \{(x_1, x_2, x_3)^t \mid 0 \le x_i \le 1.5, \text{ for } i = 1, 2, 3\}$

(c) $x_1^3 + x_1^2 x_2 - x_1 x_3 + 6 = 0$
$e^{x_1} + e^{x_2} - x_3 = 0$
$x_2^2 - 2x_1 x_3 = 4$

$D = \{(x_1, x_2, x_3)^t \mid -2 \le x_1, x_2 \le -1, 0 \le x_3 \le 1\}$

(d) $\sin x_1 + \sin(x_1x_2) + \sin(x_1x_3) = 0$
$\sin x_2 + \sin(x_2x_3) = 0$
$\sin(x_1x_2x_3) = 0$

$D = \{(x_1, x_2, x_3)^t \mid -2 \le x_1, x_2 \le -1, -6 \le x_3 \le -4\}$

5. Use Newton's method to solve the following nonlinear systems. Iterate until successive approximations agree to within 10^{-5} in the l_∞ norm.

(a)
$$6x_1 - 2\cos(x_2 x_3) - 1 = 0$$
$$9x_2 + \sqrt{x_1^2 + \sin x_3 + 1.06} + 0.9 = 0$$
$$60x_3 + 3e^{-x_1 x_2} + 10\pi - 3 = 0$$

(b)
$$\cos x_1 + \cos x_2 + \cos x_3 = 1$$
$$\cos(x_1 x_2) + \cos(x_2 x_3) + \cos(x_1 x_3) = 0$$
$$\cos(x_1 x_2 x_3) = -1$$

(c)
$$2x_1 + x_2 + x_3 + x_4 - 5 = 0$$
$$x_1 + 2x_2 + x_3 + x_4 - 5 = 0$$
$$x_1 + x_2 + 2x_3 + x_4 - 5 = 0$$
$$x_1 x_2 x_3 x_4 - 1 = 0$$

(Find a solution other than $(1, 1, 1)'$.)

(d)
$$4x_1 - x_2 + x_3 = x_1 x_4$$
$$-x_1 + 3x_2 - 2x_3 = x_2 x_4$$
$$x_1 - 2x_2 + 3x_3 = x_3 x_4$$
$$x_1^2 + x_2^2 + x_3^2 = 1$$

(Find three solutions.)

6. The following nonlinear systems have singular Jacobian matrices at the solution. Can Newton's method still be used? How is the rate of convergence affected?

(a)
$$3x_1 - \cos(x_2 x_3) - 0.5 = 0$$
$$x_1^2 - 625x_2^2 = 0$$
$$e^{-x_1 x_2} + 20x_3 + \frac{10\pi - 3}{3} = 0$$

(b)
$$x_1 - 10x_2 + 9 = 0$$
$$\sqrt{3}(x_3 - x_4) = 0$$
$$(x_2 - 2x_3 + 1)^2 = 0$$
$$\sqrt{2}(x_1 - x_4)^2 = 0$$

7. In Exercise 7 of Section 5.7, we considered the problem of predicting the population of two species that compete for the same food supply. In the problem, we made the assumption that the populations could be predicted by solving the system of equations

$$\frac{dx_1(t)}{dt} = x_1(t)(4 - 0.0003x_1(t) - 0.0004x_2(t))$$

and

$$\frac{dx_2(t)}{dt} = x_2(t)(2 - 0.0002x_1(t) - 0.0001x_2(t)).$$

In this exercise, we would like to consider the problem of determining equilibrium populations of the two species. The mathematical criteria that must be satisfied in order for the populations to be at equilibrium is that, simultaneously,

$$\frac{dx_1(t)}{dt} = 0 \qquad \text{and} \qquad \frac{dx_2(t)}{dt} = 0.$$

This occurs when the first species is extinct and the second species has a population of 20,000 or when the second species is extinct and the first species has a population of 13,333. Can an equilibrium occur in any other situation?

8. The amount of pressure required to sink a large heavy object in a soft homogeneous soil that lies above a hard base soil can be predicted by the amount of pressure required to sink smaller objects in the same soil. Specifically, the amount of pressure p required to sink a circular plate of radius r a distance d in the soft soil, where the hard base soil lies a distance $D > d$ below the surface, can be approximated by an equation of the form

$$p = k_1 e^{k_2 r} + k_3 r,$$

where k_1, k_2, and k_3 are constants, with $k_2 > 0$, depending on d and the consistency of the soil but not on the radius of the plate.

   (a) Find the values of k_1, k_2, and k_3 if we assume that a plate of radius 1 in. requires a pressure of 10 lb/in.^{2} to sink 1 ft in a muddy field, a plate of radius 2 in. requires a pressure of 12 lb/in.^{2} to sink 1 ft, and a plate of radius 3 in. requires a pressure of 15 lb/in.^{2} to sink this distance (assuming that the mud is more than 1 ft deep).

   (b) Use your calculations from part (a) to predict the minimal size of a circular plate that would be required to sustain a load of 500 lb on this field with sinkage of less than 1 ft.

10.3 Quasi–Newton Methods

A significant weakness of Newton's method for solving systems of nonlinear equations lies in the requirement that, at each iteration, a Jacobian matrix be computed and an $n \times n$ linear system solved that involves this matrix. To illustrate the magnitude of this weakness, let us consider the amount of computation associated with one iteration of Newton's method. The Jacobian matrix associated with a system of n nonlinear equations written in the form $\mathbf{F}(\mathbf{x}) = \mathbf{0}$ requires that the n^2 partial derivatives of the n component functions of $\mathbf{F}$ be determined and evaluated. In most situations, the exact evaluation of the partial derivatives is inconvenient and in many applications impossible. This difficulty can be overcome by using finite-difference approximations to the partial derivatives. For example,

$$\frac{\partial f_j}{\partial x_k}(\mathbf{x}^{(i)}) \approx \frac{f_j(\mathbf{x}^{(i)} + h\mathbf{e}_k) - f_j(\mathbf{x}^{(i)})}{h},$$

where h is small in absolute value and $\mathbf{e}_k$ is the vector whose only nonzero entry is a 1 in the kth coordinate. This approximation, however, still requires that at least n^2 scalar functional evaluations be performed to approximate the Jacobian and does not decrease the amount of calculation, in general $O(n^3)$, required for solving the linear system involving this approximate Jacobian. The total computational effort for just one iteration of Newton's method is consequently at least $n^2 + n$ scalar functional evaluations (n^2 for the evaluation of the Jacobian matrix and n for the evaluation of $\mathbf{F}$) together with $O(n^3)$ arithmetic operations to solve the linear system. This amount of computational effort is prohibitive except for relatively small values of n and easily evaluated scalar functions.

In this section, we consider a generalization of the Secant method to systems of nonlinear equations, in particular, a technique known as **Broyden's method**. The method requires only n scalar functional evaluations per iteration and also reduces the number of arithmetic calculations to $O(n^2)$. It belongs to a class of methods known as *least-change secant updates* that produce algorithms called *quasi-Newton*. These methods replace the Jacobian matrix in Newton's method with an approximation matrix that is updated at each iteration. The disadvantage to this method is that the quadratic convergence of Newton's method is lost, being replaced in general by a convergence called *superlinear*. This implies that

$$\lim_{i \to \infty} \frac{\|\mathbf{p}^{(i+1)} - \mathbf{p}\|}{\|\mathbf{p}^{(i)} - \mathbf{p}\|} = 0,$$

where $\mathbf{p}$ denotes the solution to $\mathbf{F}(\mathbf{x}) = \mathbf{0}$, and $\mathbf{p}^{(i)}$ and $\mathbf{p}^{(i+1)}$ are consecutive approximations to $\mathbf{p}$. In most applications, the reduction to superlinear convergence is a more than acceptable trade-off for the decrease in the amount of computation.

An additional disadvantage of quasi-Newton methods is that, unlike Newton's method, they are not self-correcting. Newton's method, for example, will generally correct for round-off error with successive iterations, but unless special safeguards are incorporated, Broyden's method will not.

Suppose that an initial approximation $\mathbf{p}^{(0)}$ is given to the solution $\mathbf{p}$ of $\mathbf{F}(\mathbf{x}) = \mathbf{0}$. We calculate the next approximation $\mathbf{p}^{(1)}$ in the same manner as Newton's method or, if it is inconvenient to determine $J(\mathbf{p}^{(0)})$ exactly, we can use difference equations to approximate the partial derivatives. To compute $\mathbf{p}^{(2)}$, however, we depart from Newton's method and examine the Secant method for a single nonlinear equation. The Secant method uses the approximation

$$f'(p_1) \approx \frac{f(p_1) - f(p_0)}{p_1 - p_0}$$

as a replacement for $f'(p_1)$ in Newton's method. For nonlinear systems, $\mathbf{p}^{(1)} - \mathbf{p}^{(0)}$ is a vector, and the corresponding quotient is undefined. However, the method proceeds similarly in that we replace the matrix $J(\mathbf{p}^{(1)})$ in Newton's method by a matrix A_1 with the property that

$$A_1(\mathbf{p}^{(1)} - \mathbf{p}^{(0)}) = \mathbf{F}(\mathbf{p}^{(1)}) - \mathbf{F}(\mathbf{p}^{(0)}).$$

This equation does not define a unique matrix, because it does not describe how A_1 operates on vectors that are not parallel to $\mathbf{p}^{(1)} - \mathbf{p}^{(0)}$. We can uniquely determine a matrix if we describe its operation on vectors that are orthogonal to $\mathbf{p}^{(1)} - \mathbf{p}^{(0)}$. Since no information is available about the change in $\mathbf{F}$ in a direction orthogonal to $\mathbf{p}^{(1)} - \mathbf{p}^{(0)}$, we require that

$$A_1 \mathbf{z} = J(\mathbf{p}^{(0)})\mathbf{z} \qquad \text{whenever} \qquad (\mathbf{p}^{(1)} - \mathbf{p}^{(0)})^t \mathbf{z} = 0.$$

This specifies that any vector orthogonal to $\mathbf{p}^{(1)} - \mathbf{p}^{(0)}$ is unaffected by the update from $J(\mathbf{p}^{(0)})$, which was used to compute $\mathbf{p}^{(1)}$, to A_1, which is used in the determination of $\mathbf{p}^{(2)}$.

These conditions uniquely define A_1 as

$$A_1 = J(\mathbf{p}^{(0)}) + \frac{[\mathbf{F}(\mathbf{p}^{(1)}) - \mathbf{F}(\mathbf{p}^{(0)}) - J(\mathbf{p}^{(0)})(\mathbf{p}^{(1)} - \mathbf{p}^{(0)})](\mathbf{p}^{(1)} - \mathbf{p}^{(0)})^t}{\|\mathbf{p}^{(1)} - \mathbf{p}^{(0)}\|_2^2}.$$

It is this matrix that is used in place of $J(\mathbf{p}^{(1)})$ to determine $\mathbf{p}^{(2)}$:

$$\mathbf{p}^{(2)} = \mathbf{p}^{(1)} - A_1^{-1}\mathbf{F}(\mathbf{p}^{(1)}).$$

Once $\mathbf{p}^{(2)}$ has been determined, the method is repeated to determine $\mathbf{p}^{(3)}$, with A_1 used in place of $A_0 \equiv J(\mathbf{p}^{(0)})$ and with $\mathbf{p}^{(2)}$ and $\mathbf{p}^{(1)}$ in place of $\mathbf{p}^{(1)}$ and $\mathbf{p}^{(0)}$, respectively. In general, once $\mathbf{p}^{(i)}$ has been determined, $\mathbf{p}^{(i+1)}$ is computed by

$$A_i = A_{i-1} + \frac{\mathbf{y}_i - A_{i-1}\mathbf{s}_i}{\|\mathbf{s}_i\|_2^2}\mathbf{s}_i^t,$$

and

$$\mathbf{p}^{(i+1)} = \mathbf{p}^{(i)} - A_i^{-1}\mathbf{F}(\mathbf{p}^{(i)}),$$

where the notation $\mathbf{s}_i = \mathbf{p}^{(i)} - \mathbf{p}^{(i-1)}$ and $\mathbf{y}_i = \mathbf{F}(\mathbf{p}^{(i)}) - \mathbf{F}(\mathbf{p}^{(i-1)})$ is introduced to simplify the equations.

If the method is performed as just outlined, the number of scalar functional evaluations is reduced from $n^2 + n$ to n (those required for evaluating $\mathbf{F}(\mathbf{p}^{(i)})$), but the method still requires $O(n^3)$ calculations to solve the associated $n \times n$ linear system

$$A_i\mathbf{y}_i = -\mathbf{F}(\mathbf{p}^{(i)}).$$

Employing the method in this form would not be justified because of the reduction to superlinear convergence from the quadratic convergence of Newton's method.

A significant improvement can be incorporated with a matrix inversion formula.

Sherman-Morrison Formula

If A is a nonsingular matrix and $\mathbf{x}$ and $\mathbf{y}$ are vectors, then $A + \mathbf{x}\mathbf{y}^t$ is nonsingular provided that $\mathbf{y}^t A^{-1}\mathbf{x} \neq -1$ and

$$(A + \mathbf{x}\mathbf{y}^t)^{-1} = A^{-1} - \frac{A^{-1}\mathbf{x}\mathbf{y}^t A^{-1}}{1 + \mathbf{y}^t A^{-1}\mathbf{x}}.$$

This formula permits A_i^{-1} to be computed directly from A_{i-1}^{-1}, eliminating the need for a matrix inversion with each iteration. By letting $A = A_{i-1}, \mathbf{x} = (\mathbf{y}_i - A_{i-1}\mathbf{s}_i)/\|\mathbf{s}_i\|_2^2$, and $\mathbf{y} = \mathbf{s}_i$, the Sherman–Morrison formula implies that

$$
\begin{aligned}
A_i^{-1} &= \left(A_{i-1} + \frac{\mathbf{y}_i - A_{i-1}\mathbf{s}_i}{\|\mathbf{s}_i\|_2^2}\mathbf{s}_i^t\right)^{-1} \\
&= A_{i-1}^{-1} - \frac{A_{i-1}^{-1}\left(\dfrac{\mathbf{y}_i - A_{i-1}\mathbf{s}_i}{\|\mathbf{s}_i\|_2^2}\mathbf{s}_i^t\right)A_{i-1}^{-1}}{1 + \mathbf{s}_i^t A_{i-1}^{-1}\left(\dfrac{\mathbf{y}_i - A_{i-1}\mathbf{s}_i}{\|\mathbf{s}_i\|_2^2}\right)} \\
&= A_{i-1}^{-1} - \frac{(A_{i-1}^{-1}\mathbf{y}_i - \mathbf{s}_i)\mathbf{s}_i^t A_{i-1}^{-1}}{\|\mathbf{s}_i\|_2^2 + \mathbf{s}_i^t A_{i-1}^{-1}\mathbf{y}_i - \|\mathbf{s}_i\|_2^2} \\
&= A_{i-1}^{-1} + \frac{(\mathbf{s}_i - A_{i-1}^{-1}\mathbf{y}_i)\mathbf{s}_i^t A_{i-1}^{-1}}{\mathbf{s}_i^t A_{i-1}^{-1}\mathbf{y}_i}.
\end{aligned}
$$

This computation involves only matrix-vector multiplicaton at each step and therefore requires only $O(n^2)$ arithmetic calculations. The calculation of A_i is bypassed, as is the necessity of solving the linear system. The program BROYM102 is used to implement this technique.

EXAMPLE 1   The nonlinear system

$$3x_1 - \cos(x_2 x_3) - \tfrac{1}{2} = 0,$$

$$x_1^2 - 81(x_2 + 0.1)^2 + \sin x_3 + 1.06 = 0,$$

$$e^{-x_1 x_2} + 20x_3 + \frac{10\pi - 3}{3} = 0$$

was solved by Newton's method in Example 1 of Section 10.2. The Jacobian matrix for this system is

$$J(x_1, x_2, x_3) = \begin{bmatrix} 3 & x_3 \sin x_2 x_3 & x_2 \sin x_2 x_3 \\ 2x_1 & -162(x_2 + 0.1) & \cos x_3 \\ -x_2 e^{-x_1 x_2} & -x_1 e^{-x_1 x_2} & 20 \end{bmatrix}.$$

With $\mathbf{p}^{(0)} = (0.1, 0.1, -0.1)^t$,

$$\mathbf{F}(\mathbf{p}^{(0)}) = \begin{bmatrix} -1.199949 \\ -2.269832 \\ 8.462026 \end{bmatrix}.$$

Since
$$A_0 = J(p_1^{(0)}, p_2^{(0)}, p_3^{(0)})$$

$$= \begin{bmatrix} 3 & 9.999836 \times 10^{-4} & -9.999836 \times 10^{-4} \\ 0.2 & -32.4 & 0.9950041 \\ -9.900498 \times 10^{-2} & -9.900498 \times 10^{-2} & 20 \end{bmatrix},$$

we have

$$A_0^{-1} = J(p_1^{(0)}, p_2^{(0)}, p_3^{(0)})^{-1}$$

$$= \begin{bmatrix} 0.3333331 & 1.023852 \times 10^{-5} & 1.615703 \times 10^{-5} \\ 2.108606 \times 10^{-3} & -3.086882 \times 10^{-2} & 1.535838 \times 10^{-3} \\ 1.660522 \times 10^{-3} & -1.527579 \times 10^{-4} & 5.000774 \times 10^{-2} \end{bmatrix},$$

$$\mathbf{p}^{(1)} = \mathbf{p}^{(0)} - A_0^{-1}\mathbf{F}(\mathbf{p}^{(0)}) = \begin{bmatrix} 0.4998693 \\ 1.946693 \times 10^{-2} \\ -0.5215209 \end{bmatrix},$$

$$\mathbf{F}(\mathbf{p}^{(1)}) = \begin{bmatrix} -3.404021 \times 10^{-4} \\ -0.3443899 \\ 3.18737 \times 10^{-2} \end{bmatrix},$$

$$\mathbf{y}_1 = \mathbf{F}(\mathbf{p}^{(1)}) - \mathbf{F}(\mathbf{p}^{(0)}) = \begin{bmatrix} 1.199608 \\ 1.925442 \\ -8.430152 \end{bmatrix},$$

$$\mathbf{s}_1 = \begin{bmatrix} 0.3998693 \\ -8.053307 \times 10^{-2} \\ -0.4215209 \end{bmatrix},$$

$$\mathbf{s}_1^t A_0^{-1} \mathbf{y}_1 = 0.3424604,$$

$$A_1^{-1} = A_0^{-1} + (1/0.3424604)[(\mathbf{s}_1 - A_0^{-1}\mathbf{y}_1)\mathbf{s}_1^t A_0^{-1}]$$

$$= \begin{bmatrix} 0.3333781 & 1.11077 \times 10^{-5} & 8.944584 \times 10^{-6} \\ -2.021271 \times 10^{-3} & -3.094847 \times 10^{-2} & 2.196909 \times 10^{-3} \\ 1.022381 \times 10^{-3} & -1.650679 \times 10^{-4} & 5.010987 \times 10^{-2} \end{bmatrix},$$

and
$$\mathbf{p}^{(2)} = \mathbf{p}^{(1)} - A_1^{-1}\mathbf{F}(\mathbf{p}^{(1)}) = \begin{bmatrix} 0.4999863 \\ 8.737888 \times 10^{-3} \\ -0.5231746 \end{bmatrix}.$$

Additional iterations are listed in Table 10.2. The fifth iteration of Broyden's method is slightly less accurate than was the fourth iteration of Newton's method given in the example at the end of the preceding section.   ■ ■ ■

Table 10.2

k	$p_1^{(k)}$	$p_2^{(k)}$	$p_3^{(k)}$	$\|\mathbf{p}^{(k)} - \mathbf{p}^{(k-1)}\|_\infty$
3	0.5000066	8.672215×10^{-4}	-0.5236918	7.87×10^{-3}
4	0.5000005	6.087473×10^{-5}	-0.5235954	8.06×10^{-4}
5	0.5000002	-1.445223×10^{-6}	-0.5235989	6.23×10^{-5}

EXERCISE SET 10.3

1. Use Broyden's method to approximate the two solutions to the nonlinear system

$$x_1(1 - x_1) + 4x_2 = 12$$
$$(x_1 - 2)^2 + (2x_2 - 3)^2 = 25$$

to within 10^{-5} in the l_∞ norm. Compare the number of iterations required for this accuracy to the number required in Exercise 1 of Section 10.2.

2. Use Broyden's method to approximate the solution near $(\frac{1}{4}, \frac{1}{4})^t$ to the nonlinear system

$$5x_1^2 - x_2^2 = 0$$
$$x_2 - 0.25(\sin x_1 + \cos x_2) = 0$$

accurate to within 10^{-5} in the l_∞ norm. Compare the number of iterations required for this accuracy to the number required in Exercise 2 of Section 10.2.

3. Use Broyden's method to find a solution to the following nonlinear systems. Iterate until successive approximations agree to within 10^{-6} in the l_∞ norm.

(a) $4x_1^2 - 20x_1 + \frac{1}{4}x_2^2 + 8 = 0$
$\frac{1}{2}x_1x_2^2 + 2x_1 - 5x_2 + 8 = 0$

(b) $3x_1^2 - x_2^2 = 0$
$3x_1x_2^2 - x_1^3 - 1 = 0$

(c) $\ln(x_1^2 + x_2^2) - \sin(x_1x_2) = \ln 2 + \ln \pi$
$e^{x_1 - x_2} + \cos(x_1x_2) = 0$

(d) $\sin(4\pi x_1 x_2) - 2x_2 - x_1 = 0$
$\left(\dfrac{4\pi - 1}{4\pi}\right)(e^{2x_1} - e) + 4ex_2^2 - 2ex_1 = 0$

4. Use Broyden's method to find a solution to the following nonlinear systems in the given domain. Iterate until successive approximations agree to within 10^{-6} in the l_∞ norm. Compare the number of iterations required for this accuracy to the number required in Exercise 4 of Section 10.2.

(a) $15x_1 + x_2^2 - 4x_3 = 13$
$x_1^2 + 10x_2 - x_3 = 11$
$x_2^3 - 25x_3 = -22$
$D = \{(x_1, x_2, x_3)^t \mid 0 \le x_1 \le 2, 0 \le x_2 \le 2, 0 \le x_3 \le 2\}$

(b) $x_1 + \cos(x_1x_2x_3) - 1 = 0$
$(1 - x_1)^{\frac{1}{4}} + x_2 + 0.05x_3^2 - 0.15x_3 - 1 = 0$
$-x_1^2 - 0.1x_2^2 + 0.01x_2 + x_3 - 1 = 0$
$D = \{(x_1, x_2, x_3)^t \mid 0 \le x_i \le 1.5, \text{ for } i = 1, 2, 3\}$

(c) $x_1^3 + x_1^2x_2 - x_1x_3 + 6 = 0$
$e^{x_1} + e^{x_2} - x_3 = 0$
$x_2^2 - 2x_1x_3 = 4$
$D = \{(x_1, x_2, x_3)^t \mid -2 \le x_1, x_2 \le -1, 0 \le x_3 \le 1\}$

(d) $\sin x_1 + \sin(x_1x_2) + \sin(x_1x_3) = 0$
$\sin x_2 + \sin(x_2x_3) = 0$
$\sin(x_1x_2x_3) = 0$
$D = \{(x_1, x_2, x_3)^t \mid -2 \le x_1, x_2 \le -1, -6 \le x_3 \le -4\}$

5. Use Broyden's method to find a solution to the following nonlinear systems. Iterate until successive approximations agree to within 10^{-5} in the l_∞ norm. Compare results with those obtained in Exercise 5 of Section 10.2.

(a) $6x_1 - 2\cos(x_2x_3) - 1 = 0$
$9x_2 + \sqrt{x_1^2 + \sin x_3 + 1.06} + 0.9 = 0$
$60x_3 + 3e^{-x_1x_2} + 10\pi - 3 = 0$

(b) $\cos x_1 + \cos x_2 + \cos x_3 = 1$
$\cos(x_1x_2) + \cos(x_2x_3) + \cos(x_1x_3) = 0$
$\cos(x_1x_2x_3) = -1$

(c) $2x_1 + x_2 + x_3 + x_4 - 5 = 0$

$\quad x_1 + 2x_2 + x_3 + x_4 - 5 = 0$

$\quad x_1 + x_2 + 2x_3 + x_4 - 5 = 0$

$\quad\quad\quad\quad x_1 x_2 x_3 x_4 - 1 = 0$

(Find a solution other than $(1, 1, 1)^t$.)

(d) $4x_1 - \quad x_2 + \quad x_3 = x_1 x_4$

$\quad -x_1 + 3x_2 - 2x_3 = x_2 x_4$

$\quad\quad x_1 - 2x_2 + 3x_3 = x_3 x_4$

$\quad\quad x_1^2 + \quad x_2^2 + \quad x_3^2 = 1$

(Find three solutions.)

6. The following nonlinear systems have singular Jacobian matrices as the solution. Can Broyden's method still be used? How is the rate of convergence affected?

(a) $\quad 3x_1 - \cos(x_2 x_3) - 0.5 = 0$

$\quad\quad\quad\quad x_1^2 - 625x_2^2 = 0$

$\quad e^{-x_1 x_2} + 20x_3 + \dfrac{10\pi - 3}{3} = 0$

(b) $x_1 - 10x_2 + 9 = 0$

$\quad \sqrt{3}(x_3 - x_4) = 0$

$\quad (x_2 - 2x_3 + 1)^2 = 0$

$\quad \sqrt{2}(x_1 - x_4)^2 = 0$

10.4   The Steepest-Descent Method

The advantage of the Newton and quasi-Newton methods for solving systems of nonlinear equations is their speed of convergence once a sufficiently accurate approximation is known. A weakness of these methods is that an accurate initial approximation to the solution is needed to ensure convergence. The **method of Steepest Descent** will generally converge only linearly to the solution, but it is global in nature. This method is often used to find sufficiently accurate starting approximations for the Newton-based techniques, in the same way the Bisection method is used for a single equation.

The method of Steepest Descent determines a local minimum for a multivariable function of the form $g : \mathbb{R}^n \to \mathbb{R}$. The method is valuable quite apart from the application as a starting method for solving nonlinear systems.

The connection between the minimization of a function from $\mathbb{R}^n$ to $\mathbb{R}$ and the solution of a system of nonlinear equations is due to the fact that a system of the form

$$f_1(x_1, x_2, \ldots, x_n) = 0$$
$$f_2(x_1, x_2, \ldots, x_n) = 0$$
$$\vdots$$
$$f_n(x_1, x_2, \ldots, x_n) = 0$$

has a solution at $\mathbf{x} = (x_1, x_2, \ldots, x_n)^t$ precisely when the function g defined by

$$g(x_1, x_2, \ldots, x_n) = \sum_{i=1}^{n} [f_i(x_1, x_2, \ldots, x_n)]^2$$

has the minimal value zero.

The method of Steepest Descent for finding a local minimum for an arbitrary function g from $\mathbb{R}^n$ into $\mathbb{R}$ can be intuitively described as follows:

1. Evaluate g at an initial approximation $\mathbf{p}^{(0)} = (p_1^{(0)}, p_2^{(0)}, \ldots, p_n^{(0)})^t$.

2. Determine a direction from $\mathbf{p}^{(0)}$ that results in a decrease in the value of g.

3. Move an appropriate amount in this direction and call the new value $\mathbf{p}^{(1)}$.

4. Repeat the steps with $\mathbf{p}^{(0)}$ replaced by $\mathbf{p}^{(1)}$.

Before describing how to choose the correct direction and the appropriate distance to move in this direction, we need to review some results from calculus. The Extreme Value Theorem implies that a differentiable single-variable function can have a relative minimum only when the derivative is zero. To extend this result to multivariable functions, we need the following definition.

If $g : \mathbb{R}^n \to \mathbb{R}$, the **gradient** of g at $\mathbf{x} = (x_1, x_2, \ldots, x_n)^t$ is denoted $\nabla g(\mathbf{x})$ and is defined by

$$\nabla g(\mathbf{x}) = \left(\frac{\partial g}{\partial x_1}(\mathbf{x}), \frac{\partial g}{\partial x_2}(\mathbf{x}), \ldots, \frac{\partial g}{\partial x_n}(\mathbf{x}) \right)^t.$$

The gradient for a multivariable function is analogous to the derivative of a single-variable function in the sense that a differentiable multivariable function can have a relative minimum at $\mathbf{x}$ only when the gradient at $\mathbf{x}$ is the zero vector.

The gradient has another important property connected with the minimization of multivariable functions. Suppose that $\mathbf{v} = (v_1, v_2, \ldots, v_n)^t$ is a vector in $\mathbb{R}^n$, with

$$\|\mathbf{v}\|_2^2 = \sum_{i=1}^{n} v_i^2 = 1.$$

The **directional derivative** of g at $\mathbf{x}$ in the direction of $\mathbf{v}$ is defined by

$$D_\mathbf{v} g(\mathbf{x}) = \lim_{h \to 0} \frac{1}{h} [g(\mathbf{x} + h\mathbf{v}) - g(\mathbf{x})].$$

The directional derivative of g at $\mathbf{x}$ in the direction of $\mathbf{v}$ measures the change in the value of the function g relative to the change in the variable in the direction of $\mathbf{v}$. (See Figure 10.2 for an illustration when g is a function of two variables.)

A standard result from the calculus of multivariable functions states that the direction that produces the maximum magnitude for the directional derivative occurs when $\mathbf{v}$ is chosen to be parallel to $\nabla g(\mathbf{x})$, provided that $\nabla g(\mathbf{x}) \neq \mathbf{0}$. The direction of greatest decrease in the value of g at $\mathbf{x}$ is the direction given by $-\nabla g(\mathbf{x})$. The object is to reduce $g(\mathbf{x})$ to its minimal value of zero, so given the initial approximation $\mathbf{p}^{(0)}$, we choose

$$\mathbf{p}^{(1)} = \mathbf{p}^{(0)} - \alpha \nabla g(\mathbf{p}^{(0)})$$

for some constant $\alpha > 0$.

Figure 10.2

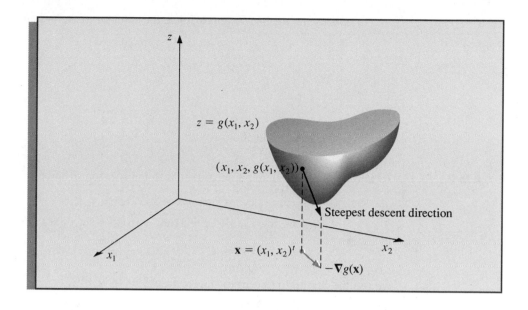

The problem now reduces to choosing α so that $g(\mathbf{p}^{(1)})$ will be significantly less than $g(\mathbf{p}^{(0)})$. To determine an appropriate choice for the value α we consider the single-variable function

$$h(\alpha) = g(\mathbf{p}^{(0)} - \alpha \nabla g(\mathbf{p}^{(0)})).$$

The value of α that minimizes h is the value used to determine $\mathbf{p}^{(1)}$.

Finding a minimal value for h directly requires differentiating h and then solving a root-finding problem to determine the critical points of h. This is generally too costly a procedure. Instead, we interpolate h using a quadratic polynomial P and three numbers α_1, α_2, and α_3 that we hope are close to the minimum value of h.

We define $\hat{\alpha}$ so that the absolute minimum of P on the smallest closed interval that contains α_1, α_2, and α_3 is $|P(\hat{\alpha})|$ and use $P(\hat{\alpha})$ to approximate the minimum value of $h(\alpha)$. Then $\hat{\alpha}$ is used to determine the new iterate for approximating the minimal value of g:

$$\mathbf{p}^{(1)} = \mathbf{p}^{(0)} - \hat{\alpha} \nabla g(\mathbf{p}^{(0)}).$$

Since $g(\mathbf{p}^{(0)})$ is available, we first choose $\alpha_1 = 0$. Next, a number α_3 is found with $h(\alpha_3) < h(\alpha_1)$. (Since α_1 does not minimize h, such a number α_3 does exist.) Finally, α_2 is chosen to be $\alpha_3/2$.

The minimum value $\hat{\alpha}$ of P on $[\alpha_1, \alpha_3]$ occurs either at the only critical point of P or at the right endpoint α_3, for, by assumption, $P(\alpha_3) = h(\alpha_3) < h(\alpha_1) = P(\alpha_1)$. The critical point is easily determined, since P is a quadratic polynomial.

The program STPDC103 applies the method of Steepest Descent to approximate the minimal value of $g(\mathbf{x})$. To begin each iteration, the value 0 is assigned to α_1, and the value 1 is assigned to α_3. If $h(\alpha_3) \geq h(\alpha_1)$, then successive divisions of α_3 by 2 are performed and the value of α_3 is reassigned until $h(\alpha_3) \leq h(\alpha_1)$.

To employ the method to approximate the solution to the system

$$f_1(x_1, x_2, \ldots, x_n) = 0$$
$$f_2(x_1, x_2, \ldots, x_n) = 0$$
$$\vdots$$
$$f_n(x_1, x_2, \ldots, x_n) = 0,$$

we simply replace the function g with $\sum_{i=1}^{n} f_i^2$.

EXAMPLE 1   To find a reasonable initial approximation to the solution of the nonlinear system

$$f_1(x_1, x_2, x_3) = 3x_1 - \cos(x_2 x_3) - \tfrac{1}{2} = 0$$
$$f_2(x_1, x_2, x_3) = x_1^2 - 81(x_2 + 0.1)^2 + \sin x_3 + 1.06 = 0$$
$$f_3(x_1, x_2, x_3) = e^{-x_1 x_2} + 20x_3 + \frac{10\pi - 3}{3} = 0$$

we use the Steepest-Descent method with $\mathbf{p}^{(0)} = (0, 0, 0)^t$.

Let $g(x_1, x_2, x_3) = [f_1(x_1, x_2, x_3)]^2 + [f_2(x_1, x_2, x_3)]^2 + [f_3(x_1, x_2, x_3)]^2$.

Then

$$\nabla g(x_1, x_2, x_3) \equiv \nabla g(\mathbf{x}) = \left(2f_1(\mathbf{x})\frac{\partial f_1}{\partial x_1}(\mathbf{x}) + 2f_2(\mathbf{x})\frac{\partial f_2}{\partial x_1}(\mathbf{x}) + 2f_3(\mathbf{x})\frac{\partial f_3}{\partial x_1}(\mathbf{x}), \right.$$
$$2f_1(\mathbf{x})\frac{\partial f_1}{\partial x_2}(\mathbf{x}) + 2f_2(\mathbf{x})\frac{\partial f_2}{\partial x_2}(\mathbf{x}) + 2f_3(\mathbf{x})\frac{\partial f_3}{\partial x_2}(\mathbf{x}),$$
$$\left. 2f_1(\mathbf{x})\frac{\partial f_1}{\partial x_3}(\mathbf{x}) + 2f_2(\mathbf{x})\frac{\partial f_2}{\partial x_3}(\mathbf{x}) + 2f_3(\mathbf{x})\frac{\partial f_3}{\partial x_3}(\mathbf{x}) \right)$$
$$= 2\mathbf{J}(\mathbf{x})^t \mathbf{F}(\mathbf{x}).$$

For $\mathbf{p}^{(0)} = (0, 0, 0)^t$, we have

$$g(\mathbf{p}^{(0)}) = 111.975, \qquad z_0 = \|\mathbf{z}\|_2 = 419.554$$

and

$$\mathbf{z} = (-0.0214514, -0.0193062, 0.999583)^t,$$

so
$$\alpha_1 = 0, \qquad g_1 = 111.975,$$
$$\alpha_2 = 0.5, \qquad g_2 = 2.53557, \qquad h_1 = -218.878,$$
$$\alpha_3 = 1, \qquad g_3 = 93.5649, \qquad h_2 = 182.059, \qquad h_3 = 400.937,$$

and

$$P(\alpha) = 111.975 - 218.878\alpha + 400.937\alpha(\alpha - 0.5).$$

Hence,

$$\alpha_0 = 0.522959 \qquad \text{and} \qquad g_0 = 2.32762.$$

(Note that $P(\alpha_0) = 2.32424$). Thus, $\alpha = 0.522959$,

$$\mathbf{p}^{(1)} = \mathbf{p}^{(0)} - 0.522959\mathbf{z} = (0.0112182, 0.0100964, -0.522741)^t,$$

and

$$g(\mathbf{p}^{(1)}) = 2.32762.$$

Table 10.3 contains the remainder of the results. A solution to the nonlinear system is $(0.5, 0, -0.5235988)^t$, so $\mathbf{p}^{(2)}$ would be adequate as an initial approximation for Newton's method or Broyden's method. ■ ■ ■

Table 10.3

k	$p_1^{(k)}$	$p_2^{(k)}$	$p_3^{(k)}$	$g(p_1^{(k)}, p_2^{(k)}, p_3^{(k)})$
0	0.0	0.0	0.0	111.975
1	0.0112182	0.0100964	-0.522741	2.32762
2	0.137860	-0.205453	-0.522059	1.27406
3	0.266959	0.00551102	-0.558494	1.06813
4	0.272734	-0.00811751	-0.522006	0.468309

EXERCISE SET 10.4

1. Use the method of Steepest Descent to approximate the solutions of the following nonlinear systems. Iterate until successive approximations agree to within 0.005 in the l_∞ norm.

    (a) $4x_1^2 - 20x_1 + \frac{1}{4}x_2^2 + 8 = 0$
         $\frac{1}{2}x_1x_2^2 + 2x_1 - 5x_2 + 8 = 0$

    (b) $3x_1^2 - x_2^2 = 0$
         $3x_1x_2^2 - x_1^3 - 1 = 0$

    (c) $\ln(x_1^2 + x_2^2) - \sin(x_1x_2) = \ln 2 + \ln \pi$
         $e^{x_1-x_2} + \cos(x_1x_2) = 0$

    (d) $\sin(4\pi x_1x_2) - 2x_2 - x_1 = 0$
         $\left(\frac{4\pi - 1}{4\pi}\right)(e^{2x_1} - e) + 4ex_2^2 - 2ex_1 = 0$

2. Use the results in Exercise 1 and Newton's method to approximate the solutions of the nonlinear systems in Exercise 1. Iterate until successive approximations agree to within 10^{-5} in the l_∞ norm.

3. Use the method of Steepest Descent to approximate the solutions of the following nonlinear systems. Iterate until successive approximations agree to within 0.005 in the l_∞ norm.

    (a) $15x_1 + x_2^2 - 4x_3 = 13$
         $x_1^2 + 10x_2 - x_3 = 11$
         $x_2^3 - 25x_3 = -22$

    (b) $x_1 + \cos(x_1x_2x_3) - 1 = 0$
         $(1 - x_1)^{\frac{1}{4}} + x_2 + 0.05x_3^2 - 0.15x_3 - 1 = 0$
         $-x_1^2 - 0.1x_2^2 + 0.01x_2 + x_3 - 1 = 0$

(c) $x_1^3 + x_1^2 x_2 - x_1 x_3 + g = 0$

$e^{x_1} + e^{x_2} - x_3 = 0$

$x_2^2 - 2x_1 x_3 = 4$

(d) $\sin x_1 + \sin(x_1 x_2) + \sin(x_1 x_3) = 0$

$\sin x_2 + \sin(x_2 x_3) = 0$

$\sin(x_1 x_2 x_3) = 0$

(e) $\cos x_1 + \cos x_2 + \cos x_3 = 1$

$\cos(x_1 x_2) + \cos(x_2 x_3) + \cos(x_1 x_3) = 0$

$\cos(x_1 x_2 x_3) = -1$

(f) $4x_1 - x_2 + x_3 = x_1 x_4$

$-x_1 + 3x_2 - 2x_3 = x_2 x_4$

$x_1 - 2x_2 + 3x_3 = x_3 x_4$

$x_1^2 + x_2^2 + x_3^2 = 1$

(Find three solutions.)

4. Use the results of Exercise 3 and Newton's method to approximate the solutions for the nonlinear systems in Exercise 3. Iterate until successive approximations agree to within 10^{-5} in the l_∞ norm.

5. Use the method of Steepest Descent to approximate minima for the following functions. Iterate until successive approximations agree to within 0.005 in the l_∞ norm.
    (a) $g(x_1, x_2) = \cos(x_1 + x_2) + \sin x_1 + \cos x_2$
    (b) $g(x_1, x_2) = 100(x_1^2 - x_2)^2 + (1 - x_1)^2$
    (c) $g(x_1, x_2, x_3) = x_1^2 + 2x_2^2 + x_3^2 - 2x_1 x_2 + 2x_1 - 2.5x_2 - x_3 + 2$
    (d) $g(x_1, x_2, x_3) = x_1^4 + 2x_2^4 + 3x_3^4 + 1.01$

10.5 Survey of Methods and Software

In this chapter we considered methods to approximate solutions to nonlinear systems

$$f_1(x_1, x_2, \ldots, x_n) = 0$$
$$f_2(x_1, x_2, \ldots, x_n) = 0$$

$$\vdots$$

$$f_n(x_1, x_2, \ldots, x_n) = 0.$$

Newton's method for systems requires a good initial approximation $(p_1^{(0)}, p_2^{(0)}, \ldots, p_n^{(0)})^t$ and generates a sequence

$$\mathbf{p}^{(k)} = \mathbf{p}^{(k-1)} - J(\mathbf{p}^{(k-1)})^{-1} F(\mathbf{p}^{(k-1)}),$$

that converges rapidly to a solution $\mathbf{p}$ if $\mathbf{p}^{(0)}$ is sufficiently close to $\mathbf{p}$. However, Newton's method requires evaluating, or approximating, n^2 partial derivatives and solving an n by n linear system at each step.

Broyden's method reduces the amount of computation at each step without significantly degrading the speed of convergence. This technique replaces the Jacobian matrix

J with a matrix A_{k-1} whose inverse is directly determined at each step. This reduces the arithmetic computations from $O(n^3)$ to $O(n^2)$. Moreover, the only scalar function evaluations required are those involving the f_i, saving n^2 scalar function evaluations per step. Broyden's method also requires a good initial approximation.

The Steepest-Descent method was presented as a way of obtaining good initial approximations for Newton's and Broyden's methods. Steepest-Descent does not give a rapidly convergent sequence, but it also does not require a good initial approximation. The Steepest-Descent method approximates a minimum of a multivariable function g. For our application we chose

$$g(x_1, x_2, \ldots, x_n) = \sum_{i=1}^{n} f_i(x_1, x_2, \ldots, x_n)^2.$$

The minimum of g is zero, which occurs when the functions f_i are simultaneously zero.

Homotopy and continuation methods are also used for nonlinear systems and are the subject of much current research. In these methods, a given problem

$$\mathbf{F(x)} = \mathbf{0}$$

is embedded in a one-parameter family of problems using a parameter λ that assumes values in $[0, 1]$. The original problem corresponds to $\lambda = 1$ and a problem with a known solution corresponds to $\lambda = 0$. For example, the set of problems

$$G(\mathbf{x}, \lambda) = \mathbf{F(x)} + (\lambda - 1)\mathbf{F(x_0)} = \mathbf{0}, \quad \text{where } 0 \leq \lambda \leq 1,$$

for fixed $\mathbf{x_0} \in \mathbb{R}^n$ forms a homotopy. When $\lambda = 0$ the solution is $\mathbf{x}(\lambda = 0) = \mathbf{x_0}$. The solution to the original problem corresponds to $\mathbf{x}(\lambda = 1)$. A continuation method attempts to determine $\mathbf{x}(\lambda = 1)$ by solving the sequence of problems corresponding to $\lambda_0 = 0 < \lambda_1 < \lambda_2 < \ldots < \lambda_m = 1$. The initial approximation to the solution of

$$\mathbf{F(x)} + (\lambda_i - 1)\mathbf{F(x_0)} = \mathbf{0}$$

is the solution $\mathbf{x}(\lambda = \lambda_{i-1})$ to the problem

$$\mathbf{F(x)} + (\lambda_i - 1)\mathbf{F(x_0)} = \mathbf{0}$$

The methods in the IMSL and NAG libraries are based on two subroutines HYBRDI and HYBRDJ contained in MINPACK, a public domain package. Both methods use the Levenberg–Marquardt method, which is a weighted average of Newton's method and the Steepest-Descent method. The weight is biased toward the Steepest-Descent method until convergence is detected, at which time the weight is shifted toward the more rapidly convergent Newton's method. The subroutine HYBRDI uses a finite-difference approximation to the Jacobian, and HYBRDJ requires a user-supplied subroutine to compute the Jacobian.

The IMSL subroutine NEQNF solves a nonlinear system without a user-supplied Jacobian. The subroutine NEQNJ is similar, except that the user must supply a subroutine to calculate the Jacobian.

In the NAG library, C05NBF is similar to HYBRDI and C05PBF is based on HYBRDJ in MINPACK. NAG also contains other modifications of the Levenberg–Marquardt method.

Boundary-Value Problems for Ordinary Differential Equations

The differential equations in the first sections of Chapter 5 are of first order and have one initial condition to satisfy. Later in the chapter we saw that the techniques could be extended to systems of equations and then to higher-order equations, but all the specified conditions had to be on the same endpoint. These are initial-value problems. In this chapter we show how to approximate the solution to **boundary-value** problems, differential equations where conditions are imposed at different points. For first-order differential equations only one condition is specified, so there is no distinction between initial-value and boundary-value problems. The differential equations whose solutions we will approximate are of second order, specifically of the form

$$y'' = f(x, y, y'), \qquad \text{for} \qquad a \le x \le b,$$

with the boundary conditions on the solution prescribed by

$$y(a) = \alpha \qquad \text{and} \qquad y(b) = \beta,$$

for some constants α and β. Such a problem has a unique solution provided that f and its partial derivatives with respect to y and y' are continuous, and that the partial derivative with respect to y is positive and with respect to y' is bounded. These are all reasonable conditions for boundary-value problems representing physical problems.

11.2 The Linear Shooting Method

A boundary-value problem is *linear* when the function f has the form

$$f(x, y, y') = p(x)y' + q(x)y + r(x).$$

Linear problems occur frequently in applications and are much easier to solve than nonlinear equations, since adding any solution to the *inhomogeneous* differential equation

$$y'' = p(x)y' + q(x)y + r(x)$$

to the complete solution of the *homogeneous* differential equation

$$y'' = p(x)y' + q(x)y$$

gives all the solutions to the inhomogeneous problem. The solutions of the homogeneous problem are easier to determine than are those of the inhomogeneous. Moreover, to show that a linear problem has a unique solution, we need only show that p, q, and r are continuous and that the values of q are positive.

To approximate the unique solution to the linear boundary-value problem, let us first consider the two initial-value problems:

(11.1)
$$y'' = p(x)y' + q(x)y + r(x), \quad \text{for } a \le x \le b, \quad \text{where} \quad y(a) = \alpha \quad \text{and} \quad y'(a) = 0,$$

and

(11.2)
$$y'' = p(x)y' + q(x)y, \quad \text{for } a \le x \le b, \quad \text{where} \quad y(a) = 0 \quad \text{and} \quad y'(a) = 1,$$

both of which have unique solutions. Suppose that $y_1(x)$ is the solution to Eq. (11.1) and $y_2(x)$ is the solution to Eq. (11.2). Then

(11.3)
$$y(x) = y_1(x) + \frac{\beta - y_1(b)}{y_2(b)} y_2(x)$$

is the unique solution to the linear boundary-value problem

(11.4)
$$y'' = p(x)y' + q(x)y + r(x), \quad \text{for} \quad a \le x \le b, \quad \text{with} \quad y(a) = \alpha \quad \text{and} \quad y(b) = \beta.$$

To verify this, first note that

$$y'' - p(x)y' - q(x)y = y_1'' - p(x)y_1' - q(x)y_1 + \frac{\beta - y_1(b)}{y_2(b)} [y_2'' - p(x)y_2' - q(x)y_2]$$

$$= r(x) + \frac{\beta - y_1(b)}{y_2(b)} \cdot 0 = r(x).$$

Moreover,

$$y(a) = y_1(a) + \frac{\beta - y_1(b)}{y_2(b)} y_2(a) = y_1(a) + \frac{\beta - y_1(b)}{y_2(b)} \cdot 0 = \alpha$$

and

$$y(b) = y_1(b) + \frac{\beta - y_1(b)}{y_2(b)} y_2(b) = y_1(b) + \beta - y_1(b) = \beta.$$

The Linear Shooting method is based on the replacement of the boundary-value problem by the two initial-value problems (11.1) and (11.2). Numerous methods are available from Chapter 5 for approximating the solutions $y_1(x)$ and $y_2(x)$, and once these approximations are available, the solution to the boundary-value problem is approximated using the weighted sum in Eq. (11.3). Graphically, the method has the appearance shown in Figure 11.1.

Figure 11.1

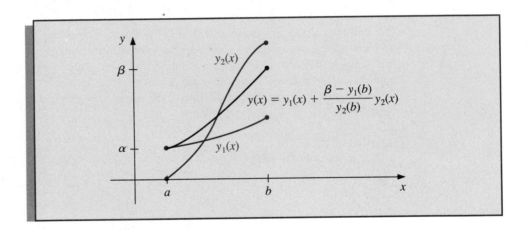

The program LINST111 incorporates the Runge–Kutta method of order four to find the approximations to $y_1(x)$ and $y_2(x)$, but any technique for approximating the solutions to initial-value problems could be substituted. The program has the additional feature of obtaining approximations for the derivative of the solution to the boundary-value problem as well as to the solution of the problem itself.

EXAMPLE 1 The boundary-value problem

$$y'' = -\frac{2}{x}y' + \frac{2}{x^2}y + \frac{\sin(\ln x)}{x^2}, \quad \text{for } 1 \le x \le 2, \quad \text{where } y(1) = 1 \text{ and } y(2) = 2,$$

has the exact solution

$$y = c_1 x + \frac{c_2}{x^2} - \frac{3}{10}\sin(\ln x) - \frac{1}{10}\cos(\ln x),$$

where

$$c_2 = \tfrac{1}{70}[8 - 12\sin(\ln 2) - 4\cos(\ln 2)]$$

and

$$c_1 = \tfrac{11}{10} - c_2.$$

Applying the Linear Shooting method to this problem requires approximating the solutions to the initial-value problems

$$y_1'' = -\frac{2}{x}y_1' + \frac{2}{x^2}y_1 + \frac{\sin(\ln x)}{x^2}, \quad \text{for } 1 \le x \le 2, \quad \text{where } y_1(1) = 1 \quad \text{and } y_1'(1) = 0,$$

and

$$y_2'' = -\frac{2}{x}y_2' + \frac{2}{x^2}y_2, \quad \text{for } 1 \le x \le 2, \quad \text{where } y_2(1) = 0 \quad \text{and } y_2'(1) = 1.$$

The results of the calculations are given in Table 11.1. The value listed as $u_{1,i}$ approximates $y_1(x_i)$, the value of $v_{1,i}$ approximates $y_2(x_i)$, and w_i approximates $y(x_i)$.

■　■　■

Table 11.1

x_i	$u_{1,i}$	$v_{1,i}$	w_i	$y(x_i)$	$|y(x_i) - w_i|$
1.0	1.00000000	0.00000000	1.00000000	1.00000000	
1.1	1.00896058	0.09117986	1.09262917	1.09262930	1.43×10^{-7}
1.2	1.03245472	0.16851175	1.18708471	1.18708484	1.34×10^{-7}
1.3	1.06674375	0.23608704	1.28338227	1.28338236	9.78×10^{-8}
1.4	1.10928795	0.29659067	1.38144589	1.38144595	6.02×10^{-8}
1.5	1.15830000	0.35184379	1.48115939	1.48115942	3.06×10^{-8}
1.6	1.21248372	0.40311695	1.58239245	1.58239246	1.08×10^{-8}
1.7	1.27087454	0.45131840	1.68501396	1.68501396	5.43×10^{-10}
1.8	1.33273851	0.49711137	1.78889854	1.78889853	5.05×10^{-9}
1.9	1.39750618	0.54098928	1.89392951	1.89392951	4.41×10^{-9}
2.0	1.46472815	0.58332538	2.00000000	2.00000000	

The accuracy exhibited in this example is expected because the Runge–Kutta method of order four gives $O(h^4)$ accuracy to the solutions of the initial-value problems. Unfortunately, there can be round-off error problems hidden in this technique. If $y_1(x)$ rapidly increases as x goes from a to b, then $u_{1,N} \approx y_1(b)$ will be large. Should β be small in magnitude compared to $u_{1,N}$, the term $(\beta - u_{1,N})/v_{1,N}$ will be approximately $-u_{1,N}/v_{1,N}$. So the approximations

$$y(x_i) \approx w_i = u_{1,i} - \left(\frac{\beta - u_{1,N}}{v_{1,N}}\right)v_{1,i} \approx u_{1,i} - \left(\frac{u_{1,N}}{v_{1,N}}\right)v_{1,i}$$

allow the possibility of a loss of significant digits due to cancellation. However, since $u_{1,i}$ is an approximation to $y_1(x_i)$, the behavior of y_1 can easily be monitored, and if $u_{1,i}$ increases rapidly from a to b, the shooting technique can be employed in the other direction—that is, solving instead the initial-value problems

$$y'' = p(x)y' + q(x)y + r(x), \quad \text{for } a \le x \le b, \quad \text{where } y(b) = \beta \quad \text{and } y'(b) = 0,$$

and

$$y'' = p(x)y' + q(x)y, \quad \text{for } a \le x \le b, \quad \text{where } y(b) = 0 \quad \text{and } y'(b) = 1.$$

If the reverse shooting technique still gives cancellation of significant digits and if increased precision does not yield greater accuracy, other techniques must be employed. In general, however, if $u_{1,i}$ and $v_{1,i}$ are $O(h^n)$ approximations to $y_1(x_i)$ and $y_2(x_i)$, respectively, for each $i = 0, 1, \ldots, N$, then $w_{1,i}$ will be an $O(h^n)$ approximation to $y(x_i)$.

EXERCISE SET 11.2

1. The boundary-value problem

   $$y'' = 4(y - x), \qquad 0 \le x \le 1, \qquad y(0) = 0, \qquad y(1) = 2$$

   has the solution $y(x) = \dfrac{e^2}{e^4 - 1}(e^{2x} - e^{-2x}) + x$. Use the Linear Shooting method to approximate the solution and then compare results to the actual solution in each case.
   (a) With $h = \frac{1}{3}$
   (b) With $h = \frac{1}{4}$

2. The boundary-value problem

   $$y'' = y' + 2y + \cos x, \qquad 0 \le x \le \frac{\pi}{2}, \qquad y(0) = -0.3, \qquad y(\frac{\pi}{2}) = -0.1$$

   has the solution $y(x) = -\frac{1}{10}[\sin x + 3\cos x]$. Use the Linear Shooting method to approximate the solution and compare the results to the actual solution in each case.
   (a) With $h = \frac{\pi}{4}$
   (b) With $h = \frac{\pi}{6}$

3. Use the Linear Shooting method to approximate the solution to the following boundary-value problems.
   (a) $y'' = -3y' + 2y + 2x + 3, \quad 0 \le x \le 1, \ y(0) = 2, \ y(1) = 1; \quad$ use $h = 0.1$.
   (b) $y'' = -4y' + 4y, \quad 0 \le x \le 5, \ y(0) = 1, \ y(5) = 0; \quad$ use $h = 0.2$.
   (c) $y'' = -2(2xy' - y + \ln x)/x^2, \quad 1 \le x \le 2, \ y(1) = -\frac{1}{2}, \ y(2) = \ln 2;$
        use $h = 0.05$.
   (d) $y'' = -(x + 1)y' + 2y + (1 - x^2)e^{-x}, \quad 0 \le x \le 1, \ y(0) = -1, \ y(1) = 0;$
        use $h = 0.1$.
   (e) $y'' = (xy' + 3y + x \ln x - x^2)/x^2, \quad 1 \le x \le 2, \ y(1) = y(2) = 0; \quad$ use $h = 0.1$.
   (f) $y'' = (-y' + (x + 1)y + 2e^x)/x, \quad 1 \le x \le e, \ y(1) = 0, \ y(e) = e^e; \quad$ use $h = \frac{1}{10}e$.

4. Although $q(x) < 0$ in the following boundary-value problems, unique solutions exist and are given. Use the Linear Shooting method to approximate the solutions to the following problems and compare the results to the actual solutions.
   (a) $y'' + y = 0, \quad 0 \le x \le \frac{\pi}{4}, \ y(0) = 1, \ y(\frac{\pi}{4}) = 1; \quad$ use $h = \frac{\pi}{20}; \quad$ actual solution
        $y(x) = \cos x + (\sqrt{2} - 1)\sin x$.
   (b) $y'' + 4y = \cos x, \quad 0 \le x \le \frac{\pi}{4}, \ y(0) = 0, \ y(\frac{\pi}{4}) = 0; \quad$ use $h = \frac{\pi}{20};$ actual solution
        $y(x) = -\frac{1}{3}\cos 2x - \frac{\sqrt{2}}{6}\sin 2x + \frac{1}{3}\cos x$.
   (c) $y'' = -2(2xy' + y - \ln x)/x^2, \quad 1 \le x \le 2, \ y(1) = \frac{1}{2}, \ y(2) = \ln 2; \quad$ use $h = 0.05;$
        actual solution $y(x) = 4x^{-1} - 2x^{-2} + \ln x - \frac{3}{2}$.
   (d) $y'' = 2y' - y + xe^x - x, \quad 0 \le x \le 2, \ y(0) = 0, \ y(2) = -4; \quad$ use $h = 0.2; \quad$ actual
        solution $y(x) = \frac{1}{6}x^3e^x - \frac{5}{3}xe^x + 2e^x - x - 2$.

5. Use the Linear Shooting method to approximate the solution $y = e^{-10x}$ to the boundary-value problem

$$y'' = 100y, \qquad 0 \le x \le 1, \qquad y(0) = 1, \qquad y(1) = e^{-10}.$$

Use $h = 0.1$ and 0.05. Can you explain the consequences?

6. Write the second-order initial-value problems (11.1) and (11.2) as first-order systems, and derive the equations necessary to solve the systems using the Runge–Kutta method of order four for systems.

7. The boundary-value problem governing the deflection of a beam with supported ends subject to uniform loading is

$$\frac{d^2 w}{dx^2} = \frac{S}{EI}w + \frac{qx}{2EI}(x - l), \qquad 0 < x < l,$$

with boundary conditions $w(0) = 0$ and $w(l) = 0$.

   Suppose the beam has the following characteristics: length $l = 120$ in., intensity of uniform load $q = 100$ lb/ft, modulus of elasticity $E = 3.0 \times 10^7$ lb/in.^{2}, stress at ends $S = 1000$ lb, and central moment of inertia $I = 625$ in.^{4}

   (a) Use the Linear Shooting method to approximate the deflection $w(x)$ of the beam at every 6-in. interval.

   (b) The actual relationship is given by

$$w(x) = c_1 e^{ax} + c_2 e^{-ax} + b(x - l)x + c,$$

   where $c_1 = 7.7042537 \times 10^4$, $c_2 = 7.9207462 \times 10^4$, $a = 2.3094010 \times 10^{-4}$, $b = -4.1666666 \times 10^{-3}$, and $c = -1.5625 \times 10^5$. Is the maximum error on the interval within 0.2 in.?

   (c) State law in Ohio requires that $\max_{0<x<1} w(x) < \frac{1}{300}$. Does this beam meet Ohio state code?

8. Let u represent the electrostatic potential between two concentric metal spheres of radii R_1 and R_2 $(R_1 < R_2)$, such that the potential of the inner sphere is kept constant at V_1 volts and the potential of the outer sphere is 0 volts. The potential in the region between the two spheres is governed by Laplace's equation, which, in this particular application, reduces to

$$\frac{d^2 u}{dr^2} + \frac{2}{r}\frac{du}{dr} = 0, \qquad R_1 \le r \le R_2, \qquad u(R_1) = V_1, \qquad u(R_2) = 0.$$

   Suppose $R_1 = 2$ in., $R_2 = 4$ in., and $V_1 = 110$ volts.

   (a) Use the Linear Shooting method with $N = 20$ to approximate $u(3)$.

   (b) Use the Linear Shooting method with $N = 40$ to approximate $u(3)$.

   (c) Compare the results of parts (a) and (b) with the actual potential $u(3)$, where

$$u(r) = \frac{V_1 R_1}{r}\left(\frac{R_2 - r}{R_2 - R_1}\right).$$

11.3 Linear Finite Difference Methods

The Shooting method often has round-off error instability. The methods we present in this section have better stability characteristics, but they generally require more computation to obtain a specified accuracy.

Methods involving finite differences for solving boundary-value problems replace each of the derivatives in the differential equation with an appropriate difference-quotient approximation of the type considered in Section 4.9. The particular difference quotient is chosen to maintain a specified order of error.

The linear second-order boundary-value problem,

$$y'' = p(x)y' + q(x)y + r(x), \quad \text{for } a \le x \le b, \quad \text{where } y(a) = \alpha \text{ and } y(b) = \beta,$$

requires that difference-quotient approximations be used for approximating both y' and y''. First, we select an integer $N > 0$ and divide the interval $[a, b]$ into $(N + 1)$ equal subintervals, whose endpoints are the mesh points $x_i = a + ih$, for $i = 0, 1, \ldots, N+1$, where $h = (b - a)/(N + 1)$.

At the interior mesh points, x_i, for $i = 1, 2, \ldots, N$, the differential equation to be approximated is

(11.5)
$$y''(x_i) = p(x_i)y'(x_i) + q(x_i)y(x_i) + r(x_i).$$

Expanding y in a third-degree Taylor polynomial about x_i evaluated at x_{i+1} and x_{i-1}, we have:

$$y(x_{i+1}) = y(x_i + h) = y(x_i) + hy'(x_i) + \frac{h^2}{2}y''(x_i) + \frac{h^3}{6}y'''(x_i) + \frac{h^4}{24}y^{(4)}(\xi_i^+),$$

for some ξ_i^+ in (x_i, x_{i+1}), and

$$y(x_{i-1}) = y(x_i - h) = y(x_i) - hy'(x_i) + \frac{h^2}{2}y''(x_i) - \frac{h^3}{6}y'''(x_i) + \frac{h^4}{24}y^{(4)}(\xi_i^-),$$

for some ξ_i^- in (x_{i-1}, x_i), assuming $y \in C^4[x_{i-1}, x_{i+1}]$. If these equations are added, the terms involving $y'(x_i)$ and $y'''(x_i)$ are eliminated, and a simple algebraic manipulation gives

$$y''(x_i) = \frac{1}{h^2}[y(x_{i+1}) - 2y(x_i) + y(x_{i-1})] - \frac{h^2}{24}[y^{(4)}(\xi_i^+) + y^{(4)}(\xi_i^-)].$$

The Intermediate Value Theorem can be used to simplify this even further.

Centered-Difference Formula for $y''(x_i)$

$$y''(x_i) = \frac{1}{h^2}[y(x_{i+1}) - 2y(x_i) + y(x_{i-1})] - \frac{h^2}{12}y^{(4)}(\xi_i),$$

for some ξ_i in (x_{i-1}, x_{i+1}).

A centered-difference formula for $y'(x_i)$ is obtained in a similar manner.

Centered-Difference Formula for $y'(x_i)$

$$y'(x_i) = \frac{1}{2h}[y(x_{i+1}) - y(x_{i-1})] - \frac{h^2}{6}y'''(\eta_i),$$

for some η_i in (x_{i-1}, x_{i+1}).

The use of these centered-difference formulas in Eq. (11.5) results in the equation

$$\frac{y(x_{i+1}) - 2y(x_i) + y(x_{i-1})}{h^2} = p(x_i)\left[\frac{y(x_{i+1}) - y(x_{i-1})}{2h}\right] + q(x_i)y(x_i)$$

$$+ r(x_i) - \frac{h^2}{12}[2p(x_i)y'''(\eta_i) - y^{(4)}(\xi_i)].$$

A Finite-Difference method with truncation error of order $O(h^2)$ results from using this equation together with the boundary conditions $y(a) = \alpha$ and $y(b) = \beta$ to define

$$w_0 = \alpha, \qquad w_{N+1} = \beta,$$

and

$$\left(\frac{2w_i - w_{i+1} - w_{i-1}}{h^2}\right) + p(x_i)\left(\frac{w_{i+1} - w_{i-1}}{2h}\right) + q(x_i)w_i = -r(x_i),$$

for each $i = 1, 2, \ldots, N$.

In the form we will consider, the equation is rewritten as

$$-\left(1 + \frac{h}{2}p(x_i)\right)w_{i-1} + (2 + h^2 q(x_i))w_i - \left(1 - \frac{h}{2}p(x_i)\right)w_{i+1} = -h^2 r(x_i),$$

and the resulting system of equations is expressed in the tridiagonal $N \times N$ matrix form $A\mathbf{w} = \mathbf{b}$, where

$$A = \begin{bmatrix} 2 + h^2 q(x_1) & -1 + \frac{h}{2}p(x_1) & 0 & \cdots & \cdots & 0 \\ -1 - \frac{h}{2}p(x_2) & 2 + h^2 q(x_2) & -1 + \frac{h}{2}p(x_2) & & & \vdots \\ 0 & & & & & 0 \\ \vdots & & & & & -1 + \frac{h}{2}p(x_{N-1}) \\ 0 & \cdots & \cdots & 0 & -1 + \frac{h}{2}p(x_N) & 2 + h^2 q(x_N) \end{bmatrix},$$

$$\mathbf{w} = \begin{bmatrix} w_1 \\ w_2 \\ \vdots \\ w_{N-1} \\ w_N \end{bmatrix}, \quad \text{and} \quad \mathbf{b} = \begin{bmatrix} -h^2 r(x_1) + \left(1 + \frac{h}{2}p(x_1)\right)w_0 \\ -h^2 r(x_2) \\ \vdots \\ -h^2 r(x_{N-1}) \\ -h^2 r(x_N) + \left(1 - \frac{h}{2}p(x_N)\right)w_{N+1} \end{bmatrix}.$$

This system has a unique solution provided that p, q, and r are continuous on $[a, b]$, that $q(x) \geq 0$ on $[a, b]$, and that $h < 2/L$, where $L = \max_{a \leq x \leq b} |p(x)|$.

The program LINFD112 implements the Linear Finite Difference method.

EXAMPLE 1 The Linear Finite-Difference method will be used to approximate the solution to the linear boundary-value problem

$$y'' = -\frac{2}{x}y' + \frac{2}{x^2}y + \frac{\sin(\ln x)}{x^2}, \quad \text{for } 1 \leq x \leq 2, \quad \text{where } y(1) = 1 \text{ and } y(2) = 2,$$

which was also approximated by the Shooting method in Example 1 of Section 11.2. For this example, we use the same spacing as in Example 1 of Section 11.2. The results are listed in Table 11.2. ■ ■ ■

Table 11.2

x_i	w_i	$y(x_i)$	$\|w_i - y(x_i)\|$
1.0	1.00000000	1.00000000	
1.1	1.09260052	1.09262930	2.88×10^{-5}
1.2	1.18704313	1.18708484	4.17×10^{-5}
1.3	1.28333687	1.28338236	4.55×10^{-5}
1.4	1.38140205	1.38144595	4.39×10^{-5}
1.5	1.48112026	1.48115942	3.92×10^{-5}
1.6	1.58235990	1.58239246	3.26×10^{-5}
1.7	1.68498902	1.68501396	2.49×10^{-5}
1.8	1.78888175	1.78889853	1.68×10^{-5}
1.9	1.89392110	1.89392951	8.41×10^{-6}
2.0	2.00000000	2.00000000	

Note that these results are considerably less accurate than those obtained in Example 1 of Section 11.2 and listed in Table 11.1 on page 367. This is because the method used in that example involved a Runge–Kutta technique with error of order $O(h^4)$, whereas the difference method used here has error of order $O(h^2)$.

To obtain a difference method with greater accuracy, we can proceed in a number of ways. Using fifth-order Taylor series for approximating $y''(x_i)$ and $y'(x_i)$ results in an error term involving h^4. However, this requires using multiples not only of $y(x_{i+1})$ and $y(x_{i-1})$, but also $y(x_{i+2})$ and $y(x_{i-2})$ in the approximation formulas for $y''(x_i)$ and $y'(x_i)$. This leads to difficulty at $i = 0$ and $i = N$. Moreover, the resulting system of equations is not in tridiagonal form, and the solution to the system requires many more calculations.

Instead of obtaining a difference method with a higher-order error term in this manner, it is generally more satisfactory to consider a reduction in step size. In addition, the Richardson's extrapolation technique can be used effectively for this method, since the error term is expressed in even powers of h with coefficients independent of h, provided y is sufficiently differentiable.

EXAMPLE 2 Richardson's extrapolation for approximating the solution to the boundary-value problem

$$y'' = -\frac{2}{x}y' + \frac{2}{x^2}y + \frac{\sin(\ln x)}{x^2}, \quad \text{for } 1 \leq x \leq 2, \quad \text{where } y(1) = 1 \quad \text{and } y(2) = 2,$$

with $h = 0.1, 0.05,$ and $0.025,$ gives the results listed in Table 11.3. The first extrapolation is

$$\text{Ext}_{1i} = \frac{4w_i(h = 0.05) - w_i(h = 0.1)}{3};$$

the second extrapolation is

$$\text{Ext}_{2i} = \frac{4w_i(h = 0.025) - w_i(h = 0.05)}{3};$$

and the final extrapolation is

$$\text{Ext}_{3i} = \frac{16\text{Ext}_{2i} - \text{Ext}_{1i}}{15}.$$

All the results of Ext_{3i} are correct to the decimal places listed. In fact, if sufficient digits are maintained, this approximation gives results that agree with the exact solution with a maximum error of 6.3×10^{-11}.   ■ ■ ■

Table 11.3

x_i	$w_i(h = 0.1)$	$w_i(h = 0.05)$	$w_i(h = 0.025)$	Ext_{1i}	Ext_{2i}	Ext_{3i}
1.0	1.00000000	1.00000000	1.00000000	1.00000000	1.00000000	1.00000000
1.1	1.09260052	1.09262207	1.09262749	1.09262925	1.09262930	1.09262930
1.2	1.18704313	1.18707436	1.18708222	1.18708477	1.18708484	1.18708484
1.3	1.28333687	1.28337094	1.28337950	1.28338230	1.28338236	1.28338236
1.4	1.38140205	1.38143493	1.38144319	1.38144589	1.38144595	1.38144595
1.5	1.48112026	1.48114959	1.48115696	1.48115937	1.48115941	1.48115942
1.6	1.58235990	1.58238429	1.58239042	1.58239242	1.58239246	1.58239246
1.7	1.68498902	1.68500770	1.68501240	1.68501393	1.68501396	1.68501396
1.8	1.78888175	1.78889432	1.78889748	1.78889852	1.78889853	1.78889853
1.9	1.89392110	1.89392740	1.89392898	1.89392950	1.89392951	1.89392951
2.0	2.00000000	2.00000000	2.00000000	2.00000000	2.00000000	2.00000000

EXERCISE SET 11.3

1. The boundary-value problem

$$y'' = 4(y - x), \qquad 0 \le x \le 1, \qquad y(0) = 0, \qquad y(1) = 2$$

has the solution $y(x) = \dfrac{e^2}{e^4 - 1}(e^{2x} - e^{-2x}) + x$. Use the Linear Finite-Difference method to approximate the solution and compare the results to the actual solution in each case.

   (a) With $h = \frac{1}{3}$
   (b) With $h = \frac{1}{4}$

2. The boundary-value problem

$$y'' = y' + 2y + \cos x, \qquad 0 \le x \le \tfrac{\pi}{2}, \qquad y(0) = -0.3, \qquad y(\tfrac{\pi}{2}) = -0.1$$

has the solution $y(x) = -\frac{1}{10}[\sin x + 3\cos x]$. Use the Linear Finite-Difference method to approximate the solution and compare the results to the actual solution in each case.

(a) With $h = \frac{\pi}{4}$

(b) With $h = \frac{\pi}{6}$

3. Use the Linear Finite-Difference method to approximate the solution to the following boundary-value problems:

(a) $y'' = -3y' + 2y + 2x + 3$, $\quad 0 \le x \le 1$, $y(0) = 2$, $y(1) = 1$; use $h = 0.1$.

(b) $y'' = -4y' + 4y$, $\quad 0 \le x \le 5$, $y(0) = 1$, $y(5) = 0$; $\quad$ use $h = 0.2$.

(c) $y'' = \dfrac{-2(2xy' - y + \ln x)}{x^2}$, $\quad 1 \le x \le 2$, $y(1) = -\frac{1}{2}$, $y(2) = \ln 2$; $\quad$ use $h = 0.05$.

(d) $y'' = -(x + 1)y' + 2y + (1 - x^2)e^{-x}$, $\quad 0 \le x \le 1$, $y(0) = -1$, $y(1) = 0$; use $h = 0.1$.

(e) $y'' = \dfrac{(xy' + 3y + x\ln x - x^2)}{x^2}$, $\quad 1 \le x \le 2$, $y(1) = y(2) = 0$; $\quad$ use $h = 0.1$.

(f) $y'' = \dfrac{(-y' + (x + 1)y + 2e^x)}{x}$, $\quad 1 \le x \le e$, $y(1) = 0$, $y(e) = e^e$; $\quad$ use $h = \frac{1}{10}e$.

4. Although $q(x) < 0$ in the following boundary-value problems, unique solutions exist and are given. Use the Linear Finite-Difference method to approximate the solutions and compare the results to the actual solutions.

(a) $y'' + y = 0$, $\quad 0 \le x \le \frac{\pi}{4}$, $y(0) = 1$, $y(\frac{\pi}{4}) = 1$; $\quad$ use $h = \frac{\pi}{20}$; $\quad$ actual solution $y(x) = \cos x + (\sqrt{2} - 1)\sin x$.

(b) $y'' + 4y = \cos x$, $\quad 0 \le x \le \frac{\pi}{4}$, $y(0) = 0$, $y(\frac{\pi}{4}) = 0$; $\quad$ use $h = \frac{\pi}{20}$; $\quad$ actual solution $y(x) = -\frac{1}{3}\cos 2x - \frac{\sqrt{2}}{6}\sin 2x + \frac{1}{3}\cos x$.

(c) $y'' = \dfrac{-2(2xy' + y - \ln x)}{x^2}$, $\quad 1 \le x \le 2$, $y(1) = \frac{1}{2}$, $y(2) = \ln 2$; $\quad$ use $h = 0.05$; actual solution $y(x) = 4x^{-1} - 2x^{-2} + \ln x - \frac{3}{2}$.

(d) $y'' = 2y' - y + xe^x - x$, $\quad 0 \le x \le 2$, $y(0) = 0$, $y(2) = -4$; $\quad$ use $h = 0.2$; $\quad$ actual solution $y(x) = \frac{1}{6}x^3e^x - \frac{5}{3}xe^x + 2e^x - x - 2$.

5. Use the Linear Finite-Difference method to approximate the solution $y = e^{-10x}$ to the boundary-value problem

$$y'' = 100y, \qquad 0 \le x \le 1, \qquad y(0) = 1, \qquad y(1) = e^{-10}.$$

Use $h = 0.1$ and $h = 0.05$. Compare the results to the actual solution and to Exercise 5 in Section 11.2.

6. Repeat Exercises 3 (a), (b), and (c) using the extrapolation discussed in Example 2.

7. The deflection of a uniformly loaded, long rectangular plate under an axial tension force is governed by a second-order differential equation. Let S represent the axial force and q the intensity of the uniform load. The deflection W along the elemental length is given by:

$$W''(x) - \frac{S}{D}W(x) = \frac{-ql}{2D}x + \frac{q}{2D}x^2, \quad 0 \le x \le l, \ W(0) = W(l) = 0,$$

where l is the length of the plate and D is the flexural rigidity of the plate. Let $q = 200$ lb/in.^{2}, $S = 100$ lb/in., $D = 8.8 \times 10^7$ lb/in., and $l = 50$ in. Use the Linear Finite-Difference method to approximate the deflection at 1-in. intervals.

11.4 The Nonlinear Shooting Method

The shooting technique for the nonlinear second-order boundary-value problem

(11.6) $y'' = f(x, y, y')$, for $a \leq x \leq b$, where $y(a) = \alpha$ and $y(b) = \beta$,

is similar to the linear method, except that the solution to a nonlinear problem cannot be expressed as a linear combination of the solutions to two initial-value problems. Instead, we need to use the solutions to a sequence of initial-value problems of the form

(11.7) $y'' = f(x, y, y')$, for $a \leq x \leq b$, where $y(a) = \alpha$ and $y'(a) = t$,

involving a parameter t, to approximate the solution to the boundary-value problem. We do this by choosing the parameters $t = t_k$ in a manner to ensure that

$$\lim_{k \to \infty} y(b, t_k) = y(b) = \beta,$$

where $y(x, t_k)$ denotes the solution to the initial-value problem (11.7) with $t = t_k$ and $y(x)$ denotes the solution to the boundary-value problem (11.6).

This technique is called a "shooting" method, by analogy to the procedure of firing objects at a stationary target. (See Figure 11.2.) We start with a parameter t_0 that determines the initial elevation at which the object is fired from the point (a, α) along the curve described by the solution to the initial-value problem:

$$y'' = f(x, y, y'), \quad \text{for } a \leq x \leq b, \quad \text{where } y(a) = \alpha \text{ and } y'(a) = t_0.$$

Figure 11.2

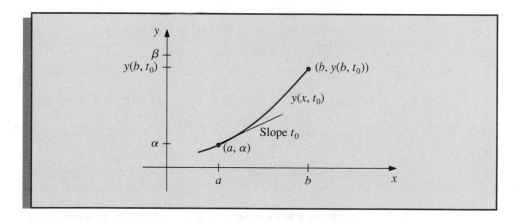

If $y(b, t_0)$ is not sufficiently close to β, we correct our approximation by choosing elevations t_1, t_2, and so on, until $y(b, t_k)$ is sufficiently close to "hitting" β. (See Figure 11.3, page 376.)

The problem is to determine the parameter t in the initial-value problem so that

$$y(b, t) - \beta = 0.$$

Since this is a nonlinear equation of the type considered in Chapter 2, a number of

Figure 11.3

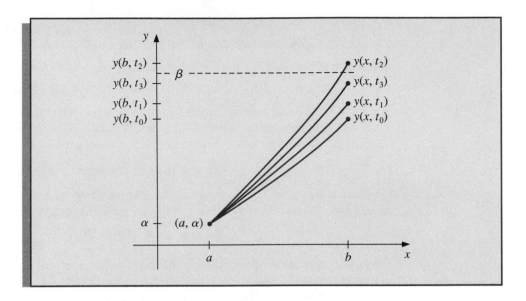

methods are available. To employ the Secant method to solve the problem, we choose initial approximations t_0 and t_1 to t and then generate the remaining terms of the sequence by using the following procedure:

Secant Method Solution

Suppose that t_0 and t_1 are given initial approximations to parameter t that solves the nonlinear boundary-value problem. For each successive $k = 2, 3, \ldots$, solve the initial-value problem

$$y'' = f(x, y, y'), \quad \text{for } a \leq x \leq b, \quad \text{where } y(a) = \alpha,$$

first with $y'(a) = t_{k-2}$ and then with $y'(a) = t_{k-1}$. Then define

$$t_k = t_{k-1} - \frac{(y(b, t_{k-1}) - \beta)(t_{k-1} - t_{k-2})}{y(b, t_{k-1}) - y(b, t_{k-2})}$$

and repeat the process with t_{k-1} replacing t_{k-2} and t_k replacing t_{k-1}.

To use the more powerful Newton's method to generate the sequence $\{t_k\}$, only one initial approximation, t_0, is needed. However, the iteration has the form

$$t_k = t_{k-1} - \frac{y(b, t_{k-1}) - \beta}{(dy/dt)(b, t_{k-1})}, \quad \text{where } (dy/dt)(b, t_{k-1}) \equiv \frac{dy}{dt}(b, t_{k-1}),$$

and requires the knowledge of $(dy/dt)(b, t_{k-1})$. This presents a difficulty, since an explicit representation for $y(b, t)$ is not known; we know only the values $y(b, t_0)$, $y(b, t_1)$, $\ldots, y(b, t_{k-1})$.

To overcome this difficulty, first rewrite the initial-value problem, emphasizing that the solution depends on both x and t:

$$y''(x, t) = f(x, y(x, t), y'(x, t)), \text{ for } a \le x \le b, \text{ where } y(a, t) = \alpha \text{ and } y'(a, t) = t,$$

retaining the prime notation to indicate differentiation with respect to x. Since we are interested in determining $(dy/dt)(b, t)$ when $t = t_{k-1}$, we take the partial derivative with respect to t. This implies that

$$\frac{\partial y''}{\partial t}(x, t) = \frac{\partial f}{\partial t}(x, y(x, t), y'(x, t))$$

$$= \frac{\partial f}{\partial x}(x, y(x, t), y'(x, t))\frac{\partial x}{\partial t} + \frac{\partial f}{\partial y}(x, y(x, t), y'(x, t))\frac{\partial y}{\partial t}(x, t)$$

$$+ \frac{\partial f}{\partial y'}(x, y(x, t), y'(x, t))\frac{\partial y'}{\partial t}(x, t)$$

or, since x and t are independent,

(11.8) $\quad \dfrac{\partial y''}{\partial t}(x, t) = \dfrac{\partial f}{\partial y}(x, y(x, t), y'(x, t))\dfrac{\partial y}{\partial t}(x, t) + \dfrac{\partial f}{\partial y'}(x, y(x, t), y'(x, t))\dfrac{\partial y'}{\partial t}(x, t)$

for $a \le x \le b$. The initial conditions give

$$\frac{\partial y}{\partial t}(a, t) = 0 \qquad \text{and} \qquad \frac{\partial y'}{\partial t}(a, t) = 1.$$

If we simplify the notation by using $z(x, t)$ to denote $(\partial y/\partial t)(x, t)$ and assume that the order of differentiation of x and t can be reversed, Eq. (11.8) becomes the linear initial-value problem

$$z'' = \frac{\partial f}{\partial y}(x, y, y')z + \frac{\partial f}{\partial y'}(x, y, y')z', \text{ for } a \le x \le b, \text{ where } z(a, t) = 0 \text{ and } z'(a, t) = 1.$$

Newton's method therefore requires that two initial-value problems be solved for each iteration.

Newton's Method Solution

Suppose that t_0 is a given initial approximation to the parameter t that solves the nonlinear boundary-value problem. For each successive $k = 1, 2, \ldots$, solve the initial-value problems

$$y'' = f(x, y, y'), \qquad \text{for } a \le x \le b, \qquad \text{where } y(a) = \alpha \text{ and } y'(a) = t_{k-1}$$

and

$$z'' = f_y(x, y, y')z + f_{y'}(x, y, y')z', \qquad \text{for } a \le x \le b,$$

where $\qquad z(a, t) = 0$ and $z'(a, t) = 1$.

Then define

$$t_k = t_{k-1} - \frac{y(b, t_{k-1}) - \beta}{z(b, t_{k-1})}$$

and repeat the process with t_k replacing t_{k-1}.

In practice, none of these initial-value problems is likely to be solved exactly; instead the solutions are approximated by one of the methods discussed in Chapter 5. The program NLINS113 uses the Runge–Kutta method of order four to approximate both solutions required for Newton's method.

EXAMPLE 1 Consider the boundary-value problem

$$y'' = \tfrac{1}{8}(32 + 2x^3 - yy'), \qquad \text{for } 1 \le x \le 3, \qquad \text{where } y(1) = 17 \text{ and } y(3) = \tfrac{43}{3},$$

which has the exact solution $y(x) = x^2 + 16/x$.

Applying the Shooting method to this problem requires approximating the initial-value problems

$$y'' = \tfrac{1}{8}(32 + 2x^3 - yy'), \qquad \text{for } 1 \le x \le 3, \qquad \text{where } y(1) = 17 \text{ and } y'(1) = t_k,$$

and

$$z'' = \frac{\partial f}{\partial y} z + \frac{\partial f}{\partial y'} z' = -\tfrac{1}{8}(y'z + yz'), \text{ for } 1 \le x \le 3, \text{ where } z(1) = 0 \text{ and } z'(1) = 1,$$

at each step in the iteration. If the stopping technique requires 10^{-5} accuracy at the right endpoints, this problem requires four iterations and $t_4 = -14.000203$. The results obtained for this value of t are shown in Table 11.4.   ■ ■ ■

Table 11.4

x_i	$w_{1,i}$	$y(x_i)$	$\lvert w_{1,i} - y(x_i)\rvert$	x_i	$w_{1,i}$	$y(x_i)$	$\lvert w_{1,i} - y(x_i)\rvert$
1.0	17.000000	17.000000		2.0	12.000023	12.000000	2.32×10^{-5}
1.1	15.755495	15.755455	4.06×10^{-5}	2.1	12.029066	12.029048	1.84×10^{-5}
1.2	14.773389	14.773333	5.60×10^{-5}	2.2	12.112741	12.112727	1.40×10^{-5}
1.3	13.997752	13.997692	5.94×10^{-5}	2.3	12.246532	12.246522	1.01×10^{-5}
1.4	13.388629	13.388571	5.71×10^{-5}	2.4	12.426673	12.426667	6.68×10^{-6}
1.5	12.916719	12.916667	5.23×10^{-5}	2.5	12.650004	12.650000	3.61×10^{-6}
1.6	12.560046	12.560000	4.64×10^{-5}	2.6	12.913847	12.913845	9.17×10^{-7}
1.7	12.301805	12.301765	4.02×10^{-5}	2.7	13.215924	13.215926	1.43×10^{-6}
1.8	12.128923	12.128889	3.14×10^{-5}	2.8	13.554282	13.554286	3.46×10^{-6}
1.9	12.031081	12.031053	2.84×10^{-5}	2.9	13.927236	13.927241	5.21×10^{-6}
				3.0	14.333327	14.333333	6.69×10^{-6}

Although Newton's method used with the shooting technique requires the solution of an additional initial-value problem, it will generally be faster than the Secant method. Both methods are only locally convergent, since they require good initial approximations.

EXERCISE SET 11.4

1. Use the Nonlinear Shooting method and the Newton method solution with $h = 0.5$ to approximate the solution to the boundary-value problem

    $$y'' = -(y')^2 - y + \ln x, \qquad \text{for} \qquad 1 \le x \le 2, \qquad \text{where } y(1) = 0 \text{ and } y(2) = \ln 2.$$

    Compare your results to the actual solution $y = \ln x$.

2. Use the Nonlinear Shooting method and the Newton method solution with $h = 0.25$ to approximate the solution to the boundary-value problem

    $$y'' = 2y^3, \qquad \text{for} \qquad 1 \le x \le 2, \qquad \text{where } y(1) = \frac{1}{4} \text{ and } y(2) = \frac{1}{5}.$$

    Compare your results to the actual solution $y(x) = (x + 3)^{-1}$.

3. Use the Nonlinear Shooting method with Newton's method and $TOL = 10^{-4}$ to approximate the solution to the following boundary-value problems. The actual solution is given for comparison to your results.

    (a) $y'' = y^3 - yy'$, $1 \le x \le 2$, $y(1) = \frac{1}{2}$, $y(2) = \frac{1}{3}$; use $h = 0.1$ and compare the results to $y(x) = (x + 1)^{-1}$.

    (b) $y'' = 2y^3 - 6y - 2x^3$, $1 \le x \le 2$, $y(1) = 2$, $y(2) = \frac{5}{2}$; use $h = 0.1$ and compare the results to $y(x) = x + x^{-1}$.

    (c) $y'' = y' + 2(y - \ln x)^3 - x^{-1}$, $1 \le x \le 2$, $y(1) = 1$, $y(2) = \frac{1}{2} + \ln 2$; use $h = 0.1$ and compare the results to $y(x) = x^{-1} + \ln x$.

    (d) $y'' = 2y' - (y')^2 - 2 + e^y - e^x(\cos x + \sin x)$, $0 \le x \le \frac{\pi}{2}$, $y(0) = 0$, $y(\frac{\pi}{2}) = \frac{\pi}{2}$; use $h = \frac{\pi}{20}$ and compare the results to $y(x) = \ln(e^x \cos x + e^x \sin x)$.

    (e) $y'' = \frac{1}{2}y^3$, $1 \le x \le 2$, $y(1) = -\frac{2}{3}$, $y(2) = -1$; use $h = 0.05$ and compare the results to $y(x) = 2(x - 4)^{-1}$.

    (f) $y'' = [x^2(y')^2 - 9y^2 + 4x^4]/x^5$, $1 \le x \le 2$, $y(1) = 0$, $y(2) = \ln 256$; use $h = 0.05$ and compare the results to $y(x) = x^3 \ln x$.

4. Use the Nonlinear Shooting method and the Secant method instead of Newton's method with $t_0 = (\beta - \alpha)/(b - a)$ and $t_1 = t_0 + (\beta - y(b, t_0))/(b - a)$ to solve the problems given in Exercise 3.

11.5 Nonlinear Finite-Difference Methods

The difference method for the general nonlinear boundary-value problem

$$y'' = f(x, y, y'), \qquad \text{for } a \le x \le b, \qquad \text{where } y(a) = \alpha \text{ and } y(b) = \beta,$$

is similar to the method applied to linear problems in Section 11.3. Here, however, the system of equations will not be linear, so an iterative process is required to solve it.

As in the linear case, we divide $[a, b]$ into $(N+1)$ equal subintervals whose endpoints are at $x_i = a + ih$ for $i = 0, 1, \ldots, N + 1$. Assuming that the exact solution has a bounded fourth derivative allows us to replace $y''(x_i)$ and $y'(x_i)$ in each of the equations by the appropriate centered-difference formula to obtain, for each $i = 1, 2, \ldots, N$,

$$\frac{y(x_{i+1}) - 2y(x_i) + y(x_{i-1})}{h^2} = f\left(x_i, y(x_i), \frac{y(x_{i+1}) - y(x_{i-1})}{2h} - \frac{h^2}{6}y'''(\eta_i)\right) + \frac{h^2}{12}y^{(4)}(\xi_i),$$

for some ξ_i and η_i in the interval (x_{i-1}, x_{i+1}).

As in the linear case, the difference method results when the error terms are deleted and the boundary conditions added. This produces the $N \times N$ nonlinear system

$$2w_1 - w_2 + h^2 f\left(x_1, w_1, \frac{w_2 - \alpha}{2h}\right) - \alpha = 0,$$

$$-w_1 + 2w_2 - w_3 + h^2 f\left(x_2, w_2, \frac{w_3 - w_1}{2h}\right) = 0,$$

$$\vdots$$

$$-w_{N-2} + 2w_{N-1} - w_N + h^2 f\left(x_{N-1}, w_{N-1}, \frac{w_N - w_{N-2}}{2h}\right) = 0,$$

$$-w_{N-1} + 2w_N + h^2 f\left(x_N, w_N, \frac{\beta - w_{N-1}}{2h}\right) - \beta = 0.$$

To approximate the solution to this system, we use Newton's method for nonlinear systems, as discussed in Section 10.2. A sequence of iterates $\{(w_1^{(k)}, w_2^{(k)}, \ldots, w_N^{(k)})^t\}$ is generated that converges to the solution of the system, provided that the initial approximation $(w_1^{(0)}, w_2^{(0)}, \ldots, w_N^{(0)})^t$ is sufficiently close to the solution, $(w_1, w_2, \ldots, w_N)^t$.

Newton's method for nonlinear systems requires solving, at each iteration, an $N \times N$ linear system involving the Jacobian matrix. In our case, the Jacobian matrix is tridiagonal, and Crout factorization can be applied. The initial approximations $w_i^{(0)}$ to w_i, for each $i = 1, 2, \ldots, N$, are obtained by passing a straight line through (a, α) and (b, β) and evaluating at x_i.

Since a good initial approximation may be required, an upper bound for k should be specified and, if exceeded, a new initial approximation or a reduction in step size considered.

The program NLFDM114 can be used to employ the Nonlinear Finite-Difference method.

EXAMPLE 1 Applying the Nonlinear Finite-Difference method, with $h = 0.1$, to the nonlinear boundary-value problem

$$y'' = \tfrac{1}{8}(32 + 2x^3 - yy'), \qquad \text{for } 1 \le x \le 3, \qquad \text{where } y(1) = 17 \text{ and } y(3) = \tfrac{43}{3}$$

gives the results in Table 11.5. The stopping procedure used in this example was to iterate until values of successive iterates differed by less than 10^{-8}. This was accomplished with four iterations. The problem in this example is the same as that considered for the Nonlinear Shooting method, Example 1 of Section 11.4. ■ ■ ■

Table 11.5

x_i	w_i	$y(x_i)$	$\|w_i - y(x_i)\|$	x_i	w_i	$y(x_i)$	$\|w_i - y(x_i)\|$
1.0	17.000000	17.000000		2.0	11.997915	12.000000	2.085×10^{-3}
1.1	15.754503	15.755455	9.520×10^{-4}	2.1	12.027142	12.029048	1.905×10^{-3}
1.2	14.771740	14.773333	1.594×10^{-3}	2.2	12.111020	12.112727	1.707×10^{-3}
1.3	13.995677	13.997692	2.015×10^{-3}	2.3	12.245025	12.246522	1.497×10^{-3}
1.4	13.386297	13.388571	2.275×10^{-3}	2.4	12.425388	12.426667	1.278×10^{-3}
1.5	12.914252	12.916667	2.414×10^{-3}	2.5	12.648944	12.650000	1.056×10^{-3}
1.6	12.557538	12.560000	2.462×10^{-3}	2.6	12.913013	12.913846	8.335×10^{-4}
1.7	12.299326	12.301765	2.438×10^{-3}	2.7	13.215312	13.215926	6.142×10^{-4}
1.8	12.126529	12.128889	2.360×10^{-3}	2.8	13.553885	13.554286	4.006×10^{-4}
1.9	12.028814	12.031053	2.239×10^{-3}	2.9	13.927046	13.927241	1.953×10^{-4}
				3.0	14.333333	14.333333	

Richardson's extrapolation can also be used for the Nonlinear Finite-Difference method. Table 11.6 lists the results when this method is applied to our example using $h = 0.1, 0.05$, and 0.025, with four iterations in each case. The notation is the same as in Example 1 of Section 11.4, and the values of EXT_{3i} are all accurate to the places listed, with an actual maximum error of 3.68×10^{-10}. The values of $w_i(h = 0.1)$ are omitted from the table, since they were listed previously.

Table 11.6

x_i	$w_i\ (h = 0.05)$	$w_i\ (h = 0.025)$	Ext_{1i}	Ext_{2i}	Ext_{3i}
1.0	17.00000000	17.00000000	17.00000000	17.00000000	17.00000000
1.1	15.75521721	15.75539525	15.75545543	15.75545460	15.75545455
1.2	14.77293601	14.77323407	14.77333479	14.77333342	14.77333333
1.3	13.99718996	13.99756690	13.99769413	13.99769242	13.99769231
1.4	13.38800424	13.38842973	13.38857346	13.38857156	13.38857143
1.5	12.91606471	12.91651628	12.91666881	12.91666680	12.91666667
1.6	12.55938618	12.55984665	12.56000217	12.56000014	12.56000000
1.7	12.30115670	12.30161280	12.30176684	12.30176484	12.30176471
1.8	12.12830042	12.12874287	12.12899094	12.12888902	12.12888889
1.9	12.03049438	12.03091316	12.03105457	12.03105275	12.03105263
2.0	11.99948020	11.99987013	12.00000179	12.00000011	12.00000000
2.1	12.02857252	12.02892892	12.02902924	12.02904772	12.02904762
2.2	12.11230149	12.11262089	12.11272872	12.11272736	12.11272727
2.3	12.24614846	12.24642848	12.24652299	12.24652182	12.24652174
2.4	12.42634789	12.42658702	12.42666773	12.42666673	12.42666667
2.5	12.64973666	12.64993420	12.65000086	12.65000005	12.65000000
2.6	12.91362828	12.91379422	12.91384683	12.91384620	12.91384615
2.7	13.21577275	13.21588765	13.21592641	13.21592596	13.21592593
2.8	13.55418579	13.55426075	13.55428603	13.55428573	13.55428571
2.9	13.92719268	13.92722921	13.92724153	13.92724139	13.92724138
3.0	14.33333333	14.33333333	14.33333333	14.33333333	14.33333333

EXERCISE SET 11.5

1. Use the Nonlinear Finite-Difference method with $h = 0.5$ to approximate the solution to the boundary-value problem

$$y'' = -(y')^2 - y + \ln x, \qquad 1 \le x \le 2, \qquad y(1) = 0, \qquad y(2) = \ln 2.$$

Compare your results to the actual solution $y = \ln x$.

2. Use the Nonlinear Finite-Difference method with $h = 0.25$ to approximate the solution to the boundary-value problem

$$y'' = 2y^3, \qquad 1 \le x \le 2, \qquad y(1) = \tfrac{1}{4}, \qquad y(2) = \tfrac{1}{5}.$$

Compare your results to the actual solution $y(x) = (x + 3)^{-1}$.

3. Use the Nonlinear Finite-Difference method with $TOL = 10^{-4}$ to approximate the solution to the following boundary-value problems. The actual solution is given for comparison to your results.

   (a) $y'' = y^3 - yy'$, $\;1 \le x \le 2$, $\;y(1) = \tfrac{1}{2}$, $\;y(2) = \tfrac{1}{3}$; use $h = 0.1$ and compare the results to $y(x) = (x + 1)^{-1}$.

   (b) $y'' = 2y^3 - 6y - 2x^3$, $\;1 \le x \le 2$, $\;y(1) = 2$, $\;y(2) = \tfrac{5}{2}$; use $h = 0.1$ and compare the results to $y(x) = x + x^{-1}$.

   (c) $y'' = y' + 2(y - \ln x)^3 - x^{-1}$, $\;1 \le x \le 2, y(1) = 1, y(2) = \tfrac{1}{2} + \ln 2$; use $h = 0.1$ and compare the results to $y(x) = x^{-1} + \ln x$.

   (d) $y'' = 2y' - (y')^2 - 2 + e^y - e^x(\cos x + \sin x)$, $\;0 \le x \le \tfrac{\pi}{2}$, $\;y(0) = 0$, $\;y(\tfrac{\pi}{2}) = \tfrac{\pi}{2}$; use $h = \tfrac{\pi}{20}$ and compare the results to $y(x) = \ln(e^x \cos x + e^x \sin x)$.

   (e) $y'' = \tfrac{1}{2}y^3$, $\;1 \le x \le 2$, $\;y(1) = -\tfrac{2}{3}, y(2) = -1$; use $h = 0.05$ and compare the results to $y(x) = 2(x - 4)^{-1}$.

   (f) $y'' = [x^2(y')^2 - 9y^2 + 4x^4]/x^5$, $\;1 \le x \le 2, y(1) = 0$, $\;y(2) = \ln 256$; use $h = 0.05$ and compare results to $y(x) = x^3 \ln x$.

4. Repeat Exercises 3 (a) and (b) using extrapolation.

11.6 Variational Techniques

The Shooting method for approximating the solution to a boundary-value problem replaced it with an initial-value problem. The finite-difference approach replaces the continuous operation of differentiation with the discrete operation of finite differences. The Rayleigh–Ritz method is a variational technique that attacks the problem from a third approach. The boundary-value problem is first reformulated as a problem of choosing, from the set of all sufficiently differentiable functions satisfying the boundary conditions, the function to minimize a certain integral. Then the set of feasible functions is reduced in size, to result in an approximation to the solution to the minimization problem and (as a consequence) an approximation to the solution to the boundary-value problem.

To describe the Rayleigh–Ritz method, we consider approximating the solution to a linear two-point boundary-value problem from beam stress analysis. This boundary-value problem is described by the differential equation

$$-\frac{d}{dx}\left(p(x)\frac{dy}{dx}\right) + q(x)y = f(x) \qquad \text{for } 0 \le x \le 1,$$

with the boundary conditions

$$y(0) = y(1) = 0.$$

The differential equation describes the deflection $y(x)$ on a beam of length one with variable cross section given by $q(x)$. The deflection is due to the added stresses $p(x)$ and $f(x)$.

As is the case with many boundary-value problems that describe physical phenomena, the solution to the beam equation satisfies a variational property. The solution to the beam equation is the function that minimizes a certain integral over all functions in $C_0^2[0, 1]$, the set of those functions u on $[0, 1]$ that have two continuous derivatives and satisfy $u(0) = u(1) = 0$.

Variational Property for the Beam Equation

The function $y \in C_0^2[0, 1]$ is the unique solution to the boundary-value problem

$$-\frac{d}{dx}\left(p(x)\frac{dy}{dx}\right) + q(x)y = f(x), \qquad \text{for } 0 \le x \le 1,$$

if and only if y is the unique function in $C_0^2[0, 1]$ that minimizes the integral

$$I[u] = \int_0^1 \{p(x)[u'(x)]^2 + q(x)[u(x)]^2 - 2f(x)u(x)\}dx.$$

The Rayleigh–Ritz method approximates the solution y by minimizing the integral, not over all the functions in $C_0^2[0, 1]$, but over a smaller set of functions consisting of linear combinations of certain basis functions $\phi_1, \phi_2, \ldots, \phi_n$. The basis functions must be linearly independent and satisfy

$$\phi_i(0) = \phi_i(1) = 0, \qquad \text{for each } i = 1, 2, \ldots, n.$$

An approximation $\phi(x) = \sum_{i=1}^n c_i \phi_i(x)$ to the solution $y(x)$ is obtained by finding constants $c_1, c_2, \ldots, c_n$ to minimize $I\left[\sum_{i=1}^n c_i \phi_i\right]$.

From the variational property,

$$I[\phi] = I\left[\sum_{i=1}^n c_i \phi_i\right]$$

$$= \int_0^1 \left\{ p(x)\left[\sum_{i=1}^n c_i \phi_i'(x)\right]^2 + q(x)\left[\sum_{i=1}^n c_i \phi_i(x)\right]^2 - 2f(x)\sum_{i=1}^n c_i \phi_i(x) \right\} dx,$$

and, for a minimum to occur, it is necessary to have

$$\frac{\partial I}{\partial c_j} = 0, \quad \text{for each } j = 1, 2, \ldots, n.$$

Differentiating with respect to the coefficients gives

$$\frac{\partial I}{\partial c_j} = \int_0^1 \left\{ 2p(x) \sum_{i=1}^n c_i \phi_i'(x)\phi_j'(x) + 2q(x) \sum_{i=1}^n c_i \phi_i(x)\phi_j(x) - 2f(x)\phi_j(x) \right\} dx,$$

so $\quad 0 = \sum_{i=1}^n \left[\int_0^1 \{ p(x)\phi_i'(x)\phi_j'(x) + q(x)\phi_i(x)\phi_j(x) \} dx \right] c_i - \int_0^1 f(x)\phi_j(x) \, dx,$

for each $j = 1, 2, \ldots, n$. These *normal equations* produce an $n \times n$ linear system $A\mathbf{c} = \mathbf{b}$ in the variables $c_1, c_2, \ldots, c_n$, where the symmetric matrix A is given by

$$a_{ij} = \int_0^1 [p(x)\phi_i'(x)\phi_j'(x) + q(x)\phi_i(x)\phi_j(x)] \, dx,$$

and $\mathbf{b}$ has coordinates

$$b_i = \int_0^1 f(x)\phi_i(x) \, dx.$$

The most elementary choice of basis functions involves piecewise linear polynomials. The first step is to form a partition of $[0, 1]$ by choosing points $x_0, x_1, \ldots, x_{n+1}$ with

$$0 = x_0 < x_1 < \cdots < x_n < x_{n+1} = 1.$$

Let $h_i = x_{i+1} - x_i$ for each $i = 0, 1, \ldots, n$, and define the basis functions $\phi_1(x)$, $\phi_2(x), \ldots, \phi_n(x)$ by

$$\phi_i(x) = \begin{cases} 0, & 0 \le x \le x_{i-1}, \\ \dfrac{(x - x_{i-1})}{h_{i-1}}, & x_{i-1} < x \le x_i, \\ \dfrac{(x_{i+1} - x)}{h_i}, & x_i < x \le x_{i+1}, \\ 0, & x_{i+1} < x \le 1, \end{cases}$$

for each $i = 1, 2, \ldots, n$. (See Figure 11.4.)

Figure 11.4

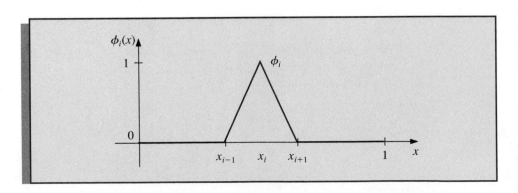

Since the functions ϕ_i are piecewise linear, the derivatives ϕ_i', while not continuous, are constant on the open subinterval (x_j, x_{j+1}) for each $j = 0, 1, \ldots, n$. Thus, we have

$$\phi_i'(x) = \begin{cases} 0, & 0 < x < x_{i-1}, \\ \dfrac{1}{h_{i-1}}, & x_{i-1} < x < x_i, \\ -\dfrac{1}{h_i}, & x_i < x < x_{i+1}, \\ 0, & x_{i+1} < x < 1, \end{cases}$$

for each $i = 1, 2, \ldots, n$. Because ϕ_i and ϕ_i' are nonzero only on (x_{i-1}, x_{i+1}),

$$\phi_i(x)\phi_j(x) \equiv 0 \qquad \text{and} \qquad \phi_i'(x)\phi_j'(x) \equiv 0,$$

except when j is $i - 1, i$, or $i + 1$. As a consequence, the linear system reduces to an $n \times n$ tridiagonal linear system. The nonzero entries in A are

$$a_{ii} = \int_0^1 \{p(x)[\phi_i'(x)]^2 + q(x)[\phi_i(x)]^2\}\, dx$$

$$= \int_{x_{i-1}}^{x_i} \left(\frac{1}{h_{i-1}}\right)^2 p(x)\, dx + \int_{x_i}^{x_{i+1}} \left(\frac{-1}{h_i}\right)^2 p(x)\, dx$$

$$+ \int_{x_{i-1}}^{x_i} \left(\frac{1}{h_{i-1}}\right)^2 (x - x_{i-1})^2 q(x)\, dx + \int_{x_i}^{x_{i+1}} \left(\frac{1}{h_i}\right)^2 (x_{i+1} - x)^2 q(x)\, dx,$$

for each $i = 1, 2, \ldots, n$;

$$a_{i,i+1} = \int_0^1 \{p(x)\phi_i'(x)\phi_{i+1}'(x) + q(x)\phi_i(x)\phi_{i+1}(x)\}\, dx$$

$$= \int_{x_i}^{x_{i+1}} -\left(\frac{1}{h_i}\right)^2 p(x)\, dx + \int_{x_i}^{x_{i+1}} \left(\frac{1}{h_i}\right)^2 (x_{i+1} - x)(x - x_i) q(x)\, dx,$$

for each $i = 1, 2, \ldots, n - 1$; and

$$a_{i,i-1} = \int_0^1 \{p(x)\phi_i'(x)\phi_{i-1}'(x) + q(x)\phi_i(x)\phi_{i-1}(x)\}\, dx$$

$$= \int_{x_{i-1}}^{x_i} -\left(\frac{1}{h_{i-1}}\right)^2 p(x)\, dx + \int_{x_{i-1}}^{x_i} \left(\frac{1}{h_{i-1}}\right)^2 (x_i - x)(x - x_{i-1}) q(x)\, dx,$$

for each $i = 2, \ldots, n$. The entries in **b** are

$$b_i = \int_0^1 f(x)\phi_i(x)\, dx = \int_{x_{i-1}}^{x_i} \frac{1}{h_{i-1}}(x - x_{i-1})f(x)\, dx + \int_{x_i}^{x_{i+1}} \frac{1}{h_i}(x_{i+1} - x)f(x)\, dx,$$

for each $i = 1, 2, \ldots, n$.

Although it appears as if there are ten types of integrals to be evaluated, there are actually only six types:

$$Q_{1,i} = \left(\frac{1}{h_i}\right)^2 \int_{x_i}^{x_{i+1}} (x_{i+1} - x)(x - x_i)q(x)\,dx, \qquad \text{for each } i = 1, 2, \ldots, n-1,$$

$$Q_{2,i} = \left(\frac{1}{h_{i-1}}\right)^2 \int_{x_{i-1}}^{x_i} (x - x_{i-1})^2 q(x)\,dx, \qquad \text{for each } i = 1, 2, \ldots, n,$$

$$Q_{3,i} = \left(\frac{1}{h_i}\right)^2 \int_{x_i}^{x_{i+1}} (x_{i+1} - x)^2 q(x)\,dx, \qquad \text{for each } i = 1, 2, \ldots, n,$$

$$Q_{4,i} = \left(\frac{1}{h_{i-1}}\right)^2 \int_{x_{i-1}}^{x_i} p(x)\,dx, \qquad \text{for each } i = 1, 2, \ldots, n+1,$$

$$Q_{5,i} = \frac{1}{h_{i-1}} \int_{x_{i-1}}^{x_i} (x - x_{i-1})f(x)\,dx, \qquad \text{for each } i = 1, 2, \ldots, n,$$

and

$$Q_{6,i} = \frac{1}{h_i} \int_{x_i}^{x_{i+1}} (x_{i+1} - x)f(x)\,dx, \qquad \text{for each } i = 1, 2, \ldots, n.$$

Then

$$a_{i,i} = Q_{4,i} + Q_{4,i+1} + Q_{2,i} + Q_{3,i}, \qquad \text{for each } i = 1, 2, \ldots, n,$$
$$a_{i,i+1} = -Q_{4,i+1} + Q_{1,i}, \qquad \text{for each } i = 1, 2, \ldots, n-1,$$
$$a_{i,i-1} = -Q_{4,i} + Q_{1,i-1}, \qquad \text{for each } i = 2, 3, \ldots, n,$$

and $\quad b_i = Q_{5,i} + Q_{6,i}, \qquad \text{for each } i = 1, 2, \ldots, n.$

The entries in $\mathbf{c}$ are the unknown coefficients $c_1, c_2, \ldots, c_n$, from which the Rayleigh–Ritz approximation ϕ, given by $\phi(x) = \sum_{i=1}^n c_i \phi_i(x)$, is constructed.

A practical difficulty with this method is the necessity of evaluating $6n$ integrals. The integrals can be evaluated either directly or by a quadrature formula such as Simpson's method.

An alternative approach for the integral evaluation is to approximate each of the functions p, q, and f with its piecewise linear interpolating polynomial and then integrate the approximation. Consider, for example, the integral $Q_{1,i}$. The piecewise linear interpolation of q is given by

$$P_q(x) = \sum_{i=0}^{n+1} q(x_i)\phi_i(x),$$

where $\phi_1, \ldots, \phi_n$ are the piecewise linear basis functions and

$$\phi_0(x) = \begin{cases} \dfrac{x_1 - x}{x_1}, & 0 \le x \le x_1, \\ 0, & \text{elsewhere,} \end{cases} \quad \text{and} \quad \phi_{n+1}(x) = \begin{cases} \dfrac{x - x_n}{1 - x_n}, & x_n \le x \le 1, \\ 0, & \text{elsewhere.} \end{cases}$$

since the interval of integration is $[x_i, x_{i+1}]$, P_q reduces to

$$P_q(x) = q(x_i)\phi_i(x) + q(x_{i+1})\phi_{i+1}(x).$$

This is the first-degree interpolating polynomial studied in Section 3.2, with error

$$|q(x) - P_q(x)| = O(h_i^2), \qquad \text{when } x_i \le x \le x_{i+1},$$

if $q \in C^2[x_i, x_{i+1}]$. For $i = 1, 2, \ldots, n - 1$, the approximation to $Q_{1,i}$ is obtained by integrating the approximation to the integrand

$$Q_{1,i} = \left(\frac{1}{h_i}\right)^2 \int_{x_i}^{x_{i+1}} (x_{i+1} - x)(x - x_i)q(x)\, dx$$

$$\approx \left(\frac{1}{h_i}\right)^2 \int_{x_i}^{x_{i+1}} (x_{i+1} - x)(x - x_i)\left[\frac{q(x_i)(x_{i+1} - x)}{h_i} + \frac{q(x_{i+1})(x - x_i)}{h_i}\right] dx$$

$$= \frac{h_i}{12}[q(x_i) + q(x_{i+1})],$$

with

$$\left| Q_{1,i} - \frac{h_i}{12}[q(x_i) + q(x_{i+1})] \right| = O(h_i^3).$$

Approximations to the other integrals are derived in a similar manner and are given by

$$Q_{2,i} \approx \frac{h_{i-1}}{12}[3q(x_i) + q(x_{i-1})], \qquad Q_{3,i} \approx \frac{h_i}{12}[3q(x_i) + q(x_{i+1})],$$

$$Q_{4,i} \approx \frac{h_{i-1}}{2}[p(x_i) + p(x_{i-1})], \qquad Q_{5,i} \approx \frac{h_{i-1}}{6}[2f(x_i) + f(x_{i-1})],$$

and

$$Q_{6,i} \approx \frac{h_i}{6}[2f(x_i) + f(x_{i+1})].$$

The program PLRRM115 sets up the tridiagonal linear system and incorporates Crout factorization for tridiagonal systems to solve the system. The integrals $Q_{1,i}, \ldots, Q_{6,i}$ can be computed by one of the methods just mentioned. Because of the elementary nature of the following example, the integrals were found directly.

EXAMPLE 1 Consider the boundary-value problem

$$-y'' + \pi^2 y = 2\pi^2 \sin(\pi x), \qquad \text{for } 0 \le x \le 1, \qquad \text{where } y(0) = y(1) = 0.$$

Let $h_i = h = 0.1$, so that $x_i = 0.1i$ for each $i = 0, 1, \ldots, 9$. The integrals are

$$Q_{1,i} = 100 \int_{0.1i}^{0.1i+0.1} (0.1i + 0.1 - x)(x - 0.1i)\pi^2\, dx = \frac{\pi^2}{60},$$

$$Q_{2,i} = 100 \int_{0.1i-0.1}^{0.1i} (x - 0.1i + 0.1)^2\pi^2\, dx = \frac{\pi^2}{30},$$

$$Q_{3,i} = 100 \int_{0.1i}^{0.1i+0.1} (0.1i + 0.1 - x)^2\pi^2\, dx = \frac{\pi^2}{30},$$

$$Q_{4,i} = 100 \int_{0.1i-0.1}^{0.1i} dx = 10,$$

$$Q_{5,i} = 10 \int_{0.1i-0.1}^{0.1i} (x - 0.1i + 0.1)2\pi^2 \sin \pi x\, dx$$

$$= -2\pi \cos 0.1\pi i + 20[\sin(0.1\pi i) - \sin((0.1i - 0.1)\pi)],$$

and
$$Q_{6,i} = 10 \int_{0.1i}^{0.1i+0.1} (0.1i + 0.1 - x)2\pi^2 \sin \pi x \ dx$$
$$= 2\pi \cos 0.1\pi i - 20[\sin((0.1i + 0.1)\pi) - \sin(0.1\pi i)].$$

The linear system $A\mathbf{c} = \mathbf{b}$ has

$$a_{i,i} = 20 + \frac{\pi^2}{15}, \quad \text{for each } i = 1, 2, \ldots, 9,$$

$$a_{i,i+1} = -10 + \frac{\pi^2}{60}, \quad \text{for each } i = 1, 2, \ldots, 8,$$

$$a_{i,i-1} = -10 + \frac{\pi^2}{60}, \quad \text{for each } i = 2, 3, \ldots, 9,$$

and
$$b_i = 40 \sin(0.1\pi i)[1 - \cos 0.1\pi], \quad \text{for each } i = 1, 2, \ldots, 9.$$

The solution to the tridiagonal linear system is

$$\begin{array}{lll}
c_9 = 0.3102866742, & c_6 = 0.9549641893, & c_3 = 0.8123410598, \\
c_8 = 0.5902003271, & c_5 = 1.004108771, & c_2 = 0.5902003271, \\
c_7 = 0.8123410598, & c_4 = 0.9549641893, & c_1 = 0.3102866742.
\end{array}$$

The piecewise linear approximation is

$$\phi(x) = \sum_{i=1}^{9} c_i \phi_i(x).$$

The actual solution to the boundary-value problem is

$$y(x) = \sin \pi x.$$

Table 11.7 lists the error in the approximation at x_i for each $i = 1, \ldots, 9$.

◼ ◼ ◼

Table 11.7

i	x_i	$\phi(x_i)$	$y(x_i)$	$|\phi(x_i) - y(x_i)|$
1	0.1	0.3102866742	0.3090169943	0.00127
2	0.2	0.5902003271	0.5877852522	0.00242
3	0.3	0.8123410598	0.8090169943	0.00332
4	0.4	0.9549641896	0.9510565162	0.00391
5	0.5	1.0041087710	1.0000000000	0.00411
6	0.6	0.9549641893	0.9510565162	0.00391
7	0.7	0.8123410598	0.8090169943	0.00332
8	0.8	0.5902003271	0.5877852522	0.00242
9	0.9	0.3102866742	0.3090169943	0.00127

The tridiagonal matrix A given by the piecewise linear basis functions is positive definite, so the linear system is stable with respect to round-off error and

$$|\phi(x) - y(x)| = O(h^2), \qquad 0 \le x \le 1.$$

The use of piecewise linear basis functions results in an approximate solution that is continuous but not differentiable on [0, 1]. A more complicated set of basis functions is required to construct an approximation that belongs to $C_0^2[0, 1]$. These basis functions are similar to the cubic interpolatory splines discussed in Section 3.6.

The cubic *interpolatory* spline S on the five nodes x_0, x_1, x_2, x_3, and x_4 for a function f is defined as follows:

(a) S is a cubic polynomial, denoted by S_j, on $[x_j, x_{j+1}]$ for $j = 0, 1, 2, 3$. (*This gives 16 selectable constants for S, 4 for each cubic.*)

(b) $S(x_j) = f(x_j)$, for $j = 0, 1, 2, 3, 4$ (5 *specified conditions*).

(c) $S_{j+1}(x_{j+1}) = S_j(x_{j+1})$, for $j = 0, 1, 2$ (3 *specified conditions*).

(d) $S'_{j+1}(x_{j+1}) = S'_j(x_{j+1})$, for $j = 0, 1, 2$ (3 *specified conditions*).

(e) $S''_{j+1}(x_{j+1}) = S''(x_{j+1})$, for $j = 0, 1, 2$ (3 *specified conditions*).

(f) One of the following boundary conditions is satisfied:

   i. Free: $S''(x_0) = S''(x_4) = 0$ (2 *specified conditions*).

   ii. Clamped: $S'(x_0) = f''(x_0)$ and $S'(x_4) = f'(x_4)$ (2 *specified conditions*).

Since uniqueness of solution requires that the number of constants in (a), 16, must equal the number of conditions in (b) through (f), only *one* of the boundary conditions in (f) can be specified for the interpolatory cubic splines. The cubic spline functions we use for our basis functions are called **B-splines**, or *bell-shaped splines*. These differ from interpolatory splines in that both sets of boundary conditions in (f) are satisfied. This requires the relaxation of two of the conditions in (b) through (e). The spline must have two continuous derivatives on $[x_0, x_4]$, so we are forced to delete two of the interpolation conditions. In particular, we modify condition (b) to

(b') $S(x_j) = f(x_j)$, for $j = 0, 2, 4$.

The basic B-spline S defined next uses the equally spaced nodes $x_0 = -2$, $x_1 = -1$, $x_2 = 0$, $x_3 = 1$, and $x_4 = 2$. It satisfies the interpolatory conditions

(b') $S(x_0) = 0$, $S(x_2) = 1$, $S(x_4) = 0$

as well as both sets of conditions

i. $S''(x_0) = S''(x_4) = 0$ and   **ii.** $S'(x_0) = S'(x_4) = 0$.

As a consequence, S has two continuous derivatives and is defined by

$$S(x) = \begin{cases} 0, & x \le -2; \\ \frac{1}{4}[(2-x)^3 - 4(1-x)^3 - 6x^3 + 4(1+x)^3], & -2 \le x \le -1; \\ \frac{1}{4}[(2-x)^3 - 4(1-x)^3 - 6x^3], & -1 < x \le 0; \\ \frac{1}{4}[(2-x)^3 - 4(1-x)^3], & 0 < x \le 1; \\ \frac{1}{4}[(2-x)^3], & 1 < x \le 2; \\ 0, & 2 < x. \end{cases}$$

To construct the basis functions ϕ_i in $C_0^2[0, 1]$, we first partition $[0, 1]$ by choosing a positive integer n and defining $h = 1/(n+1)$. This produces the equally spaced nodes $x_i = ih$, for each $i = 0, 1, \ldots, n + 1$. We then define S_i by

$$S_i(x) = S\left(\frac{x - x_i}{h}\right)$$

for each $i = 0, 1, \ldots, n + 1$. The graph of a typical S_i is shown in Figure 11.6, which is a simple translation and expansion or compression of the graph of S shown in Figure 11.5.

Figure 11.5

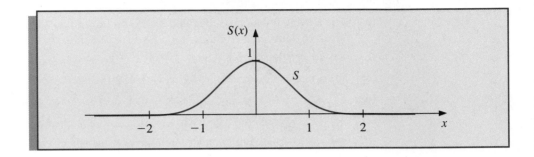

When $2 \leq i \leq n - 1$, the graph of ϕ_i is the same as the graph of S_i shown in Figure 11.6. The graphs of ϕ_i for $i = 0, 1, n$, and $n + 1$ are shown in Figure 11.7.

Figure 11.6

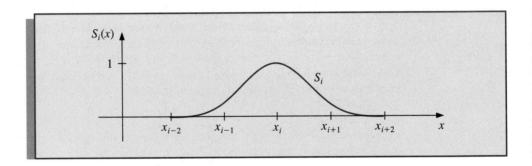

For the linearly independent set $\{S_i\}_{i=0}^{n+1}$ to satisfy the boundary conditions $\phi_i(0) = \phi_i(1) = 0$, we must modify S_0, S_1, S_n, and S_{n+1}. The basis with this modification is defined by

$$\phi_i(x) = \begin{cases} S_0(x) - 4S\left(\dfrac{x+h}{h}\right), & \text{for } i = 0, \\[2mm] S_1(x) - S\left(\dfrac{x+h}{h}\right), & \text{for } i = 1, \\[2mm] S_i(x), & \text{for } 2 \leq i \leq n - 1, \\[2mm] S_n(x) - S\left(\dfrac{x - (n+2)h}{h}\right), & \text{for } i = n, \\[2mm] S_{n+1}(x) - 4S\left(\dfrac{x - (n+2)h}{h}\right), & \text{for } i = n + 1. \end{cases}$$

Figure 11.7

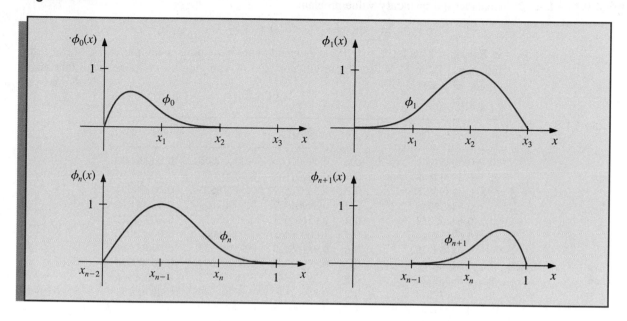

Since $\phi_i(x)$ and $\phi_i'(x)$ are nonzero only for $x_{i-2} \le x \le x_{i+2}$, the matrix in the Rayleigh–Ritz approximation is a band matrix with band width at most seven:

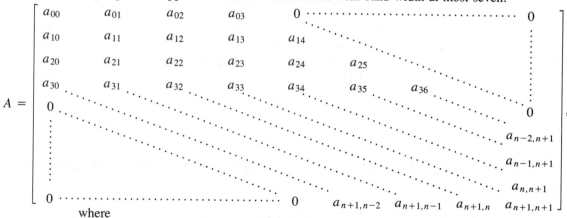

where

$$a_{ij} = \int_0^1 \{p(x)\phi_i'(x)\phi_j'(x) + q(x)\phi_i(x)\phi_j(x)\}\,dx,$$

for each $i, j = 0, 1, \ldots, n + 1$. The matrix A is positive definite, so the linear system can be quickly solved by Choleski's method or by Gaussian elimination. The program CSRRM116 constructs the cubic spline approximation described by the Rayleigh–Ritz technique.

EXAMPLE 2 Consider the boundary-value problem

$$-y'' + \pi^2 y = 2\pi^2 \sin(\pi x), \qquad \text{for } 0 \le x \le 1, \qquad \text{where } y(0) = y(1) = 0.$$

In Example 1 we let $h = 0.1$ and generated approximations using piecewise linear basis functions. Table 11.8 lists the results obtained by applying the B-splines with this same choice of nodes. ■ ■ ■

Table 11.8

i	c_i	x_i	$\phi(x_i)$	$y(x_i)$	$\|y(x_i) - \phi(x_i)\|$
0	$0.50964361 \times 10^{-5}$	0	0.00000000	0.00000000	0.00000000
1	0.20942608	0.1	0.30901644	0.30901699	0.00000055
2	0.39835678	0.2	0.58778549	0.58778525	0.00000024
3	0.54828946	0.3	0.80901687	0.80901699	0.00000012
4	0.64455358	0.4	0.95105667	0.95105652	0.00000015
5	0.67772340	0.5	1.00000002	1.00000000	0.00000020
6	0.64455370	0.6	0.95105713	0.95105652	0.00000061
7	0.54828951	0.7	0.80901773	0.80901699	0.00000074
8	0.39835730	0.8	0.58778690	0.58778525	0.00000165
9	0.20942593	0.9	0.30901810	0.30901699	0.00000111
10	$0.74931285 \times 10^{-5}$	1.0	0.00000000	0.00000000	0.00000000

The integrations should be performed in two steps, as was done in the Piecewise Linear method. First, construct cubic spline interpolatory polynomials for p, q, and f using the methods presented in Section 3.6. Then approximate the integrands by products of cubic splines or derivatives of cubic splines. Since these integrands are piecewise polynomials, they can be integrated exactly on each subinterval and then summed.

In general, this technique produces approximations $\phi(x)$ to $y(x)$ that satisfy

$$\left\{ \int_0^1 |y(x) - \phi(x)|^2 \, dx \right\}^{\frac{1}{2}} = O(h^4), \qquad 0 \le x \le 1.$$

EXERCISE SET 11.6

1. Use the Piecewise Linear method to approximate the solution to the boundary-value problem

$$y'' + \frac{\pi^2}{4} y = \frac{\pi^2}{16} \cos \frac{\pi}{4} x, \qquad 0 \le x \le 1, \qquad y(0) = y(1) = 0,$$

using $x_0 = 0$, $x_1 = 0.3$, $x_2 = 0.7$, $x_3 = 1$ and compare the results to the actual solution

$$y(x) = -\frac{1}{3} \cos \frac{\pi}{2} x - \frac{\sqrt{2}}{6} \sin \frac{\pi}{2} x + \frac{1}{3} \cos \frac{\pi}{4} x.$$

2. Use the Piecewise Linear method to approximate the solution to the boundary-value problem

$$-\frac{d}{dx}(xy') + 4y = 4x^2 - 8x + 1, \qquad 0 \le x \le 1, \qquad y(0) = y(1) = 0,$$

using $x_0 = 0$, $x_1 = 0.4$, $x_2 = 0.8$, $x_3 = 1$ and compare the results to the actual solution $y(x) = x^2 - x$.

3. Use the Piecewise Linear method to approximate the solutions to the following boundary-value problems and compare the results to the actual solution:

   (a) $-y'' + y = x$, $0 \le x \le 1$, $y(0) = y(1) = 0$; use $h = 0.1$; actual solution
   $$y(x) = x + \left(\frac{e}{e^2 - 1} \right) (e^{-x} - e^x).$$

   (b) $-x^2 y'' - 2xy' + 2y = -4x^2$, $0 \le x \le 1$, $y(0) = y(1) = 0$; use $h = 0.1$; actual solution $y(x) = x^2 - x$.

   (c) $-\dfrac{d}{dx}(e^x y') + e^x y = x + (2 - x)e^x$, $0 \le x \le 1$, $y(0) = y(1) = 0$; use $h = 0.1$; actual solution $y(x) = (x - 1)(e^{-x} - 1)$.

   (d) $-\dfrac{d}{dx}(e^{-x} y') + e^{-x} y = (x - 1) - (x + 1)e^{-(x-1)}$, $0 \le x \le 1$, $y(0) = y(1) = 0$; use $h = 0.05$; actual solution $y(x) = x(e^x - e)$.

   (e) $-(x+1)y'' - y' + (x+2)y = (2 - (x+1)^2) e \ln 2 - 2e^x$, $0 \le x \le 1$, $y(0) = y(1) = 0$; use $h = 0.05$; actual solution $y(x) = e^x \ln(x + 1) - (e \ln 2)x$.

   (f) $-(x + 1)^2 y'' - 2(x + 1)y' = -1$, $0 \le x \le 1$, $y(0) = y(1) = 0$; use $h = 0.1$; actual solution $y(x) = -2x(x + 1)^{-1} \ln 2 + \ln(x + 1)$.

4. Use the Cubic Spline method with $n = 3$ to approximate the solution to each of the following boundary-value problems and compare the results to the actual solutions:

   (a) $y'' + \dfrac{\pi^2}{4} y = \dfrac{\pi^2}{16} \cos \dfrac{\pi}{4} x$, $0 \le x \le 1$, $y(0) = 0$, $y(1) = 0$.

   (b) $-\dfrac{d}{dx}(xy') + 4y = 4x^2 - 8x + 1$, $0 \le x \le 1$, $y(0) = 0$, $y(1) = 0$.

5. Repeat Exercise 3 using the Cubic Spline method.

6. Show that the boundary-value problem

$$-\frac{d}{dx}(p(x)y') + q(x)y = f(x), \qquad 0 \le x \le 1, \qquad y(0) = \alpha, \qquad y(1) = \beta$$

   can be transformed by the change of variable

$$z = y - \beta x - (1 - x)\alpha$$

   into the form

$$-\frac{d}{dx}(p(x)z') + q(x)z = F(x), \qquad 0 \le x \le 1, \qquad z(0) = 0, \qquad z(1) = 0.$$

7. Use Exercise 6 and the Piecewise Linear method with $n = 9$ to approximate the solution to the boundary-value problem

$$-y'' + y = x, \qquad 0 \le x \le 1, \qquad y(0) = 1, \qquad y(1) = 1 + e^{-1}.$$

8. Repeat Exercise 7 using the Cubic Spline method.

9. Show that the boundary-value problem

$$-\frac{d}{dx}(p(x)y') + q(x)y = f(x), \qquad a \le x \le b, \qquad y(a) = \alpha, \qquad y(b) = \beta$$

   can be transformed into the form

$$-\frac{d}{dw}(p(w)z') + q(w)z = F(w), \qquad 0 \le w \le 1, \qquad z(0) = 0, \qquad z(1) = 0.$$

   by a method similar to that given in Exercise 6.

11.7 Survey of Methods and Software

In this chapter we discussed methods for approximating solutions to boundary-value problems. For the linear boundary-value problem

$$y'' = p(x)y' + q(x)y + r(x), \quad \text{for } a \le x \le b, \quad \text{where } y(a) = \alpha \text{ and } y(b) = \beta,$$

we considered both a Linear Shooting method and a Finite-Difference method to approximate the solution. The shooting method uses an initial-value technique to solve the problems

$$y'' = p(x)y' + q(x)y + r(x), \quad \text{for } a \le x \le b, \quad \text{where } y(a) = \alpha \text{ and } y'(a) = 0,$$

and

$$y'' = p(x)y' + q(x)y, \quad \text{for } a \le x \le b, \quad \text{where } y(a) = 0 \text{ and } y'(a) = 1.$$

In the finite difference method, we replace y'' and y' with difference approximations and solve a linear system. Although the approximations may not be as accurate as the shooting method, there is less sensitivity to round-off error. Higher-order difference methods are available, or extrapolation can be used to improve accuracy.

For the nonlinear boundary problem

$$y'' = f(x, y, y'), \quad \text{for } a \le x \le b, \quad \text{where } y(a) = \alpha \text{ and } y(b) = \beta,$$

we also presented two methods. The Nonlinear Shooting method requires the solution of the initial-value problem

$$y'' = f(x, y, y'), \quad \text{for } a \le x \le b, \quad \text{where } y(a) = \alpha \text{ and } y'(a) = t,$$

for an initial choice of t. We improve on the choice by using Newton's method to approximate the solution, t, to $y(b, t) = \beta$. This method requires solving two initial-value problems at each iteration. The accuracy is dependent on the choice of method for solving the initial-value problems. The Nonlinear Finite-Difference method requires the replacement of y'' and y' by difference quotients, which results in a nonlinear system. This system is solved using Newton's method. Higher-order differences or extrapolation can be used to improve accuracy. Finite-difference methods tend to be less sensitive to round-off error than shooting methods.

The Rayleigh–Ritz method was introduced to approximate the solution to the boundary-value problem

$$-\frac{d\left(p(x)\dfrac{dy}{dx}\right)}{dx} + q(x)y = f(x), \quad \text{for } 0 \le x \le 1, \quad \text{where } y(0) = y(1) = 0.$$

A piecewise linear approximation or a cubic spline approximation can be obtained.

Most of the material concerning second-order boundary-value problems can be extended to problems with boundary conditions of the form

$$\alpha_1 y(a) + \beta_1 y'(a) = \alpha \quad \text{and} \quad \alpha_2 y(b) + \beta_2 y'(b) = \beta,$$

where $|\alpha_1| + |\beta_1| \neq 0$ and $|\alpha_2| + |\beta_2| \neq 0$, but some of the techniques become quite complicated.

We mention only two of the methods given in the IMSL library for solving boundary-value problems. The subroutine BVPFD is based on finite differences and BVPMS is based on multiple shooting using IVPRK, a Runge–Kutta–Verner method for initial-value problems. Both methods can be used for systems of parameterized boundary-value problems.

The NAG library has a multitude of subroutines for solving boundary-value problems. The subroutine D02HAF is a shooting method using the Runge–Kutta–Merson initial-value method in conjunction with Newton's method. The subroutine D02GAF uses the finite-difference method with Newton's method to solve the nonlinear system. The subroutine D02GBF is a linear finite difference method.

Numerical Methods for Partial-Differential Equations

12.1 Introduction

Physical problems involving more than one variable are often expressed using equations involving partial derivatives. In this chapter, we present a brief introduction to some of the techniques available for approximating the solution to partial-differential equations involving two variables by showing how these techniques can be applied to certain standard physical problems.

The partial-differential equation we consider in Section 12.2 is an **elliptic** equation known as the **Poisson equation**:

$$\frac{\partial^2 u}{\partial x^2}(x, y) + \frac{\partial^2 u}{\partial y^2}(x, y) = f(x, y).$$

In this equation we assume that f describes the input to the problem on a plane region R with boundary S. Equations of this type arise in the study of various time-independent physical problems such as the steady-state distribution of heat in a plane region, the potential energy of a point in a plane acted on by gravitational forces in the plane, and two-dimensional steady-state problems involving incompressible fluids.

Additional constraints must be imposed to obtain a unique solution to the Poisson equation. For example, the study of the steady-state distribution of heat in a plane region requires that $f(x, y) \equiv 0$, resulting in a simplification to

$$\frac{\partial^2 u}{\partial x^2}(x, y) + \frac{\partial^2 u}{\partial y^2}(x, y) = 0,$$

which is called **Laplace's equation**. If the temperature within the region is determined by the temperature distribution on the boundary of the region, the constraints are called the **Dirichlet boundary conditions**, given by

$$u(x, y) = g(x, y)$$

for all (x, y) on S, the boundary of the region R. (See Figure 12.1.)

Figure 12.1

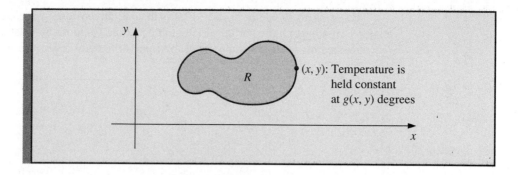

In Section 12.3 we consider the numerical solution to a problem involving a **parabolic** partial-differential equation of the form

$$\frac{\partial u}{\partial t}(x, t) - \alpha^2 \frac{\partial^2 u}{\partial x^2}(x, t) = 0.$$

The physical problem considered here concerns the flow of heat along a rod of length l (see Figure 12.2), which has a uniform temperature within each cross-sectional element. This requires the rod to be perfectly insulated on its lateral surface. The constant α is determined by the heat-conductive properties of the material of which the rod is composed and is assumed to be independent of the position in the rod.

Figure 12.2

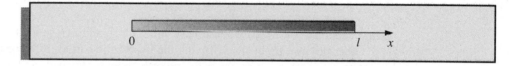

One of the typical sets of constraints for a heat-flow problem of this type is to specify the initial heat distribution in the rod,

$$u(x, 0) = f(x),$$

and to describe the behavior at the ends of the rod. For example, if the ends are held at constant temperatures U_1 and U_2, the boundary conditions have the form

$$u(0, t) = U_1 \quad \text{and} \quad u(l, t) = U_2,$$

and the heat distribution approaches the limiting temperature distribution

$$\lim_{t \to \infty} u(x, t) = U_1 + \frac{U_2 - U_1}{l} x.$$

If, instead, the rod is insulated so that no heat flows through the ends, the boundary conditions are

$$\frac{\partial u}{\partial x}(0, t) = 0 \quad \text{and} \quad \frac{\partial u}{\partial x}(l, t) = 0,$$

resulting in a constant temperature in the rod as the limiting case. The parabolic partial-differential equation is also of importance in the study of gas diffusion; in fact, it is known in some circles as the **diffusion equation**.

The problem studied in Section 12.4 is the one-dimensional **wave equation** and is an example of a **hyperbolic** partial-differential equation. Suppose an elastic string of length l is stretched between two supports at the same horizontal level (see Figure 12.3).

Figure 12.3

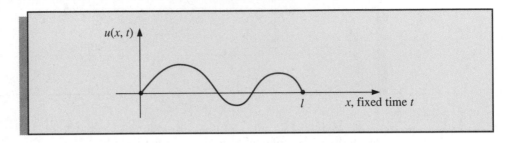

If the string is set to vibrate in a vertical plane, the vertical displacement $u(x, t)$ of a point x at time t satisfies the partial-differential equation

$$\alpha^2 \frac{\partial^2 u}{\partial x^2}(x, t) = \frac{\partial^2 u}{\partial t^2}(x, t), \qquad \text{for } 0 < x < l, \qquad \text{and} \qquad 0 < t,$$

provided that damping effects are neglected and the amplitude is not too large. To impose constraints on this problem, assume that the initial position and velocity of the string are given by

$$u(x, 0) = f(x) \qquad \text{and} \qquad \frac{\partial u}{\partial t}(x, 0) = g(x), \qquad \text{for } 0 \le x \le l.$$

If the endpoints are fixed, we also have $u(0, t) = 0$ and $u(l, t) = 0$.

Other physical problems involving the hyperbolic partial-differential equation occur in the study of vibrating beams with one or both ends clamped, and in the transmission of electricity over a long line where there is some leakage of current to the ground.

12.2 Finite-Difference Methods for Elliptic Problems

The *elliptic* partial-differential equation we consider is the Poisson equation,

$$\nabla^2 u(x, y) \equiv \frac{\partial^2 u}{\partial x^2}(x, y) + \frac{\partial^2 u}{\partial y^2}(x, y) = f(x, y)$$

on $R = \{(x, y) | a < x < b, c < y < d\}$, with

$$u(x, y) = g(x, y) \qquad \text{for } (x, y) \in S,$$

where S denotes the boundary of R. If both f and g are continuous on their domains, then there is a unique solution to this equation.

The method used here is an adaptation of the finite-difference method for linear boundary-value problems, which was discussed in Section 11.3. The first step is to choose integers n and m and define step sizes h and k by $h = (b - a)/n$ and $k = (d - c)/m$. Partition the interval $[a, b]$ into n equal parts of width h and the interval $[c, d]$ into m equal parts of width k. This provides a grid on the rectangle R

by drawing vertical and horizontal lines through the points with coordinates (x_i, y_j), where

$$x_i = a + ih \qquad \text{and} \qquad y_j = b + jk$$

for each $i = 0, 1, \ldots, n$ and $j = 0, 1, \ldots, m$ (see Figure 12.4).

Figure 12.4

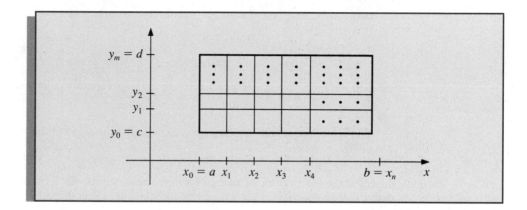

The lines $x = x_i$ and $y = y_j$ are **grid lines**, and their intersections are the **mesh points** of the grid. For each mesh point in the interior of the grid, we use a Taylor polynomial in the variable x about x_i to generate the central-difference formula

$$\frac{\partial^2 u}{\partial x^2}(x_i, y_j) = \frac{u(x_{i+1}, y_j) - 2u(x_i, y_j) + u(x_{i-1}, y_j)}{h^2} - \frac{h^2}{12} \frac{\partial^4 u}{\partial x^4}(\xi_i, y_j),$$

for some ξ_i in (x_{i-1}, x_{i+1}), and Taylor polynomial in the variable y about y_j to generate the central-difference formula

$$\frac{\partial^2 u}{\partial y^2}(x_i, y_j) = \frac{u(x_i, y_{j+1}) - 2u(x_i, y_j) + u(x_i, y_{j-1})}{k^2} - \frac{k^2}{12} \frac{\partial^4 u}{\partial y^4}(x_i, \eta_j),$$

for some η_j in (y_{j-1}, y_{j+1}).

Using these formulas in the Poisson equation produces the following equations:

$$\frac{u(x_{i+1}, y_j) - 2u(x_i, y_j) + u(x_{i-1}, y_j)}{h^2} + \frac{u(x_i, y_{j+1}) - 2u(x_i, y_j) + u(x_i, y_{j-1})}{k^2}$$

$$= f(x_i, y_j) + \frac{h^2}{12} \frac{\partial^4 u}{\partial x^4}(\xi_i, y_j) + \frac{k^2}{12} \frac{\partial^4 u}{\partial y^4}(x_i, \eta_j),$$

for each $i = 1, 2, \ldots, (n-1)$ and $j = 1, 2, \ldots, (m-1)$, and the boundary conditions

$$u(x_i, y_0) = g(x_i, y_0) \qquad \text{and} \qquad u(x_i, y_m) = g(x_i, y_m),$$

for each $i = 1, 2, \ldots, n-1$, and

$$u(x_0, y_j) = g(x_0, y_j) \qquad \text{and} \qquad u(x_n, y_j) = g(x_n, y_j),$$

for each $j = 0, 1, \ldots, m$.

In difference-equation form, this results in the *central-difference* method, with error of order $O(h^2 + k^2)$:

Central-Difference Method

$$2\left[\left(\frac{h}{k}\right)^2 + 1\right]w_{ij} - (w_{i+1,j} + w_{i-1,j}) - \left(\frac{h}{k}\right)^2(w_{i,j+1} + w_{i,j-1}) = -h^2 f(x_i, y_j),$$

for each $i = 1, 2, \ldots, n - 1$ and $j = 1, 2, \ldots, m - 1$, and

$$w_{0j} = g(x_0, y_j), \qquad\qquad w_{nj} = g(x_n, y_j),$$
$$w_{i0} = g(x_i, y_0), \qquad \text{and} \qquad w_{im} = g(x_i, y_m),$$

for each $i = 1, 2, \ldots, n - 1$ and $j = 0, 1, \ldots, m$, where w_{ij} approximates $u(x_i, y_j)$.

The typical equation involves approximations to $u(x, y)$ at the points

$$(x_{i-1}, y_j), \qquad (x_i, y_j), \qquad (x_{i+1}, y_j), \qquad (x_i, y_{j-1}), \qquad \text{and} \qquad (x_i, y_{j+1}).$$

Reproducing the portion of the grid where these points are located (see Figure 12.5) shows that each equation involves approximations in a star-shaped region about (x_i, y_j).

Figure 12.5

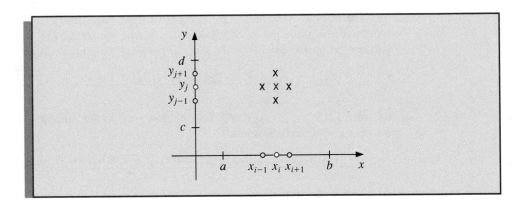

If we use the information from the boundary conditions, whenever appropriate, in the system given by the central-difference method (that is, at all points (x_i, y_j) that are adjacent to a boundary mesh point) we have an $(n - 1)(m - 1) \times (n - 1)(m - 1)$ linear system with the unknowns being the approximations w_{ij} to $u(x_i, y_j)$ at the interior mesh points. The linear system involving these unknowns is expressed for matrix calculations more efficiently if the interior mesh points are relabeled by letting

$$P_l = (x_i, y_j) \qquad \text{and} \qquad w_l = w_{ij},$$

where $l = i + (m - 1 - j)(n - 1)$ for each $i = 1, 2, \ldots, n - 1$ and $j = 1, 2, \ldots, m - 1$. This, in effect, labels the mesh points consecutively from left to right and top to bottom. For example, with $n = 4$ and $m = 5$ the relabeling results in a grid whose points are shown in Figure 12.6. Labeling the points in this manner ensures that the system needed to determine the w_{ij} is a banded matrix with band width at most $2n - 1$.

Figure 12.6

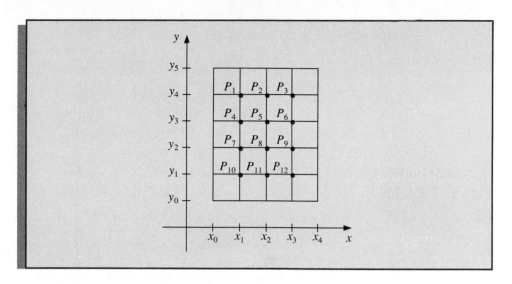

EXAMPLE 1 Consider the problem of determining the steady-state heat distribution in a thin square metal plate 0.5 m on a side. Two adjacent boundaries are held at 0° C, while the heat on the other boundaries increases linearly from 0° at one corner to 100° where the sides meet. If we place the sides with the zero boundary conditions along the x- and y-axes, the problem is expressed as

$$\frac{\partial^2 u}{\partial x^2}(x, y) + \frac{\partial^2 u}{\partial y^2}(x, y) = 0,$$

for (x, y) in the set $R = \{(x, y)|0 < x < 0.5\,;0 < y < 0.5\}$, with the boundary conditions

$$u(0, y) = 0, \qquad u(x, 0) = 0, \qquad u(x, 0.5) = 200x, \qquad u(0.5, y) = 200y.$$

If $n = m = 4$, the problem has the grid in Figure 12.7 and the difference equation

$$4w_{i,j} - w_{i+1,j} - w_{i-1,j} - w_{i,j-1} - w_{i,j+1} = 0,$$

for each $i = 1, 2, 3$, and $j = 1, 2, 3$.

Figure 12.7

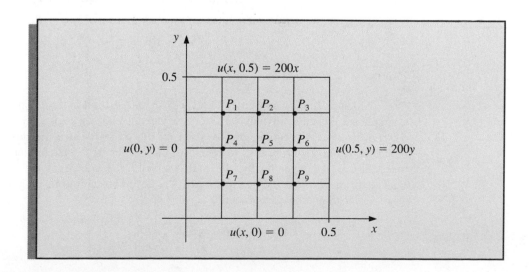

Expressing this in terms of the relabeled interior grid points $w_i = u(P_i)$ implies that the equations at the points P_i are

$$P_1: \qquad 4w_1 - w_2 - w_4 = w_{0,3} + w_{1,4},$$
$$P_2: \qquad 4w_2 - w_3 - w_1 - w_5 = w_{2,3},$$
$$P_3: \qquad 4w_3 - w_2 - w_6 = w_{4,3} + w_{3,4},$$
$$P_4: \quad 4w_4 - w_5 - w_1 - w_7 - w_8 = w_{0,2},$$
$$P_5: \quad 4w_5 - w_6 - w_4 - w_2 - w_8 = 0,$$
$$P_6: \qquad 4w_6 - w_5 - w_3 - w_9 = w_{4,2},$$
$$P_7: \qquad 4w_7 - w_8 - w_4 = w_{0,1} + w_{1,0},$$
$$P_8: \qquad 4w_8 - w_9 - w_7 - w_5 = w_{2,0},$$
$$P_9: \qquad 4w_9 - w_8 - w_6 = w_{3,0} + w_{4,1},$$

where the right sides of the equations are obtained from the boundary conditions. In fact, the boundary conditions imply that $w_{1,0} = w_{2,0} = w_{3,0} = w_{0,1} = w_{0,2} = w_{0,3} = 0$, $w_{1,4} = w_{4,1} = 25$, $w_{2,4} = w_{4,2} = 50$, and $w_{3,4} = w_{4,3} = 75$.

The linear system associated with this problem has the form

$$
\begin{bmatrix}
4 & -1 & 0 & -1 & 0 & 0 & 0 & 0 & 0 \\
-1 & 4 & -1 & 0 & -1 & 0 & 0 & 0 & 0 \\
0 & -1 & 4 & 0 & 0 & -1 & 0 & 0 & 0 \\
-1 & 0 & 0 & 4 & -1 & 0 & -1 & 0 & 0 \\
0 & -1 & 0 & -1 & 4 & -1 & 0 & -1 & 0 \\
0 & 0 & -1 & 0 & -1 & 4 & 0 & 0 & -1 \\
0 & 0 & 0 & -1 & 0 & 0 & 4 & -1 & 0 \\
0 & 0 & 0 & 0 & -1 & 0 & -1 & 4 & -1 \\
0 & 0 & 0 & 0 & 0 & -1 & 0 & -1 & 4
\end{bmatrix}
\begin{bmatrix}
w_1 \\ w_2 \\ w_3 \\ w_4 \\ w_5 \\ w_6 \\ w_7 \\ w_8 \\ w_9
\end{bmatrix}
=
\begin{bmatrix}
25 \\ 50 \\ 150 \\ 0 \\ 0 \\ 50 \\ 0 \\ 0 \\ 25
\end{bmatrix}.
$$

The values of $w_1, w_2, \ldots, w_9$, found by applying the Gauss–Seidel method to this matrix, are given in Table 12.1.

Table 12.1

i	1	2	3	4	5	6	7	8	9
w_i	18.75	37.50	56.25	12.50	25.00	37.50	6.25	12.50	18.75

These answers are exact, since the true solution, $u(x, y) = 400xy$, has

$$\frac{\partial^4 u}{\partial x^4} = \frac{\partial^4 u}{\partial y^4} = 0,$$

so the error is zero at each step. ■ ■ ■

Although the Gauss–Seidel iterative procedure is used in the program POIFD121 for simplicity, it is generally advisable to use a direct technique such as Gaussian elimination when the system is small, on the order of 100 or less, since the positive definiteness ensures stability with respect to round-off errors. For large systems, an iterative method should be used, specifically, the SOR method. The choice of the optimal ω in this situation comes from the fact that when A is decomposed into its diagonal D and upper- and lower-triangular parts U and L,

$$A = D - L - U,$$

and B is the matrix for the Jacobi method,

$$B = D^{-1}(L + U),$$

then the spectral radius of B is

$$\rho(B) = \frac{1}{2}\left[\cos\left(\frac{\pi}{m}\right) + \cos\left(\frac{\pi}{n}\right)\right]$$

and the value of ω to be used is

$$\omega = \frac{2}{1 + \sqrt{1 - [\rho(B)]^2}} = \frac{4}{2 + \sqrt{4 - \left[\cos\left(\frac{\pi}{m}\right) + \cos\left(\frac{\pi}{n}\right)\right]^2}}.$$

EXAMPLE 2 Consider Poisson's equation

$$\frac{\partial^2 u}{\partial x^2}(x, y) + \frac{\partial^2 u}{\partial y^2}(x, y) = xe^y, \qquad \text{for } 0 < x < 2 \text{ and } 0 < y < 1,$$

with the boundary conditions

$$u(0, y) = 0, \qquad u(2, y) = 2e^y, \qquad \text{for } 0 \le y \le 1,$$
$$u(x, 0) = x, \qquad u(x, 1) = ex, \qquad \text{for } 0 \le x \le 2.$$

We will use the Finite-Difference method to approximate the exact solution $u(x, y) = xe^y$ with $n = 6$ and $m = 5$. The stopping criterion for the Gauss–Seidel method in the program POIFD121 requires that

$$\left|w_{ij}^{(l)} - w_{ij}^{(l-1)}\right| \le 10^{-10},$$

for each $i = 1, \ldots, 5$ and $j = 1, \ldots, 4$. So the solution to the difference equation was accurately obtained, and the procedure stopped at $l = 61$. The results, along with the correct values, are presented in Table 12.2. ■ ■ ■

Table 12.2

i	j	x_i	y_i	$w_{i,j}^{(61)}$	$u(x_i, y_i)$	$\left|u(x_i, y_j) - w_{i,j}^{(61)}\right|$
1	1	0.3333	0.2000	0.40726	0.40713	1.30×10^{-4}
1	2	0.3333	0.4000	0.49748	0.49727	2.08×10^{-4}
1	3	0.3333	0.6000	0.60760	0.60737	2.23×10^{-4}
1	4	0.3333	0.8000	0.74201	0.74185	1.60×10^{-4}
2	1	0.6667	0.2000	0.81452	0.81427	2.55×10^{-4}
2	2	0.6667	0.4000	0.99496	0.99455	4.08×10^{-4}
2	3	0.6667	0.6000	1.2152	1.2147	4.37×10^{-4}
2	4	0.6667	0.8000	1.4840	1.4837	3.15×10^{-4}
3	1	1.0000	0.2000	1.2218	1.2214	3.64×10^{-4}
3	2	1.0000	0.4000	1.4924	1.4918	5.80×10^{-4}
3	3	1.0000	0.6000	1.8227	1.8221	6.24×10^{-4}
3	4	1.0000	0.8000	2.2260	2.2255	4.51×10^{-4}
4	1	1.3333	0.2000	1.6290	1.6285	4.27×10^{-4}
4	2	1.3333	0.4000	1.9898	1.9891	6.79×10^{-4}
4	3	1.3333	0.6000	2.4302	2.4295	7.35×10^{-4}
4	4	1.3333	0.8000	2.9679	2.9674	5.40×10^{-4}
5	1	1.6667	0.2000	2.0360	2.0357	3.71×10^{-4}
5	2	1.6667	0.4000	2.4870	2.4864	5.84×10^{-4}
5	3	1.6667	0.6000	3.0375	3.0369	6.41×10^{-4}
5	4	1.6667	0.8000	3.7097	3.7092	4.89×10^{-4}

EXERCISE SET 12.2

1. Use the Finite-Difference method to approximate the solution to the elliptic partial-differential equation

$$\frac{\partial^2 u}{\partial x^2} + \frac{\partial^2 u}{\partial y^2} = 4, \qquad 0 < x < 1, \qquad 0 < y < 2;$$

$$u(x, 0) = x^2, \qquad u(x, 2) = (x - 2)^2, \qquad 0 \le x \le 1;$$

$$u(0, y) = y^2, \qquad u(1, y) = (y - 1)^2, \qquad 0 \le y \le 2.$$

Use $h = 0.25$ and $k = 0.5$ and compare the results to the actual solution $u(x, y) = (x - y)^2$.

2. Use the Finite-Difference method to approximate the solution to the elliptic partial-differential equation

$$\frac{\partial^2 u}{\partial x^2} + \frac{\partial^2 u}{\partial y^2} = 0, \qquad 1 < x < 2, \qquad 0 < y < 1;$$

$$u(x, 0) = 2 \ln x, \qquad u(x, 1) = \ln(x^2 + 1), \qquad 0 \le x \le 2;$$

$$u(1, y) = \ln(y^2 + 1), \qquad u(2, y) = \ln(y^2 + 4), \qquad 0 \le y \le 1.$$

Use $h = k = \frac{1}{3}$ and compare the results to the actual solution $u(x, y) = \ln(x^2 + y^2)$.

3. Use the Finite-Difference method to approximate the solutions to the following elliptic partial-differential equations.

(a) $\dfrac{\partial^2 u}{\partial x^2} + \dfrac{\partial^2 u}{\partial y^2} = 0, \qquad 0 < x < 1, \ 0 < y < 1;$

$$u(x, 0) = 0, \qquad u(x, 1) = x, \qquad 0 \le x \le 1;$$

$$u(0, y) = 0, \qquad u(1, y) = y, \qquad 0 \le y \le 1.$$

Use $h = k = 0.2$ and compare the results with the solution $u(x, y) = xy$.

(b) $\dfrac{\partial^2 u}{\partial x^2} + \dfrac{\partial^2 u}{\partial y^2} = -2, \qquad 0 < x < 1, \ 0 < y < 1;$

$$u(0, y) = 0, \qquad u(1, y) = \sinh(\pi) \sin \pi y, \qquad 0 \le y \le 1;$$

$$u(x, 0) = u(x, 1) = x(1 - x), \qquad 0 \le x \le 1.$$

Use $h = k = 0.2$, and compare the results with the solution $u(x, y) = \sinh(\pi x) \sin(\pi y) + x(1 - x)$.

(c) $\dfrac{\partial^2 u}{\partial x^2} + \dfrac{\partial^2 u}{\partial y^2} = (x^2 + y^2)e^{xy}, \qquad 0 < x < 2, \ 0 < y < 1;$

$$u(0, y) = 1, \qquad u(2, y) = e^{2y}, \qquad 0 \le y \le 1;$$

$$u(x, 0) = 1, \qquad u(x, 1) = e^x, \qquad 0 \le x \le 2.$$

Use $h = 0.2$ and $k = 0.1$, and compare the results with the solution $u(x, y) = e^{xy}$.

(d) $\dfrac{\partial^2 u}{\partial x^2} + \dfrac{\partial^2 u}{\partial y^2} = -(\cos(x + y) + \cos(x - y)), \qquad 0 < x < \pi, \ 0 < y < \dfrac{\pi}{2};$

$$u(0, y) = \cos y, \qquad u(\pi, y) = -\cos y, \qquad 0 \le y \le \tfrac{\pi}{2};$$

$$u(x, 0) = \cos x, \qquad u\left(x, \tfrac{\pi}{2}\right) = 0, \qquad 0 \le x \le \pi.$$

Use $h = \pi/5$ and $k = \pi/10$ and compare the results with the solution $u(x, y) = \cos x \cos y$.

(e) $\dfrac{\partial^2 u}{\partial x^2} + \dfrac{\partial^2 u}{\partial y^2} = \dfrac{x}{y} + \dfrac{y}{x}$, $1 < x < 2,\ 1 < y < 2$;

$u(x, 1) = x \ln x$, $u(x, 2) = x \ln(4x^2)$, $1 \le x \le 2$;

$u(1, y) = y \ln y$, $u(2, y) = 2y \ln(2y)$, $1 \le y \le 2$.

Use $h = k = 0.1$ and compare the results with the solution $u(x, y) = xy \ln xy$.

(f) $\dfrac{\partial^2 u}{\partial x^2} + \dfrac{\partial^2 u}{\partial y^2} = (x^2 + y^2) \cos xy - \cos \pi x$, $0 < x < 1,\ 0 < y < 1$;

$u(x, 0) = \pi^{-2} \cos \pi x - 1$, $u(x, 1) = \pi^{-2} \cos \pi x - \cos x$, $0 \le x \le 1$;

$u(0, y) = \pi^{-2} - 1$, $u(1, y) = -(\pi^{-2} + \cos y)$, $0 \le y \le 1$.

Use $h = k = 0.1$ and compare the results with the solution $u(x, y) = \pi^{-2} \cos \pi x - \cos xy$.

4. A coaxial cable is made of a 0.1-in.-square inner conductor and a 0.5-in.-square outer conductor. The potential at a point in the cross section of the cable is described by Laplace's equation. Suppose the inner conductor is kept at 0 volts, whereas the outer conductor is kept at 110 volts. Find the potential between the two conductors by placing a grid with horizontal mesh spacing $h = 0.1$ in. and vertical mesh spacing $k = 0.1$ in. on the region

$$D = \{(x, y) \mid 0 \le x, y \le 0.5\}.$$

Approximate the solution to Laplace's equation at each grid point, and use the two sets of boundary conditions to derive a linear system to be solved by the Gauss–Seidel method.

5. A 6-cm by 5-cm rectangular silver plate has heat being uniformly generated at each point at the rate $q = 1.5\text{cal/cm}^3 \cdot$ s. Let x represent the distance along the edge of the plate of length 6 cm and y be the distance along the edge of the plate of length 5 cm. Suppose the temperature u along the edges is kept at

$$u(x, 0) = x(6 - x), \qquad u(x, 5) = 0, \quad 0 \le x \le 6,$$
$$u(0, y) = y(5 - y), \qquad u(6, y) = 0, \quad 0 \le y \le 5,$$

where the origin lies at a corner of the plate with coordinates $(0, 0)$ and the edges lie along the positive x- and y-axes. The steady-state temperature $u = u(x, y)$ satisfies Poisson's equation:

$$\frac{\partial^2 u}{\partial x^2}(x, y) + \frac{\partial^2 u}{\partial y^2}(x, y) = \frac{-q}{K}, \qquad 0 < x < 6, \qquad 0 < y < 5,$$

where K, the thermal conductivity, is 1.04 cal/cm $\cdot\ °C \cdot$ s. Use the Finite-Difference method with $h = 0.4$ and $k = \frac{1}{3}$ to approximate the temperature $u(x, y)$.

12.3 Finite-Difference Methods for Parabolic Problems

The *parabolic* partial-differential equation we will study is the heat or diffusion equation

$$\frac{\partial u}{\partial t}(x, t) = \alpha^2 \frac{\partial^2 u}{\partial x^2}(x, t), \qquad \text{for } 0 < x < l \text{ and } t > 0,$$

subject to the conditions

$u(0, t) = u(l, t) = 0$ for $t > 0$,   and $u(x, 0) = f(x)$ for $0 \le x \le l$.

The approach we use to approximate the solution to this problem involves finite differences similar to those in Section 12.2. First select two mesh constants h and k, with the stipulation that $m = l/h$ is an integer. The grid points for this situation are (x_i, t_j), where $x_i = ih$, for $i = 0, 1, \ldots, m$, and $t_j = jk$, for $j = 0, 1, \ldots$.

We obtain the difference method by using a Taylor polynomial in t to form the difference quotient

$$\frac{\partial u}{\partial t}(x_i, t_j) = \frac{u(x_i, t_j + k) - u(x_i, t_j)}{k} - \frac{k}{2}\frac{\partial^2}{\partial t^2}(x_i, \mu_j),$$

for some μ_j in (t_j, t_{j+1}), and a Taylor polynomial in x to form the difference quotient

$$\frac{\partial^2 u}{\partial x^2}(x_i, t_j) = \frac{u(x_i + h, t_j) - 2u(x_i, t_j) + u(x_i - h, t_j)}{h^2} - \frac{h^2}{12}\frac{\partial^4 u}{\partial x^4}(\xi_i, t_j),$$

for some ξ_i in (x_{i-1}, x_{i+1}).

The parabolic partial-differential equation implies that at the interior gridpoint (x_i, t_j) we have

$$\frac{\partial u}{\partial t}(x_i, t_j) - \alpha^2 \frac{\partial^2 u}{\partial x^2}(x_i, t_j) = 0,$$

so the difference method using the two difference quotients is

$$\frac{w_{i,j+1} - w_{ij}}{k} - \alpha^2 \frac{w_{i+1,j} - 2w_{ij} + w_{i-1,j}}{h^2} = 0,$$

where w_{ij} approximates $u(x_i, t_j)$. The error for this difference equation is

$$\tau_{ij} = \frac{k}{2}\frac{\partial^2 u}{\partial t^2}(x_i, \mu_j) - \alpha^2 \frac{h^2}{12}\frac{\partial^4 u}{\partial x^4}(\xi_i, t_j).$$

Solving the difference equation for $w_{i,j+1}$ gives

$$w_{i,j+1} = \left(1 - \frac{2\alpha^2 k}{h^2}\right)w_{ij} + \alpha^2 \frac{k}{h^2}(w_{i+1,j} + w_{i-1,j}),$$

for each $i = 1, 2, \ldots, (m - 1)$ and $j = 1, 2, \ldots$. Since the initial condition $u(x, 0) = f(x)$ implies that $w_{i0} = f(x_i)$, for each $i = 0, 1, \ldots, m$, these values can be used in the difference equation to find the value of w_{i1} for each $i = 1, 2, \ldots, (m - 1)$. The additional conditions $u(0, t) = 0$ and $u(l, t) = 0$ imply that $w_{01} = w_{m1} = 0$, so all the entries of the form w_{i1} can be determined. If the procedure is reapplied once all the approximations w_{i1}, are known, the values of $w_{i2}, w_{i3}, \ldots, w_{i,m-1}$ can be obtained in a similar manner.

The explicit nature of the difference method implies that the $(m - 1)$ by $(m - 1)$ matrix associated with this system can be written in the tridiagonal form

$$A = \begin{bmatrix} (1 - 2\lambda) & \lambda & 0 & \cdots\cdots\cdots & 0 \\ \lambda & (1 - 2\lambda) & \lambda & & \vdots \\ 0 & & & & 0 \\ \vdots & & & & \lambda \\ 0 & \cdots\cdots\cdots & 0 & \lambda & (1 - 2\lambda) \end{bmatrix},$$

where $\lambda = \alpha^2(k/h^2)$. If we let

$$\mathbf{w}^{(0)} = (f(x_1), f(x_2), \ldots, f(x_{m-1}))^t$$

and

$$\mathbf{w}^{(j)} = (w_{1j}, w_{2j}, \ldots, w_{m-1,j})^t, \qquad \text{for each } j = 1, 2, \ldots,$$

then the approximate solution is given by

$$\mathbf{w}^{(j)} = A\mathbf{w}^{(j-1)}, \qquad \text{for each } j = 1, 2, \ldots.$$

This is known as the **Forward-Difference method**. It is of order $O(k + h^2)$.

EXAMPLE 1 Consider the heat equation

$$\frac{\partial u}{\partial t}(x, t) - \frac{\partial^2 u}{\partial x^2}(x, t) = 0, \qquad \text{for } 0 < x < 1 \text{ and } 0 \le t,$$

with boundary conditions

$$u(0, t) = u(1, t) = 0, \qquad \text{for } 0 < t,$$

and initial conditions

$$u(x, 0) = \sin(\pi x), \qquad \text{for } 0 \le x \le 1.$$

It is easily verified that the solution to this problem is $u(x, t) = e^{-\pi^2 t} \sin(\pi x)$. The solution at $t = 0.5$ will be approximated using the Forward-Difference method, first with $h = 0.1$, $k = 0.0005$, and $\lambda = 0.05$ and then with $h = 0.1$, $k = 0.01$, and $\lambda = 1$. The results are presented in Table 12.3. ■ ■ ■

Table 12.3

x_i	$u(x_i, 0.5)$	$w_{i,1000}$ $k = 0.0005$	$\|u(x_i, 0.5) - w_{i,1000}\|$	$w_{i,50}$ $k = 0.01$	$\|u(x_i, 0.5) - w_{i,50}\|$
0.0	0	0		0	
0.1	0.00222241	0.00228652	6.411×10^{-5}	8.19876×10^7	8.199×10^7
0.2	0.00422728	0.00434922	1.219×10^{-4}	-1.55719×10^8	1.557×10^8
0.3	0.00581836	0.00598619	1.678×10^{-4}	2.13833×10^8	2.138×10^8
0.4	0.00683989	0.00703719	1.973×10^{-4}	-2.50642×10^8	2.506×10^8
0.5	0.00719188	0.00739934	2.075×10^{-4}	2.62685×10^8	2.627×10^8
0.6	0.00683989	0.00703719	1.973×10^{-4}	-2.49015×10^8	2.490×10^8
0.7	0.00581836	0.00598619	1.678×10^{-4}	2.11200×10^8	2.112×10^8
0.8	0.00422728	0.00434922	1.219×10^{-4}	-1.53086×10^8	1.531×10^8
0.9	0.00222241	0.00228652	6.511×10^{-5}	8.03604×10^7	8.036×10^7
1.0	0	0		0	

An error of order $O(k + h^2)$ is expected in Example 1. This is obtained with $h = 0.1$ and $k = 0.0005$, but it is certainly not obtained when $h = 0.1$ and $k = 0.01$. To explain the difficulty, we must look at the stability of the Forward-Difference method.

If the error $\mathbf{e}^{(0)} = (e_1^{(0)}, e_2^{(0)}, \ldots, e_{m-1}^{(0)})^t$ is made in representing the initial data $\mathbf{w}^{(0)} = (f(x_1), f(x_2), \ldots, f(x_{m-1}))^t$, or in any particular step (the choice of the initial step is simply for convenience), an error of $A\mathbf{e}^{(0)}$ propagates in $\mathbf{w}^{(1)}$, since

$$\mathbf{w}^{(1)} = A(\mathbf{w}^{(0)} + \mathbf{e}^{(0)}) = A\mathbf{w}^{(0)} + A\mathbf{e}^{(0)}.$$

This process continues. At the nth time step, the error in $\mathbf{w}^{(n)}$ due to $\mathbf{e}^{(0)}$ is $A^n \mathbf{e}^{(0)}$. Consequently, the method is stable precisely when these errors do not grow as n increases—that is, if and only if $\|A^n \mathbf{e}^{(0)}\| \leq \|\mathbf{e}^{(0)}\|$ for all n. This implies that $\|A^n\| \leq 1$, a condition that requires that the spectral radius $\rho(A^n) = (\rho(A))^n \leq 1$. The Forward-Difference method is, therefore, stable only if $\rho(A) \leq 1$.

The eigenvalues of A are

$$\mu_i = 1 - 4\lambda \left(\sin\left(\frac{i\pi}{2m} \right) \right)^2, \qquad \text{for each } i = 1, 2, \ldots, (m-1).$$

So the condition for stability consequently reduces to determining whether

$$\rho(A) = \max_{1 \leq i \leq m-1} \left| 1 - 4\lambda \left(\sin\left(\frac{i\pi}{2m} \right) \right)^2 \right| \leq 1,$$

which simplifies to

$$0 \leq \lambda \left(\sin\left(\frac{i\pi}{2m} \right) \right)^2 \leq \frac{1}{2}, \qquad \text{for each } i = 1, 2, \ldots, m-1.$$

Since stability requires that this inequality condition hold as $h \to 0$ or, equivalently, as $m \to \infty$, the fact that

$$\lim_{m \to \infty} \left[\sin\left(\frac{(m-1)\pi}{2m} \right) \right]^2 = 1$$

means that stability will occur only if $0 \leq \lambda \leq \frac{1}{2}$. Since $\lambda = \alpha^2 (k/h^2)$, this inequality requires that h and k be chosen so that

$$\alpha^2 \frac{k}{h^2} \leq \frac{1}{2}.$$

This condition was satisfied in our example when $h = 0.1$ and $k = 0.0005$; but when k was increased to 0.01 with no corresponding increase in h, the ratio was

$$\frac{0.01}{(0.1)^2} = 1 > \frac{1}{2},$$

and stability problems became apparent.

To obtain a more stable method, we consider an implicit-difference method that results from using the backward-difference quotient for $(\partial u / \partial t)(x_i, t_j)$ in the form

$$\frac{\partial u}{\partial t}(x_i, t_j) = \frac{u(x_i, t_j) - u(x_i, t_{j-1})}{k} + \frac{k}{2} \frac{\partial^2 u}{\partial t^2}(x_i, \mu_j),$$

for some μ_j in (t_{j-1}, t_j). Substituting this equation, together with the centered-difference formula for $\partial^2 u / \partial x^2$, into the partial-differential equation gives

$$\frac{u(x_i, t_j) - u(x_i, t_{j-1})}{k} - \alpha^2 \frac{u(x_{i+1}, t_j) - 2u(x_i, t_j) + u(x_{i-1}, t_j)}{h^2}$$

$$= -\frac{k}{2} \frac{\partial^2 u}{\partial t^2}(x_i, \mu_j) - \frac{h^2}{12} \frac{\partial^4 u}{\partial x^4}(\xi_i, t_j),$$

for some ξ_i in (x_{i-1}, x_{i+1}). The difference method this produces is called the *backward-difference method*.

Backward-Difference Method

$$\frac{w_{ij} - w_{i,j-1}}{k} - \alpha^2 \frac{w_{i+1,j} - 2w_{ij} + w_{i-1,j}}{h^2} = 0,$$

for each $i = 1, 2, \ldots, m-1$, and $j = 1, 2, \ldots$.

This method involves, at a typical step, the mesh points

$$(x_i, t_j), \qquad (x_i, t_{j-1}), \qquad (x_{i-1}, t_j), \qquad \text{and} \qquad (x_{i+1}, t_j),$$

and, in grid form, involves approximations at the points marked with ×'s in Figure 12.8.

Figure 12.8

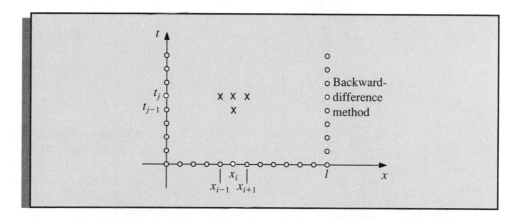

Since the boundary and initial conditions associated with the problem give information at the circled mesh points, the figure shows that no explicit procedures can be used for the Backward-Difference method.

Recall that in the Forward-Difference method (see Figure 12.9, page 410), approximations at

$$(x_{i-1}, t_j), \qquad (x_i, t_j), \qquad (x_i, t_{j+1}), \qquad \text{and} \qquad (x_{i+1}, t_j)$$

Figure 12.9

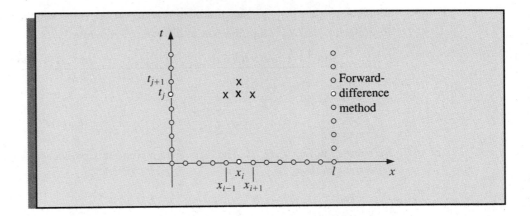

were used, so an explicit method for finding the approximations, based on the information from the initial and boundary conditions, was available.

If we again let λ denote the quantity $\alpha^2(k/h^2)$, the Backward-Difference method becomes

$$(1 + 2\lambda)w_{ij} - \lambda w_{i+1,j} - \lambda w_{i-1,j} = w_{i,j-1},$$

for each $i = 1, 2, \ldots, m - 1$, and $j = 1, 2, \ldots$. Using the knowledge that $w_{i0} = f(x_i)$ for each $i = 1, 2, \ldots, m - 1$, and $w_{mj} = w_{0j} = 0$ for each $j = 1, 2, \ldots$, this difference method has the matrix representation:

$$
\begin{bmatrix}
(1 + 2\lambda) & -\lambda & 0 & \cdots\cdots\cdots & 0 \\
-\lambda & \ddots & \ddots & & \vdots \\
0 & \ddots & \ddots & \ddots & 0 \\
\vdots & & \ddots & \ddots & -\lambda \\
0 & \cdots\cdots\cdots & 0 & -\lambda & (1 + 2\lambda)
\end{bmatrix}
\begin{bmatrix}
w_{1j} \\
w_{2j} \\
\vdots \\
w_{m-1,j}
\end{bmatrix}
=
\begin{bmatrix}
w_{1,j-1} \\
w_{2,j-1} \\
\vdots \\
w_{m-1,j-1}
\end{bmatrix},
$$

or $A\mathbf{w}^{(j)} = \mathbf{w}^{(j-1)}$. Since $\lambda > 0$, the matrix A is positive definite and strictly diagonally dominant, as well as being tridiagonal. We can use either Crout factorization for tridiagonal linear systems or the SOR method to solve this system. The program HEBDM122 uses Crout factorization, which is an acceptable method unless m is large.

EXAMPLE 2   The Backward-Difference method with $h = 0.1$ and $k = 0.01$ will be used to approximate the solution to the heat equation

$$\frac{\partial u}{\partial t}(x, t) - \frac{\partial^2 u}{\partial x^2}(x, t) = 0, \qquad 0 < x < 1, \qquad 0 < t,$$

subject to the constraints

$$u(0, t) = u(1, t) = 0, \qquad 0 < t, \qquad u(x, 0) = \sin \pi x, \qquad 0 \le x \le 1,$$

which was considered in Example 1. To demonstrate the stability of the Backward-Difference method, we again compare $w_{i,50}$ to $u(x_i, 0.5)$, where $i = 0, 1, \ldots, 10$. The results listed in Table 12.4 should be compared with the fifth and sixth columns of Table 12.3. ■ ■ ■

Table 12.4

x_i	$w_{i,50}$	$u(x_i, 0.5)$	$|w_{i,50} - u(x_i, 0.5)|$
0.0	0	0	
0.1	0.00289802	0.00222241	6.756×10^{-4}
0.2	0.00551236	0.00422728	1.285×10^{-3}
0.3	0.00758711	0.00581836	1.769×10^{-3}
0.4	0.00891918	0.00683989	2.079×10^{-3}
0.5	0.00937818	0.00719188	2.186×10^{-3}
0.6	0.00891918	0.00683989	2.079×10^{-3}
0.7	0.00758711	0.00581836	1.769×10^{-3}
0.8	0.00551236	0.00422728	1.285×10^{-3}
0.9	0.00289802	0.00222241	6.756×10^{-4}
1.0	0	0	

The reason that the Backward-Difference method does not have the stability problems of the Forward-Difference method can be seen by analyzing the eigenvalues of the matrix A. For the Backward-Difference method the eigenvalues are

$$\mu_i = 1 + 4\lambda \left[\sin\left(\frac{i\pi}{2m}\right) \right]^2, \qquad \text{for each } i = 1, 2, \ldots, (m-1);$$

and since $\lambda > 0$, we have $\mu_i > 1$, for all $i = 1, 2, \ldots, (m-1)$. This implies that A^{-1} exists, since zero cannot be an eigenvalue of A. An error $\mathbf{e}^{(0)}$ in the initial data produces an error $(A^{-1})^n \mathbf{e}^{(0)}$ at the nth step. Since the eigenvalues of A^{-1} are the reciprocals of the eigenvalues of A, the spectral radius of A^{-1} is bounded above by 1 and the method is stable, independent of the choice of $\lambda = \alpha^2(k/h^2)$. The error for the method is of order $O(k + h^2)$, provided the solution of the differential equation satisfies the usual differentiability conditions.

The weakness in the Backward-Difference method results from the fact that the error has a portion with order $O(k)$, requiring that time intervals be made much smaller than spatial intervals. It would clearly be desirable to have a procedure with error of order $O(k^2 + h^2)$. A method with this error term is derived by averaging the Forward-Difference method at the jth step in t,

$$\frac{w_{i,j+1} - w_{ij}}{k} - \alpha^2 \frac{w_{i+1,j} - 2w_{ij} + w_{i-1,j}}{h^2} = 0,$$

which has error $\dfrac{k}{2} \dfrac{\partial^2 u}{\partial t^2}(x_i, \mu_j) + O(h^2)$, and the Backward-Difference method at the $(j+1)$st step in t,

$$\frac{w_{i,j+1} - w_{ij}}{k} - \alpha^2 \frac{w_{i+1,j+1} - 2w_{i,j+1} + w_{i-1,j+1}}{h^2} = 0,$$

which has error $-\dfrac{k}{2}\dfrac{\partial^2 u}{\partial t^2}(x_i, \hat{u}_j) + O(h^2)$. If we assume that

$$\frac{\partial^2 u}{\partial t^2}(x_i, \hat{\mu}_j) \approx \frac{\partial^2 u}{\partial t^2}(x_i, \mu_j),$$

then the averaged-difference method,

$$\frac{w_{i,j+1} - w_{ij}}{k} - \frac{\alpha^2}{2}\left[\frac{w_{i+1,j} - 2w_{ij} + w_{i-1,j}}{h^2} + \frac{w_{i+1,j+1} - 2w_{i,j+1} + w_{i-1,j+1}}{h^2}\right] = 0,$$

has error of order $O(k^2 + h^2)$, provided, of course, that the usual differentiability conditions are satisfied. This is known as the **Crank–Nicolson** method and is represented in the matrix form $A\mathbf{w}^{(j+1)} = B\mathbf{w}^{(j)}$ for each $j = 0, 1, 2, \ldots$, where

$$\lambda = \alpha^2 \frac{k}{h^2}, \qquad \mathbf{w}^{(j)} = (w_{1j}, w_{2j}, \ldots, w_{m-1,j})^t,$$

and the matrices A and B are given by

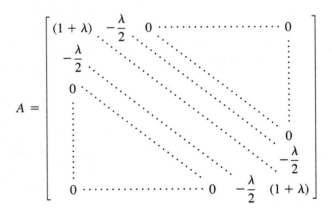

and

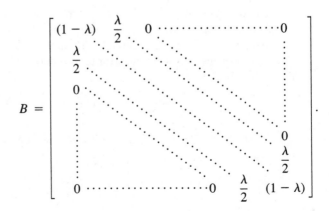

Since A is a positive definite, strictly diagonal dominant, and a tridiagonal matrix, it is nonsingular. Either Crout factorization for tridiagonal linear systems or the SOR method can be used to obtain $\mathbf{w}^{(j+1)}$ from $\mathbf{w}^{(j)}$, for each $j = 0, 1, 2, \ldots$. The program HECNM123 incorporates Crout factorization into the Crank–Nicolson technique.

EXAMPLE 3   The Crank–Nicolson method will be used to approximate the solution to the problem in Examples 1 and 2, consisting of the equation

$$\frac{\partial u}{\partial t}(x, t) - \frac{\partial^2 u}{\partial x^2}(x, t) = 0, \qquad \text{for } 0 < x < 1 \text{ and } 0 < t,$$

subject to the conditions

$$u(0, t) = u(1, t) = 0, \qquad 0 < t, \qquad \text{and} \qquad u(x, 0) = \sin(\pi x), \qquad 0 \le x \le 1.$$

The choices $m = 10$, $h = 0.1$, $k = 0.01$, and $\lambda = 1$ are used, as they were in the previous examples. The results in Table 12.5 indicate the increase in accuracy of the Crank–Nicolson method over the Backward-Difference method, the best of the two previously discussed techniques.   ■ ■ ■

Table 12.5

x_i	$w_{i,50}$	$u(x_i, 0.5)$	$|w_{i,50} - u(x_i, 0.5)|$
0.0	0	0	
0.1	0.00230512	0.00222241	8.271×10^{-5}
0.2	0.00438461	0.00422728	1.573×10^{-4}
0.3	0.00603489	0.00581836	2.165×10^{-4}
0.4	0.00709444	0.00683989	2.546×10^{-4}
0.5	0.00745954	0.00719188	2.677×10^{-4}
0.6	0.00709444	0.00683989	2.546×10^{-4}
0.7	0.00603489	0.00581836	2.165×10^{-4}
0.8	0.00438461	0.00422728	1.573×10^{-4}
0.9	0.00230512	0.00222241	8.271×10^{-5}
1.0	0	0	

EXERCISE SET 12.3

1. Approximate the solution to the following partial-differential equations using the Backward-Difference method.

   (a) $\dfrac{\partial u}{\partial t} - \dfrac{\partial^2 u}{\partial x^2} = 0,$ $0 < x < 2,\ 0 < t;$

   $u(0, t) = u(2, t) = 0,$ $0 < t;$

   $u(x, 0) = \sin \dfrac{\pi}{2} x,$ $0 \le x \le 2.$

   Use $m = 4, T = 0.1,$ and $N = 2$ and compare your answers to the actual solution $u(x, t) = e^{-\frac{\pi^2}{4} t} \sin \dfrac{\pi}{2} x.$

   (b) $\dfrac{\partial u}{\partial t} - \dfrac{1}{16} \dfrac{\partial^2 u}{\partial x^2} = 0,$ $0 < x < 1,\ 0 < t;$

   $u(0, t) = u(1, t) = 0,$ $0 < t;$

   $u(x, 0) = 2 \sin 2\pi x,$ $0 \le x \le 1.$

   Use $m = 3, T = 0.1,$ and $N = 2$ and compare your answers to the actual solution $u(x, t) = 2e^{-\frac{\pi^2}{4} t} \sin 2\pi x.$

2. Repeat Exercise 1 using the Crank–Nicolson method.

3. Use the Forward-Difference method to approximate the solution to the following parabolic partial-differential equations.

   (a) $\dfrac{\partial u}{\partial t} - \dfrac{\partial^2 u}{\partial x^2} = 0,$ $0 < x < 2,\ 0 < t;$

   $u(0, t) = u(2, t) = 0,$ $0 < t;$

   $u(x, 0) = \sin 2\pi x,$ $0 \le x \le 2.$

   Use $h = 0.1$ and $k = 0.01,$ and compare your answers at $t = 0.5$ to the actual solution $u(x, t) = e^{-4\pi^2 t} \sin 2\pi x.$ Then use $h = 0.1$ and $k = 0.005,$ and compare the answers.

   (b) $\dfrac{\partial u}{\partial t} - \dfrac{\partial^2 u}{\partial x^2} = 0,$ $0 < x < \pi,\ 0 < t;$

   $u(0, t) = u(\pi, t) = 0,$ $0 < t;$

   $u(x, 0) = \sin x,$ $0 \le x \le \pi.$

   Use $h = \dfrac{\pi}{10}$ and choose an appropriate value for k. Compare your answers to the actual solution $u(x, t) = e^{-t} \sin x$ at $t = 0.5.$

   (c) $\dfrac{\partial u}{\partial t} - \dfrac{4}{\pi^2} \dfrac{\partial^2 u}{\partial x^2} = 0,$ $0 < x < 4,\ 0 < t;$

   $u(0, t) = u(4, t) = 0,$ $0 < t;$

   $u(x, 0) = \sin \dfrac{\pi}{4} x \left(1 + 2 \cos \dfrac{\pi}{4} x\right),$ $0 \le x \le 4.$

   Use $h = 0.2$ and $k = 0.04.$ Compare your answers to the actual solution $u(x, t) = e^{-t} \sin \dfrac{\pi}{2} x + e^{-t/4} \sin \dfrac{\pi}{4} x$ at $t = 0.4.$

(d) $\dfrac{\partial u}{\partial t} - \dfrac{1}{\pi^2} \dfrac{\partial^2 u}{\partial x^2} = 0,$ $0 < x < 1,\ 0 < t;$

$u(0, t) = u(1, t) = 0,$ $0 < t;$

$u(x, 0) = \cos \pi(x - \tfrac{1}{2}),$ $0 \le x \le 1.$

Use $h = 0.1$ and $k = 0.04$. Compare your answers to the actual solution $u(x, t) = e^{-t} \cos \pi(x - \tfrac{1}{2})$ at $t = 0.4$.

4. Repeat Exercise 3 using the Backward-Difference method.

5. Repeat Exercise 3 using the Crank–Nicolson method.

6. Modify the Backward-Difference method to include the parabolic partial-differential equation

$$\frac{\partial u}{\partial t} - \frac{\partial^2 u}{\partial x^2} = F(x), \qquad 0 < x < l,\ 0 < t;$$

$$u(0, t) = u(l, t) = 0, \qquad 0 < t;$$

$$u(x, 0) = f(x), \qquad 0 \le x \le l.$$

7. Use the results of Exercise 6 to approximate the solution to

$$\frac{\partial u}{\partial t} - \frac{\partial^2 u}{\partial x^2} = 2, \qquad 0 < x < 1, \qquad 0 < t;$$

$$u(0, t) = u(1, t) = 0, \qquad 0 < t;$$

$$u(x, 0) = \sin \pi x + x(1 - x),$$

with $h = 0.1$ and $k = 0.01$. Compare your answer to the actual solution $u(x, t) = e^{-\pi^2 t} \sin \pi x + x(1 - x)$ at $t = 0.25$.

8. Modify the Backward-Difference method to accommodate the partial-differential equation

$$\frac{\partial u}{\partial t} - \alpha^2 \frac{\partial^2 u}{\partial x^2} = 0, \qquad 0 < x < l,\ 0 < t;$$

$$u(0, t) = \phi(t),\ u(l, t) = \psi(t), \qquad 0 < t;$$

$$u(x, 0) = f(x), \qquad 0 \le x \le l,$$

where $f(0) = \phi(0)$ and $f(l) = \psi(0)$.

9. The temperature $u(x, t)$ of a long thin rod of constant cross section and homogeneous conducting material is governed by the one-dimensional heat equation. If heat is generated in the material, for example by resistance to current or nuclear reaction, the heat equation becomes

$$\frac{\partial^2 u}{\partial x^2} + \frac{Kr}{\rho C} = K \frac{\partial u}{\partial t}, \qquad 0 < x < l, \qquad 0 < t,$$

where l is the length, ρ the density, C the specific heat, and K the thermal diffusivity of the rod. The function $r = r(x, t, u)$ represents the heat generated per unit volume. Suppose that

$$l = 1.5 \text{ cm}, \qquad K = 1.04 \ \frac{\text{cal}}{\text{cm} \cdot {}^\circ C \cdot \text{s}},$$

$$\rho = 10.6 \text{ g/cm}^3, \qquad C = 0.056 \ \frac{\text{cal}}{\text{g} \cdot {}^\circ C},$$

and

$$r(x, t, u) = 5.0 \ \frac{\text{cal}}{\text{s} \cdot \text{cm}^3}.$$

If the ends of the rod are kept at 0°C, then

$$u(0, t) = u(l, t) = 0, \qquad t > 0.$$

Suppose the initial temperature distribution is given by

$$u(x, 0) = \sin \frac{\pi x}{l}, \qquad 0 \le x \le l.$$

Use the results of Exercise 6 to approximate the temperature distribution with $h = 0.15$ and $k = 0.0225$.

10. In analyzing the stress-strain relationships and material properties of a cylinder alternately subjected to heating and cooling, V. Sagar and D. J. Payne consider the equation

$$\frac{\partial^2 T}{\partial r^2} + \frac{1}{r} \frac{\partial T}{\partial r} = \frac{1}{4K} \frac{\partial T}{\partial t}, \qquad \frac{1}{2} < r < 1, \ 0 < T,$$

where $T = T(r, t)$ is the temperature, r is the radial distance from the center of the cylinder, t is time, and K is a diffusivity coefficient.

(a) Find approximations to $T(r, 10)$ for a cylinder with outside radius 1, given the initial and boundary conditions:

$$T(l, t) = 100 + 40t, \qquad 0 \le t \le 10,$$

$$T(\tfrac{1}{2}, t) = t, \qquad 0 \le t \le 10,$$

$$T(r, 0) = 200(r - 0.5), \qquad 0.5 \le r \le 1.$$

Use a modification of the Backward-Difference method with $K = 0.1$, $k = 0.5$, and $h = \Delta r = 0.1$.

(b) Use the temperature distribution of part (a) to calculate the strain I by approximating the integral

$$I = \int_{0.5}^{1} \alpha T(r, t) r \, dr,$$

where $\alpha = 10.7$ and $t = 10$. Use the Composite Trapezoidal method with $n = 5$.

11. An equation describing one-dimensional, single-phase, slightly compressible flow in a producing petroleum reservoir is given, for $0 < x < 1000$ and $0 < t$, by

$$\frac{\phi \mu C}{K} \frac{\partial p}{\partial t}(x, t) = \frac{\partial^2 p}{\partial x^2}(x, t) - \begin{cases} 0, & \text{if } x \ne 500, \\ 1000, & \text{if } x = 500, \end{cases}$$

where it has been assumed that the porous medium and the reservoir are homogeneous, that the liquid is ideal, and that gravitational effects are negligible. The symbols are defined as x representing distance (in feet), t the time (in days), p the pressure (in pounds per square inch), ϕ the dimensionless constant porosity of the medium, μ the viscosity (in centipoise), K the permeability of the medium (in millidarcies), C the compressibility (in [pounds per square inch]$^{-1}$). Assume that $\alpha = \phi \mu C/K = 0.00004$ days/ft^2 and that the following conditions hold:

$$p(x, 0) = 2.5 \times 10^7, \qquad 0 \le x \le 1000,$$

$$\frac{1}{K} \frac{\partial p}{\partial x}(0, t) = \frac{\partial p}{\partial x}(1000, t) = 0, \qquad 0 < t.$$

Use the Crank–Nicolson method with $k = \Delta t = 0.5$ and $h = \Delta x = 100$ to find the pressure p at $t = 5$.

12.4 Finite-Difference Methods for Hyperbolic Problems

In this section, we consider the numerical solution to the **wave equation**, an example of a *hyperbolic* partial-differential equation. The wave equation is given by the differential equation

$$\frac{\partial^2 u}{\partial t^2}(x, t) - \alpha^2 \frac{\partial^2 u}{\partial x^2}(x, t) = 0, \qquad \text{for } 0 < x < l \text{ and } t > 0,$$

subject to the conditions

$$u(0, t) = u(l, t) = 0, \qquad \text{for } t > 0,$$

$$u(x, 0) = f(x), \qquad \text{and} \qquad \frac{\partial u}{\partial t}(x, 0) = g(x), \qquad \text{for } 0 \leq x \leq l,$$

where α is a constant. Select an integer $m > 0$ and time-step size $k > 0$. With $h = l/m$, the mesh points (x_i, t_j) are defined by

$$x_i = ih, \qquad \text{and} \qquad t_j = jk,$$

for each $i = 0, 1, \ldots, m$ and $j = 0, 1, \ldots$. At any interior mesh point (x_i, t_j), the wave equation becomes

$$\frac{\partial^2 u}{\partial t^2}(x_i, t_j) - \alpha^2 \frac{\partial^2 u}{\partial x^2}(x_i, t_j) = 0.$$

The difference method is obtained using the centered-difference quotient for the second partial derivatives given by

$$\frac{\partial^2 u}{\partial t^2}(x_i, t_j) = \frac{u(x_i, t_{j+1}) - 2u(x_i, t_j) + u(x_i, t_{j-1})}{k^2} - \frac{k^2}{12} \frac{\partial^4 u}{\partial t^4}(x_i, \mu_j),$$

for some μ_j in (t_{j-1}, t_{j+1}) and

$$\frac{\partial^2 u}{\partial x^2}(x_i, t_j) = \frac{u(x_{i+1}, t_j) - 2u(x_i, t_j) + u(x_{i-1}, t_j)}{h^2} - \frac{h^2}{12} \frac{\partial^4 u}{\partial x^4}(\xi_i, t_j),$$

for some ξ_i in (x_{i-1}, x_{i+1}). Substituting these into the wave equation gives

$$\frac{u(x_i, t_{j+1}) - 2u(x_i, t_j) + u(x_i, t_{j-1})}{k^2} - \alpha^2 \frac{u(x_{i+1}, t_j) - 2u(x_i, t_j) + u(x_{i-1}, t_j)}{h^2}$$

$$= \frac{1}{12} \left[k^2 \frac{\partial^4 u}{\partial t^4}(x_i, \mu_j) - \alpha^2 h^2 \frac{\partial^4 u}{\partial x^4}(\xi_i, t_j) \right].$$

Neglecting the error term

$$\tau_{ij} = \frac{1}{12}\left[k^2 \frac{\partial^4 u}{\partial t^4}(x_i, \mu_j) - \alpha^2 h^2 \frac{\partial^4 u}{\partial x^4}(\xi_i, t_j) \right]$$

leads to the difference equation

$$\frac{w_{i,j+1} - 2w_{ij} + w_{i,j-1}}{k^2} - \alpha^2 \frac{w_{i+1,j} - 2w_{ij} + w_{i-1,j}}{h^2} = 0.$$

With $\lambda = \alpha k/h$, we can solve for $w_{i,j+1}$, the most advanced time-step approximation, to obtain

$$w_{i,j+1} = 2(1 - \lambda^2)w_{ij} + \lambda^2(w_{i+1,j} + w_{i-1,j}) - w_{i,j-1}.$$

This equation holds for each $i = 1, 2, \ldots, (m - 1)$, and $j = 1, 2, \ldots$. The boundary conditions give

$$w_{0j} = w_{mj} = 0, \qquad \text{for each } j = 1, 2, 3, \ldots,$$

and the initial condition implies that

$$w_{i0} = f(x_i), \qquad \text{for each } i = 1, 2, \ldots, m - 1.$$

Writing this set of equations in matrix form gives

$$
\begin{bmatrix} w_{i,j+1} \\ w_{2,j+1} \\ \vdots \\ w_{m-1,j+1} \end{bmatrix}
=
\begin{bmatrix}
2(1-\lambda^2) & \lambda^2 & 0 & \cdots\cdots & 0 \\
\lambda^2 & 2(1-\lambda^2) & \lambda^2 & & \vdots \\
0 & & & & 0 \\
\vdots & & & & \lambda^2 \\
0 & \cdots\cdots & 0 & \lambda^2 & 2(1-\lambda^2)
\end{bmatrix}
\begin{bmatrix} w_{1j} \\ w_{2j} \\ \vdots \\ w_{m-1,j} \end{bmatrix}
-
\begin{bmatrix} w_{1,j-1} \\ w_{2,j-1} \\ \vdots \\ w_{m-1,j-1} \end{bmatrix}.
$$

To determine $w_{i,j+1}$ requires values from the jth and $(j - 1)$st time steps. (See Figure 12.10.)

Figure 12.10

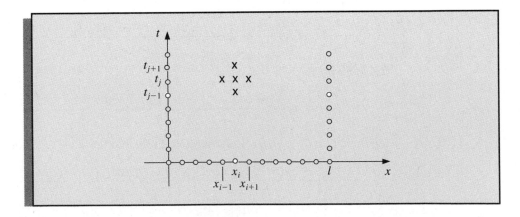

There is a minor starting problem since values for $j = 0$ are given in the initial conditions, but values for $j = 1$, which are needed to compute w_{i2}, must be obtained from the initial-velocity condition

$$\frac{\partial u}{\partial t}(x,0) = g(x), \qquad \text{for } 0 \le x \le l.$$

One approach is to replace $\partial u / \partial t$ by a forward-difference approximation,

$$\frac{\partial u}{\partial t}(x_i,0) = \frac{u(x_i,t_1) - u(x_i,0)}{k} - \frac{k}{2}\frac{\partial^2 u}{\partial t^2}(x_1,\tilde{\mu}_i),$$

for some $\tilde{\mu}_i$ in $(0, t_1)$. Solving for $u(x_i, t_1)$ gives

$$u(x_i,t_1) = u(x_i,0) + k\frac{\partial u}{\partial t^2}(x_1,0) + \frac{k^2}{2}\frac{\partial^2 u}{\partial t^2}(x_i,\tilde{\mu}_i)$$

$$= u(x_i,0) + kg(x_i) + \frac{k^2}{2}\frac{\partial^2 u}{\partial t^2}(x_i,\tilde{\mu}_i).$$

As a consequence,

$$w_{i1} = w_{i0} + kg(x_i), \qquad \text{for each } i = 1,\dots,m-1.$$

However, this gives an approximation that has error of only $O(k)$. A better approximation to $u(x_i, 0)$ can be obtained.

Consider the equation

$$u(x_i,t_1) = u(x_i,0) + k\frac{\partial u}{\partial t}(x_i,0) + \frac{k^2}{2}\frac{\partial^2 u}{\partial t^2}(x_i,0) + \frac{k^3}{6}\frac{\partial^3 u}{\partial t^3}(x_i,\hat{u}_i)$$

for some $\hat{\mu}_i$ in $(0, t_1)$, which comes from expanding $u(x_i, t_1)$ in a second Maclaurin polynomial in t.

If f'' exists, then

$$\frac{\partial^2 u}{\partial t^2}(x_i,0) = \alpha^2 \frac{\partial^2 u}{\partial x^2}(x_i,0) = \alpha^2 \frac{d^2 f}{dx^2}(x_i) = \alpha^2 f''(x_i)$$

and

$$u(x_i,t_1) = u(x_i,0) + kg(x_i) + \frac{\alpha^2 k^2}{2}f''(x_i) + \frac{k^3}{6}\frac{\partial^3 u}{\partial t^3}(x_i,\hat{\mu}_i),$$

producing an approximation with error $O(k^2)$:

$$w_{i1} = w_{i0} + kg(x_i) + \frac{\alpha^2 k^2}{2}f''(x_i).$$

If $f''(x_i)$ is not readily available, we can use a central-difference equation to write

$$f''(x_i) = \frac{f(x_{i+1}) - 2f(x_i) + f(x_{i-1})}{h^2} - \frac{h^2}{12}f^{(4)}(\tilde{\xi}_i),$$

for some $\tilde{\xi}_i$ in (x_{i-1}, x_{i+1}). The approximation then becomes

$$\frac{u(x_i,t_1) - u(x_i,0)}{k} = g(x_i) + \frac{k\alpha^2}{2h^2}[f(x_{i+1}) - 2f(x_i) + f(x_{i-1})] + O(k^2 + h^2 k)$$

or, letting $\lambda = \dfrac{k\alpha}{h}$,

$$u(x_i, t_1) = u(x_i, 0) + kg(x_i) + \frac{\lambda^2}{2}[f(x_{i+1}) - 2f(x_i) + f(x_{i-1})] + O(k^3 + h^2 k^2)$$

$$= (1 - \lambda^2)f(x_i) + \frac{\lambda^2}{2}f(x_{i+1}) + \frac{\lambda^2}{2}f(x_{i-1}) + kg(x_i) + O(k^3 + h^2 k^2).$$

So, the difference equation

$$w_{i1} = (1 - \lambda^2)f(x_i) + \frac{\lambda^2}{2}f(x_{i+1}) + \frac{\lambda^2}{2}f(x_{i-1}) + kg(x_i)$$

can be used to find w_{i1} for each $i = 1, 2, \ldots, m - 1$.

The program WVFDM124 uses this equation to approximate w_{i1}. It is assumed that there is an upper bound for the value of t, to be used in the stopping technique.

EXAMPLE 1   Consider the hyperbolic problem

$$\frac{\partial^2 u}{\partial t}(x, t) - 4\frac{\partial^2 u}{\partial x^2}(x, t) = 0, \qquad \text{for } 0 < x < 1 \text{ and } 0 < t,$$

with boundary conditions

$$u(0, t) = u(1, t) = 0, \qquad \text{for } 0 < t,$$

and initial conditions

$$u(x, 0) = \sin(\pi x), \qquad 0 \le x \le 1, \qquad \text{and} \qquad \frac{\partial u}{\partial t}(x, 0) = 0, \qquad \text{for } 0 \le x \le 1.$$

It is easily verified that the solution to this problem is

$$u(x, t) = \sin(\pi x)\cos(2\pi t).$$

The Finite-Difference method is used in this example with $m = 10, T = 1$, and $N = 20$, which implies that $h = 0.1, k = 0.05$, and $\lambda = 1$. The following table lists the results of the approximation, w_{iN}, for $i = 0, 1, \ldots, 10$. The values listed in Table 12.6 are correct to the places given.   ▪   ▪   ▪

Table 12.6

x_i	$w_{i,20}$
0.0	0.0000000000
0.1	0.3090169944
0.2	0.5877852523
0.3	0.8090169944
0.4	0.9510565163
0.5	1.0000000000
0.6	0.9510565163
0.7	0.8090169944
0.8	0.5877852523
0.9	0.3090169944
1.0	0.0000000000

EXERCISE SET 12.4

1. Use the Finite-Difference method with $m = 4$, $N = 4$, and $T = 1$ to approximate the solution to the wave equation

$$\frac{\partial^2 u}{\partial t^2} - \frac{\partial^2 u}{\partial x^2} = 0, \qquad 0 < x < 1, \ 0 < t;$$

$$u(0, t) = u(1, t) = 0, \qquad 0 < t,$$

$$u(x, 0) = \sin \pi x, \qquad 0 \le x \le 1,$$

$$\frac{\partial u}{\partial t}(x, 0) = 0, \qquad 0 \le x \le 1,$$

and compare your results to the actual solution $u(x, t) = \cos \pi t \sin \pi x$ at $t = 0.5$.

2. Use the Finite-Difference method with $m = 4$, $N = 4$, and $T = 0.5$ to approximate the solution to the wave equation

$$\frac{\partial^2 u}{\partial t^2} - \frac{1}{16\pi^2} \frac{\partial^2 u}{\partial x^2} = 0, \qquad 0 < x < 0.5, \ 0 < t;$$

$$u(0, t) = u(0.5, t) = 0, \qquad 0 < t,$$

$$u(x, 0) = 0, \qquad 0 < x < 0.5,$$

$$\frac{\partial u}{\partial t}(x, 0) = \sin 4\pi x, \qquad 0 \le x \le 0.5,$$

and compare your results to the actual solution $u(x, t) = \sin t \sin 4\pi x$ at $t = 0.5$.

3. Use the Finite-Difference method with $h = \dfrac{\pi}{10}$ and $k = 0.05$, with $h = \dfrac{\pi}{20}$ and $k = 0.1$, and then with $h = \dfrac{\pi}{20}$ and $k = 0.05$ to approximate the solution $u(x, t) = \sin \pi x \cos \pi t$ to the wave equation

$$\frac{\partial^2 u}{\partial t^2} - \frac{\partial^2 u}{\partial x^2} = 0, \qquad 0 < x < \pi, \ 0 < t;$$

$$u(0, t) = u(\pi, t) = 0, \qquad 0 < t,$$

$$u(x, 0) = \sin x, \qquad 0 \le x \le \pi,$$

$$\frac{\partial u}{\partial t}(x, 0) = 0, \qquad 0 \le x \le \pi,$$

and compare your results to the actual solution $u(x, t) = \cos t \sin x$ at $t = 0.5$.

4. Use the Finite-Difference method with $h = 0.1$ and $k = 0.1$ to approximate the solution to the wave equation

$$\frac{\partial^2 u}{\partial t^2} - \frac{\partial^2 u}{\partial x^2} = 0, \qquad 0 < x < 1, \ 0 < t;$$

$$u(0, t) = u(1, t) = 0, \qquad 0 < t,$$

$$u(x, 0) = \sin 2\pi x, \qquad 0 \le x \le 1,$$

$$\frac{\partial u}{\partial t}(x, 0) = 2\pi \sin 2\pi x, \qquad 0 \le x \le 1,$$

and compare your results to the actual solution $u(x, t) = \sin 2\pi x (\cos 2\pi t + \sin 2\pi t)$ at $t = 0.3$.

5. Use the Finite-Difference method with $h = 0.1$ and $k = 0.1$ to approximate the solution to the wave equation

$$\frac{\partial^2 u}{\partial t^2} - \frac{\partial^2 u}{\partial x^2} = 0, \qquad 0 < x < 1, \quad 0 < t;$$

$$u(0, t) = u(1, t) = 0, \qquad 0 < t,$$

$$u(x, 0) = \begin{cases} 1, & 0 \le x \le \frac{1}{2}, \\ -1, & \frac{1}{2} < x \le 1, \end{cases}$$

$$\frac{\partial u}{\partial t}(x, 0) = 0, \qquad 0 \le x \le 1.$$

6. The air pressure $p(x, t)$ in an organ pipe is governed by the wave equation

$$\frac{\partial^2 p}{\partial x^2} = \frac{1}{c^2}\frac{\partial^2 p}{\partial t^2}, \qquad 0 < x < 1, \quad 0 < t,$$

where l is the length of the pipe and c is a physical constant. If the pipe is open, the boundary conditions are given by

$$p(0, t) = p_0 \qquad \text{and} \qquad p(l, t) = p_0.$$

If the pipe is closed at the end where $x = l$, the boundary conditions are

$$p(0, t) = p_0 \qquad \text{and} \qquad \frac{\partial p}{\partial x}(l, t) = 0.$$

Assume that $c = 1, l = 1$, and the initial conditions are

$$p(x, 0) = p_0 \cos 2\pi x, \qquad \text{and} \qquad \frac{\partial p}{\partial t}(x, 0) = 0, \qquad 0 \le x \le 1.$$

(a) Use the Finite-Difference method with $h = k = 0.1$ to approximate the pressure for an open pipe with $p_0 = 0.9$ at $x = \frac{1}{2}$ for $t = 0.5$ and $t = 1$.
(b) Modify the Finite-Difference method for the closed pipe problem with $p_0 = 0.9$, and approximate $p(0.5, 0.5)$ and $p(0.5, 1)$, using $h = k = 0.1$.

7. In an electric transmission line of length l that carries alternating current of high frequency (called a "lossless" line), the voltage V and current i are described by

$$\frac{\partial^2 V}{\partial x^2} = LC\frac{\partial^2 V}{\partial t^2}, \qquad 0 < x < l, \ 0 < t,$$

$$\frac{\partial^2 i}{\partial x^2} = LC\frac{\partial^2 i}{\partial t^2}, \qquad 0 < x < l, \ 0 < t,$$

where L is the inductance per unit length and C is the capacitance per unit length. Suppose the line is 200 ft long and the constants C and L are given by

$$C = 0.1 \text{ farads/ft} \qquad \text{and} \qquad L = 0.3 \text{ henries/ft}.$$

Suppose the voltage and current also satisfy

$$V(0, t) = V(200, t) = 0, \quad 0 < t,$$

$$V(x, 0) = 110 \sin \frac{\pi x}{200}, \quad 0 \le x \le 200,$$

$$\frac{\partial V}{\partial t}(x, 0) = 0, \quad 0 \le x \le 200,$$

$$i(0, t) = i(200, t) = 0, \quad 0 < t,$$

$$i(x, 0) = 5.5 \cos \frac{\pi x}{200}, \quad 0 \le x \le 200,$$

and

$$\frac{\partial i}{\partial t}(x, 0) = 0, \quad 0 \le x \le 200.$$

Use the Finite-Difference method with $h = 10$ and $k = 0.1$ to approximate the voltage and current at $t = 0.2$ and $t = 0.5$.

12.5 Introduction to the Finite-Element Method

The **Finite-Element method** for partial-differential equations is similar to the Rayleigh–Ritz method for approximating the solution to two-point boundary-value problems. It was originally developed for use in civil engineering, but it is now used for approximating the solutions to partial-differential equations that arise in all areas of applied mathematics.

One advantage of the finite-element method over finite-difference methods is the relative ease with which the boundary conditions of the problem are handled. Many physical problems have boundary conditions involving derivatives and irregularly shaped boundaries. Boundary conditions of this type are difficult to handle using finite-difference techniques, since each boundary condition involving a derivative must be approximated by a difference quotient at the grid points, and irregular shaping of the boundary makes placing the grid points difficult. The finite-element method includes the boundary conditions as integrals in a functional that is being minimized, so the construction procedure is independent of the particular boundary conditions of the problem.

In our discussion, we consider the partial-differential equation

$$\frac{\partial}{\partial x}\left(p(x, y)\frac{\partial u}{\partial x}\right) + \frac{\partial}{\partial y}\left(q(x, y)\frac{\partial u}{\partial y}\right) + r(x, y)u(x, y) = f(x, y),$$

with (x, y) in $\mathcal{D}$, where $\mathcal{D}$ is a plane region with boundary $\mathcal{S}$. Boundary conditions of the form

$$u(x, y) = g(x, y)$$

are imposed on a portion $\mathcal{S}_1$ of the boundary. On the remainder of the boundary,

Figure 12.11

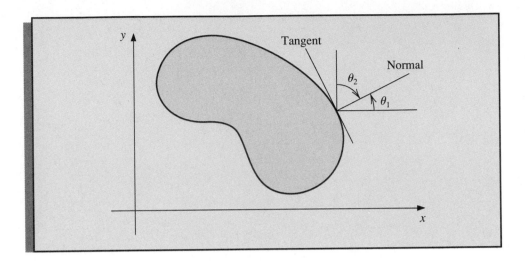

$\mathcal{S}_2$, $u(x, y)$ is required to satisfy

$$p(x, y)\frac{\partial u}{\partial x}(x, y)\cos\theta_1 + q(x, y)\frac{\partial u}{\partial y}(x, y)\cos\theta_2 + g_1(x, y)u(x, y) = g_2(x, y),$$

where θ_1 and θ_2 are the direction angles of the outward normal to the boundary at the point (x, y). (See Figure 12.11.) Physical problems in the areas of solid mechanics and elasticity have associated partial-differential equations of this type. The solution to such a problem is typically the minimization of a certain functional, involving integrals, over a class of functions determined by the problem.

Suppose p, q, r, and f are all continuous in $\mathcal{D} \cup \mathcal{S}$, p and q have continuous first partial derivatives, and g_1 and g_2 are continuous on $\mathcal{S}_2$. Suppose, in addition, that $p(x, y) > 0$, $q(x, y) > 0$, $r(x, y) \le 0$, and $g_1(x, y) > 0$. Then a solution to our problem uniquely minimizes the functional

$$I[w] = \iint_{\mathcal{D}} \left\{ \frac{1}{2} \left[p(x, y)\left(\frac{\partial w}{\partial x}\right)^2 + q(x, y)\left(\frac{\partial w}{\partial y}\right)^2 - r(x, y)w^2 \right] + f(x, y)w \right\} \, dx \, dy$$

$$+ \int_{\mathcal{S}_2} \left\{ -g_2(x, y)w + \frac{1}{2}g_1(x, y)w^2 \right\} \, dS$$

over all twice continuously differentiable functions w satisfying $w(x, y) = g(x, y)$ on $\mathcal{S}_1$. The finite-element method approximates this solution by minimizing the functional I over a smaller class of functions, just as the Rayleigh-Ritz method did for the boundary-value problem considered in Section 11.6.

The first step is to divide the region into a finite number of sections, or elements, of a regular shape, either rectangles or triangles. (See Figure 12.12.)

The set of functions used for approximation is generally a set of piecewise polynomials of fixed degree in x and y, and the approximation requires that the polynomials be pieced together in such a manner that the resulting function is continuous with an integrable or continuous first or second derivative on the entire region. Polynomials of linear type in x and y,

$$\phi(x, y) = a + bx + cy,$$

Figure 12.12

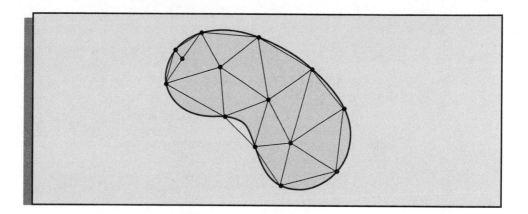

are commonly used with triangular elements, whereas polynomials of bilinear type in x and y,

$$\phi(x, y) = a + bx + cy + dxy,$$

are used with rectangular elements.

For our discussion, suppose that the region $\mathcal{D}$ has been subdivided into triangular elements. The collection of triangles is denoted D, and the vertices of these triangles are called **nodes**. The method seeks an approximation of the form

$$\phi(x, y) = \sum_{i=1}^{m} \gamma_i \phi_i(x, y),$$

where $\phi_1, \phi_2, \ldots, \phi_m$ are linearly independent piecewise-linear polynomials and γ_1, $\gamma_2, \ldots, \gamma_m$ are constants. Some of these constants, say, $\gamma_{n+1}, \gamma_{n+2}, \ldots, \gamma_m$, are used to ensure that the boundary condition

$$\phi(x, y) = g(x, y)$$

is satisfied on $\mathcal{S}_1$, and the remaining constants $\gamma_1, \gamma_2, \ldots, \gamma_n$ are used to minimize the functional $I\left[\sum_{i=1}^{m} \gamma_i \phi_i\right]$.

Since the functional is of the form

$$I[\phi] = I\left[\sum_{i=1}^{m} \gamma_i \phi_i\right]$$

$$= \iint_{\mathcal{D}} \left(\frac{1}{2} \left\{ p(x, y) \left[\sum_{i=1}^{m} \gamma_i \frac{\partial \phi_i}{\partial x}(x, y) \right]^2 + q(x, y) \left[\sum_{i=1}^{m} \gamma_i \frac{\partial \phi_i}{\partial y}(x, y) \right]^2 \right. \right.$$

$$\left. - r(x, y) \left[\sum_{i=1}^{m} \gamma_i \phi_i(x, y) \right]^2 \right\} + f(x, y) \sum_{i=1}^{m} \gamma_i \phi_i(x, y) \bigg) \, dy \, dx$$

$$+ \int_{\mathcal{S}_2} \left\{ -g_2(x, y) \sum_{i=1}^{m} \gamma_i \phi_i(x, y) + \frac{1}{2} g_1(x, y) \left[\sum_{i=1}^{m} \gamma_i \phi_i(x, y) \right]^2 \right\} \, dS,$$

for a minimum to occur, considering I as a function of $\gamma_1, \gamma_2, \ldots, \gamma_n$, it is necessary to have

$$\frac{\partial I}{\partial \gamma_j} = 0, \qquad \text{for each } j = 1, 2, \ldots, n.$$

Performing the partial differentiation allows us to write this set of equations as a linear system,

$$A\mathbf{c} = \mathbf{b},$$

where $\mathbf{c} = (\gamma_1, \ldots, \gamma_n)^t$ and where $A = (\alpha_{ij})$ and $\mathbf{b} = (\beta_1, \ldots, \beta_n)^t$ are defined by

$$\alpha_{ij} = \iint_{\mathcal{D}} \left[p(x, y)\frac{\partial \phi_i}{\partial x}(x, y)\frac{\partial \phi_j}{\partial x}(x, y) + q(x, y)\frac{\partial \phi_i}{\partial y}(x, y)\frac{\partial \phi_j}{\partial y}(x, y) \right.$$

$$\left. - r(x, y)\phi_i(x, y)\phi_j(x, y) \right] dx \, dy + \int_{\mathcal{S}_2} g_1(x, y)\phi_i(x, y)\phi_j(x, y) \, dS,$$

for each $i = 1, 2, \ldots, n$ and $j = 1, 2, \ldots, m$, and

$$\beta_i = -\iint_{\mathcal{D}} f(x, y)\phi_i(x, y) \, dx \, dy + \int_{\mathcal{S}_2} g_2(x, y)\phi_i(x, y) \, dS - \sum_{k=n+1}^{m} \alpha_{ik}\gamma_k,$$

for each $i = 1, \ldots, n$.

The particular choice of basis functions is important, since the appropriate choice can often make the matrix A positive definite and banded. For our second-order problem, we assume that $\mathcal{D}$ is polygonal and that $\mathcal{S}_1$ is a contiguous set of straight lines so that $\mathcal{D} = D$. To begin the procedure, we divide the region D into a collection of triangles $T_1, T_2, \ldots, T_M$, with the ith triangle having three vertices, or nodes, denoted

$$V_j^{(i)} = (x_j^{(i)}, y_j^{(i)}), \qquad \text{for } j = 1, 2, 3.$$

To simplify the notation, we write $V_j^{(i)}$ simply as $V_j = (x_j, y_j)$ when working with the fixed triangle T_i. With each vertex V_j we associate a linear polynomial

$$N_j^{(i)} \equiv N_j = a_j + b_j x + c_j y, \qquad \text{where} \quad N_j^{(i)}(x_k, y_k) = \begin{cases} 1, & \text{if } j = k \\ 0, & \text{if } j \neq k. \end{cases}$$

This produces linear systems of the form

$$\begin{bmatrix} 1 & x_1 & y_1 \\ 1 & x_2 & y_2 \\ 1 & x_3 & y_3 \end{bmatrix} \begin{bmatrix} a_j \\ b_j \\ c_j \end{bmatrix} = \begin{bmatrix} 0 \\ 1 \\ 0 \end{bmatrix},$$

with the element 1 occurring in the jth row in the vector on the right.

Let $E_1, \ldots, E_n$ be a labeling of the nodes lying in $D \cup \mathcal{S}$ in a left-to-right, top-to-bottom fashion. With each node E_k, we associate a function ϕ_k that is linear on each triangle, has the value 1 at E_k, and is 0 at each of the other nodes. This choice makes ϕ_k identical to $N_j^{(i)}$ on triangle T_i when the node E_k is the vertex denoted $V_j^{(i)}$.

Figure 12.13

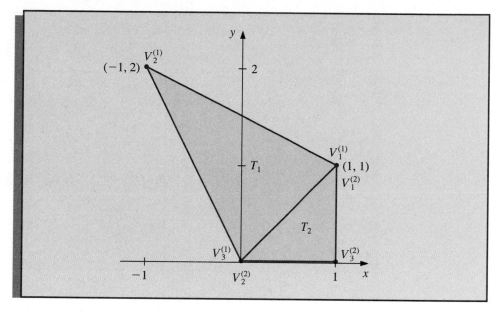

EXAMPLE 1   Suppose that a finite-element problem contains the triangles T_1 and T_2 shown in Figure 12.13. The linear function $N_1^{(1)}(x, y)$ that assumes the value 1 at $(1, 1)$ and 0 at both $(0, 0)$ and $(-1, 2)$ satisfies

$$a_1^{(1)} + b_1^{(1)}(1) + c_1^{(1)}(1) = 1,$$
$$a_1^{(1)} + b_1^{(1)}(-1) + c_1^{(1)}(2) = 0,$$
$$a_1^{(1)} + b_1^{(1)}(0) + c_1^{(1)}(0) = 0,$$

so

$$a_1^{(1)} = 0, b_1^{(1)} = \tfrac{2}{3}, c_1^{(1)} = \tfrac{1}{3},$$

and

$$N_1^{(1)}(x, y) = \tfrac{2}{3}x + \tfrac{1}{3}y.$$

In a similar manner, the linear function $N_1^{(2)}(x, y)$ that assumes the value 1 at $(1, 1)$ and 0 at both $(0, 0)$ and $(1, 0)$ satisfies

$$a_1^{(2)} + b_1^{(2)}(1) + c_1^{(2)}(1) = 1,$$
$$a_1^{(2)} + b_1^{(2)}(0) + c_1^{(2)}(0) = 0,$$
$$a_1^{(2)} + b_1^{(2)}(1) + c_1^{(2)}(0) = 0,$$

so $a_1^{(2)} = 0$, $b_1^{(2)} = 0$, and $c_1^{(2)} = 1$. As a consequence, $N_1^{(2)}(x, y) = y$. Note that on the common boundary of T_1 and T_2 we have $N_1^{(1)}(x, y) = N_1^{(2)}(x, y)$, since $y = x$.

■   ■   ■

Consider Figure 12.14, the upper left portion of the region shown in Figure 12.12 on page 425. We will generate the entries in the matrix A that correspond to the nodes shown in this figure.

For simplicity, we assume that E_1 is not one of the nodes on $\mathcal{S}_2$. The relationship between the nodes and the vertices of the triangles for this portion is

$$E_1 = V_3^{(1)} = V_1^{(2)}, \qquad E_4 = V_2^{(2)}, \qquad E_3 = V_2^{(1)} = V_3^{(2)}, \qquad \text{and} \qquad E_2 = V_1^{(1)}.$$

Figure 12.14

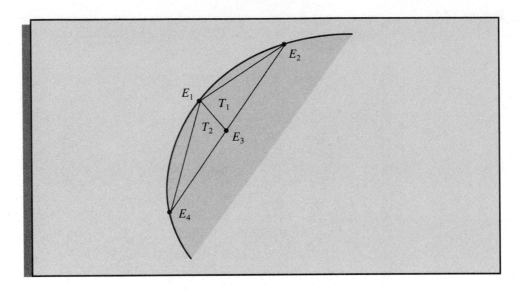

Since ϕ_1 and ϕ_3 are both nonzero on T_1 and T_2, the entries $\alpha_{1,3} = \alpha_{3,1}$ are computed by

$$\alpha_{1,3} = \iint_D \left[p \frac{\partial \phi_1}{\partial x} \frac{\partial \phi_3}{\partial x} + q \frac{\partial \phi_1}{\partial y} \frac{\partial \phi_3}{\partial y} - r \phi_1 \phi_3 \right] dx \, dy$$

$$= \iint_{T_1} \left[p \frac{\partial \phi_1}{\partial x} \frac{\partial \phi_3}{\partial x} + q \frac{\partial \phi_1}{\partial y} \frac{\partial \phi_3}{\partial y} - r \phi_1 \phi_3 \right] dx \, dy$$

$$+ \iint_{T_2} \left[p \frac{\partial \phi_1}{\partial x} \frac{\partial \phi_3}{\partial x} + q \frac{\partial \phi_1}{\partial y} \frac{\partial \phi_3}{\partial y} - r \phi_1 \phi_3 \right] dx \, dy.$$

On triangle T_1,

$$\phi_1(x, y) = N_3^{(1)} = a_3^{(1)} + b_3^{(1)} x + c_3^{(1)} y$$

and

$$\phi_3(x, y) = N_2^{(1)} = a_2^{(1)} + b_2^{(1)} x + c_2^{(1)} y,$$

so

$$\frac{\partial \phi_1}{\partial x} = b_3^{(1)}, \qquad \frac{\partial \phi_1}{\partial y} = c_3^{(1)}, \qquad \frac{\partial \phi_3}{\partial x} = b_2^{(1)}, \qquad \text{and} \qquad \frac{\partial \phi_3}{\partial y} = c_2^{(1)}.$$

Similarly, on T_2,

$$\phi_1(x, y) = N_1^{(2)} = a_1^{(2)} + b_1^{(2)} x + c_1^{(2)} y$$

and

$$\phi_3(x, y) = N_3^{(2)} = a_3^{(2)} + b_3^{(2)} x + c_3^{(2)} y,$$

so

$$\frac{\partial \phi_1}{\partial x} = b_1^{(2)}, \qquad \frac{\partial \phi_1}{\partial y} = c_1^{(2)}, \qquad \frac{\partial \phi_3}{\partial x} = b_3^{(2)}, \qquad \text{and} \qquad \frac{\partial \phi_3}{\partial y} = c_3^{(2)}.$$

Thus,

$$\alpha_{13} = b_3^{(1)} b_2^{(1)} \iint_{T_1} p \; dx \; dy + c_3^{(1)} c_2^{(1)} \iint_{T_1} q \; dx \; dy$$

$$- \iint_{T_1} r(a_3^{(1)} + b_3^{(1)} x + c_3^{(1)} y)(a_2^{(1)} + b_2^{(1)} x + c_2^{(1)} y) \; dx \; dy$$

$$+ b_1^{(2)} b_3^{(2)} \iint_{T_2} p \; dx \; dy + c_1^{(2)} c_3^{(2)} \iint_{T_2} q \; dx \; dy$$

$$- \iint_{T_2} r(a_1^{(2)} + b_1^{(2)} x + c_1^{(2)} y)(a_3^{(2)} + b_3^{(2)} x + c_3^{(2)} y) \; dx \; dy.$$

All the double integrals over D reduce to double integrals over triangles. The usual procedure is to compute all possible integrals over the triangles and accumulate them into the correct entry α_{ij} in A.

Similarly, the double integrals of the form

$$\iint_D f(x, y)\phi_i(x, y) \; dx \; dy$$

are computed over triangles and then accumulated into the correct entry b_i. For example, to determine b_i we need

$$- \iint_D f(x, y)\phi_1(x, y) \; dx \; dy = - \iint_{T_1} f(x, y)[a_3^{(1)} + b_3^{(1)} x + c_3^{(1)} y] \; dx \; dy$$

$$- \iint_{T_2} f(x, y)[a_1^{(2)} + b_1^{(2)} x + c_1^{(2)} y] \; dx \; dy.$$

Part of b_1 is contributed by ϕ_1 restricted to T_1, and the remainder by ϕ_1 restricted to T_2, since E_1 is a vertex of both T_1 and T_2. In addition, nodes that lie on $\mathscr{S}_2$ have line integrals added to their entries in A and $\mathbf{b}$.

The program LINFE125 performs the Finite-Element method on a second-order elliptic differential equation. In the program, all values of the matrix A and vector $\mathbf{b}$ are initially set to zero and after all the integrations have been performed on all the triangles these values are added to the appropriate entries in A and $\mathbf{b}$.

EXAMPLE 2 The temperature, $u(x, y)$, in a two-dimensional region D satisfies Laplace's equation

$$\frac{\partial^2 u}{\partial x^2}(x, y) + \frac{\partial^2 u}{\partial y^2}(x, y) = 0 \qquad \text{on } D.$$

Consider the region D shown in Figure 12.15 and suppose that the following boundary conditions are given:

$$u(x, y) = 4, \qquad \text{for } (x, y) \text{ on } L_6 \text{ and for } (x, y) \text{ on } L_7,$$

$$\frac{\partial u}{\partial n}(x, y) = x, \qquad \text{for } (x, y) \text{ on } L_2 \text{ and for } (x, y) \text{ on } L_4,$$

$$\frac{\partial u}{\partial n}(x, y) = y, \qquad \text{for } (x, y) \text{ on } L_5,$$

$$\frac{\partial u}{\partial n}(x, y) = \frac{x + y}{\sqrt{2}}, \qquad \text{for } (x, y) \text{ on } L_1 \text{ and for } (x, y) \text{ on } L_3,$$

where $\partial u / \partial n$ denotes the directional derivative in the direction of the normal to the boundary of the region D at the point (x, y).

Figure 12.15

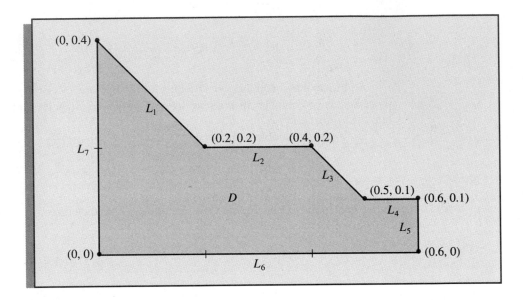

For this example, $\mathcal{S}_1 = L_6 \cup L_7$ and $\mathcal{S}_2 = L_1 \cup L_2 \cup L_3 \cup L_4 \cup L_5$. We first subdivide D into triangles with the labeling shown in Figure 12.16.

Figure 12.16

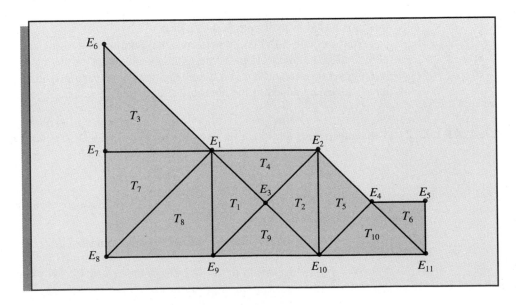

The boundary condition $u(x, y) = 4$ on L_6 and L_7 implies that $\gamma_t = 4$ when $t = 6, 7, \ldots, 11$. To determine the values of γ_l for $l = 1, 2, \ldots, 5$, generate the matrix and vector

$$
A = \begin{bmatrix} 2.5 & 0 & -1 & 0 & 0 \\ 0 & 1.5 & -1 & -0.5 & 0 \\ -1 & -1 & 4 & 0 & 0 \\ 0 & -0.5 & 0 & 2.5 & -0.5 \\ 0 & 0 & 0 & -0.5 & 1 \end{bmatrix} \quad \text{and} \quad \mathbf{b} = \begin{bmatrix} 6.066\bar{6} \\ 0.063\bar{3} \\ 8.0000 \\ 6.056\bar{6} \\ 2.031\bar{6} \end{bmatrix}.
$$

The solution to the equation $A\mathbf{c} = \mathbf{b}$ is

$$
\mathbf{c} = \begin{bmatrix} \gamma_1 \\ \gamma_2 \\ \gamma_3 \\ \gamma_4 \\ \gamma_5 \end{bmatrix} = \begin{bmatrix} 4.0383 \\ 4.0782 \\ 4.0291 \\ 4.0496 \\ 4.0565 \end{bmatrix},
$$

which gives the following approximation to the solution of Laplace's equation and the boundary conditions on the respective triangles:

$T_1:$ $\phi(x, y) = 4.0383(1 - 5x + 5y) + 4.0291(-2 + 10x) + 4(2 - 5x - 5y),$

$T_2:$ $\phi(x, y) = 4.0782(-2 + 5x + 5y) + 4.0291(4 - 10x) + 4(-1 + 5x - 5y),$

$T_3:$ $\phi(x, y) = 4(-1 + 5y) + 4(2 - 5x - 5y) + 4.0383(5x),$

$T_4:$ $\phi(x, y) = 4.0383(1 - 5x + 5y) + 4.0782(-2 + 5x + 5y) + 4.0291(2 - 10y),$

$T_5:$ $\phi(x, y) = 4.0782(2 - 5x + 5y) + 4.0496(-4 + 10x) + 4(3 - 5x - 5y),$

$T_6:$ $\phi(x, y) = 4.0496(6 - 10x) + 4.0565(-6 + 10x + 10y) + 4(1 - 10y),$

$T_7:$ $\phi(x, y) = 4(-5x + 5y) + 4.0383(5x) + 4(1 - 5y),$

$T_8:$ $\phi(x, y) = 4.0383(5y) + 4(1 - 5x) + 4(5x - 5y),$

$T_9:$ $\phi(x, y) = 4.0291(10y) + 4(2 - 5x - 5y) + 4(-1 + 5x - 5y),$

$T_{10}:$ $\phi(x, y) = 4.0496(10y) + 4(3 - 5x - 5y) + 4(-2 + 5x - 5y).$

The actual solution to the boundary value problem is $u(x, y) = xy + 4$. Table 12.7 compares the value of u to the value of ϕ at E_i, for each $i = 1, \ldots, 5$. ■ ■ ■

Table 12.7

x	y	$\phi(x, y)$	$u(x, y)$	$\lvert \phi(x, y) - u(x, y) \rvert$
0.2	0.2	4.0383	4.04	0.0017
0.4	0.2	4.0782	4.08	0.0018
0.3	0.1	4.0291	4.03	0.0009
0.5	0.1	4.0496	4.05	0.0004
0.6	0.1	4.0565	4.06	0.0035

Typically, the error for elliptic second-order problems with smooth coefficient functions is $O(h^2)$, where h is the maximum diameter of the triangular elements. Piecewise bilinear basis functions on rectangular elements are also expected to give $O(h^2)$ results, where h is the maximum diagonal length of the rectangular elements. Other classes of basis functions can be used to give $O(h^4)$ results, but the construction is more complex. Efficient error theorems for finite-element methods are difficult to state and apply because the accuracy of the approximation depends on the continuity properties of the solution and the regularity of the boundary.

EXERCISE SET 12.5

1. Use the Finite-Element method to approximate the solution to the following partial-differential equation (see the figure).

$$\frac{\partial}{\partial x}\left(y^2\frac{\partial u}{\partial x}(x, y)\right) + \frac{\partial}{\partial y}\left(y^2\frac{\partial u}{\partial y}(x, y)\right) - yu(x, y) = -x, \qquad (x, y) \in D,$$

$$u(x, 0.5) = 2x, \qquad 0 \le x \le 0.5, \qquad u(0, y) = 0, \qquad 0.5 \le y \le 1,$$

$$y^2\frac{\partial u}{\partial x}\cos\theta_1 + y^2\frac{\partial u}{\partial y}(x, y)\cos\theta_2 = \frac{\sqrt{2}}{2}(y - x) \qquad \text{for } (x, y) \in \mathcal{S}_2.$$

Let $M = 2$; T_1 have vertices $(0, 0.5), (0.25, 0.75), (0, 1)$ and T_2 have vertices $(0, 0.5),$ $(0.5, 0.5),$ and $(0.25, 0.75)$.

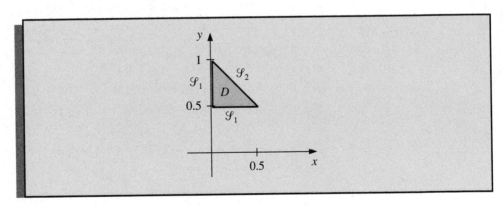

2. Repeat Exercise 1 using the following triangles:

$$T_1: \qquad (0, 0.75), (0, 1), (0.25, 0.75);$$
$$T_2: \qquad (0.25, 0.5), (0.25, 0.75), (0.5, 0.5);$$
$$T_3: \qquad (0, 0.5), (0, 0.75), (0.25, 0.75);$$
$$T_4: \qquad (0, 0.5), (0.25, 0.05), (0.25, 0.75).$$

3. Approximate the solution to the partial-differential equation

$$\frac{\partial^2 u}{\partial x^2}(x, y) + \frac{\partial^2 u}{\partial y^2}(x, y) - 12.5\pi^2 u(x, y) = -25\pi^2 \sin\frac{5\pi}{2}x \sin\frac{5\pi}{2}y, \qquad 0 < x, \ y < 0.4,$$

subject to the Dirichlet boundary condition

$$u(x, y) = 0,$$

using the Finite-Element method with the elements given in the figure. Compare the approximate solution to the actual solution

$$u(x, y) = \sin \frac{5\pi}{2} x \sin \frac{5\pi}{2} y$$

at the interior vertices and at the points $(0.125, 0.125)$, $(0.125, 0.25)$, $(0.25, 0.125)$, and $(0.25, 0.25)$.

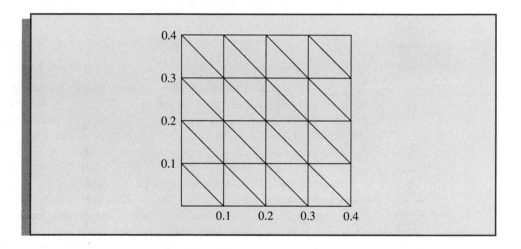

4. Repeat Exercise 3, with $f(x, y) = -25\pi^2 \cos \frac{5\pi}{2} x \cos \frac{5\pi}{2} y$, using the Neumann boundary condition

$$\frac{\partial u}{\partial n}(x, y) = 0.$$

The actual solution for this problem is

$$u(x, y) = \cos \frac{5\pi}{2} x \cos \frac{5\pi}{2} y.$$

5. A silver plate in the shape of a trapezoid shown in the accompanying figure has heat being uniformly generated at each point at the rate $q = 1.5 \text{cal/cm}^3 \cdot \text{s}$. The steady-state temperature

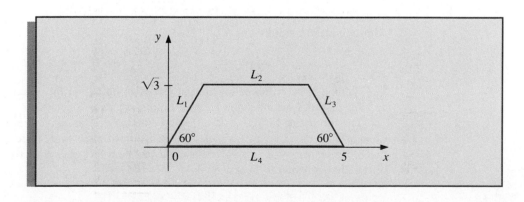

$u(x, y)$ of the plate satisfies the Poisson equation

$$\frac{\partial^2 u}{\partial x^2}(x, y) + \frac{\partial^2 u}{\partial y^2}(x, y) = \frac{-q}{K},$$

where K, the thermal conductivity, is 1.04 cal/cm · °C · s. Assume that the temperature is held at 15°C on L_2, that heat is lost on the slanted edges L_1 and L_3 according to the boundary condition $\partial u / \partial n = 4$, and that no heat is lost on L_4; that is, $\partial u / \partial n = 0$. Use the Finite-Element method to approximate the temperature of the plate at $(1, 0)$, $(4, 0)$, and $(\frac{5}{2}, \sqrt{3}/2)$.

12.6 Survey of Methods and Software

In this chapter, we considered methods to approximate solutions to partial-differential equations. We restricted attention to Poisson's equation as an example of an elliptic partial-differential equation, the heat or diffusion equation as an example of a parabolic partial-differential equation, and the wave equation as an example of a hyperbolic partial-differential equation. Finite difference approximations were discussed for these three examples.

Poisson's equation on a rectangle required the solution of a large sparse linear system, for which iterative techniques, such as the SOR method, are recommended. Three finite-difference methods were presented for the heat equation. The forward-difference method has stability problems, so the backward-difference method and the Crank–Nicolson methods were introduced. Although a tridiagonal linear system must be solved at each time step with these implicit methods, they are more stable than the explicit forward-difference method. The finite-difference method for the wave equation is explicit and can also have stability problems for certain choice of time and space discretizations.

In the last section of the chapter we presented an introduction to the finite-element method for a self-adjoint elliptic partial-differential equation on a polygonal domain. Although our methods will work adequately for the problems and examples in the textbook, more powerful generalizations and modifications of these techniques are required for sophisticated applications.

We consider two subroutines from the IMSL library. The subroutine MOLCH is used to solve the partial-differential equation

$$\frac{\partial u}{\partial t} = F\left(x, t, u, \frac{\partial u}{\partial x}, \frac{\partial^2 u}{\partial x^2}\right)$$

with boundary conditions

$$\alpha(x, t)u(x, t) + \beta(x, t)\frac{\partial u}{\partial x}(x, t) = \gamma(x, t).$$

The method is based on collocation at Gaussian points on the x-axis for each value of t. The solution uses the cubic Hermite splines as basis functions.

The subroutine FPS2H is used to solve Poisson's equation on a rectangle. The method of solution is based on a choice of second- or fourth-order finite differences on a uniform mesh. The linear system is solved using a fast Fourier transform technique. The partial-differential equation to be solved is

$$u_{xx} + u_{yy} + cu = P.$$

The NAG library has a number of subroutines for partial differential equations. The subroutine D03EAF is used for Laplace's equation on an arbitrary domain in the xy-plane. The subroutine D03PAF is used to solve a single parabolic partial-differential equation by the method of lines.

There are specialized packages, such as NASTRAN, consisting of codes for the finite element method. These packages are very popular in engineering applications. General codes for partial-differential equations are difficult to write because of the problem of specifying domains other than common geometrical figures. Research in the area of solution of partial-differential equations is currently very active.

BIBLIOGRAPHY

Aiken, R. C. (ed.) *Stiff Computation*. Oxford University Press, Oxford, 1985. 462 pp. QA371.S8474.

Allgower, E., and K. Georg. *Numerical Continuation Methods: An Introduction*. Springer-Verlag, Berlin, 1990. 388 pp. QA377.A562.

Anderson, E., et al. *LAPACK User's Guide*. SIAM Publications, Philadelphia, 1992. 270 pp.

Atkinson, K. E. *An Introduction to Numerical Analysis* (second edition). John Wiley & Sons, New York, 1989. 693 pp. QA297.A84.

Axelson, O., and V. A. Barker. *Finite Element Solution of Boundary Value Problems*. Academic Press, New York, 1984. 432 pp. QA379.A9.

Bailey, P. B., L. F. Shampine, and P. E. Waltman. *Nonlinear Two-Point Boundary-Value Problems*. Academic Press, New York, 1968. 171 pp. QA372.B27.

Bekker, M. G. *Introduction to Terrain Vehicle Systems*. University of Michigan Press, Ann Arbor, Mich., 1969. 846 pp. TL243.B39.

Bernadelli, H. "Population Waves," *Journal of the Burma Research Society* **31** (1941): 1–18.

Birkhoff, G., and G. Rota. *Ordinary Differential Equations*. John Wiley & Sons, New York, 1978. 342 pp. QA372.B58.

Bracewell, R. *The Fourier Transform and Its Application* (second edition). McGraw-Hill, New York, 1978. 444 pp. QA403.5.B7.

Brent, R. *Algorithms for Minimization without Derivatives*. Prentice Hall, Englewood Cliffs, N.J., 1973. 195 pp. QA402.5.B74.

Brigham, E. O. *The Fast Fourier Transform*. Prentice Hall, Englewood Cliffs, N.J., 1974. 252 pp. QA403.B74.

Broyden, C. G. "A class of methods for solving nonlinear simultaneous equations," *Mathematics of Computation* **19** (1965): 577–593, QA1.M4144.

Burden, R. C., and J. D. Faires. *Numerical Analysis* (fifth edition). PWS-KENT, Boston, 1993. 760 pp. QA297.B84.

Cody, W. J., and W. Waite. *Software Manual for Elementary Functions*. Prentice Hall, Englewood Cliffs, N.J., 1980. 269 pp. QA331.C635.

Coleman, T. F., and C. Van Loan, *Handbook for Matrix Computations*. SIAM Publications, Philadelphia, 1988. 264 pp. QA188.C65.

Cooley, J. W., and J. W. Tukey, "An algorithm for the machine calculation of complex Fourier series," *Mathematics of Computation* **19,** no. 90 (1965): 297–301. QA1.M4144.

Crowell, W. (ed.) *Sources and Development of Mathematical Software*. Prentice Hall, Englewood Cliffs, N.J., 1984. 404 pp. QA76.95.S68.

Dahlquist, G., and Å. Björck (translated by N. Anderson). *Numerical Methods*. Prentice Hall, Englewood Cliffs, N.J., 1974. 573 pp. QA297.D33l3.

Davis, P. J., and P. Rabinowitz. *Methods of Numerical Integration*. Academic Press, New York, 1975. 459 pp. QA299.3.D28.

DeBoor, C., *A Practical Guide to Splines*. Springer-Verlag, New York, 1978. 392 pp. QA1.A647.

Dennis, J. E., Jr., and R. B. Schnabel. *Numerical Methods for Unconstrained Optimization and Nonlinear Equations*. Prentice Hall, Englewood Cliffs, N.J., 1983. 378 pp. QA402.5.D44.

Dongarra, J. J., J. R. Bunch, C. B. Moler, and G. W. Stewart. *LINPACK Users Guide*. SIAM Publications, Philadelphia, 1979. 367 pp. QA214.L56.

Dorn, G. L., and A. B. Burdick. "On the recombinational structure of complementation relationships in the *m–dy* complex of the *Drosophila melanogaster,*" *Genetics* **47** (1962): 503–518, QH431.G43.

Engels, H. *Numerical Quadrature and Cubature*. Academic Press, New York, 1980. 441 pp. QA299.3.E5.

Faires, J. D., and B. T. Faires. *Calculus* (second edition). Random House, New York, 1988. 1028 pp. QA303.F294.

Forsythe, G. E., M. A. Malcolm, and C. B. Moler. *Computer Methods for Mathematical Computations*. Prentice Hall, Englewood Cliffs, N.J., 1977. 259 pp. QA297.F568.

Forsythe, G. E., and C. B. Moler. *Computer Solution of Linear Algebraic Systems*. Prentice Hall, Englewood Cliffs, N.J., 1967. 148 pp. QA297.F57.

Garbow, B. S., J. M. Boyle, J. J. Dongarra, and C. B. Moler. *Matrix Eigensystem Routines: EISPACK Guide Extension*. Springer-Verlag, New York, 1977. 343 pp. QA193.M38.

Gear, C. W. *Numerical Initial-Value Problems in Ordinary Differential Equations*. Prentice Hall, Englewood Cliffs, N.J., 1971. 253 pp. QA372.G4.

George, J. A., and J. W. H. Liu. *Computer Solution of Large Sparse Positive Definite Systems*. Prentice Hall, Englewood Cliffs, N.J., 1981. 324 pp. QA188.G46.

Golub, G. H., and C. F. Van Loan. *Matrix Computations* (second edition). Johns Hopkins University Press, Baltimore, 1989. 642 pp. QA188.G65.

Gragg, W. B. "On extrapolation algorithms for ordinary inital-value problems," *SIAM Journal on Numerical Analysis* **2** (1965): 384–403, QA297.A1S2.

Hageman, L. A., and D. M. Young. *Applied Iterative Methods*. Academic Press, New York, 1981. 386 pp. QA297.8.H34.

Hamming, R. W. *Numerical Methods for Scientists and Engineers* (second edition). McGraw-Hill, New York, 1973. 721 pp. QA297.H28.

Henrici, P. *Discrete Variable Methods in Ordinary Differential Equations*. John Wiley & Sons, New York, 1962. 407 pp. QA372.H48.

Henrici, P. *Elements of Numerical Analysis*. John Wiley & Sons, New York, 1964. 328 pp. QA297.H4.

Hildebrand, F. B. *Introduction to Numerical Analysis* (second edition). McGraw-Hill, New York, 1974. 669 pp. QA297.H54.

Hill, D. R. *Experiments in Computational Matrix Algebra*. Random House, New York, 1988. 446 pp. QA188.H55.

Householder, A. S. *The Numerical Treatment of a Single Nonlinear Equation*. McGraw-Hill, New York, 1970. 216 pp. QA218.H68.

Isaacson, E., and H. B. Keller. *Analysis of Numerical Methods*. John Wiley & Sons, New York, 1966. 541 pp. QA297.I8.

Jain, M. K. *Numerical Solution of Differential Equations* (second edition). John Wiley & Sons, New York, 1984. 698 pp. QA371.J34.

Johnston, R. L. *Numerical Methods: A Software Approach*. John Wiley & Sons, New York, 1982. 276 pp. QA297.J64.

Kahaner, D., C. Moler, and S. Nash. *Numerical Methods and Software*. Prentice Hall, Englewood Cliffs, N.J., 1989. 495 pp. TA345.K34.

Keller, H. B. *Numerical Methods for Two-Point Boundary-Value Problems*. Blaisdell, Waltham, MA, 1968. 184 pp. QA372.K42.

Kincaid, D., and W. Cheney. *Numerical Analysis: Mathematics of Scientific Computing*. Brooks/Cole Publishing Company, Pacific Grove, Calif., 1991. 690 pp. QA297.K563.

Lapidus, L., and G. F. Pinder. *Numerical Solution of Partial Differential Equations in Science and Engineering*. John Wiley & Sons, New York, 1982. 677 pp. Q172.L36.

Moler, C. B. "Demonstration of a matrix laboratory," *Lecture Notes in Mathematics*. Springer-Verlag, Berlin, 1982. pages 84–98.

Noble, B., and J. W. Daniel. *Applied Linear Algebra* (second edition). Prentice Hall, Englewood Cliffs, N.J., 1977. 477 pp. QA184.N6.

Ortega, J. M. *Introduction to Parallel and Vector Solution of Linear Systems*. Plenum Press, New York, 1988. 305 pp. QA218.O78.

Ortega, J. M. *Numerical Analysis—A Second Course*. Academic Press, New York, 1972. 201 pp. QA297.O78.

Ortega, J. M., and W. G. Poole, Jr. *An Introduction to Numerical Methods for Differential Equations*. Pitman Press, Marshfield, Mass., 1981. 329 pp. QA371.O65.

Ortega, J. M., and W. C. Rheinboldt. *Iterative Solution of Nonlinear Equations in Several Variables*. Academic Press, New York, 1970. 572 pp. QA297.8.O77.

Parlett, B. *The Symmetric Eigenvalue Problem*. Prentice Hall, Englewood Cliffs, N.J., 1980. 348 pp. QA188.P37.

Phillips, J. *The NAG Library: A Beginner's Guide*. Clarendon Press, Oxford, 1986. 245 pp. QA297.P35.

Piessens, R., E. de Doncker-Kapenga, C. W. Überhuber, and D. K. Kahaner. *QUAD-PACK: A Subroutine Package for Automatic Integration*. Springer-Verlag, New York, 1983. 301 pp. QA299.3.Q36.

Pissantsky, S. *Sparse Matrix Technology*. Acadamic Press, New York, 1984. 321 pp. QA188.P57.

Powell, M. J. D. *Approximation Theory and Methods*. Cambridge University Press, Cambridge, 1981. 339 pp. QA221.P65.

Press, W. H., B. P. Flannery, S.A. Teukolsky, and W. T. Vetterling. *Numerical Recipes: The Art of Scientific Computing*. Cambridge University Press, Cambridge, 1986. 818 pp. QA297.N866.

Ralston, A., and P. Rabinowitz. *A First Course in Numerical Analysis* (second edition). McGraw-Hill, New York, 1978. 556 pp. QA297.R3.

Rice, J. R., *Matrix Computations and Mathematical Software*. McGraw-Hill, New York, 1981. 248 pp. QA188.R52.

Rice, J. R., *Numerical Methods, Software, and Analysis: IMSL Reference Edition*. McGraw-Hill, New York, 1983. 661 pp. QA297.R49.

Rice, J. R., and R. F. Boisvert. *Solving Elliptic Problems Using ELLPACK*. Springer-Verlag, New York, 1985. 497 pp. QA377.R53.

Sagar, V., and D. J. Payne. "Incremental collapse of thick-walled circular cylinders under steady axial tension and torsion loads and cyclic transient heating," *Journal of the Mechanics and Physics of Solids* **21,** no. 1 (1975): 39–54, TA350.J68.

Schendel, U. *Introduction to Numerical Methods for Parallel Computers*. John Wiley & Sons, New York, 1984. 151 pp. QA297.S3813.

Schroeder, L. A. "Energy budget of the larvae of the moth *Pachysphinx modesta*," *Oikos* **24** (1973): 278–281, QH540.O35.

Schultz, M. H. *Spline Analysis,* Prentice Hall, Englewood Cliffs, N.J., 1973. 156 pp. QA211.S33.

Shampine, L. F., and M. K. Gordon. *Computer Solution of Ordinary Differential Equations: The Initial Value Problem*. W. H. Freeman, San Francisco, 1975. 318 pp. QA372.S416.

Simon, B., and R. M. Wilson. "Supercalculators on the PC," *Notices of the American Mathematical Society* **35,** no. 7 (1988): 978–1001. QA1.A5213.

Singh, V. P. "Investigations of attentuation and internal friction of rocks by ultrasonics," *International Journal of Rock Mechanics and Mining Sciences* (1976): 69–72. TA706.I45.

Smith, B. T., J. M. Boyle, J. J. Dongarra, B. S. Garbow, Y. Ikebe, V. C. Klema, and C. B. Moler. *Matrix Eigensystem Routines: EISPACK Guide* (second edition). Springer-Verlag, New York, 1976. 551 pp. QA193.M37.

Stetter, H. J. *Analysis of Discretization Methods for Ordinary Differential Equations. From Tracts in Natural Philosophy*. Springer-Verlag, Berlin, 1973. 388 pp. QA372.S84.

Stewart, G. W. *Introduction to Matrix Computations*. Academic Press, New York, 1973. 441 pp. QA188.S7.

Strang, W. G., and G. J. Fix. *An Analysis of the Finite Element Method*. Prentice Hall, Englewood Cliffs, N.J., 1973. 306 pp. TA335.S77.

Strikwerda, J. C. *Finite Difference Schemes and Partial Differential Equations*. Brooks/ Cole Publishing Company, Pacific Grove, Calif., 1989. 386 pp. QA374.S88.

Stroud, A. H. *Approximate Calculations of Multiple Integrals*. Prentice Hall, Englewood Cliffs, N.J., 1971. 431 pp. QA311.S85.

Stroud, A. H., and D. Secrest. *Gaussian Quadrature Formulas*. Prentice Hall, Englewood Cliffs, N.J., 1966. 374 pp. QA299.4.G4S7.

Varga, R. S. *Matrix Iterative Analysis*. Prentice Hall, Englewood Cliffs, N.J., 1962. 322 pp. QA263.V3.

Vichnevetsky, R. *Computer Methods for Partial Differential Equations,* vol. 1. Prentice Hall, Englewood Cliffs, N.J., 1981. 400 pp. QA374.V53.

Wait, R., and A. R. Mitchel. *Finite Element Analysis and Applications*. John Wiley & Sons, New York, 1985. 260 pp. TA374.F5W35.

Watkins, D. S. "Undertanding the QR algorithm," *SIAM Review* **24,** no. 4 (1982): 427–440. QA1.S2.

Wendroff, B. *Theoretical Numerical Analysis*. Academic Press, New York, 1966. 239 pp. QA297.W43.

Wilkinson, J. H. *The Algebraic Eigenvalue Problem*. Clarendon Press, Oxford, 1965. 662 pp. QA218.W5.

Wilkinson, J. H., *Rounding Errors in Algebraic Processes*. H. M. Stationary Office, London, 1963. 161 pp. QA76.5.W53.

Wilkinson, J. H., and C. Reinsch. *Handbook for Automatic Computation,* vol. 2: *Linear Algebra*. Springer-Verlag, Berlin, 1971. 439 pp. QA251.W67.

Young, D. M., *Iterative Solution of Large Linear Systems*. Academic Press, New York, 1971. 570 pp. QA195.Y68.

Young, D. M., and R. T. Gregory. *A Survey of Numerical Mathematics,* vol. 1. Addison-Wesley, Reading, Mass., 1972. 533 pp. QA297.Y63.

Zienkiewicz, O. C., and K. Morgan. *Finite Elements and Approximation*. John Wiley & Sons, New York, 1983. 328 pp. QA297.5.Z53.

Answers to Odd-Numbered Exercises

Exercise Set 1.2 (page 9)

1. For each part, $f \in C[a, b]$ on the given interval. Since $f(a)$ and $f(b)$ are of opposite sign, the Intermediate Value Theorem implies a number c exists with $f(c) = 0$.

3. The maximum value for $|f(x)|$ is **(a)** 0.4620981   **(b)** 1.333333   **(c)** 5.164000   **(d)** 1.582572   **(e)** 160
   (f) 10.79869

5. **(a)** $P_3(x) = -1 + 4x - 2x^2 - \frac{1}{2}x^3$ and $P_3(0.25) = -0.1328125$
   (b) $|R_3(0.25)| \leq 2.005 \times 10^{-4}$ and $|f(0.25) - P_3(0.25)| = 4.061 \times 10^{-5}$
   (c) $P_7(x) = -1 + 4x - 2x^2 - \frac{1}{2}x^3$ and $P_4(0.25) = -0.1328125$
   (d) $|R_4(0.25)| \leq 4.069 \times 10^{-5}$ and $|f(0.25) - P_4(0.25)| = 4.061 \times 10^{-5}$

7. $P_4(x) = 1 + 2x + \frac{3x^2}{2} + \frac{x^3}{3} - \frac{7x^4}{24}$; $|P_4(x)| \leq 0.03753$

9. $P_3(x) = \frac{\sqrt{3}}{2} - \frac{1}{2}(x - \frac{\pi}{6}) - \frac{\sqrt{3}}{4}(x - \frac{\pi}{6})^2 + \frac{1}{12}(x - \frac{\pi}{6})^3$ and $|\cos 32° - P_3(0.5585054)| \leq 5.4 \times 10^{-8}$.

11. **(a)** Expanding about $x_0 = 0$, we have $P_1(x) = 7 - 2x$, $P_2(x) = 7 - 2x$, $P_3(x) = 7 - 2x + 5x^3$, and $P_4(x) = 7 - 2x + 5x^3$.
    (b) Expanding about $x_0 = 1$, we have $P_1(x) = 10 + 13(x - 1)$, $P_2(x) = 10 + 13(x - 1) + 15(x - 1)^2$, $P_3(x) = 10 + 13(x - 1) + 15(x - 1)^2 + 5(x - 1)^3$, and $P_4(x) = 10 + 13(x - 1) + 15(x - 1)^2 + 5(x - 1)^3$.

13. 0.176

Exercise Set 1.3 (page 14)

1. The machine numbers are equivalent to   **(a)** 2707   **(b)** -2707 **(c)** 0.0174560546875
   (d) 0.0174560584127903

3.

	Absolute Error	Relative Error
(a)	0.001264	4.025×10^{-4}
(b)	7.346×10^{-6}	2.338×10^{-6}
(c)	2.818×10^{-4}	1.037×10^{-4}
(d)	2.136×10^{-4}	1.510×10^{-4}
(e)	2.647×10^{1}	1.202×10^{-3}
(f)	1.454×10^{1}	1.050×10^{-2}
(g)	4.200×10^{2}	1.042×10^{-2}
(h)	3.343×10^{3}	9.213×10^{-3}

5.

	Approximation	Absolute Error	Relative Error
(a)	134	0.079	5.90×10^{-4}
(b)	133	0.499	3.77×10^{-3}
(c)	2.00	0.327	0.195
(d)	1.67	0.003	1.79×10^{-3}
(e)	1.80	0.154	0.0786
(f)	-15.1	0.0546	3.60×10^{-3}
(g)	0.286	2.86×10^{-4}	10^{-3}
(h)	0.00	0.0215	1.00

7.

	Approximation	Absolute Error	Relative Error
(a)	133.9	0.021	1.568×10^{-4}
(b)	132.5	0.001	7.55×10^{-6}
(c)	1.700	0.027	0.01614
(d)	1.673	0	0
(e)	1.986	0.03246	0.01662
(f)	-15.16	0.005377	3.548×10^{-4}
(g)	0.2857	1.429×10^{-5}	5×10^{-5}
(h)	-0.01700	0.0045	0.2092

9.

	Approximation	Absolute Error	Relative Error
(a)	133.9	0.021	1.568×10^{-4}
(b)	132.5	0.001	7.55×10^{-6}
(c)	1.600	0.073	0.04363
(d)	1.673	0	0
(e)	1.983	0.02945	0.01508
(f)	-15.15	0.004622	3.050×10^{-4}
(g)	0.2855	2.142×10^{-4}	7.5×10^{-4}
(h)	-0.01700	0.0045	0.2092

11.

	Approximation	Absolute Error	Relative Error
(a)	3.145	0.003407	1.085×10^{-3}
(b)	3.139	0.002593	8.253×10^{-4}

Exercise Set 1.4 (page 19)

1.

	x_1	Absolute Error	Relative Error	x_2	Absolute Error	Relative Error
(a)	92.26	1.542×10^{-2}	1.672×10^{-4}	0.005421	6.273×10^{-7}	1.157×10^{-4}
(b)	0.005421	1.264×10^{-6}	2.333×10^{-4}	-92.26	4.581×10^{-3}	4.966×10^{-5}
(c)	10.98	6.874×10^{-3}	6.257×10^{-4}	0.001149	7.560×10^{-8}	6.580×10^{-5}
(d)	-0.001149	7.566×10^{-8}	6.584×10^{-5}	-10.98	6.875×10^{-3}	6.257×10^{-4}

3. **(a)** -0.1000 **(b)** -0.1010

(c) Absolute error for part (a) 2.331×10^{-3} with relative error 2.387×10^{-2}. Absolute error for part (b) 3.331×10^{-3} with relative error 3.411×10^{-2}.

5. The approximate solutions to the systems are **(a)** $x = 2.451, y = -1.635$ **(b)** $x = 507.7, y = 82.00$

7.

	Approximation	Absolute Error	Relative Error
(a)	0.3743	1.011×10^{-3}	2.694×10^{-3}
(b)	0.3755	1.889×10^{-4}	5.033×10^{-4}

9. **(a)** $O\left(\frac{1}{n}\right)$ **(b)** $O\left(\frac{1}{n^2}\right)$ **(c)** $O\left(\frac{1}{n^2}\right)$ **(d)** $O\left(\frac{1}{n}\right)$

Exercise Set 2.2 (page 29)

1. The Bisection method gives **(a)** $p_7 = 0.5859$ **(b)** $p_8 = 3.002$ **(c)** $p_7 = 3.419$

3. The Bisection method gives $p_9 = 4.4932$.

5. The Bisection method gives **(a)** $p_{17} = 0.641182$ **(b)** $p_{17} = 0.257530$

(c) For the interval $[-3, -2]$ we have $p_{17} = -2.191307$ and for the interval $[-1, 0]$ we have $p_{17} = -0.798164$.

(d) For the interval $[0.2, 0.3]$ we have $p_{14} = 0.297528$ and for the interval $[1.2, 1.3]$ we have $p_{14} = 1.256622$.

7. A bound is $n \geq 14$ and $p_{14} = 1.32477$. **9.** A bound is $n \geq 7$ and $p_7 = -0.473$.

11.

Interval	Approximation
$[0.1, 3]$	$p_{15} = 2.36324$
$[3, 4]$	$p_{14} = 3.81793$
$[4, 6]$	$p_{15} = 5.83929$
$[6, 8]$	$p_{15} = 6.60309$

13. The depth of the water is 0.838 ft.

Exercise Set 2.3 (page 34)

1. Using the endpoints of the intervals as p_0 and p_1, we have   (a) $p_{11} = 2.69065$ (b) $p_7 = -2.87939$
   (c) $p_6 = 0.73909$ (d) $p_5 = 0.96433$

3. Using the endpoints of the intervals as p_0 and p_1, we have   (a) $p_7 = 1.829384$ (b) $p_9 = 1.397748$
   (c) For $[0, 1]$, $p_6 = 0.9100076$ and for $[3, 5]$, $p_{10} = 3.733079$.
   (d) For $[1, 2]$, $p_8 = 1.412391$ and for $[e, 4]$, $p_7 = 3.057104$.

5. Using the endpoints of the intervals as p_0 and p_1, we have   (a) $p_2 = -0.479$ (b) $p_3 = 0.705$

7. Using the endpoints of the intervals as p_0 and p_1, we have
   (a) For $[2, 3]$, $p_6 = 2.370687$ and for $[3, 4]$, $p_7 = 3.722113$.
   (b) For $[0.2, 0.3]$, $p_4 = 0.297530$ and for $[1.2, 1.3]$, $p_5 = 1.256623$.

9. Using the endpoints of the intervals as p_0 and p_1, we have   (a) $p_{5610} = -1.3977446$ (b) $p_{3960} = 0.6491861$

11. Using the Secant method we have the following: 13. $w_4 = -0.3170618$

    For $p_0 = -4$ and $p_1 = -3$, we have $p_7 = -3.161950$.
    For $p_0 = -3$ and $p_1 = -1$, we have $p_9 = -1.968873$.
    For $p_0 = -1$ and $p_1 = 2$, we have $p_6 = 1.968873$.
    For $p_0 = 3$ and $p_1 = 4$, we have $p_7 = 3.161950$.

Exercise Set 2.4 (page 39)

1. Newton's method gives the following approximations:
   (a) For $p_0 = 2$, we have $p_5 = 2.69065$.
   (b) For $p_0 = -3$, we have $p_3 = -2.87939$.
   (c) For $p_0 = 0$, we have $p_4 = 0.73909$.
   (d) For $p_0 = 0$, we have $p_3 = 0.96434$.

3. Newton's method gives the approximation $p_3 = 0.90479$.

5. Newton's method gives the following approximations:

    With $p_0 = 1.5$, $p_6 = 2.363171$.
    With $p_0 = 3.5$, $p_5 = 3.817926$.
    With $p_0 = 5.5$, $p_4 = 5.839252$.
    With $p_0 = 7$, $p_5 = 6.603085$.

7. With $p_0 = \frac{\pi}{2}$, we have $p_{15} = 1.895488$; $p_0 = 5\pi$ gives $p_{19} = 1.895489$; and $p_0 = 10\pi$ does not converge in 200 iterations.

9. (a) For $p_0 = -1$ and $p_1 = 0$, we have $p_{17} = -0.04065850$; and for $p_0 = 0$ and $p_1 = 1$, we have $p_9 = 0.9623984$.
   (b) For $p_0 = -1$ and $p_1 = 0$, we have $p_5 = -0.04065929$; and for $p_0 = 0$ and $p_1 = 1$, we have $p_{12} = -0.04065929$.
   (c) For $p_0 = -0.5$, we have $p_5 = -0.04065929$; and for $p_0 = 0.5$, we have $p_{21} = 0.9623989$.

11. For $f(x) = \ln(x^2 + 1) - e^{0.4x} \cos \pi x$, we have the following roots:
  (a) For $p_0 = -0.5$, we have $p_3 = -0.4341431$.
  (b) For $p_0 = 0.5$, we have $p_3 = 0.4506567$; for $p_0 = 1.5$, we have $p_3 = 1.7447381$; for $p_0 = 2.5$, we have $p_5 = 2.2383198$; and for $p_0 = 3.5$, we have $p_4 = 3.7090412$.
  (c) The initial approximation $n - 0.5$ is quite reasonable.
  (d) For $p_0 = 24.5$, we have $p_2 = 24.4998870$.

13. The borrower can afford to pay at most 9.4%.

Exercise Set 2.5 (page 44)

1. The results of Aitken's Δ^2 method are listed in the following table:

	(a)	(b)	(c)	(d)
$\hat{p}_0$	0.258684	0.907859	0.548101	0.731385
$\hat{p}_1$	0.257613	0.909568	0.547915	0.736087
$\hat{p}_2$	0.257536	0.909917	0.547847	0.737653
$\hat{p}_3$	0.257531	0.909989	0.547823	0.738469
$\hat{p}_4$	0.257530	0.910004	0.547814	0.738798
$\hat{p}_5$	0.257530	0.910007	0.547810	0.738958

3. $p_{10} = -0.169607$ and $\hat{p}_8 = -0.169607$ are accurate to within 2×10^{-4}.

5. Aitken's Δ^2 method gives   **(a)** $\hat{p}_{10} = 0.0\overline{45}$ **(b)** $\hat{p}_2 = 0.0363$

7. **(a)** $\displaystyle \lim_{n \to \infty} \frac{|p_{n+1} - 0|}{|p_n - 0|^2} = \lim_{n \to \infty} \frac{10^{-2^{n+1}}}{(10^{-2^n})^2} = \lim_{n \to \infty} \frac{10^{-2^{n+1}}}{10^{-2^{n+1}}} = 1$

Exercise Set 2.6 (page 48)

1. **(a)** For $p_0 = 1$, we have $p_{22} = 2.69065$.
  (b) For $p_0 = 1$, we have $p_5 = 0.53209$; for $p_0 = -1$, we have $p_3 = -0.65270$; and for $p_0 = -3$, we have $p_3 = -2.87939$.
  (c) For $p_0 = 1$, we have $p_5 = 1.32472$.
  (d) For $p_0 = 1$, we have $p_4 = 1.12412$; and for $p_0 = 0$, we have $p_8 = -0.87605$.
  (e) For $p_0 = 0$, we have $p_6 = -0.47006$; for $p_0 = -1$, we have $p_4 = -0.88533$; and for $p_0 = -3$, we have $p_4 = -2.64561$.
  (f) For $p_0 = 0$, we have $p_{10} = 1.49819$.

3. The following table lists the initial approximation and the roots found by Müller's method:

p_0	p_1	p_2	Approximate Roots	Complex Conjugate Roots
(a) -1	0	1	$p_7 = -0.34532 - 1.31873i$	$-0.34532 + 1.31873i$
0	1	2	$p_6 = 2.69065$	
(b) 0	1	2	$p_6 = 0.53209$	
1	2	3	$p_9 = -0.65270$	
-2	-3	-2.5	$p_4 = -2.87939$	

(Continued)

	p_0	p_1	p_2	Approximate Roots	Complex Conjugate Roots
(c)	0	1	2	$p_5 = 1.32472$	
	-2	-1	0	$p_7 = -0.66236 - 0.56228i$	$-0.66236 + 0.56228i$
(d)	0	1	2	$p_5 = 1.12412$	
	2	3	4	$p_{12} = -0.12403 + 1.74096i$	$-0.12403 - 1.74096i$
	-2	0	-1	$p_5 = -0.87605$	
(e)	0	1	2	$p_{10} = -0.88533$	
	1	0	-0.5	$p_5 = -0.47006$	
	-1	-2	-3	$p_5 = -2.64561$	
(f)	0	1	2	$p_6 = 1.49819$	
	-1	-2	-3	$p_{10} = -0.51363 - 1.09156i$	$-0.51363 + 1.09156i$
	1	0	-1	$p_8 = 0.26454 - 1.32837i$	$0.26454 + 1.32837i$

5. **(a)** The zeros are 1.244, 8.847, and -1.091, and the critical points are 0 and 6.
   (b) The zeros are 0.5798, 1.521, 2.332, and -2.432, and the critical points are 1, 2.001, and -1.5.

7. $W = 16.2121$ ft.

Exercise Set 3.2 (page 61)

1. (a)

n	$x_0, x_1, \ldots, x_n$	$P_n(8.4)$
1	8.3, 8.6	17.87833
2	8.3, 8.6, 8.7	17.87716
3	8.3, 8.6, 8.7, 8.1	17.81000
4	8.3, 8.6, 8.7, 8.1, 8.0	17.60859

(b)

n	$x_0, x_1, \ldots, x_n$	$P_n\left(-\frac{1}{3}\right)$
1	$-0.5, -0.25$	0.21504167
2	$-0.5, -0.25, 0.0$	0.16988889
3	$-0.5, -0.25, 0.0, -0.75$	0.17451852
4	$-0.5, -0.25, 0.0, -0.75, -1.0$	0.17451852

(c)

n	$x_0, x_1, \ldots, x_n$	$P_n(0.25)$
1	0.2, 0.3	-0.13869287
2	0.2, 0.3, 0.4	-0.13259734
3	0.2, 0.3, 0.4, 0.1	-0.13277477
4	0.2, 0.3, 0.4, 0.1, 0.0	-0.13277246

(d)

n	$x_0, x_1, \ldots, x_n$	$P_n(0.9)$
1	0.8, 1.0	0.44086280
2	0.8, 1.0, 0.7	0.43841352
3	0.8, 1.0, 0.7, 0.6	0.44198500
4	0.8, 1.0, 0.7, 0.6, 0.5	0.44325624

(e)

n	$x_0, x_1, \ldots, x_n$	$P_n(\pi)$
1	3.1, 3.2	-3.122526
2	3.1, 3.2, 3.0	-3.141593
3	3.1, 3.2, 3.0, 2.9	-3.141686
4	3.1, 3.2, 3.0, 2.9, 3.4	-3.141590

(f)

n	$x_0, x_1, \ldots, x_n$	$P_n\left(\frac{\pi}{2}\right)$
1	1.5, 1.4	19.98002
2	1.5, 1.4, 1.3	23.67269
3	1.5, 1.4, 1.3, 1.2	26.36966
4	1.5, 1.4, 1.3, 1.2, 1.1	28.49372

(g) n	$x_0, x_1, \ldots, x_n$	$P_n(1.15)$	(h) n	$x_0, x_1, \ldots, x_n$	$P_n(4.1)$
1	1.1, 1.2	2.074637	1	4.0, 4.2	-0.0262639
2	1.1, 1.2, 1.0	2.076485	2	4.0, 4.2, 4.4	0.0005223
3	1.1, 1.2, 1.0, 1.3	2.076244	3	4.0, 4.2, 4.4, 3.8	-0.0022188
4	1.1, 1.2, 1.0, 1.3, 1.4	2.076216	4	4.0, 4.2, 4.4, 3.8, 3.6	-0.0024334

3. $\sqrt{3} \approx P_4(\frac{1}{2}) = 1.708\overline{3}$

5. (a)

n	Actual Error	Error Bound
1	0.00118	0.00120
2	1.367×10^{-5}	1.452×10^{-5}
3	3.830×10^{-6}	2.823×10^{-7}
4	3.470×10^{-6}	8.789×10^{-9}

(b)

n	Actual Error	Error Bound
1	4.0523×10^{-2}	4.5153×10^{-2}
2	4.6296×10^{-3}	4.6296×10^{-3}
3	0	0
4	0	0

(c)

n	Actual Error	Error Bound
1	5.9210×10^{-3}	6.0971×10^{-3}
2	1.7455×10^{-4}	1.8128×10^{-4}
3	2.8798×10^{-6}	4.5143×10^{-6}
4	5.6320×10^{-7}	5.8594×10^{-7}

(d)

n	Actual Error	Error Bound
1	2.7296×10^{-3}	1.4080×10^{-2}
2	5.1789×10^{-3}	9.2215×10^{-3}

7. $f(1.09) \approx 0.2826$

9. **(a)** $f(1.03) \approx P_{0,1,2} = 0.8094418$ **(b)** $f(1.03) \approx P_{0,1,2,3} = 0.8092831$

11. **(a)** $P_2(x) = -11.22017728x^2 + 3.808210517x + 1$, and an error bound is 0.11371294.

  (b) $P_2(x) = -0.1306344167x^2 + 0.8969979335x - 0.63249693$, and an error bound is 9.45762×10^{-4}.

  (c) $P_3(x) = 0.1970056667x^3 - 1.06259055x^2 + 2.532453189x - 1.666868305$, and an error bound is 10^{-4}.

  (d) $P_3(x) = -0.07932x^3 - 0.545506x^2 + 1.0065992x + 1$, and an error bound is 1.591376×10^{-3}.

13. The first 10 terms of the sequence are 0.038462, 0.333671, 0.116605, -0.371760, -0.0548919, 0.605935, 0.190249, -0.513353, -0.0668173, 0.448335. Since $f(1 + \sqrt{10}) = 0.0545716$, the sequence does not appear to converge.

Exercise Set 3.3 (page 68)

1. Newton's interpolatory divided-difference formula gives the following results:

  (a) $P_1(x) = 16.63553 + 9.7996(x - 8); P_1(8.4) = 20.55537$
  $P_2(x) = P_1(x) - 33.50817(x - 8)(x - 8.1); P_2(8.4) = 16.53439$
  $P_3(x) = P_2(x) + 67.13678(x - 8)(x - 8.1)(x - 8.3); P_3(8.4) = 17.34003$
  $P_4(x) = P_3(x) - 111.8981(x - 8)(x - 8.1)(x - 8.3)(x - 8.6); P_4(8.4) = 17.60859$

(b) $P_1(x) = -0.3440987 + 1.671541(x - 0.5)$; $P_1(0.9) = 0.3245177$
$P_2(x) = P_1(x) + 1.177137(x - 0.5)(x - 0.6)$; $P_2(0.9) = 0.4657741$
$P_3(x) = P_2(x) - 0.7263767(x - 0.5)(x - 0.6)(x - 0.7)$; $P_3(0.9) = 0.4483411$
$P_4(x) = P_3(x) - 2.118729(x - 0.5)(x - 0.6)(x - 0.7)(x - 0.8)$; $P_4(0.9) = 0.4432562$

(c) $P_1(x) = -4.827866 + 5.87808(x - 2.9)$; $P_1(\pi) = -3.407765$
$P_2(x) = P_1(x) + 7.76705(x - 2.9)(x - 3.0)$; $P_2(\pi) = -3.142072$
$P_3(x) = P_2(x) + 0.2711667(x - 2.9)(x - 3.0)(x - 3.1)$; $P_3(\pi) = -3.141686$
$P_4(x) = P_3(x) - 1.159392(x - 2.9)(x - 3.0)(x - 3.1)(x - 3.2)$; $P_4(\pi) = -3.141590$

3. In the following equations we have $s = \frac{1}{h}(x - x_n)$:

(a) $P_1(s) = 1.101 + 0.7660625s$; $f(-\frac{1}{3}) \approx P_1(-\frac{4}{3}) = 0.07958333$
$P_2(s) = P_1(s) + 0.406375s\frac{(s+1)}{2}$; $f(-\frac{1}{3}) \approx P_2(-\frac{4}{3}) = 0.1698889$
$P_3(s) = P_2(s) + 0.09375s\frac{(s+1)(s+2)}{6}$; $f(-\frac{1}{3}) \approx P_3(-\frac{4}{3}) = 0.1745185$
$P_4(s) = P_3(s)$; $f(-\frac{1}{3}) \approx P_4(-\frac{4}{3}) = 0.1745185$

(b) $P_1(s) = 0.2484244 + 0.2418235s$; $f(0.25) \approx P_1(-1.5) = -0.1143108$
$P_2(s) = P_1(s) - 0.04876419s\frac{(s+1)}{2}$; $f(0.25) \approx P_2(-1.5) = -0.1325973$
$P_3(s) = P_2(s) - 0.00283893s\frac{(s+1)(s+2)}{6}$; $f(0.25) \approx P_3(-1.5) = -0.1327748$
$P_4(s) = P_3(s) + 0.0009881s\frac{(s+1)(s+2)(s+3)}{24}$; $f(0.25) \approx P_4(-1.5) = -0.1327725$

(c) $P_1(s) = 14.10142 + 8.303536s$; $f(\frac{\pi}{2}) \approx P_1(5\pi - 15) = 19.98002$
$P_2(s) = P_1(s) + 6.107754s\frac{(s+1)}{2}$; $f(\frac{\pi}{2}) \approx P_2(5\pi - 15) = 23.67269$
$P_3(s) = P_2(s) + 4.941922s\frac{(s+1)(s+2)}{6}$; $f(\frac{\pi}{2}) \approx P_3(5\pi - 15) = 26.36966$
$P_4(s) = P_3(s) + 4.198648s\frac{(s+1)(s+2)(s+3)}{24}$; $f(\frac{\pi}{2}) \approx P_4(5\pi - 15) = 28.49372$

5. (a) $f(0.05) \approx 1.05126$ (b) $f(0.65) \approx 1.9155505$

Exercise Set 3.4 (page 72)

1. The coefficients for the polynomials in divided-difference form are given in the following tables. For example, the polynomial in part (a) is

$$H_3(x) = 17.56492 + 1.116256(x - 8.3) + 6.726147(x - 8.3)^2 - 44.44647(x - 8.3)^2(x - 8.6).$$

(a)	(b)	(c)	(d)
17.56492	22.363362	-0.02475	-4.240058
1.116256	2.1691753	0.751	6.64986
6.726147	0.01558225	2.751	7.8163
-44.44647	-3.2177925	1	0.2733
			-1.128
			-0.18

(e)	(f)	(g)	(h)
−0.62049958	2.572152	1.684370	1.16164956
3.5850208	7.615964	2.742245	−1.5072822
−2.1989182	26.83536	−0.9117500	−1.4545692
−0.490447	99.21040	0.9499000	−0.56972300
0.037205	580.7080	−0.8815000	−0.17751125
0.040475	3400.600	0.8600000	−0.0446421875
−0.0025277777	48026.08	−0.8722222	−0.0097682292
0.0029629628	678365.5	1.037037	−0.0016232639
		0.04629630	−0.00071614579
		−15.85648	0.00021927433

3. For 2(a) we have an error bound of 5.9×10^{-8}, and for 2(e) we have an error bound of 2.7×10^{-13}. The error bound for 2(c) is 0 since $f^{(n)}(x) \equiv 0$ for $n > 3$.

5. (a) We have $\sin 0.34 \approx H_5(0.34) = 0.33349$.
   (b) The formula gives an error bound of 3.05×10^{-14}, but the actual error is 2.91×10^{-6}. The discrepancy is due to the fact that the data are only given to 5 decimal places.
   (c) We have $\sin 0.34 \approx H_7(0.34) = 0.33350$. Although the error bound is now 5.4×10^{-24}, the accuracy of the given data dominates the calculations. This result is actually less accurate than the approximation in part (b), since $\sin 0.34 = 0.333487$.

7. The Hermite polynomial generated from these data is

$$H_9(x) = 75x + 0.222222x^2(x-3) - 0.0311111x^2(x-3)^2$$
$$- 0.00644444x^2(x-3)^2(x-5) + 0.00226389x^2(x-3)^2(x-5)^2$$
$$- 0.000913194x^2(x-3)^2(x-5)^2(x-8)$$
$$+ 0.000130527x^2(x-3)^2(x-5)^2(x-8)^2$$
$$- 0.0000202236x^2(x-3)^2(x-5)^2(x-8)^2(x-13).$$

   (a) The Hermite polynomial predicts a position of $H_9(10) = 743$ ft and a speed of $H_9'(10) = 36$ ft/s. Although the position approximation is reasonable, the low speed prediction is suspect.
   (b) To find the first time the speed exceeds 55 mi/hr, we solve for the smallest value of t in the equation $55 = H_9'(x)$. This gives $x \approx 5.9119932$.
   (c) The estimated maximum speed is $H_9'(6.5009714) = 81.0004518$ ft/s ≈ 55.228 mi/hr.

Exercise Set 3.5 (page 81)

1. The equations of the respective free cubic splines are given by

$$S(x) = S_i(x) = a_i + b_i(x - x_i) + c_i(x - x_i)^2 + d_i(x - x_i)^3,$$

   for x in $[x_i, x_{i+1}]$, and the coefficients in the following tables:

(a)
i	a_i	b_i	c_i	d_i
0	17.564920	3.13410000	0.00000000	0.00000000

(b)

i	a_i	b_i	c_i	d_i
0	0.22363362	2.17229175	0.00000000	0.00000000

(c)

i	a_i	b_i	c_i	d_i
0	−0.02475000	1.03237500	0.00000000	6.50200000
1	0.33493750	2.25150000	4.87650000	−6.50200000

(d)

i	a_i	b_i	c_i	d_i
0	−4.24005800	7.03907000	0.00000000	39.242000
1	−3.49690900	8.21633000	11.77260000	−39.242000

(e)

i	a_i	b_i	c_i	d_i
0	−0.62049958	3.45508693	0.00000000	−8.9957933
1	−0.28398668	3.18521313	−2.69873800	−0.94630333
2	0.00660095	2.61707643	−2.98262900	9.9420966

(f)

i	a_i	b_i	c_i	d_i
0	2.57215200	11.26245066	0.00000000	−96.295066
1	3.60210200	8.37359866	−28.88852000	1647.3073
2	5.79788400	52.01511466	465.30368000	−1551.0122

(g)

i	a_i	b_i	c_i	d_i
0	1.68437000	2.68435107	0.00000000	−3.32810714
1	1.94947700	2.58450785	−0.99843214	1.85253571
2	2.19979600	2.44039750	−0.44267142	−0.22003571
3	2.43918900	2.34526214	−0.50868214	1.69560714

(h)

i	a_i	b_i	c_i	d_i
0	1.16164956	−1.65814692	0.00000000	−3.50122696
1	0.80201036	−2.07829415	−2.10073617	0.53954232
2	0.30663842	−2.85384355	−1.77701078	−2.99443232
3	−0.35916618	−3.92397974	−3.57367017	5.95611696

3. The equations of the respective clamped cubic splines are given by

$$s(x) = s_i(x) = a_i + b_i(x - x_i) + c_i(x - x_i)^2 + d_i(x - x_i)^3,$$

for x in $[x_i, x_{i+1}]$, and the coefficients in the following tables:

(a)

i	a_i	b_i	c_i	d_i
0	17.564920	1.1162560	20.060086	−44.446466

(b)

i	a_i	b_i	c_i	d_i
0	0.22363362	2.1691753	0.65914075	−3.2177925

(c)

i	a_i	b_i	c_i	d_i
0	−0.02475000	0.75100000	2.5010000	1.0000000
1	0.33493750	2.18900000	3.2510000	1.0000000

(d)

i	a_i	b_i	c_i	d_i
0	−4.2400580	6.6498600	7.7887900	0.27510000
1	−3.4969090	8.2158710	7.8713200	−0.18330000

(e)

i	a_i	b_i	c_i	d_i
0	−0.62049958	3.5850208	−2.1498407	−0.49077413
1	−0.28398668	3.1403294	−2.2970730	−0.47458360
2	0.006600950	2.6666773	−2.4394481	−0.44980146

(f)

i	a_i	b_i	c_i	d_i
0	2.5721520	7.6159640	−4.3305760	311.65936
1	3.6021020	16.099629	89.167232	−305.85328
2	5.7978840	24.757477	−2.5887520	5853.6757

(g)

i	a_i	b_i	c_i	d_i
0	1.684370	2.742245	−1.005926	0.9417607
1	1.949477	2.569313	−0.723398	0.6217179
2	2.199796	2.443285	−0.536883	0.4333679
3	2.439189	2.348909	−0.406872	0.3128107

(h)

i	a_i	b_i	c_i	d_i
0	1.16164956	-1.50728217	-1.34091868	-0.56825233
1	0.80201036	-2.11183992	-1.68187008	-0.71614399
2	0.30663842	-2.87052523	-2.11155648	-0.90466168
3	-0.35916618	-3.82370723	-2.65435349	-1.14727927

5. **(a)** $S(x) = 0.29552 + 0.95430(x - 0.3) - 4.5(x - 0.3)^3$ on $[0.30, 0.32]$, and
     $S(x) = 0.31457 + 0.94973(x - 0.32) - 0.27(x - 0.32)^2 + 3(x - 0.32)^3$ on $[0.32, 0.35]$;
     $S(0.34) = 0.33348$.

     (b) 7.09×10^{-6}

     (c) $s(x) = 0.29552 + 0.95433(x - 0.3) - 0.1105(x - 0.3)^2 - 1.5583(x - 0.3)$ on $[0.30, 0.32]$ and
     $s(x) = 0.31457 + 0.94987(x - 0.32) - 0.204(x - 0.32)^2 + 0.64078(x - 0.32)^3$ on $[0.32, 0.35]$;
     $s(0.34) = 0.33349$.

     (d) 2.91×10^{-6} **(e)** $S'(0.34) = 0.94253$ **(f)** $s'(0.34) = 0.94248$ **(g)** $\int_{0.30}^{0.35} S(x)\, dx = 0.015964$

     (h) $\int_{0.30}^{0.35} s(x)\, dx = 0.015964$

7. Using the portion of the spline on $[1.02, 1.04]$, we have $S(1.03) = 0.809324$ with an error bound of 1.9×10^{-7} and an actual error of 3.8×10^{-7}.

9. On $[0, 0.05]$, $F(x) = 20(e^{0.1} - 1)x + 1$; and on $(0.05, 1]$, $F(x) = 20(e^{0.2} - e^{0.1})x + 2e^{0.1} - e^{0.2}$. So $\int_0^{0.1} F(x)\, dx = 0.1107936$.

11. The spline has equation $S(x) = S_i(x) = a_i + b_i(x - x_i) + c_i(x - x_i)^2 + d_i(x - x_i)^3$ on $[x_i, x_{i+1}]$, where the coefficients are in the following table:

x_i	a_i	b_i	c_i	d_i
0	0	75	-0.659292	0.219764
3	225	76.99779	1.31858	-0.153761
5	383	80.4071	0.396018	-0.177237
8	623	77.9978	-1.19912	0.0799115

The spline predicts a position of $S(10) = 774.84$ ft and a speed of $S'(10) = 74.16$ ft/s. To maximize the speed, we find the single critical points of $S'(x)$ and compare the values of $S(x)$ at these points and the endpoints. We find that max $S'(x) = S'(5.7448) = 80.7$ ft/s $= 55.02$ mi/hr. The speed 55 mi/hr was first exceeded at approximately 5.5 s.

13. **(a)** $S(x) = S_i(x) = a_i + b_i(x - x_i) + c_i(x - x_i)^2 + d_i(x - x_i)^3$ on $[x_i, x_{i+1}]$, where the coefficients are in the following table:

x_i	a_i	b_i	c_i	d_i
0	0	103.0425	0	-23.0820
0.25	25.4	98.7147	-17.3114	25.8098
0.5	49.4	94.8984	2.04590	1.91475
1.0	97.6	98.3803	4.91803	-6.55738
1.25				

     (b) $1:13\frac{7}{25}$ **(c)** Starting speed $\approx 9.7047 \times 10^{-3}$ mi/s $= 34.94$ mi/h. Ending speed ≈ 36.14 mi/h.

Exercise Set 3.6 (page 89)

1. The parametric cubic Hermite approximations are
   (a) $x(t) = -10t^3 + 14t^2 + t$; $y(t) = -2t^3 + 3t^2 + t$
   (b) $x(t) = -10t^3 + 14.5t^2 + 0.5t$; $y(t) = -3t^3 + 4.5t^2 + 0.5t$
   (c) $x(t) = -10t^3 + 14t^2 + t$; $y(t) = -4t^3 + 5t^2 + t$
   (d) $x(t) = -10t^3 + 13t^2 + 2t$; $y(t) = 2t$

3. The Bézier approximations are
   (a) $x(t) = -11.5t^3 + 15t^2 + 1.5t + 1$; $y(t) = -4.25t^3 + 4.5t^2 + 0.75t + 1$
   (b) $x(t) = -6.25t^3 + 10.5t^2 + 0.75t + 1$; $y(t) = -3.5t^3 + 3t^2 + 1.5t + 1$
   (c) Between (0,0) and (4,6), we have

   $$x(t) = -5t^3 + 7.5t^2 + 1.5t; \ y(t) = -13.5t^3 + 18t^2 + 1.5t,$$

   and between (4,6) and (6,1), we have

   $$x(t) = -5.5t^3 + 6t^2 + 1.5t + 4; \ y(t) = 4t^3 - 6t^2 - 3t + 6.$$

   (d) Between (0,0) and (2,1), we have

   $$x(t) = -5.5t^3 + 6t^2 + 1.5t; \ y(t) = -1.25t^3 + 1.5t^2 + 0.75t,$$

   between (2,1) and (4,0), we have

   $$x(t) = -4t^3 + 3t^2 + 3t + 2; \ y(t) = -t^3 + 1,$$

   and between (4,0) and (6,−1), we have

   $$x(t) = -8.5t^3 + 13.5t^2 - 3t + 4; \ y(t) = -3.25t^3 + 5.25t^2 - 3t.$$

Exercise Set 4.2 (page 98)

1. The Midpoint rule gives the following approximations: **(a)** 0.608197 **(b)** 0.0 **(c)** 0.8 **(d)** 0.0 **(e)** 2.08549 **(f)** 0.5 **(g)** 0.577350 **(h)** 0.787179

3. The Trapezoidal rule gives the following approximations: **(a)** 0.693147 **(b)** 116.060 **(c)** 0.75 **(d)** −15.5031 **(e)** −15.2556 **(f)** 0.430769 **(g)** 0.665431 **(h)** 1.42209

5. Simpson's rule gives the following approximations: **(a)** 0.636514 **(b)** 38.6865 **(c)** 0.783333 **(d)** −5.16771 **(e)** −3.69486 **(f)** 0.476923 **(g)** 0.606711 **(h)** 0.998816

7. The Midpoint rule gives the following approximations: **(a)** 0.174331 **(b)** 0.151633 **(c)** −0.176350 **(d)** 0.0463500 **(e)** 1.80391 **(f)** −0.675325 **(g)** 0.634335 **(h)** 0.670379

9. The Trapezoidal rule gives the following approximations: **(a)** 0.228074 **(b)** 0.183940 **(c)** −0.177764 **(d)** 0.171287 **(e)** 4.14326 **(f)** −0.866667 **(g)** 0.640046 **(h)** 0.589049

11. Simpson's rule gives the following approximations: **(a)** 0.192245 **(b)** 0.162402 **(c)** −0.176822 **(d)** 0.0879957 **(e)** 2.58370 **(f)** −0.739105 **(g)** 0.636239 **(h)** 0.643269

13.

	Approximation	Error Bound	Actual Error
(a)	0.636394	3.08642×10^{-4}	9.94862×10^{-5}
(b)	29.6524	177.484	9.73151
(c)	0.784615	0.00370370	7.82775×10^{-4}
(d)	-5.81368	0.871151	0.469508
(e)	-9.39559	22.1008	4.81839
(f)	0.477413	0.00155496	3.43065×10^{-4}
(g)	0.605504	0.0165370	0.00112880
(h)	0.980303	0.899961	0.0197555

15.

	Approximation	Error Bound	Actual Error
(a)	0.617476	0.277778	0.0188183
(b)	0.849966	577.988	19.0709
(c)	0.796154	0.0555556	0.0107557
(d)	-2.58386	6.77798	3.69933
(e)	-7.44227	156.747	6.77171
(f)	0.492960	0.0391111	0.0152047
(g)	0.585529	0.0874551	0.0188469
(h)	0.833041	1.49750	0.0127507

17.

	Approximation	Error Bound	Actual Error
(a)	0.636162	4.22222×10^{-4}	1.32382×10^{-4}
(b)	9.57271	242.798	10.3481
(c)	0.786733	0.00506667	0.00133531
(d)	-6.97811	1.19173	0.694927
(e)	-20.6282	30.2339	6.41419
(f)	0.478281	0.00212719	5.25342×10^{-4}
(g)	0.603029	0.0226226	0.00134693
(h)	0.939762	1.23115	0.0207852

Exercise Set 4.3 (page 105)

1. The Composite Trapezoidal rule approximations are (a) 0.639900 (b) 31.3653 (c) 0.784241 (d) -6.42872 (e) -13.5760 (f) 0.476977 (g) 0.605498 (h) 0.970926

3. The Composite Simpson's rule approximations are (a) 0.6363098 (b) 22.47713 (c) 0.7853980 (d) -6.274868 (e) -14.18334 (f) 0.4777547 (g) 0.6043941 (h) 0.9610554

5. (a) The Composite Trapezoidal rule requires $h < 0.00092295$ and $n \geq 2167$.
(b) The Composite Midpoint rule requires $h < 0.00065216$ and $n \geq 3066$.
(c) The Composite Simpson's rule requires $h < 0.037658$ and $n \geq 54$.

7. **(a)** The Composite Trapezoidal rule requires $h < 0.04382$ and $n \geq 46$. The approximation is 0.405471.
   (b) The Composite Midpoint rule requires $h < 0.03098$ and $n \geq 64$. The approximation is 0.405460.
   (c) The Composite Simpson's rule requires $h < 0.44267$ and $n \geq 6$. The approximation is 0.405466.

9. The integral is approximately 58.47047.   **11.** The area of the region is approximately 0.68271097.

13. Composite Simpson's rule with $h = 0.25$ gives 2.61972 seconds.

Exercise Set 4.4 (page 111)

1. Gaussian quadrature gives **(a)** 0.1922687   **(b)** 0.1594104   **(c)** -0.1768190 **(d)** 0.08926302
   (e) 2.5913247   **(f)** -0.7307230 **(g)** 0.6361966   **(h)** 0.6423172

3. Gaussian quadrature gives **(a)** 0.1922594   **(b)** 0.1606028   **(c)** -0.1768200 **(d)** 0.08875529
   (e) 2.5886327   **(f)** -0.7339604 **(g)** 0.6362133   **(h)** 0.6426991

Exercise Set 4.5 (page 117)

1. Romberg integration gives $R_{3,3}$ as follows: **(a)** 0.1922593   **(b)** 0.1606105   **(c)** -0.1768200
   (d) 0.08875677   **(e)** 2.5879685   **(f)** -0.7341567 **(g)** 0.6362135   **(h)** 0.6426970

3. Romberg integration gives   **(a)** $R_{4,4} = 0.63629437$ **(b)** $R_{6,6} = 19.920853$ **(c)** $R_{5,5} = 0.78539817$
   (d) $R_{5,5} = -6.2831855$ **(e)** $R_{6,6} = -14.213977$ **(f)** $R_{4,4} = 0.47775587$ **(g)** $R_{5,5} = 0.60437561$
   (h) $R_{5,5} = 0.96055026$

5. Romberg integration gives
   (a) 62.4373714, 57.2885616, 56.4437507, 56.2630547, and 56.2187727.
   (b) 55.5722917, 56.2014707, 56.2055989, and 56.2040624.
   (c) 58.3626837, 59.0773207, 59.2688746, 59.3175220, 59.3297316, and 59.3327870.
   (d) 58.4220930, 59.4707174, 58.4704791, and 58.4704691. Prediction is 58.47047.
   (e) Consider the graph of the function.

Exercise Set 4.6 (page 123)

1. Adaptive quadrature gives
   (a) $S(1, 1.5) = 0.19224530$, $S(1, 1.25) = 0.039372434$, $S(1.25, 1.5) = 0.15288602$, and the actual value is 0.19225935.
   (b) $S(0, 1) = 0.16240168$, $S(0, 0.5) = 0.028861071$, $S(0.5, 1) = 0.13186140$, and the actual value is 0.16060279.
   (c) $S(0, 0.35) = -0.17682156$, $S(0, 0.175) = -0.087724382$, $S(0.175, 0.35) = -0.089095736$, and the actual value is -0.17682002.
   (d) $S(0, \frac{\pi}{4}) = 0.087995669$, $S(0, \frac{\pi}{8}) = 0.0058315797$, $S(\frac{\pi}{8}, \frac{\pi}{4}) = 0.082877624$, and the actual value is 0.088755285.
   (e) $S(0, \frac{\pi}{4}) = 2.5836964$, $S(0, \frac{\pi}{8}) = 0.033088926$, $S(\frac{\pi}{8}, \frac{\pi}{4}) = 2.2568121$, and the actual value is 2.5886286.
   (f) $S(1, 1.6) = -0.73910533$, $S(1, 1.3) = -0.26141244$, $S(1.3, 1.6) = -0.47305351$, and the actual value is -0.73396917.

(g) $S(3, 3.5) = 0.63623873$, $S(3, 3.25) = 0.32567095$, $S(3.25, 3.5) = 0.31054412$, and the actual value is 0.63621334.

(h) $S(0, \frac{\pi}{4}) = 0.64326905$, $S(0, \frac{\pi}{8}) = 0.37315002$, $S(\frac{\pi}{8}, \frac{\pi}{4}) = 0.26958270$, and the actual value is 0.64269908.

3. Adaptive quadrature gives (a) 108.555281 (b) -1724.966983 (c) -15.306308 (d) -18.945949

5. Adaptive quadrature gives

$$\int_{0.1}^{2} \sin \frac{1}{x} \, dx \approx 1.1454 \qquad \text{and} \qquad \int_{0.1}^{2} \cos \frac{1}{x} \, dx \approx 0.67378$$

7. $\int_{1}^{3} u(t) \, dt \approx -0.1369488$

9.

t	$c(t)$	$s(t)$
0.1	0.0999975	0.000523589
0.2	0.199921	0.00418759
0.3	0.299399	0.0141166
0.4	0.397475	0.0333568
0.5	0.492327	0.0647203
0.6	0.581061	0.110498
0.7	0.659650	0.172129
0.8	0.722844	0.249325
0.9	0.764972	0.339747
1.0	0.779880	0.438245

Exercise Set 4.7 (page 132)

1. With $n = m = 2$ we have (a) 0.3115733 (b) 0.2552526 (c) 16.50864 (d) 1.476684

3. With $n = 2$ and $m = 4$, $n = 4$ and $m = 2$, and $n = m = 3$ we have
(a) 0.5119875, 0.5118533, 0.5118722 (b) 1.718857, 1.718220, 1.718385
(c) 1.001953, 1.000122, 1.000386 (d) 0.7838542, 0.7833659, 0.7834362
(e) -1.985611, -1.999182, -1.997353 (f) 2.004596, 2.000879, 2.000980
(g) 0.3084277, 0.3084562, 0.3084323 (h) -22.61612, -19.85408, -20.14117

5. With $n = m = 2$ we have (a) 0.3115733 (b) 0.2552446 (c) 16.50863 (d) 1.488875

7. With $n = m = 3$, $n = 3$ and $m = 4$, $n = 4$ and $m = 3$, and $n = m = 4$ we have
(a) 0.5118655, 0.5118445, 0.5118655, 0.5118445 (b) 1.718163, 1.718302, 1.718139, 1.718277
(c) 1.000000, 1.000000, 1.000000, 1.000000 (d) 0.7833333, 0.7833333, 0.7833333, 0.7833333
(e) -1.991878, -2.000124, -1.991878, -2.000124 (f) 2.001494, 2.000080, 2.001388, 1.999984
(g) 0.3084151, 0.3084145, 0.3084246, 0.3084245 (h) -12.74790, -21.21539, -11.83624, -20.30373

9. With $n = m = 7$ we have 0.1479099, and with $n = m = 4$ we have 0.1506823.

11. The area approximations are (a) 1.040253 (b) 1.040252

13. Gaussian quadrature with $n = m = p = 3$ gives (a) 5.206442 (b) 0.08333333 (c) 0.07166667
(d) 0.08333333 (e) 6.928161 (f) 1.474577

15. Gaussian quadrature with $n = m = p = 4$ gives 3.052125.

Exercise Set 4.8 (page 138)

1. The Composite Simpson's rule gives   (a) 0.5284163   (b) 4.266654   (c) 0.4329748   (d) 0.8802010

3. The Composite Simpson's rule gives   (a) 0.4112649   (b) 0.2440679   (c) 0.05501681   (d) 0.2903746

5. The Composite Simpson's rule gives   (a) 3.141569   (b) 0.0   (c) 1.178071   (d) 2.221548

Exercise Set 4.9 (page 146)

1. From the forward-backward difference formula, we have the following approximations:
   (a) $f'(0.5) \approx 1.67154$, $f'(0.6) \approx 1.906973$, $f'(0.7) \approx 1.906973$
   (b) $f'(0.0) \approx 3.5800665$, $f'(0.2) \approx 2.6620555$, $f'(0.4) \approx 2.6620555$

3. For the endpoints of the tables we use the three-point endpoint formula; the other approximations come from the three-point endpoint formula.
   (a) $f'(1.1) \approx 2.557820$, $f'(1.2) \approx 2.448560$, $f'(1.3) \approx 2.352640$, $f'(1.4) \approx 2.270060$
   (b) $f'(8.1) \approx 3.092050$, $f'(8.3) \approx 3.116150$, $f'(8.5) \approx 3.139975$, $f'(8.7) \approx 3.163525$
   (c) $f'(2.9) \approx 5.101375$, $f'(3.0) \approx 6.654785$, $f'(3.1) \approx 8.216330$, $f'(3.2) \approx 9.786010$
   (d) $f'(3.6) \approx -1.45886415$, $f'(3.8) \approx -2.13752785$, $f'(4.0) \approx -2.90294135$, $f'(4.2) \approx -3.75510465$

5. The approximations and the formulas used are
   (a) $f'(2.1) \approx 3.899344$ from the five-point endpoint formula
       $f'(2.2) \approx 2.876876$ from the five-point endpoint formula
       $f'(2.3) \approx 2.249704$ from the five-point midpoint formula
       $f'(2.4) \approx 1.837756$ from the five-point midpoint formula
       $f'(2.5) \approx 1.544210$ from the five-point endpoint formula
       $f'(2.6) \approx 1.355496$ from the five-point endpoint formula

   (b) $f'(-3.0) \approx -5.877358$ from the five-point endpoint formula
       $f'(-2.8) \approx -5.468933$ from the five-point endpoint formula
       $f'(-2.6) \approx -5.059884$ from the five-point midpoint formula
       $f'(-2.4) \approx -4.650223$ from the five-point midpoint formula
       $f'(-2.2) \approx -4.239911$ from the five-point endpoint formula
       $f'(-2.0) \approx -3.828853$ from the five-point endpoint formula

7. From the backward-forward formula we have
   (a) $f'(0.5) \approx 1.672$, $f'(0.6) \approx 1.907$, $f'(0.7) \approx 1.907$
   (b) $f'(0.0) \approx 3.580$, $f'(0.2) \approx 2.662$, $f'(0.4) \approx 2.662$

9. For the endpoints of the tables we use the five-point endpoint formula; the other approximations come from the five-point midpoint formula.

   (a) $f'(2.1) \approx 3.884$ (b) $f'(-3.0) \approx -5.883$
       $f'(2.2) \approx 2.896$ $f'(-2.8) \approx -5.467$
       $f'(2.3) \approx 2.249$ $f'(-2.6) \approx -5.059$
       $f'(2.4) \approx 1.836$ $f'(-2.4) \approx -4.650$
       $f'(2.5) \approx 1.550$ $f'(-2.2) \approx -4.208$
       $f'(2.6) \approx 1.348$ $f'(-2.0) \approx -3.875$

11. The approximation is -3.10457 with an error bound of 3.98×10^{-2}.

Exercise Set 5.2 (page 159)

1. Euler's method gives the approximations in the following tables:

(a)

i	t_i	w_i	$y(t_i)$	$\|y(t_i) - w_i\|$
0	0.000	0.0000000	0.0000000	0.0000000
1	0.500	0.0000000	0.4160531	0.4160531
2	1.000	1.1204223	3.2678200	2.1473978

(b)

i	t_i	w_i	$y(t_i)$	$\|y(t_i) - w_i\|$
0	2.000	1.0000000	1.0000000	0.0000000
1	2.500	2.0000000	1.8333333	0.1666667
2	3.000	2.6250000	2.5000000	0.1250000

(c)

i	t_i	w_i	$y(t_i)$	$\|y(t_i) - w_i\|$
0	1.000	2.0000000	2.0000000	0.0000000
1	1.250	2.7500000	2.7789294	0.0289294
2	1.500	3.5500000	3.6081977	0.0581977
3	1.750	4.3916667	4.4793276	0.0876610
4	2.000	5.2690476	5.3862944	0.1172467

(d)

i	t_i	w_i	$y(t_i)$	$\|y(t_i) - w_i\|$
0	0.000	1.0000000	1.0000000	0.0000000
1	0.250	1.2500000	1.3291498	0.0791498
2	0.500	1.6398053	1.7304898	0.0906844
3	0.750	2.0242547	2.0414720	0.0172174
4	1.000	2.2364573	2.1179795	0.1184777

3. Euler's method gives the approximations in the following tables:

(a)

i	t_i	w_i	$y(t_i)$	$\|y(t_i) - w_i\|$
0	1.000	1.0000000	1.0000000	0.0000000
2	1.200	1.0082645	1.0149523	0.0066879
4	1.400	1.0385147	1.0475339	0.0090192
6	1.600	1.0784611	1.0884327	0.0099716
8	1.800	1.1232621	1.1336536	0.0103915
10	2.000	1.1706516	1.1812322	0.0105806

(b)

i	t_i	w_i	$y(t_i)$	$|y(t_i) - w_i|$
0	1.000	0.0000000	0.0000000	0.0000000
2	1.400	0.4388889	0.4896817	0.0507928
4	1.800	1.0520380	1.1994386	0.1474006
6	2.200	1.8842608	2.2135018	0.3292410
8	2.600	3.0028372	3.6784753	0.6756382
10	3.000	4.5142774	5.8741000	1.3598226

(c)

i	t_i	w_i	$y(t_i)$	$|y(t_i) - w_i|$
0	0.000	−2.0000000	−2.0000000	0.0000000
2	0.400	−1.6080000	−1.6200510	0.0120510
4	0.800	−1.3017370	−1.3359632	0.0342263
6	1.200	−1.1274909	−1.1663454	0.0388544
8	1.600	−1.0491191	−1.0783314	0.0292124
10	2.000	−1.0181518	−1.0359724	0.0178206

(d)

i	t_i	w_i	$y(t_i)$	$|y(t_i) - w_i|$
0	0.0	0.3333333	0.3333333	0.0000000
2	0.2	0.1083333	0.1626265	0.0542931
4	0.4	0.1620833	0.2051118	0.0430284
6	0.6	0.3455208	0.3765957	0.0310749
8	0.8	0.6213802	0.6461052	0.0247250
10	1.0	0.9803451	1.0022460	0.0219009

5. Taylor's method of order two gives the approximations in the following tables:

(a)

i	t_i	w_i
1	0.5	0.1250000
2	1.0	2.023239

(b)

i	t_i	w_i
1	2.5	1.750000
2	3.0	2.425781

(c)

i	t_i	w_i
1	1.25	2.781250
2	1.50	3.612500
3	1.75	4.485417
4	2.00	5.394048

(d)

i	t_i	w_i
1	0.25	1.343750
2	0.50	1.772187
3	0.75	2.110676
4	1.00	2.201644

7. Taylor's method of order four gives the approximations in the following tables:

(a)

i	t_i	w_i	$\|y(t_i) - w_i\|$
3	1.3	1.0298483	3.46×10^{-5}
6	1.6	1.0884694	3.67×10^{-5}
9	1.9	1.1572648	3.64×10^{-5}

(b)

i	t_i	w_i	$\|y(t_i) - w_i\|$
3	1.6	0.8126583	9.45×10^{-5}
6	2.2	2.2132495	2.52×10^{-4}
9	2.8	4.6578929	7.72×10^{-4}

(c)

i	t_i	w_i	$\|y(t_i) - w_i\|$
3	0.6	-1.4630012	5.08×10^{-5}
6	1.2	-1.1663295	1.59×10^{-5}
9	1.8	-1.0531846	9.40×10^{-6}

(d)

i	t_i	w_i	$\|y(t_i) - w_i\|$
3	0.3	0.1644651	8.84×10^{-5}
6	0.6	0.3766352	3.95×10^{-5}
9	0.9	0.8137162	1.32×10^{-5}

9. Linear interpolation gives the following results:
   - (a) $1.021957 = y(1.25) \approx 1.023555$; $1.164390 = y(1.93) \approx 1.165937$
   - (b) $1.924962 = y(2.1) \approx 1.921037$; $4.394170 = y(2.75) \approx 4.349162$
   - (c) $-1.114648 = y(1.4) \approx -1.115028$; $-1.041267 = y(1.93) \approx -1.043301$
   - (d) $0.3140018 = y(0.54) \approx 0.3210207$; $0.9108858 = y(0.94) \approx 0.8901230$

11.

j	t_j	$w_j \approx i(t_j)$
20	2	0.702938
40	4	-0.0457793
60	6	0.294870
80	8	0.341673
100	10	0.139432

13. (a)   Taylor's method of order two gives the approximations in the following table:

i	t_i	w_i
2	0.2	5.86595
5	0.5	2.82145
7	0.7	0.84926
10	1.0	-2.08606

Taylor's method of order four gives the approximations in the following table:

i	t_i	w_i
2	0.2	5.86433
5	0.5	2.81789
7	0.7	0.84455
10	1.0	-2.09015

(b)   0.8 seconds

Exercise Set 5.3 (page 166)

1. (a)

t	y(t)	Modified Euler	Error
0.5	0.2836165	0.5602111	0.2765946
1.0	3.2190993	5.3014898	2.0823905

(b)

t	y(t)	Modified Euler	Error
2.5	1.8333333	1.8125000	0.0208333
3.0	2.5000000	2.4815531	0.0184469

(c)

t	y(t)	Modified Euler	Error
1.25	2.7789294	2.7750000	0.0039294
1.50	3.6081977	3.6008333	0.0073643
1.75	4.4793276	4.4688294	0.0104983
2.00	5.3862944	5.3728586	0.0134358

(d)

t	y(t)	Modified Euler	Error
0.25	1.3291498	1.3199027	0.0092471
0.50	1.7304898	1.7070300	0.0234598
0.75	2.0414720	2.0053560	0.0361161
1.00	2.1179795	2.0770789	0.0409006

3. (a)

t	y(t)	Midpoint	Error
0.5	0.2836165	0.2646250	0.0189915
1.0	3.2190993	3.1300023	0.0890970

(b)

t	y(t)	Midpoint	Error
2.5	1.8333333	1.7812500	0.0520833
3.0	2.5000000	2.4550638	0.0449362

(c)

t	y(t)	Midpoint	Error
1.25	2.7789294	2.7777778	0.0011517
1.50	3.6081977	3.6060606	0.0021371
1.75	4.4793276	4.4763015	0.0030262
2.00	5.3862944	5.3824398	0.0038546

(d)

t	y(t)	Midpoint	Error
0.25	1.3291498	1.3337962	0.0046464
0.50	1.7304898	1.7422854	0.0117956
0.75	2.0414720	2.0596374	0.0181654
1.00	2.1179795	2.1385560	0.0205764

5. Linear interpolation on the Modified Euler method gives the following results:
   (a) $1.0221167 \approx y(1.25) = 1.0219569$, $1.1640347 \approx y(1.93) = 1.1643901$
   (b) $1.9086500 \approx y(2.1) = 1.9249616$, $4.3105913 \approx y(2.75) = 4.3941697$
   (c) $-1.1461434 \approx y(1.3) = -1.1382768$, $-1.0454854 \approx y(1.93) = -1.0412665$
   (d) $0.3271470 \approx y(0.54) = 0.3140018$, $0.8967073 \approx y(0.94) = 0.8866318$

7. Linear interpolation on Heun's method gives the following results:
   (a) $1.0225530 \approx y(1.25) = 1.0219569$, $1.1646155 \approx y(1.93) = 1.1643901$
   (b) $1.9132167 \approx y(2.1) = 1.9249616$, $4.3246152 \approx y(2.75) = 4.3941697$
   (c) $-1.1441775 \approx y(1.3) = -1.1382768$, $-1.0447403 \approx y(1.93) = -1.0412665$
   (d) $0.3251049 \approx y(0.54) = 0.3140018$, $0.8945125 \approx y(0.94) = 0.8866318$

9. Linear interpolation on the Midpoint method gives the following results:
   (a) $1.0227863 \approx y(1.25) = 1.0219569$, $1.1649247 \approx y(1.93) = 1.1643901$
   (b) $1.9153749 \approx y(2.1) = 1.9249616$, $4.3312939 \approx y(2.75) = 4.3941697$
   (c) $-1.1432070 \approx y(1.3) = -1.1382768$, $-1.0443743 \approx y(1.93) = -1.0412665$
   (d) $0.3240839 \approx y(0.54) = 0.3140018$, $0.8934152 \approx y(0.94) = 0.8866318$

11. The Runge-Kutta method of order four gives the results in the following tables:

(a)

t	Runge-Kutta	$y(t)$	Error
1.2	1.0149520	1.0149523	3.10759×10^{-7}
1.4	1.0475336	1.0475339	3.60986×10^{-7}
1.6	1.0884323	1.0884327	3.67611×10^{-7}
1.8	1.1336532	1.1336536	3.65627×10^{-7}
2.0	1.1812319	1.1812322	3.62576×10^{-7}

(b)

t	Runge-Kutta	$y(t)$	Error
1.4	0.4896842	0.4896817	2.50269×10^{-6}
1.8	1.1994320	1.1994386	6.61869×10^{-6}
2.2	2.2134693	2.2135018	3.24969×10^{-5}
2.6	3.6783790	3.6784753	9.63348×10^{-5}
3.0	5.8738386	5.8741000	2.61408×10^{-4}

(c)

t	Runge-Kutta	$y(t)$	Error
0.4	-1.6200576	-1.6200510	6.60106×10^{-6}
0.8	-1.3359824	-1.3359632	1.91530×10^{-5}
1.2	-1.1663735	-1.1663454	2.81499×10^{-5}
1.6	-1.0783582	-1.0783314	2.67598×10^{-5}
2.0	-1.0359922	-1.0359724	1.98032×10^{-5}

(d)

t	Runge-Kutta	$y(t)$	Error
0.2	0.1627655	0.1626265	1.38977×10^{-4}
0.4	0.2052405	0.2051118	1.28744×10^{-4}
0.6	0.3766981	0.3765957	1.02389×10^{-4}
0.8	0.6461896	0.6461052	8.43755×10^{-5}
1.0	1.0023207	1.0022460	7.46867×10^{-5}

13. **(a)** 6.531327 ft   **(b)** 25 min.

Exercise Set 5.4 (page 174)

1. The Adams-Bashforth methods give the results in the following tables:

(a)

t	2 step	3 step	4 step	5 step	$y(t)$
0.2	0.0268128	0.0268128	0.0268128	0.0268128	0.0268128
0.4	0.1200522	0.1507778	0.1507778	0.1507778	0.1507778
0.6	0.4153551	0.4613866	0.4960196	0.4960196	0.4960196
0.8	1.1462844	1.2512447	1.2961260	1.3308570	1.3308570
1.0	2.8241683	3.0360680	3.1461400	3.1854002	3.2190993

(b)

t	2 step	3 step	4 step	5 step	$y(t)$
2.2	1.3666667	1.3666667	1.3666667	1.3666667	1.3666667
2.4	1.6750000	1.6857143	1.6857143	1.6857143	1.6857143
2.6	1.9632431	1.9794407	1.9750000	1.9750000	1.9750000
2.8	2.2323184	2.2488759	2.2423065	2.2444444	2.2444444
3.0	2.4884512	2.5051340	2.4980306	2.5011406	2.5000000

(c)

t	2 step	3 step	4 step	5 step	$y(t)$
1.2	2.6187859	2.6187859	2.6187859	2.6187859	2.6187859
1.4	3.2734823	3.2710611	3.2710611	3.2710611	3.2710611
1.6	3.9567107	3.9514231	3.9520058	3.9520058	3.9520058
1.8	4.6647738	4.6569191	4.6582078	4.6580160	4.6580160
2.0	5.3949416	5.3848058	5.3866452	5.3862177	5.3862944

(d)

t	2 step	3 step	4 step	5 step	$y(t)$
0.2	1.2529306	1.2529306	1.2529306	1.2529306	1.2529306
0.4	1.5986417	1.5712255	1.5712255	1.5712255	1.5712255
0.6	1.9386951	1.8827238	1.8750869	1.8750869	1.8750869
0.8	2.1766821	2.0844122	2.0698063	2.0789180	2.0789180
1.0	2.2369407	2.1115540	2.0998117	2.1180642	2.1179795

3. The Adams-Bashforth methods give the results in the following tables:

(a)

t	2 step	3 step	4 step	5 step	$y(t)$
1.2	1.0161982	1.0149520	1.0149520	1.0149520	1.0149523
1.4	1.0497665	1.0468730	1.0477278	1.0475336	1.0475339
1.6	1.0910204	1.0875837	1.0887567	1.0883045	1.0884327
1.8	1.1363845	1.1327465	1.1340093	1.1334967	1.1336536
2.0	1.1840272	1.1803057	1.1815967	1.1810689	1.1812322

(b)

t	2 step	3 step	4 step	5 step	$y(t)$
1.4	0.4867550	0.4896842	0.4896842	0.4896842	0.4896817
1.8	1.1856931	1.1982110	1.1990422	1.1994320	1.1994386
2.2	2.1753785	2.2079987	2.2117448	2.2134792	2.2135018
2.6	3.5849181	3.6617484	3.6733266	3.6777236	3.6784753
3.0	5.6491203	5.8268008	5.8589944	5.8706101	5.8741000

(c)	t	2 step	3 step	4 step	5 step	$y(t)$
	0.5	−1.5357010	−1.5381988	−1.5379372	−1.5378676	−1.5378828
	1.0	−1.2374093	−1.2389605	−1.2383734	−1.2383693	−1.2384058
	1.5	−1.0952910	−1.0950952	−1.0947925	−1.0948481	−1.0948517
	2.0	−1.0366643	−1.0359996	−1.0359497	−1.0359760	−1.0359724

(d)	t	2 step	3 step	4 step	5 step	$y(t)$
	0.2	0.1739041	0.1627655	0.1627655	0.1627655	0.1626265
	0.4	0.2144877	0.2026399	0.2066057	0.2052405	0.2051118
	0.6	0.3822803	0.3747011	0.3787680	0.3765206	0.3765957
	0.8	0.6491272	0.6452640	0.6487176	0.6471458	0.6461052
	1.0	1.0037415	1.0020894	1.0064121	1.0073348	1.0022460

5. The Adams fourth-order predictor-corrector method gives the results in the following tables:

(a)	t	w	$y(t)$
	1.2	1.0149520	1.0149523
	1.4	1.0475227	1.0475339
	1.6	1.0884141	1.0884327
	1.8	1.1336331	1.1336536
	2.0	1.1812112	1.1812322

(b)	t	w	$y(t)$
	1.4	0.4896842	0.4896817
	1.8	1.1994245	1.1994386
	2.2	2.2134701	2.2135018
	2.6	3.6784144	3.6784753
	3.0	5.8739518	5.8741000

(c)	t	w	$y(t)$
	0.5	−1.5378788	−1.5378828
	1.0	−1.2384134	−1.2384058
	1.5	−1.0948609	−1.0948517
	2.0	−1.0359757	−1.0359724

(d)	t	w	$y(t)$
	0.2	0.1627655	0.1626265
	0.4	0.2048557	0.2051118
	0.6	0.3762804	0.3765957
	0.8	0.6458949	0.6461052
	1.0	1.0021372	1.0022460

7. The Milne-Simpson method gives the results in the following tables:

(a)	t	w	$y(t)$
	1.3	1.029813	1.029814
	1.6	1.088427	1.088433
	1.9	1.157226	1.157228

(b)	t	w	$y(t)$
	1.6	0.812752	0.812753
	2.2	2.213461	2.213502
	2.8	4.658474	4.658665

(c)	t	w	$y(t)$
	0.5	−1.537883	−1.537883
	1.0	−1.238408	−1.238406
	1.5	−1.094855	−1.094852
	2.0	−1.035972	−1.035972

(d)	t	w	$y(t)$
	0.3	0.164517	0.164377
	0.6	0.376374	0.376596
	0.9	0.813714	0.813703

Exercise Set 5.5 (page 178)

1. The Extrapolation method gives the results in the following tables:

(a)

t	w	h	y(t)
0.25	0.0454263	0.25	0.0454312
0.50	0.2835997	0.25	0.2836165
0.75	1.0525685	0.25	1.0525761
1.00	3.2190944	0.25	3.2190993

(b)

t	w	h	y(t)
2.25	1.4499819	0.25	1.4500000
2.50	1.8333386	0.25	1.8333333
2.75	2.1785826	0.25	2.1785714
3.00	2.5000119	0.25	2.5000000

(c)

t	w	h	y(t)
1.25	2.7789152	0.25	2.7789294
1.50	3.6081723	0.25	3.6081977
1.75	4.4792929	0.25	4.4793276
2.00	5.3862512	0.25	5.3862944

(d)

t	w	h	y(t)
0.25	1.3291498	0.25	1.3291498
0.50	1.7304898	0.25	1.7304898
0.75	2.0414720	0.25	2.0414720
1.00	2.1179795	0.25	2.1179795

3. (a)

t	w	y(t)
1.972454	-0.269257	-0.269257
2.172454	-0.398491	-0.398491
2.372454	-0.346753	-0.346753
2.572454	-0.185816	-0.185816
2.772454	-0.086328	-0.086328
2.972454	-0.170751	-0.170751
3.141593	-0.323039	-0.323039

(b)

t	w	y(t)
1.770796	1.603594	1.603594
1.970796	2.876235	2.876235
2.170796	4.753496	4.753496
2.356194	7.540468	7.540468

(c)

t	w	y(t)
2.918282	2.043602	2.043602
3.118282	2.093943	2.093943
3.318282	2.148044	2.148043
3.518282	2.204142	2.204141
3.718282	2.261173	2.261173
3.918282	2.318486	2.318486
4.118282	2.375674	2.375674
4.318282	2.432486	2.432486
4.518282	2.488767	2.488767
4.718282	2.544424	2.544424
4.918282	2.599404	2.599404
5.118282	2.653680	2.653680
5.318282	2.707245	2.707245
5.436564	2.738589	2.738589

(d)

t	w	y(t)
0.2	0.219998	0.219997
0.4	0.479830	0.479830
0.6	0.778081	0.778081
0.8	1.109507	1.109507
1.0	1.462117	1.462117
1.2	1.816909	1.816909
1.4	2.153066	2.153066
1.6	2.456485	2.456485
1.8	2.724624	2.724624
2.0	2.964028	2.964028
2.2	3.184310	3.184310
2.4	3.393718	3.393718
2.6	3.597684	3.597684
2.8	3.799213	3.799213
3.0	3.999753	3.999753

5. $P(5) \approx 56751$

Exercise Set 5.6 (page 186)

1. The Runge-Kutta-Fehlberg method gives the results in the following tables:

(a)

t	w	h	y(t)
0.2093900	0.0298184	0.2093900	0.0298337
0.3832972	0.1343260	0.1739072	0.1343488
0.5610469	0.4016438	0.1777496	0.4016860
0.7106840	0.8708372	0.1496371	0.8708882
0.8387744	1.5894061	0.1280905	1.5894600
0.9513263	2.6140226	0.1125519	2.6140771
1.0000000	3.2190497	0.0486737	3.2190993

(b)

t	w	h	y(t)
2.25	1.4499988	0.25	1.4500000
2.50	1.8333332	0.25	1.8333333
2.75	2.1785718	0.25	2.1785714
3.00	2.5000005	0.25	2.5000000

(c)

t	w	h	y(t)
1.25	2.7789299	0.25	2.7789294
1.50	3.6081985	0.25	3.6081977
1.75	4.4793288	0.25	4.4793276
2.00	5.3862958	0.25	5.3862944

(d)

t	w	h	y(t)
0.25	1.3291478	0.25	1.3291498
0.50	1.7304857	0.25	1.7304898
0.75	2.0414669	0.25	2.0414720
1.00	2.1179750	0.25	2.1179795

3. The Runge-Kutta-Fehlberg method gives the results in the following tables:

(a)

t	w	h	y(t)
1.070211	1.755041	0.0702106	1.755040
1.151431	1.545701	0.0812206	1.545701
1.233265	1.396559	0.0818337	1.396557
1.323237	1.290561	0.0899720	1.290559
1.432414	1.236108	0.1091769	1.236106
1.527855	1.264192	0.0954415	1.264189
1.612733	1.379554	0.0848773	1.379548
1.661500	1.513496	0.0487677	1.513488
1.690343	1.632540	0.0288423	1.632530
1.716154	1.777972	0.0258112	1.777960
1.736435	1.929852	0.0202807	1.929837

(b)

t	w	h	y(t)
0.2	0.310560	0.2	0.310560
0.4	0.667845	0.2	0.667841
0.5	0.854426	0.1	0.854423

(c)

t	w	h	y(t)
0.2	1.940400	0.2	1.940400
0.4	1.766384	0.2	1.766384
0.5	1.640567	0.1	1.640567

(d)

t	w	h	y(t)
0.0384750	0.276479	0.038475	0.276480
0.0795096	0.230310	0.041035	0.230310
0.3095255	0.166723	0.049442	0.166723
0.4676371	0.250851	0.054245	0.250852
0.5232524	0.298151	0.055615	0.298152
0.6379535	0.420711	0.057873	0.420712
0.7561692	0.579392	0.059467	0.579393
0.8767031	0.772768	0.060493	0.772769
1.0000000	1.002245	0.001344	1.002246

5. The Adams variable step-size predictor-corrector method gives the results in the following tables:

(a)

t	w	h	$y(t)$
1.020863	1.000210	0.020863	1.000210
1.406760	1.048812	0.031220	1.048812
1.767213	1.126041	0.048250	1.126041
2.225999	1.236525	0.072787	1.236525
2.846579	1.391210	0.111070	1.391210
3.634751	1.586853	0.121750	1.586853

(b)

t	w	h	$y(t)$
1.037039	0.037734	0.037039	0.037734
2.623606	3.783978	0.030921	3.783977
2.865890	5.029648	0.025841	5.029647

(c)

t	w	h	$y(t)$
0.033708	−1.966305	0.033708	−1.966305
0.490737	−1.545199	0.052532	−1.545199
0.690243	−1.401862	0.041910	−1.401862
1.553344	−1.085665	0.066820	−1.085664
1.828547	−1.050316	0.074745	−1.050316
2.962285	−1.005332	0.012572	−1.005332

(d)

t	w	h	$y(t)$
0.057097	0.332851	0.057097	0.332851
0.955371	0.224891	0.098912	0.224891
1.413378	0.139720	0.062361	0.139720
1.934201	0.061621	0.021933	0.061621

7. $i(2) = 8.69329$

Exercise Set 5.7 (page 193)

1. The Runge-Kutta for systems method gives the results in the following tables:

(a)

i	t_i	w_{1i}	w_{2i}
0	0.0	−1.0000000	4.0000000
1	0.5	5.7864583	8.8203125
2	1.0	48.263658	50.104513
3	1.5	342.55321	343.67019
4	2.0	2400.7289	2401.4066

(b)

i	t_i	w_{1i}	w_{2i}
0	0.0	1.0000000	1.0000000
1	0.2	2.1203658	1.5069919
2	0.4	4.4412278	3.2422402
3	0.6	9.7391333	8.1634170
4	0.8	22.676560	21.343528
5	1.0	55.661181	56.030503

(c)

i	t_i	w_{1i}	w_{2i}
5	0.5	0.9567139	−1.083820
10	1.0	1.306544	−0.8329536
15	1.5	1.344167	−0.5698033
20	2.0	1.143324	−0.3693632

(d)

i	t_i	w_{1i}	w_{2i}	w_{3i}
2	0.2	2.997466	−0.03733361	0.7813973
5	0.5	2.963536	−0.2083733	0.3981578
7	0.7	2.905641	−0.3759597	0.1206261
10	1.0	2.749648	−0.6690454	−0.3011653

(e)

i	t_i	w_{1i}	w_{2i}	w_{3i}
2	0.2	1.381651	1.007999	−0.6183311
5	0.5	1.907526	1.124998	−0.09090565
7	0.7	2.255029	1.342997	0.2634397
10	1.0	2.832112	1.999997	0.8821206

(f)

i	t_i	w_{1i}	w_{2i}	w_{3i}
0	0.1	4.6210619	−7.6818883	−1.7287952
1	0.2	7.1615826	−8.9381975	−0.9074132
2	0.3	11.261380	−11.282626	0.9731908
3	0.4	18.045437	−15.669345	4.8620475
4	0.5	29.508292	−23.861700	12.516816
5	0.6	49.209223	−39.106195	27.177722
6	0.7	83.531779	−67.363422	54.798414
7	0.8	143.96890	−119.55455	106.29239
8	0.9	251.27288	−215.65912	201.63071
9	1.0	442.99766	−392.18697	377.31385

5. The Adams fourth-order predictor-corrector method for systems method gives the results in the following tables:

(a)

i	t_i	w_{1i}	$y(t_i)$
1	0.1	0.0000090	0.0000089
2	0.2	0.0001535	0.0001535
3	0.3	0.0008343	0.0008343
4	0.4	0.0028326	0.0028323
5	0.5	0.0074311	0.0074303
6	0.6	0.0165644	0.0165626
7	0.7	0.0330013	0.0329980
8	0.8	0.0605672	0.0605619
9	0.9	0.1044131	0.1044052
10	1.0	0.1713402	0.1713288

(b)

i	t_i	w_{1i}	$y(t_i)$
5	0.5	0.71544570	0.71544565
10	1.0	0.77153962	0.77154032
15	1.5	1.1204202	1.1204222
20	2.0	1.8134261	1.8134302

(c)

i	t_i	w_{1i}	$y(t_i)$
5	1.25	0.9405698	0.94056987
10	1.50	0.7779721	0.77792370
15	1.75	0.5425639	0.54256418
20	2.00	0.2725883	0.27258872

(d)

i	t_i	w_{1i}	$y(t_i)$
2	1.4	2.4535431	2.4540442
4	1.8	6.2364230	6.2365330
6	2.2	11.8583510	11.8587121
8	2.6	19.7685613	19.7698919
10	3.0	30.3955698	30.3982630

(e)

i	t_i	w_{1i}	$y(t_i)$
2	1.1	0.2863275	0.2863276
4	1.2	0.6053290	0.6053289
6	1.3	0.9626928	0.9626923
8	1.4	1.3644628	1.3644622
10	1.5	1.8171693	1.8171685
12	1.6	2.3279441	2.3279432
14	1.7	2.9046321	2.9046311
16	1.8	3.5559032	3.5559021
18	1.9	4.2913687	4.2913673
20	2.0	5.1217050	5.1217034

(f)

i	t_i	w_{1i}	$y(t_i)$
5	1.0	3.731627	3.7317045
10	2.0	11.31425	11.314529
15	3.0	34.04396	34.045172

(g)

i	t_i	w_{1i}	$y(t_i)$
2	1.1	−1.111110	−1.111111
6	1.3	−1.428566	−1.428571
12	1.6	−2.499944	−2.500000
18	1.9	−9.967667	−10.00000

(h)

i	t_i	w_{1i}	$y(t_i)$
2	1.2	0.2727376	0.2727379
4	1.4	0.7469883	0.7469981
6	1.6	1.5162504	1.5162647
8	1.8	2.6840027	2.6840149
10	2.0	4.3615685	4.3615778

7.

i	t_i	w_{1i}	w_{2i}
6	1.2	2211	11429
12	2.4	175	17492
18	3.6	2	19704

Stable solution: $x_1 = 8000$, $x_2 = 4000$.

Exercise Set 5.8 (page 198)

1. Euler's method gives the results in the following tables:

(a)

i	t_i	w_i	$y(t_i)$
2	0.2	0.0271828	0.4493290
4	0.4	0.0002718	0.0742736
6	0.6	0.0000027	0.0122773
8	0.8	0.0000000	0.0020294
10	1.0	0.0000000	0.0003355

(b)

i	t_i	w_i	$y(t_i)$
2	0.4	1.1200000	0.4815244
4	0.8	1.0592000	0.8033231
6	1.2	1.2933120	1.2001355
8	1.6	1.6335923	1.6000055
10	2.0	2.0120932	2.0000002

(c)

i	t_i	w_i	$y(t_i)$
2	0.2	0.3733333	0.0461052
4	0.4	0.4933333	0.1601118
6	0.6	0.6933333	0.3600020
8	0.8	0.9733333	0.6400000
10	1.0	1.3333333	1.0000000

(d)

i	t_i	w_i	$y(t_i)$
2	0.50	14.005208	0.4794709
4	1.00	229.56899	0.8414710
6	1.50	3684.7269	0.9974950
8	2.00	58970.544	0.9092974

(e)

i	t_i	w_i	$y(t_i)$
2	1.2	0.9882272	0.9805185
4	1.4	0.9304751	0.9213678
6	1.6	0.8343313	0.8254933
8	1.8	0.7047064	0.6967792
10	2.0	0.5469651	0.5403338

(f)

i	t_i	w_i	$y(t_i)$
2	0.2	6.1282588	1.0000000
4	0.4	94.577575	1.0000000
6	0.6	1513.0165	1.0000000
8	0.8	24208.250	1.0000000
10	1.0	387332.00	1.0000000

3. The Adams fourth-order predictor-corrector method gives the results in the following tables:

(a)

i	t_i	w_i	$y(t_i)$
2	0.2	0.4588119	0.4493290
4	0.4	0.0366658	0.0742736
6	0.6	0.0011465	0.0122773
8	0.8	−0.0087900	0.0020294
10	1.0	0.0023604	0.0003355

(b)

i	t_i	w_i	$y(t_i)$
2	0.4	0.5462323	0.4815244
4	0.8	0.6006881	0.8033231
6	1.2	1.1619813	1.2001355
8	1.6	1.7170015	1.6000055
10	2.0	1.6477554	2.0000002

(c)

i	t_i	w_i	$y(t_i)$
2	0.2	0.0792593	0.0461052
4	0.4	0.1265329	0.1601118
6	0.6	0.2812206	0.3600020
8	0.8	0.8334432	0.6400000
10	1.0	0.7278557	1.0000000

(d)

i	t_i	w_i	$y(t_i)$
2	0.50	1.8830821×10^2	0.4794709
4	1.00	3.8932032×10^4	0.8414710
6	1.50	9.0736073×10^6	0.9974950
8	2.00	2.1157413×10^8	0.9092974

(e)

i	t_i	w_i	$y(t_i)$
2	1.2	0.9804973	0.9805185
4	1.4	0.9213599	0.9213678
6	1.6	0.8254928	0.8254933
8	1.8	0.6967793	0.6967792
10	2.0	0.5403329	0.5403338

(f)

i	t_i	w_i	$y(t_i)$
2	0.2	-2.1574586×10^2	1.0000000
4	0.4	-4.4723780×10^4	1.0000000
6	0.6	-1.0423830×10^7	1.0000000
8	0.8	-2.4305801×10^9	1.0000000
10	1.0	$-5.6675117 \times 10^{11}$	1.0000000

5. $p(50) \approx 0.10421$

Exercise Set 6.2 (page 208)

1. **(a)** Intersecting lines whose solution is $x_1 = x_2 = 1$.

   (b) Intersecting lines whose solution is $x_1 = x_2 = 0$.

   (c) One line, so there is an infinite number of solutions with $x_2 = \frac{3}{2} - \frac{1}{2}x_1$.

   (d) Parallel lines, so there is no solution.

   (e) One line, so there is an infinite number of solutions with $x_2 = -\frac{1}{2}x_1$.

   (f) Intersecting lines whose solution is $x_1 = 5$ and $x_2 = 3$.

   (g) Three lines in the plane that do not intersect in a common point.

   (h) Three lines in the plane that do not intersect in a common point.

   (i) Intersecting lines whose solution is $x_1 = \frac{2}{7}$ and $x_2 = -\frac{11}{7}$.

   (j) Two planes in space that intersect in a line with $x_1 = -\frac{5}{4}x_2$ and $x_3 = \frac{3}{2}x_2 + 1$.

3. Gaussian elimination gives the following solutions:

   (a) $x_1 = 1.1875$, $x_2 = 1.8125$, $x_3 = 0.875$ with one row interchange required;

   (b) $x_1 = 0.75$, $x_2 = 0.5$, $x_3 = -0.125$ with one row interchange required;

   (c) $x_1 = -1$, $x_2 = 0$, $x_3 = 1$ with no interchange required;

   (d) $x_1 = 1$, $x_2 = 2$, $x_3 = -1$ with no interchange required;

   (e) $x_1 = 1.5$, $x_2 = 2$, $x_3 = -1.2$, $x_4 = 3$ with no interchange required;

   (f) $x_1 = \frac{22}{9}$, $x_2 = -\frac{4}{9}$, $x_3 = \frac{4}{3}$, $x_4 = 1$ with one row interchange required;

   (g) no solution;

   (h) $x_1 = -1$, $x_2 = 2$, $x_3 = 0$, $x_4 = 1$ with one row interchange required.

5. **(a)** There is sufficient food to satisfy the average daily consumption.

   (b) We could add 200 of species 1, or 150 of species 2, or 100 of species 3, or 100 of species 4.

   (c) Assuming none of the increases indicated in part (b) was selected, species 2 could be increased by 650, or species 3 could be increased by 150, or species 4 could be increased by 150.

   (d) Assuming none of the increases indicated in parts (b) or (c) was selected, species 3 could be increased by 150 or species 4 could be increased by 150.

7. **(a)** For the Trapezoidal rule $m = n = 1$, $x_0 = 0$, $x_1 = 1$ so that for $i = 0$ and 1, we have

   $$u(x_i) = f(x_i) + \int_0^1 K(x_i, t)u(t)dt = f(x_i) + \tfrac{1}{2}[K(x_i, 0)u(0) + K(x_i, 1)u(1)].$$

   Substituting for x_i gives the desired equations.

   (b) We have $n = 4$, $h = \frac{1}{4}$, $x_0 = 0$, $x_1 = \frac{1}{4}$, $x_2 = \frac{1}{2}$, $x_3 = \frac{3}{4}$, $x_4 = 1$ so that

   $$u(x_i) = f(x_i) + \frac{h}{2}[K(x_i, 0)u(0) + K(x_i, 1)u(1) + 2K(x_i, \tfrac{1}{4})u(\tfrac{1}{4}) + 2K(x_i, \tfrac{1}{2})u(\tfrac{1}{2}) + 2K(x_i, \tfrac{3}{4})u(\tfrac{3}{4})]$$

   for $i = 0, 1, 2, 3, 4$. This gives

   $$u(x_i) = x_i^2 + \tfrac{1}{8}[e^{x_i}u(0) + e^{|x_i - 1|}u(1) + 2e^{|x_i - \frac{1}{4}|}u(\tfrac{1}{4}) + 2e^{|x_i - \frac{1}{2}|}u(\tfrac{1}{2}) + 2e^{|x_i - \frac{3}{4}|}u(\tfrac{3}{4})]$$

   for each $i = 1, \ldots, 4$. The 5 by 5 linear system has solution $u(0) = -1.154255$, $u(\tfrac{1}{4}) = -0.9093298$, $u(\tfrac{1}{2}) = -0.7153145$, $u(\tfrac{3}{4}) = -0.5472949$, and $u(1) = -0.3931261$.

(c)   The Composite Simpson's rule gives

$$\int_0^1 K(x_i, t)u(t)dt = \frac{h}{3}[K(x_i, 0)u(0) + 4K(x_i, \tfrac{1}{4})u(\tfrac{1}{4}) + 2K(x_i, \tfrac{1}{2})u(\tfrac{1}{2}) + 4K(x_i, \tfrac{3}{4})u(\tfrac{3}{4}) + K(x_i, 1)u(1)],$$

which results in the linear equations

$$u(x_i) = x_i^2 + \tfrac{1}{12}[e^{x_i}u(0) + 4e^{|x_i-\frac{1}{4}|}u(\tfrac{1}{4}) + 2e^{|x_i-\frac{1}{2}|}u(\tfrac{1}{2}) + 4e^{|x_i-\frac{3}{4}|}u(\tfrac{3}{4}) + e^{|x_i-1|}u(1)].$$

The 5 by 5 linear system has solutions $u(0) = -1.234286$, $u(\tfrac{1}{4}) = -0.9507292$, $u(\tfrac{1}{2}) = -0.7659400$, $u(\tfrac{3}{4}) = -0.5844737$, and $u(1) = -0.4484975$.

Exercise Set 6.3 (page 215)

1. Gaussian elimination with three-digit chopping arithmetic gives the following results:
   (a) $x_1 = 30.0$, $x_2 = 0.990$ (b) $x_1 = 1.00$, $x_2 = 9.98$ (c) $x_1 = -40.6$, $x_2 = -0.125$, $x_3 = 0.143$
   (d) $x_1 = 9.33$, $x_2 = 0.492$, $x_3 = -9.61$ (e) $x_1 = -0.102$, $x_2 = 1.38$, $x_3 = 2.42$
   (f) $x_1 = 57.8$, $x_2 = -258$, $x_3 = 259$ (g) $x_1 = 0.198$, $x_2 = 0.0154$, $x_3 = -0.0156$, $x_4 = -0.716$
   (h) $x_1 = 0.828$, $x_2 = -3.32$, $x_3 = 0.153$, $x_4 = 4.91$

3. Gaussian elimination with maximal column pivoting and three-digit chopping arithmetic gives the following results:
   (a) $x_1 = 10.0$, $x_2 = 1.00$ (b) $x_1 = 1.00$, $x_2 = 9.98$ (c) $x_1 = -0.160$, $x_2 = 9.98$, $x_3 = 0.142$
   (d) $x_1 = 9.33$, $x_2 = 0.492$, $x_3 = -9.61$ (e) $x_1 = -0.102$, $x_2 = 1.38$, $x_3 = 2.42$
   (f) $x_1 = 60.1$, $x_2 = -298$, $x_3 = 273$ (g) $x_1 = 0.172$, $x_2 = 0.0131$, $x_3 = -0.208$, $x_4 = -1.23$
   (h) $x_1 = 0.777$, $x_2 = -3.10$, $x_3 = 0.161$, $x_4 = 4.50$

5. Gaussian elimination with scaled column pivoting and three-digit chopping arithmetic gives the following results:
   (a) $x_1 = 10.0$, $x_2 = 1.00$ (b) $x_1 = 1.00$, $x_2 = 9.98$ (c) $x_1 = -0.160$, $x_2 = 9.98$, $x_3 = 0.142$
   (d) $x_1 = 0.987$, $x_2 = 0.500$, $x_3 = -0.997$ (e) $x_1 = -0.102$, $x_2 = 1.38$, $x_3 = 2.42$
   (f) $x_1 = 60.1$, $x_2 = -298$, $x_3 = 273$ (g) $x_1 = 0.170$, $x_2 = 0.0127$, $x_3 = -0.0217$, $x_4 = -1.28$
   (h) $x_1 = 0.837$, $x_2 = -3.31$, $x_3 = 0.158$, $x_4 = 4.92$

7. The Gaussian elimination with backward substitution method and single precision arithmetic gives the following results:
   (a) $x_1 = 10.000000$, $x_2 = 1.0000000$.
   (b) $x_1 = 1.0000000$, $x_2 = 10.000000$.
   (c) $x_1 = 0.0000000$, $x_2 = 10.000000$, $x_3 = 0.14285714$.
   (d) $x_1 = 0.99104628$, $x_2 = 0.49870656$, $x_3 = -0.99568160$.
   (e) $x_1 = -0.11108022$, $x_2 = 1.3962863$, $x_3 = 2.4190803$.
   (f) $x_1 = 54.000044$, $x_2 = -264.00023$, $x_3 = 240.00021$.
   (g) $x_1 = 0.17682530$, $x_2 = 0.012692691$, $x_3 = -0.020654050$, $x_4 = -1.1826087$.
   (h) $x_1 = 0.78842555$, $x_2 = -3.1255199$, $x_3 = 0.1676848$, $x_4 = 4.5572988$.

9. The Gaussian elimination with scaled-column pivoting method and single precision arithmetic gives the following results:

(a) $x_1 = 10.000000$, $x_2 = 1.0000000$.

(b) $x_1 = 1.0000000$, $x_2 = 10.000000$.

(c) $x_1 = 0.0000000$, $x_2 = 10.000000$, $x_3 = 0.14285714$.

(d) $x_1 = 0.99104628$, $x_2 = 0.49870656$, $x_3 = -0.99568160$.

(e) $x_1 = -0.11108022$, $x_2 = 1.3962863$, $x_3 = 2.4190803$.

(f) $x_1 = 54.000044$, $x_2 = -264.00023$, $x_3 = 240.00021$.

(g) $x_1 = 0.17682530$, $x_2 = 0.012692691$, $x_3 = -0.020654050$, $x_4 = -1.1826087$.

(h) $x_1 = 0.78842555$, $x_2 = -3.1255199$, $x_3 = 0.1676848$, $x_4 = 4.5572988$.

Exercise Set 6.4 (page 222)

1. (a) The matrix is singular.

(b) $\begin{bmatrix} -\frac{1}{4} & \frac{1}{4} & \frac{1}{4} \\ \frac{5}{8} & -\frac{1}{8} & -\frac{1}{8} \\ \frac{1}{8} & -\frac{5}{8} & \frac{3}{8} \end{bmatrix}$

(c) $\begin{bmatrix} \frac{1}{2} & 0 & 0 \\ 0 & -\frac{1}{3} & 0 \\ 0 & 0 & 1 \end{bmatrix}$

(d) The matrix is singular.   (e) The matrix is singular.   (f) The matrix is singular.

(g) $\begin{bmatrix} \frac{1}{4} & 1 & 0 & 0 \\ -\frac{3}{14} & \frac{1}{7} & 0 & 0 \\ \frac{3}{28} & -\frac{11}{7} & 1 & 0 \\ -\frac{1}{2} & 1 & -1 & 1 \end{bmatrix}$

(h) $\begin{bmatrix} 1 & 0 & 1 & -1 \\ -1 & \frac{5}{3} & \frac{5}{3} & -1 \\ -1 & \frac{2}{3} & \frac{2}{3} & 0 \\ 0 & -\frac{1}{3} & -\frac{4}{3} & 1 \end{bmatrix}$

(i) $\begin{bmatrix} 1 & 0 & 0 & 0 \\ 2 & 1 & 0 & 0 \\ 3 & 4 & 1 & 0 \\ -1 & -3 & 0 & 1 \end{bmatrix}$

(j) $\begin{bmatrix} 1 & 1 & 2 & 4 \\ 0 & 1 & 1 & 2 \\ 0 & 0 & 1 & 1 \\ 0 & 0 & 0 & 1 \end{bmatrix}$

3. The solutions to the linear systems obtained in parts (a) and (b) are, from left to right and top to bottom, $-\frac{2}{7}$, $-\frac{13}{14}$, $-\frac{3}{14}$; $\frac{17}{7}$, $-\frac{19}{14}$, $-\frac{41}{14}$; 1, 1, 1;   and $-\frac{1}{7}$, $\frac{2}{7}$, $\frac{1}{7}$.

5. The determinants of the matrices are   (a) -8 (b) 14   (c) 0   (d) 3

7. If AB is nonsingular, then $\det(AB) \neq 0$. Since $\det(AB) = \det(A)\det(B)$, $\det(A) \neq 0$ and $\det(B) \neq 0$, so A and B are nonsingular. If A and B are nonsingular, then $\det(A) \neq 0$ and $\det(B) \neq 0$. Thus, $\det(AB) \neq 0$ and AB is nonsingular.

9. (a) If $C = AB$, where A and B are lower-triangular, then $a_{ik} = 0$ if $k > i$ and $b_{kj} = 0$ if $k < j$. Thus,

$$c_{ij} = \sum_{k=1}^{n} a_{ik}b_{kj} = \sum_{k=j}^{i} a_{ik}b_{kj},$$

which will have the sum zero unless $j \leq i$. Hence, C is lower-triangular.

(b) We have $a_{ik} = 0$ if $k < i$ and $b_{kj} = 0$ if $k > j$. The steps are similar to those in part (a).

(c) Let L be a nonsingular lower-triangular matrix. To obtain the ith column of L^{-1}, solve n linear systems of the form

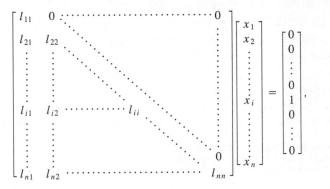

where the 1 appears in the ith position to obtain the ith column of L^{-1}.

11. (a) We have
$$\begin{bmatrix} 7 & 4 & 4 & 0 \\ -6 & -3 & -6 & 0 \\ 0 & 0 & 3 & 0 \\ 0 & 0 & 0 & 1 \end{bmatrix} \begin{bmatrix} 2(x_0 - x_1) + \alpha_0 + \alpha_1 \\ 3(x_1 - x_0) - \alpha_1 - 2\alpha_0 \\ \alpha_0 \\ x_0 \end{bmatrix} = \begin{bmatrix} 2(x_0 - x_1) + 3\alpha_0 + 3\alpha_1 \\ 3(x_1 - x_0) - 3\alpha_1 - 6\alpha_0 \\ 3\alpha_0 \\ x_0 \end{bmatrix}$$

(b)
$$B = A^{-1} = \begin{bmatrix} -1 & -\frac{4}{3} & -\frac{4}{3} & 0 \\ 2 & \frac{7}{3} & 2 & 0 \\ 0 & 0 & \frac{1}{3} & 0 \\ 0 & 0 & 0 & 1 \end{bmatrix}$$

Exercise Set 6.5 (page 229)

1. (a)
$$L = \begin{bmatrix} 1 & 0 & 0 \\ 1.5 & 1 & 0 \\ 1.5 & 1 & 1 \end{bmatrix} \quad \text{and} \quad U = \begin{bmatrix} 2 & -1 & 1 \\ 0 & 4.5 & 7.5 \\ 0 & 0 & -4 \end{bmatrix}$$

(b)
$$L = \begin{bmatrix} 1 & 0 & 0 \\ -0.5 & 1 & 0 \\ 2 & 2 & 1 \end{bmatrix} \quad \text{and} \quad U = \begin{bmatrix} 2 & -1.5 & 3 \\ 0 & -0.75 & 3.5 \\ 0 & 0 & -8 \end{bmatrix}$$

(c)
$$L = \begin{bmatrix} 1 & 0 & 0 \\ -2.106719 & 1 & 0 \\ 3.067193 & 1.19776 & 1 \end{bmatrix} \quad \text{and} \quad U = \begin{bmatrix} 1.012 & -2.132 & 3.104 \\ 0 & -0.3955249 & -0.4737443 \\ 0 & 0 & -8.939133 \end{bmatrix}$$

(d) $L = I$ and U is the original matrix.

(e)
$$L = \begin{bmatrix} 1 & 0 & 0 & 0 \\ 0.5 & 1 & 0 & 0 \\ 0 & -2 & 1 & 0 \\ 1 & -1.33333 & 2 & 1 \end{bmatrix} \quad \text{and} \quad U = \begin{bmatrix} 2 & 0 & 0 & 0 \\ 0 & 1.5 & 0 & 0 \\ 0 & 0 & 0.5 & 0 \\ 0 & 0 & 0 & 1 \end{bmatrix}$$

(f)

$$L = \begin{bmatrix} 1 & 0 & 0 & 0 \\ -1.849190 & 1 & 0 & 0 \\ -0.4596433 & -0.2501219 & 1 & 0 \\ 2.768661 & -0.3079435 & -5.35229 & 1 \end{bmatrix} \text{ and}$$

$$U = \begin{bmatrix} 2.175600 & 4.023099 & -2.173199 & 5.196700 \\ 0 & 13.43947 & -4.018660 & 10.80698 \\ 0 & 0 & -0.8929510 & 5.091692 \\ 0 & 0 & 0 & 12.03614 \end{bmatrix}$$

3. (a)

$$P'LU = \begin{bmatrix} 0 & 1 & 0 \\ 1 & 0 & 0 \\ 0 & 0 & 1 \end{bmatrix} \begin{bmatrix} 1 & 0 & 0 \\ 0 & 1 & 0 \\ 0 & -\frac{1}{2} & 1 \end{bmatrix} \begin{bmatrix} 1 & 1 & -1 \\ 0 & 2 & 3 \\ 0 & 0 & \frac{5}{2} \end{bmatrix}$$

(b)

$$P'LU = \begin{bmatrix} 1 & 0 & 0 \\ 0 & 0 & 1 \\ 0 & 1 & 0 \end{bmatrix} \begin{bmatrix} 1 & 0 & 0 \\ 2 & 1 & 0 \\ 1 & 0 & 1 \end{bmatrix} \begin{bmatrix} 1 & 2 & -1 \\ 0 & -5 & 6 \\ 0 & 0 & 4 \end{bmatrix}$$

(c)

$$P'LU = \begin{bmatrix} 1 & 0 & 0 & 0 \\ 0 & 1 & 0 & 0 \\ 0 & 0 & 0 & 1 \\ 0 & 0 & 1 & 0 \end{bmatrix} \begin{bmatrix} 1 & 0 & 0 & 0 \\ 2 & 1 & 0 & 0 \\ 3 & 4 & 1 & 0 \\ -1 & -3 & 0 & 1 \end{bmatrix} \begin{bmatrix} 1 & 1 & 0 & 3 \\ 0 & -1 & -1 & -5 \\ 0 & 0 & 3 & 13 \\ 0 & 0 & 0 & -13 \end{bmatrix}$$

(d)

$$P'LU = \begin{bmatrix} 1 & 0 & 0 & 0 \\ 0 & 0 & 0 & 1 \\ 0 & 0 & 1 & 0 \\ 0 & 1 & 0 & 0 \end{bmatrix} \begin{bmatrix} 1 & 0 & 0 & 0 \\ 2 & 1 & 0 & 0 \\ 1 & 0 & 1 & 0 \\ 1 & 0 & 0 & 1 \end{bmatrix} \begin{bmatrix} 1 & -2 & 3 & 0 \\ 0 & 5 & -3 & -1 \\ 0 & 0 & -1 & -2 \\ 0 & 0 & 0 & 1 \end{bmatrix}$$

Exercise Set 6.6 (page 236)

1. (*i*) The symmetric matrices are in (a), (b), and (f).

(*ii*) The singular matrices are in (e) and (h).

(*iii*) The strictly diagonally dominant matrices are in (a), (b), (c), and (d).

(*iv*) The positive definite matrices are in (a) and (f).

3. Choleski's method gives the following results:

(a)

$$L = \begin{bmatrix} 1.41423 & 0 & 0 \\ -0.7071069 & 1.224743 & 0 \\ 0 & -0.8164972 & 1.154699 \end{bmatrix}$$

(b)

$$L = \begin{bmatrix} 2 & 0 & 0 & 0 \\ 0.5 & 1.658311 & 0 & 0 \\ 0.5 & -0.7537785 & 1.087113 & 0 \\ 0.5 & 0.4522671 & 0.08362442 & 1.240346 \end{bmatrix}$$

(c)

$$L = \begin{bmatrix} 2 & 0 & 0 & 0 \\ 0.5 & 1.658311 & 0 & 0 \\ -0.5 & -0.4522671 & 2.132006 & 0 \\ 0 & 0 & 0.9380833 & 1.766351 \end{bmatrix}$$

(d)

$$L = \begin{bmatrix} 2.449489 & 0 & 0 & 0 \\ 0.8164966 & 1.825741 & 0 & 0 \\ 0.4082483 & 0.3651483 & 1.923538 & 0 \\ -0.4082483 & 0.1825741 & -0.4678876 & 1.606574 \end{bmatrix}$$

5. The Choleski factorization gives the following results:
    (a) $x_1 = 1$, $x_2 = -1$, $x_3 = 0$
    (b) $x_1 = 0.2$, $x_2 = -0.2$, $x_3 = -0.2$, $x_4 = 0.25$
    (c) $x_1 = 1$, $x_2 = 2$, $x_3 = -1$, $x_4 = -2$
    (d) $x_1 = -0.85863874$, $x_2 = 2.4188482$, $x_3 = -0.95811518$, $x_4 = -1.2722513$

7. We have $x_i = 1$ for each $i = 1, \ldots, 10$.

9. Only the matrix in (d) is positive definite.

11. (a) No, consider $\begin{bmatrix} -1 & 0 \\ 0 & -1 \end{bmatrix}$.   (b) Yes, since $A = A^t$.   (c) Yes, since $x^t(A + B)x = x^t Ax + x^t Bx$.

    (d) Yes, since $x^t A^2 x = x^t A^t Ax = (Ax)^t(Ax) \geq 0$ and because A is nonsingular, equality holds only if $x = 0$.

    (e) No, consider $A = \begin{bmatrix} 1 & 0 \\ 0 & 1 \end{bmatrix}$ and $B = \begin{bmatrix} 10 & 0 \\ 0 & 10 \end{bmatrix}$.

13. One example is $\begin{bmatrix} 1 & \frac{1}{5} \\ \frac{1}{10} & 1 \end{bmatrix}$.

Exercise Set 7.2 (page 249)

1. (a) We have $\| x \|_\infty = 4$ and $\| x \|_2 = 5.220153$.   (b) We have $\| x \|_\infty = 4$ and $\| x \|_2 = 5.477226$.
    (c) We have $\| x \|_\infty = 2^k$ and $\| x \|_2 = (1 + 4^k)^{\frac{1}{2}}$.
    (d) We have $\| x \|_\infty = \frac{4}{(k+1)}$ and $\| x \|_2 = (\frac{16}{(k+1)^2} + \frac{4}{k^4} + k^4 e^{-2k})^{\frac{1}{2}}$.

3. (a) We have $\lim_{k \to \infty} x^{(k)} = (0, 0, 0)^t$.   (b) We have $\lim_{k \to \infty} x^{(k)} = (0, 1, 3)^t$.
    (c) We have $\lim_{k \to \infty} x^{(k)} = (0, 0, \frac{1}{2})^t$.   (d) We have $\lim_{k \to \infty} x^{(k)} = (1, -1, 1)^t$.

5. (a) We have $\| x - \hat{x} \|_\infty = 6.67 \times 10^{-4}$ and $\| A\hat{x} - b \|_\infty = 2.06 \times 10^{-4}$.
    (b) We have $\| x - \hat{x} \|_\infty = 0.33$ and $\| A\hat{x} - b \|_\infty = 0.27$.
    (c) We have $\| x - \hat{x} \|_\infty = 0.5$ and $\| A\hat{x} - b \|_\infty = 0.3$.
    (d) We have $\| x - \hat{x} \|_\infty = 6.55 \times 10^{-2}$, and $\| A\hat{x} - b \|_\infty = 0.32$.

Exercise Set 7.3 (page 254)

1. (a) The eigenvalue $\lambda_1 = 3$ has the eigenvector $x_1 = (1, -1)^t$, and the eigenvalue $\lambda_2 = 1$ has the eigenvector $x_2 = (1, 1)^t$.
    (b) The eigenvalue $\lambda_1 = \lambda_2 = 1$ has the eigenvector $x = (1, 0)^t$.
    (c) The eigenvalue $\lambda_1 = \frac{1}{2}$ has the eigenvector $x_1 = (1, 1)^t$, and the eigenvalue $\lambda_2 = -\frac{1}{2}$ has the eigenvector $x_2 = (1, -1)^t$.
    (d) The eigenvalue $\lambda_1 = 0$ has the eigenvector $x_1 = (1, -1)^t$, and the eigenvalue $\lambda_2 = -1$ has the eigenvector $x_2 = (1, -2)^t$.
    (e) The eigenvalue $\lambda_1 = \lambda_2 = 3$ has the eigenvectors $x_1 = (0, 0, 1)^t$ and $x_2 = (1, 1, 0)^t$, and the eigenvalue $\lambda_3 = 1$ has the eigenvector $x_3 = (1, -1, 0)^t$.
    (f) The eigenvalue $\lambda_1 = 7$ has the eigenvector $x_1 = (1, 4, 4)^t$, the eigenvalue $\lambda_2 = 3$ has the eigenvector $x_2 = (1, 2, 0)^t$, and the eigenvalue $\lambda_3 = -1$ has the eigenvector $x_3 = (1, 0, 0)^t$.

(g) The eigenvalue $\lambda_1 = \lambda_2 = 1$ has the eigenvectors $x_1 = (-1, 1, 0)'$ and $x_2 = (-1, 0, 1)'$, and the eigenvalue $\lambda_3 = 5$ has the eigenvector $x_3 = (1, 2, 1)'$.

(h) The eigenvalue $\lambda_1 = 3$ has the eigenvector $x_1 = (-0.408248, 0.408248, 0.816497)'$, the eigenvalue $\lambda_2 = 4$ has the eigenvector $x_2 = (0, 0.447214, 0.447214)'$, and the eigenvalue $\lambda_3 = -2$ has the eigenvector $x_3 = (0.348743, -0.929981, -0.116248)'$.

3. Since $A_1^k = \begin{bmatrix} 1 & 0 \\ \frac{2^k - 1}{2^{k+1}} & 2^{-k} \end{bmatrix}$, we have $\lim\limits_{k \to \infty} A_1^k = \begin{bmatrix} 1 & 0 \\ \frac{1}{2} & 0 \end{bmatrix}$.

   Also, $A_2^k = \begin{bmatrix} 2^{-k} & 0 \\ \frac{16k}{2^{k-1}} & 2^{-k} \end{bmatrix}$, so $\lim\limits_{k \to \infty} A_2^k = \begin{bmatrix} 0 & 0 \\ 0 & 0 \end{bmatrix}$.

5. (a) 3   (b) 1   (c) $-\frac{1}{4}$ (d) 0   (e) 9   (f) -21 (g) 5   (h) -24

7. (a) We have the real eigenvalue $\lambda = 1$ with the eigenvector $x = (6, 3, 1)'$.
   (b) Choose any multiple of the vector $(6, 3, 1)'$.

Exercise Set 7.4 (page 259)

1. Two iterations of Jacobi's method give the following results: (a) $(0.1428571, -0.3571429, 0.4285714)'$
   (b) $(0.97, 0.91, 0.74)'$ (c) $(1.4, 2.85, 1.2, 1.5)'$ (d) $(0.075, 2.9625, -1.1875, -3.975)'$
   (e) $(2.975, -2.65, 1.75, 0.325)'$ (f) $(1.325, -1.6, 1.6, 1.675, 2.425)'$
   (g) $(-0.5208333, -0.04166667, -0.2166667, 0.4166667)'$
   (h) $(0.6875, 1.125, 0.6875, 1.375, 0.5625, 1.375)'$

3. Jacobi's method gives the following results: (a) $x^{(10)} = (0.03507839, -0.2369262, 0.6578015)'$
   (b) $x^{(6)} = (0.9957250, 0.9577750, 0.7914500)'$ (c) $x^{(5)} = (1.95, 2.9, 1.2, 1.5)'$
   (d) $x^{(13)} = (-0.08268421, 3.789723, -1.519298, -4.777234)'$ (e) Does not converge.
   (f) $x^{(12)} = (0.7870883, -1.003036, 1.866048, 1.912449, 1.989571)'$
   (g) $x^{(14)} = (-0.7529267, 0.04078538, -0.2806091, 0.6911662)'$
   (h) $x^{(17)} = (0.9996805, 1.999774, 0.9996805, 1.999840, 0.9995482, 1.999840)'$

Exercise Set 7.5 (page 263)

1. Two iterations of the SOR method give the following results: (a) $(0.05410079, -0.2115435, 0.6477159)'$
   (b) $(0.9876790, 0.9784935, 0.7899328)'$ (c) $(1.6060, 2.9645, 1.2540, 1.4850)'$
   (d) $(0.08765735, 3.818345, -1.471937, -4.824042)'$ (e) $(3.055932, -1.023653, 0.4577766, 0.6158895)'$
   (f) $(1.079675, -1.260654, 2.042489, 1.995373, 2.049536)'$
   (g) $(-0.6604902, 0.03700749, -0.2493513, 0.6561139)'$
   (h) $(0.8318750, 1.647766, 0.9189856, 1.791281, 0.8712129, 1.959155)'$

3. The SOR method gives the following results: (a) $x^{(12)} = (0.03488469, -0.2366474, 0.6579013)'$
   (b) $x^{(7)} = (0.9958341, 0.9579041, 0.7915756)'$ (c) $x^{(8)} = (1.950315, 2.899950, 1.200034, 1.499996)'$
   (d) $x^{(8)} = (-0.08276995, 3.7896231, -1.519177, -4.777632)'$ (e) Does not converge.
   (f) $x^{(10)} = (0.7866310, -1.002807, 1.866530, 1.912645, 1.989792)'$
   (g) $x^{(7)} = (-0.7534489, 0.04106617, -0.2808146, 0.6918049)'$
   (h) $x^{(7)} = (0.9999442, 1.999934, 1.000033, 1.999958, 0.9999815, 2.000007)'$

Exercise Set 7.6 (page 266)

1. The $\| \cdot \|_\infty$ condition numbers are **(a)** 50   **(b)** 241.37   **(c)** 235.23   **(d)** 60002   **(e)** 339866   **(f)** 12
    (g) 52   **(h)** 198.17

3. The matrix is ill-conditioned since $K_\infty = 60,000$. We have $\tilde{\mathbf{x}} = (-1.0000, 2.0000)^t$.

5. **(a)** We have $\tilde{\mathbf{x}} = (188.9998, 92.99998, 45.00001, 27.00001, 21.00002)^t$.
    (b) The condition number is $K_\infty = 80$.   **(c)** The exact solution is $\mathbf{x} = (189, 93, 45, 27, 21)^t$.

Exercise Set 8.2 (page 275)

1. The linear least-squares polynomial is $1.70784x + 0.89968$.

3. The least-squares polynomials with their errors are, respectively, $0.6208950 + 1.219621x$, with
    $E = 2.719 \times 10^{-5}$; $0.5965807 + 1.253293x - 0.01085343x^2$, with $E = 1.801 \times 10^{-5}$; and
    $0.6290193 + 1.185010x + 0.03533252x^2 - 0.01004723x^3$, with $E = 1.741 \times 10^{-5}$.

5. **(a)** The linear least-squares polynomial of degree one is $72.0845x - 194.138$ with an error of 329.
    (b) The least-squares polynomial of degree two is $6.61822x^2 - 1.14357x + 1.23570$ with an error of
    1.44×10^{-3}.
    (c) The least-squares polynomial of degree three is $-0.0137352x^3 + 6.84659x^2 - 2.38475x + 3.43896$ with an
    error of 5.27×10^{-4}.
    (d) The least-squares approximation of the form be^{ax} is $24.2588e^{0.372382x}$ with an error of 418.
    (e) The least-squares approximation of the form bx^a is $6.23903x^{2.01954}$ with an error of 0.00703.

7. Point average $= 0.101$ (ACT score) $+ 0.487$.

Exercise Set 8.3 (page 284)

1. The linear least-squares approximations are   **(a)** $P_1(x) = 1.833333 + 4x$
    (b) $P_1(x) = -1.600003 + 3.600003x$ **(c)** $P_1(x) = 1.140981 - 0.2958375x$
    (d) $P_1(x) = 0.1945267 + 3.000001x$ **(e)** $P_1(x) = 0.7307083 - 0.1777249x$
    (f) $P_1(x) = -1.861455 + 1.666667x$

3. The linear least-square approximations on $[-1, 1]$ are   **(a)** $P_1(x) = 3.333333 - 2x$
    (b) $P_1(x) = 0.6000025x$ **(c)** $P_1(x) = 0.5493063 - 0.2958375x$ **(d)** $P_1(x) = 1.175201 + 1.103639x$
    (e) $P_1(x) = 0.4207355 + 0.4353975x$ **(f)** $P_1(x) = 0.6479184 + 0.5281226x$

5. The errors for the approximations in Exercise 3 are   **(a)** 0.177779   **(b)** 0.0457206   **(c)** 0.00484624
    (d) 0.0526541   **(e)** 0.0153784   **(f)** 0.00363453

7. The Gram-Schmidt process produces the following collections of polynomials:
    (a) $\phi_0(x) = 1, \phi_1(x) = x - 0.5, \quad \phi_2(x) = x^2 - x + \frac{1}{6}, \quad$ and $\phi_3(x) = x^3 - 1.5x^2 + 0.6x - 0.05$.
    (b) $\phi_0(x) = 1, \phi_1(x) = x - 1, \quad \phi_2(x) = x^2 - 2x + \frac{2}{3}, \quad$ and $\phi_3(x) = x^3 - 3x^2 + \frac{12}{5}x - \frac{2}{5}$.
    (c) $\phi_0(x) = 1, \phi_1(x) = x - 2, \quad \phi_2(x) = x^2 - 4x + \frac{11}{3}, \quad$ and $\phi_3(x) = x^3 - 6x^2 + 11.4x - 6.8$.

9. The least-squares polynomials of degree two are (a) $P_2(x) = 3.83333\phi_0(x) + 4\phi_1(x) + 0.999999\phi_2(x)$
(b) $P_2(x) = 2\phi_0(x) + 3.6\phi_1(x) + 3\phi_2(x)$ (c) $P_2(x) = 0.549306\phi_0(x) - 0.295837\phi_1(x) + 0.158878\phi_2(x)$
(d) $P_2(x) = 3.19453\phi_0(x) + 3\phi_1(x) + 1.458960\phi_2(x)$
(e) $P_2(x) = 0.0656760\phi_0(x) + 0.0916711\phi_1(x) - 0.737512\phi_2(x)$
(f) $P_2(x) = 1.47188\phi_0(x) + 1.66667\phi_1(x) + 0.259771\phi_2(x)$

11. The Laguerre polynomials are $L_1(x) = x - 1$, $L_2(x) = x^2 - 4x + 2$, and $L_3(x) = x^3 - 9x^2 + 18x - 6$.

Exercise Set 8.4 (page 290)

1. The interpolating polynomials of degree two are
(a) $P_2(x) = 2.377443 + 1.590534(x - 0.8660254) + 0.5320418(x - 0.8660254)x$
(b) $P_2(x) = 0.7617600 + 0.8796047(x - 0.8660254)$
(c) $P_2(x) = 1.052926 + 0.4154370(x - 0.8660254) - 0.1384262x(x - 0.8660254)$
(d) $P_2(x) = 0.5625 + 0.649519(x - 0.8660254) + 0.75x(x - 0.8660254)$

3. The interpolating polynomials of degree three are
(a) $P_3(x) = 2.519044 + 1.945377(x - 0.9238795) + 0.7047420(x - 0.9238795)(x - 0.3826834) +$
$0.1751757(x - 0.9238795)(x - 0.3826834)(x + 0.3826834)$
(b) $P_3(x) = 0.7979459 + 0.7844380(x - 0.9238795) - 0.1464394(x - 0.9238795)(x - 0.3826834) -$
$0.1585049(x - 0.9238795)(x - 0.3826834)(x + 0.3826834)$
(c) $P_3(x) = 1.072911 + 0.3782067(x - 0.9238795) - 0.09799213(x - 0.9238795)(x - 0.3826834) +$
$0.04909073(x - 0.9238795)(x - 0.3826834)(x + 0.3826834)$
(d) $P_3(x) = 0.7285533 + 1.306563(x - 0.9238795) + 0.9999999(x - 0.9238795)(x - 0.3826834)$

5. The zeros of $\tilde{T}_3$ produce the following interpolating polynomials of degree two:
(a) $P_2(x) = 0.3489153 - 0.1744576(x - 2.866025) + 0.1538462(x - 2.866025)(x - 2)$
(b) $P_2(x) = 0.1547375 - 0.2461152(x - 1.866025) + 0.1957273(x - 1.866025)(x - 1)$
(c) $P_2(x) = 0.6166200 - 0.2370869(x - 0.9330127) - 0.7427732(x - 0.9330127)(x - 0.5)$
(d) $P_2(x) = 3.0177125 + 1.883800(x - 2.866025) + 0.2584625(x - 2.866025)(x - 2)$

7. If $i > j$, then

$$\tfrac{1}{2}[T_{i+j}(x) + T_{i-j}(x)] = \tfrac{1}{2}[\cos(i + j)\theta + \cos(i - j)\theta] = \cos i\theta \cos j\theta = T_i(x)T_j(x).$$

Exercise Set 8.5 (page 297)

1. The Padé approximations of degree two for $f(x) = e^{2x}$ are

$n = 2, m = 0 : r_{2,0}(x) = 1 + 2x + 2x^2$
$n = 1, m = 1 : r_{1,1}(x) = (1 + x)/(1 - x)$
$n = 0, m = 2 : r_{0,2}(x) = (1 - 2x + 2x^2)^{-1}$

i	x_i	$f(x_i)$	$r_{2,0}(x_i)$	$r_{1,1}(x_i)$	$r_{0,2}(x_i)$
1	0.2	1.4918	1.4800	1.5000	1.4706
2	0.4	2.2255	2.1200	2.3333	1.9231
3	0.6	3.3201	2.9200	4.0000	1.9231
4	0.8	4.9530	3.8800	9.0000	1.4706
5	1.0	7.3891	5.0000	undefined	1.0000

3. $r_{2,3}(x) = \dfrac{1 + \frac{2}{5}x + \frac{1}{20}x^2}{1 - \frac{3}{5}x + \frac{3}{20}x^2 - \frac{1}{60}x^3}$

5. $r_{3,3}(x) = \dfrac{x - \frac{7}{60}x^3}{1 + \frac{1}{20}x^2}$

7. The Padé approximations of degree five are

(a) $r_{0,5}(x) = (1 + x + \frac{1}{2}x^2 + \frac{1}{6}x^3 + \frac{1}{24}x^4 + \frac{1}{120}x^5)^{-1}$

(b) $r_{1,4}(x) = \dfrac{1 - \frac{1}{5}x}{1 + \frac{4}{5}x + \frac{3}{10}x^2 + \frac{1}{15}x^3 + \frac{1}{120}x^4}$

(c) $r_{3,2}(x) = \dfrac{1 - \frac{3}{5}x + \frac{3}{20}x^2 - \frac{1}{60}x^3}{1 + \frac{2}{5}x + \frac{1}{20}x^2}$

(d) $r_{4,1}(x) = \dfrac{1 - \frac{4}{5}x + \frac{3}{10}x^2 - \frac{1}{15}x^3 + \frac{1}{120}x^4}{1 + \frac{1}{5}x}$

9. For

8(a):	(a) 5.63,	(b) 5.63,	(c) 5.62,	exact value 5.61.
8(b):	(a) 0.303,	(b) 0.304,	(c) 0.303,	exact value 0.304.
8(c):	(a) −0.112,	(b) −0.112,	(c) −0.120,	exact value −0.113.
8(d):	(a) 0.836,	(b) 0.837,	(c) 0.836,	exact value 0.836.

11.

$r_{T_{2,0}}(x) = \dfrac{1.266066T_0(x) - 1.130318T_1(x) + 0.2714953T_2(x)}{T_0(x)}$

$r_{T_{1,1}}(x) = \dfrac{0.9945705T_0(x) - 0.4569046T_1(x)}{T_0(x) + 0.48038745T_1(x)}$

$r_{T_{0,2}}(x) = \dfrac{0.7940220T_0(x)}{T_0(x) + 0.8778575T_1(x) + 0.1774266T_2(x)}$

x	$f(x)$	$r_{T_{2,0}}(x)$	$r_{T_{1,1}}(x)$	$r_{T_{0,2}}(x)$
0.25	0.77801	0.745928	0.785954	0.746110
0.5	0.606531	0.565159	0.617741	0.588071
1.0	0.367879	0.407243	0.363193	0.386332

13. $r_{T_{2,2}}(x) = \dfrac{0.91747T_1(x)}{T_0(x) + 0.088863T_2(x)}$

Exercise Set 8.6 (page 303)

1. $S_2(x) = 2 \sin x$

3. $S_2(x) = 9.214561 - 6.515678 \cos x + 2.606271 \cos 2x + 6.515678 \sin x$

5. The trigonometric least-squares polynomials are **(a)** $S_2(x) = \cos 2x$ **(b)** $S_2(x) = 0$
    (c) $S_3(x) = 3.132905 + 0.5886815 \cos x - 0.2700642 \cos 2x + 0.2175679 \cos 3x + 0.8341640 \sin x -$
    $0.3097866 \sin 2x$
    (d) $S_3(x) = -4.092652 + 3.883872 \cos x - 2.320482 \cos 2x + 0.7310818 \cos 3x$

7. The trigonometric least-squares polynomial is $S_3(x) = -0.9937858 + 0.2391965 \cos x + 1.515393 \cos 2x +$
    $0.2391965 \cos 3x - 1.150649 \sin x$ with error $E(S_3) = 7.271197$.

9. The trigonometric least-squares polynomials and their errors are
    (a) $S_3(x) = -0.08676065 - 1.446416 \cos \pi(x-3) - 1.617554 \cos 2\pi(x-3) + 3.980729 \cos 3\pi(x-3) -$
    $2.154320 \sin \pi(x-3) + 3.907451 \sin 2\pi(x-3);$ $E(S_3) = 210.90453$
    (b) $S_3(x) = -0.0867607 - 1.446416 \cos \pi(x-3) - 1.617554 \cos 2\pi(x-3) + 3.980729 \cos 3\pi(x-3) -$
    $2.354088 \cos 4\pi(x-3) - 2.154320 \sin \pi(x-3) + 3.907451 \sin 2\pi(x-3) - 1.166181 \sin 3\pi(x-3);$
    $E(S_4) = 169.4943$

Exercise Set 8.7 (page 310)

1. The trigonometric interpolating polynomials are
    (a) $S_2(x) = -12.33701 + 4.934802 \cos x - 2.467401 \cos 2x + 4.934802 \sin x$
    (b) $S_2(x) = -6.16851 + 9.869604 \cos x - 3.701102 \cos 2x + 4.934802 \sin x$
    (c) $S_2(x) = 1.570796 - 1.570796 \cos x$ **(d)** $S_2(x) = -0.5 - 0.5 \cos 2x + \sin x$

3. The Fast Fourier Transform method gives the following trigonometric interpolating polynomials:
    (a) $S_4(x) = -11.10331 + 2.467401 \cos x - 2.467401 \cos 2x + 2.467401 \cos 3x - 1.233701 \cos 4x +$
    $5.956833 \sin x - 2.467401 \sin 2x + 1.022030 \sin 3x$
    (b) $S_4(x) = 1.570796 - 1.340756 \cos x - 0.2300378 \cos 3x$
    (c) $S_4(x) = -0.1264264 + 0.2602724 \cos x - 0.3011140 \cos 2x + 1.121372 \cos 3x + 0.04589648 \cos 4x -$
    $0.1022190 \sin x + 0.2754062 \sin 2x - 2.052955 \sin 3x$
    (d) $S_4(x) = -0.1526819 + 0.04754278 \cos x + 0.6862114 \cos 2x - 1.216913 \cos 3x + 1.176143 \cos 4x -$
    $0.8179387 \sin x + 0.1802450 \sin 2x + 0.2753402 \sin 3x$

5. The integral approximations are given in the following table:

	Approximation	Actual
(a)	-69.76412	-62.01255
(b)	9.869605	9.869604
(c)	-0.7943605	-0.2739384
(d)	-0.9593284	-0.9570636

Exercise Set 9.2 (page 317)

1. **(a)** The eigenvalues and associated eigenvectors are $\lambda_1 = 2$, $\mathbf{v}^{(1)} = (1, 0, 0)^t$; $\lambda_2 = 1$, $\mathbf{v}^{(2)} = (0, 2, 1)^t$; and $\lambda_3 = -1$, $\mathbf{v}^{(3)} = (-1, 1, 1)^t$. Yes, the set is linearly independent.
   (b) The eigenvalues and associated eigenvectors are $\lambda_1 = \lambda_2 = 2$, $\mathbf{v}^{(1)} = \mathbf{v}^{(2)} = (1, 0, 0)^t$; and $\lambda_3 = 3$, $\mathbf{v}^{(3)} = (0, 1, 1)^t$. No.
   (c) The eigenvalues and associated eigenvectors are $\lambda_1 = \lambda_2 = \lambda_3 = 2$, $\mathbf{v}^{(1)} = \mathbf{v}^{(2)} = (1, 0, 0)^t$; and $\mathbf{v}^{(3)} = (0, 1, 0)$. No.
   (d) The eigenvalues and associated eigenvectors are $\lambda_1 = \lambda_2 = \lambda_3 = 1$, $\mathbf{v}^{(1)} = \mathbf{v}^{(2)} = (1, 0, 1)^t$; and $\mathbf{v}^{(3)} = (0, 1, 1)$. No.
   (e) The eigenvalues and associated eigenvectors are $\lambda_1 = 2$, $\mathbf{v}^{(1)} = (0, 1, 0)^t$; $\lambda_2 = 3$, $\mathbf{v}^{(2)} = (1, 0, 1)^t$; and $\lambda_3 = 1$, $\mathbf{v}^{(3)} = (1, 0, -1)^t$. Yes, the set is linearly independent.
   (f) The eigenvalues and associated eigenvectors are $\lambda_1 = \lambda_2 = 3$, $\mathbf{v}^{(1)} = (1, 0, -1)^t$, $\mathbf{v}^{(2)} = (0, 1, -1)^t$; and $\lambda_3 = 0$, $\mathbf{v}^{(3)} = (1, 1, 1)^t$. Yes, the set is linearly independent.
   (g) The eigenvalues and associated eigenvectors are $\lambda_1 = 1$, $\mathbf{v}^{(1)} = (1, 0, -1)^t$; $\lambda_2 = 1 + \sqrt{2}$, $\mathbf{v}^{(2)} = (\sqrt{2}, 1, 1)^t$; and $\lambda_3 = 1 - \sqrt{2}$, $\mathbf{v}^{(3)} = (-\sqrt{2}, 1, 1)^t$. Yes, the set is linearly independent.
   (h) The eigenvalues and associated eigenvectors are $\lambda_1 = 1$, $\mathbf{v}^{(1)} = (1, 0, -1)^t$; $\lambda_2 = 1$, $\mathbf{v}^{(2)} = (1, -1, 0)^t$; and $\lambda_3 = 4$, $\mathbf{v}^{(4)} = (1, 1, 1)^t$. Yes, the set is linearly independent.

3. **(a)** The three eigenvalues are within $\{\lambda \mid |\lambda| \le 2\} \cup \{\lambda \mid |\lambda - 2| \le 2\}$.
   (b) The three eigenvalues are within $R_1 = \{\lambda \mid |\lambda - 4| \le 2\}$.
   (c) The three real eigenvalues satisfy $0 \le \lambda \le 6$.
   (d) The three real eigenvalues satisfy $1.25 \le \lambda \le 8.25$.
   (e) The four real eigenvalues satisfy $-4 \le \lambda \le 1$.
   (f) The four real eigenvalues are within $R_1 = \{\lambda \mid |\lambda - 2| \le 4\}$.

5. If $c_1\mathbf{v}_1 + \cdots + c_k\mathbf{v}_k = \mathbf{0}$, then for any j, with $1 \le j \le k$, we have $c_1\mathbf{v}_j^t\mathbf{v}_1 + \cdots + c_k\mathbf{v}_j^t\mathbf{v}_k = \mathbf{0}$. But orthogonality gives $c_j\mathbf{v}_j^t\mathbf{v}_j = 0$ and $\mathbf{v}_j^t\mathbf{v}_j \ne 0$, so $c_j = 0$.

7. Let $A\mathbf{x}^{(i)} = \lambda_i\mathbf{x}^{(i)}$ for $i = 1, 2, \ldots, n$ where the λ_i are distinct. Suppose $\{\mathbf{x}^{(i)}\}_{i=1}^{k}$ is the largest linearly independent set of eigenvectors of A where $1 \le k < n$. (Note that a reindexing may be necessary for the preceding statement to hold.) Since $\{\mathbf{x}^{(i)}\}_{i=1}^{k+1}$ is linearly dependent, there exist numbers $c_1, \ldots, c_{k+1}$, not all zero, with

$$c_1\mathbf{x}^{(1)} + \cdots + c_k\mathbf{x}^{(k)} + c_{k+1}\mathbf{x}^{(k+1)} = \mathbf{0}.$$

Since $\{\mathbf{x}^{(i)}\}_{i=1}^{k}$ is linearly independent, $c_{k+1} \ne 0$. Multiplying by A gives

$$c_1\lambda_1\mathbf{x}^{(1)} + \cdots + c_k\lambda_k\mathbf{x}^{(k)} + c_{k+1}\lambda_{k+1}\mathbf{x}^{(k+1)} = \mathbf{0}.$$

Thus,

$$c_1(\lambda_{k+1} - \lambda_1)\mathbf{x}^{(1)} + \cdots + c_k(\lambda_{k+1} - \lambda_k)\mathbf{x}^{(k)} = \mathbf{0}.$$

But $\{\mathbf{x}^{(i)}\}_{i=1}^{k}$ is linearly independent and $\mathbf{x}^{(k+1)} \ne \mathbf{0}$, so $\lambda_{k+1} = \lambda_i$ for some $1 \le i \le k$.

Exercise Set 9.3 (page 328)

1. Approximate eigenvalues and approximate eigenvectors are
   (a) $\mu^{(3)} = 3.666667$, $\mathbf{x}^{(3)} = (0.9772727, 0.9318182, 1)^t$
   (b) $\mu^{(3)} = 2.000000$, $\mathbf{x}^{(3)} = (1, 1, 0.5)^t$ (c) $\lambda = \mu^{(2)} = 5$, $\mathbf{x} = \mathbf{x}^{(2)} = (0.25, 1, 0.25)^t$
   (d) $\mu^{(3)} = 5.000000$, $\mathbf{x}^{(3)} = (-0.2578947, 1, -0.2842105)^t$
   (e) $\mu^{(3)} = 5.038462$, $\mathbf{x}^{(3)} = (1, 0.2213741, 0.3893130, 0.4045802)^t$
   (f) $\mu^{(3)} = 7.531073$, $\mathbf{x}^{(3)} = (0.6886722, -0.6706677, -0.9219805, 1)^t$
   (g) $\mu^{(3)} = 8.644444$, $\mathbf{x}^{(3)} = (0.2544987, 1, -0.2519280, -0.3161954)^t$
   (h) $\mu^{(3)} = -3.691176$, $\mathbf{x}^{(3)} = (1, -0.4462151, -0.07968127, -0.5816733)^t$

3. Approximate eigenvalues and approximate eigenvectors are
   (a) $\mu^{(3)} = 3.959538$, $\mathbf{x}^{(3)} = (0.5816124, 0.5545606, 0.5951383)^t$
   (b) $\mu^{(3)} = 2.0000000$, $\mathbf{x}^{(3)} = (-0.6666667, -0.6666667, -0.3333333)^t$
   (c) $\mu^{(3)} = 3.142857$, $\mathbf{x}^{(3)} = (0.6767155, -0.6767155, 0.2900210)^t$
   (d) $\mu^{(3)} = 7.189567$, $\mathbf{x}^{(3)} = (0.5995308, 0.7367472, 0.3126762)^t$
   (e) $\mu^{(3)} = 6.037037$, $\mathbf{x}^{(3)} = (0.5073714, 0.4878571, -0.6634857, -0.2536857)^t$
   (f) $\mu^{(3)} = 5.142562$, $\mathbf{x}^{(3)} = (0.8373051, 0.3701770, 0.1939022, 0.3525495)^t$
   (g) $\mu^{(3)} = 8.086569$, $\mathbf{x}^{(3)} = (0.2296403, 0.9023239, -0.2273207, -0.2853107)^t$
   (h) $\mu^{(3)} = 8.593142$, $\mathbf{x}^{(3)} = (-0.4134762, 0.4026664, 0.5535536, -0.6003962)^t$

5. Approximate eigenvalues and approximate eigenvectors are
   (a) $\mu^{(8)} = 4.000001$, $\mathbf{x}^{(8)} = (0.9999773, 0.99993134, 1)^t$ (b) Method fails.
   (c) $\mu^{(3)} = 5.000000$, $\mathbf{x}^{(3)} = (-0.25, 1, -0.25)^t$
   (d) $\mu^{(7)} = 5.124890$, $\mathbf{x}^{(7)} = (-0.2425938, 1, -0.3196351)^t$
   (e) $\mu^{(15)} = 5.236112$, $\mathbf{x}^{(15)} = (1, 0.6125369, 0.1217216, 0.4978318)^t$
   (f) $\mu^{(10)} = 8.999890$, $\mathbf{x}^{(10)} = (0.9944137, -0.9942148, -0.9997991, 1)^t$
   (g) Method fails. (h) Method fails.

7. Approximate eigenvalues and approximate eigenvectors are
   (a) $\mu^{(9)} = 1.000015$, $\mathbf{x}^{(9)} = (-0.1999939, 1, -0.7999909)^t$
   (b) $\mu^{(12)} = -0.4142136$, $\mathbf{x}^{(12)} = (1, -0.7070918, -0.7071217)^t$
   (c) $\mu^{(8)} = 5.000029$, $\mathbf{x}^{(8)} = (-0.2500072, 1, -0.2500072)^t$ (d) Method fails.
   (e) $\mu^{(9)} = 1.381959$, $\mathbf{x}^{(9)} = (-0.3819400, -0.2361007, 0.2360191, 1)^t$
   (f) $\mu^{(6)} = 3.999997$, $\mathbf{x}^{(6)} = (0.9999939, 0.9999999, 0.9999940, 1)^t$
   (g) $\mu^{(8)} = 3.999996$, $\mathbf{x}^{(8)} = (1, -0.4999959, -0.9999970, -0.5000023)^t$
   (h) Method fails.

9. (a) We have $|\lambda| \leq 6$ for all eigenvalues λ.
   (b) The approximate eigenvalue is $\lambda_1 = 0.6982681$ with approximate eigenvector $\mathbf{x} = (1, 0.71606, 0.25638, 0.04602)^t$.
   (c) The eigenvalues are listed in part (d).
   (d) The characteristic polynomial is $P(\lambda) = \lambda^4 - \frac{1}{4}\lambda - \frac{1}{16}$; and the eigenvalues are $\lambda_1 = 0.6976684972$, $\lambda_2 = -0.237313308$, $\lambda_3 = -0.2301775942 + 0.56965884i$, and $\lambda_4 = -0.2301775942 - 0.56965884i$.
   (e) The beetle population should approach zero since A is convergent.

Exercise Set 9.4 (page 334)

1. Householder's method produces the following tridiagonal matrices:

(a)
$$\begin{bmatrix} 12.00000 & -10.77033 & 0.0 \\ -10.77033 & 3.862069 & 5.344828 \\ 0.0 & 5.344828 & 7.137931 \end{bmatrix}$$
(b)
$$\begin{bmatrix} 2.0000000 & 1.414214 & 0.0 \\ 1.414214 & 1.000000 & 0.0 \\ 0.0 & 0.0 & 3.0 \end{bmatrix}$$

(c)
$$\begin{bmatrix} 2.0000000 & -1.414214 & 0.0 \\ -1.414214 & 3.000000 & 0.0 \\ 0.0 & 0.0 & 1.000000 \end{bmatrix}$$
(d)
$$\begin{bmatrix} 1.0000000 & -1.414214 & 0.0 \\ -1.414214 & 1.000000 & 0.0 \\ 0.0 & 0.0 & 1.000000 \end{bmatrix}$$

(e)
$$\begin{bmatrix} 2.0 & -1.0 & 0.0 \\ -1.0 & -1.0 & -2.0 \\ 0.0 & -2.0 & 3.0 \end{bmatrix}$$
(f)
$$\begin{bmatrix} 4.750000 & -2.263846 & 0.0 \\ -2.263846 & 4.475610 & -1.219512 \\ 0.0 & -1.219512 & 5.024390 \end{bmatrix}$$

Exercise Set 9.5 (page 340)

1. Two iterations of the QR method produce the following matrices:

(a)
$$A^{(3)} = \begin{bmatrix} 0.6939977 & -0.3759745 & 0.0 \\ -0.3759745 & -1.892417 & -0.03039696 \\ 0.0 & -0.03039696 & 3.413585 \end{bmatrix}$$

(b)
$$A^{(3)} = \begin{bmatrix} 4.535466 & 1.212648 & 0.0 \\ 1.212648 & 3.533242 & 3.83 \times 10^{-7} \\ 0.0 & 3.83 \times 10^{-7} & -0.06870782 \end{bmatrix}$$
(c)
$$A^{(3)} = \begin{bmatrix} 0.6939977 & 0.3759745 & 0.0 \\ 0.3759745 & 1.892417 & 0.03039696 \\ 0.0 & 0.03039696 & 3.413585 \end{bmatrix}$$

(d)
$$A^{(3)} = \begin{bmatrix} 4.679567 & -0.2969009 & 0.0 \\ -2.969009 & 3.052484 & -1.207346 \times 10^{-5} \\ 0.0 & -1.207346 \times 10^{-5} & 1.267949 \end{bmatrix}$$

(e)
$$A^{(3)} = \begin{bmatrix} 0.3862092 & 0.4423226 & 0.0 & 0.0 \\ 0.4423226 & 1.787694 & -0.3567744 & 0.0 \\ 0.0 & -0.3567744 & 3.080815 & 3.116382 \times 10^{-5} \\ 0.0 & 0.0 & 3.116382 \times 10^{-5} & 4.745201 \end{bmatrix}$$

(f)
$$A^{(3)} = \begin{bmatrix} -2.826365 & 1.130297 & 0.0 & 0.0 \\ 1.130297 & -2.429647 & -0.1734156 & 0.0 \\ 0.0 & -0.1734156 & 0.8172086 & 1.863997 \times 10^{-9} \\ 0.0 & 0.0 & 1.863997 \times 10^{-9} & 3.438803 \end{bmatrix}$$

(g)
$$A^{(3)} = \begin{bmatrix} 0.2763388 & 0.1454371 & 0.0 & 0.0 \\ 0.1454371 & 0.4543713 & 0.1020836 & 0.0 \\ 0.0 & 0.1020836 & 1.174446 & -4.36 \times 10^{-5} \\ 0.0 & 0.0 & -4.36 \times 10^{-5} & 0.9948441 \end{bmatrix}$$

(h)
$$A^{(3)} = \begin{bmatrix} 6.376729 & -1.988497 & 0.0 & 0.0 \\ -1.988497 & -1.652394 & -0.5187146 & 0.0 \\ 0.0 & -0.5187146 & 1.007133 & 1.14 \times 10^{-6} \\ 0.0 & 0.0 & 1.14 \times 10^{-6} & 2.268531 \end{bmatrix}$$

3. The matrices in Exercise 1 have the following eigenvalues, accurate to within 10^{-5}:

   (a) 3.414214, 2.000000, 0.5857864 **(b)** -0.06870782, 5.346462, 2.722246

   (c) 3.414214, 2.000000, 0.5857864 **(d)** 1.267949, 4.732051, 3.000000

   (e) 4.745281, 3.177283, 1.822717, 0.2547188 **(f)** 3.438803, 0.8275517, -1.488068, -3.778287

   (g) 0.9948440, 1.189091, 0.5238224, 0.1922421 **(h)** 2.268531, 1.084364, 6.844621, -2.197517

Exercise Set 10.2 (page 348)

1. **(b)** With $\mathbf{x}^{(0)} = (0, 0)^t$, we have $\mathbf{x}^{(10)} = (-1, 3.5)^t$.
   With $\mathbf{x}^{(0)} = (6, 3)^t$, we have $\mathbf{x}^{(6)} = (2.546947, 3.984997)^t$.

3. **(a)** With $\mathbf{x}^{(0)} = (0, 0)^t$, we have $\mathbf{x}^{(5)} = (0.5, 2.0)^t$.
   (b) With $\mathbf{x}^{(0)} = (1, 1)^t$, we have $\mathbf{x}^{(5)} = (0.5, 0.8660254)^t$.
   (c) With $\mathbf{x}^{(0)} = (2, 2)^t$, we have $\mathbf{x}^{(6)} = (1.772454, 1.772454)^t$.
   (d) With $\mathbf{x}^{(0)} = (0, 0)^t$, we have $\mathbf{x}^{(6)} = (-0.3736982, 0.05626649)^t$.

5. **(a)** With $\mathbf{x}^{(0)} = (0.1, -0.1, 0.1)^t$, we have $\mathbf{x}^{(3)} = (0.4981446, -0.1996059, -0.5288259)^t$.
   (b) With $\mathbf{x}^{(0)} = (1, 2, 3)^t$, we have $\mathbf{x}^{(16)} = (-0.4021187, -0.9302338, 8.398353)^t$.
   With $\mathbf{x}^{(0)} = (-0.5, -1, 2)^t$, we have $\mathbf{x}^{(23)} = (7.869499, 0.7214381, -4.980240)^t$.
   With $\mathbf{x}^{(0)} = (-0.5, -1, 4)^t$, we have $\mathbf{x}^{(14)} = (-0.6398107, -1.102016, 4.455568)^t$.
   With $\mathbf{x}^{(0)} = (-0.5, -1, 6)^t$, we have $\mathbf{x}^{(17)} = (-0.9302344, -0.4021213, -8.398351)^t$.
   (c) With $\mathbf{x}^{(0)} = (0.5, 0.5, 0.5, 0.5)^t$, we have $\mathbf{x}^{(14)} = (0.8688753, 0.8688760, 0.8688761, 1.524497)^t$.
   (d) With $\mathbf{x}^{(0)} = (1, 1, 1, 1)^t$, we have $\mathbf{x}^{(5)} = (0, 0.7071068, 0.7071068, 1)^t$.
   With $\mathbf{x}^{(0)} = (1, -1, 1, 5)^t$, we have $\mathbf{x}^{(6)} = (0.8164966, 0.4082483, -0.4082483, 3)^t$.
   With $\mathbf{x}^{(0)} = (1, -1, 1, 10)^t$, we have $\mathbf{x}^{(5)} = (0.5773503, -0.5773503, 0.5773503, 6)^t$.

7. Yes: $x_1 = 8000$, $x_2 = 40000$

Exercise Set 10.3 (page 355)

1. With $\mathbf{x}^{(0)} = (6, 3)^t$, we have $\mathbf{x}^{(8)} = (2.546947, 3.984997)^t$.
   With $\mathbf{x}^{(0)} = (1, 1)^t$, we have $\mathbf{x}^{(13)} = (-1, 3.5)^t$.

3. **(a)** With $\mathbf{x}^{(0)} = (0, 0)^t$, we have $\mathbf{x}^{(7)} = (0.5, 2.0)^t$.
   (b) With $\mathbf{x}^{(0)} = (1, 1)^t$, we have $\mathbf{x}^{(8)} = (0.5, 0.8660254)^t$.
   (c) With $\mathbf{x}^{(0)} = (2, 2)^t$, we have $\mathbf{x}^{(7)} = (1.772454, 1.772454)^t$.
   (d) With $\mathbf{x}^{(0)} = (0, 0)^t$, we have $\mathbf{x}^{(8)} = (-0.3736982, 0.05626649)^t$.

5. **(a)** With $\mathbf{x}^{(0)} = (0.1, -0.1, 0.1)^t$, we have $\mathbf{x}^{(4)} = (0.4981447, -0.1996059, -0.5288260)^t$.
   (b) With $\mathbf{x}^{(0)} = (8, 0.75, -5)^t$, we have $\mathbf{x}^{(32)} = (7.869503, 0.7214736, -4.980267)^t$.
   With $\mathbf{x}^{(0)} = (-0.6, -1, 4.5)^t$, we have $\mathbf{x}^{(25)} = (-0.6396095, -1.101958, 4.455394)^t$.
   (c) With $\mathbf{x}^{(0)} = (0.8, 0.8, 0.8, 1.4)^t$, we have $\mathbf{x}^{(17)} = (0.8688769, 0.8688769, 0.8688769, 1.524492)^t$.
   (d) With $\mathbf{x}^{(0)} = (1, 1, 1, 1)^t$, we have $\mathbf{x}^{(5)} = (0, 0.7071068, 0.7071068, 1)^t$.
   With $\mathbf{x}^{(0)} = (1, -1, 1, 5)^t$, we have $\mathbf{x}^{(8)} = (0.5773498, -0.5773506, 0.5773506, 6)^t$.
   With $\mathbf{x}^{(0)} = (1, 0.5, -0.5, 2)^t$, we have $\mathbf{x}^{(6)} = (0.8164966, 0.4082476, -0.4082490, 3)^t$.

Exercise Set 10.4 (page 361)

1. (a) With $x^{(0)} = (1, 1)^t$, we have $x^{(12)} = (0.498413, 1.98629)^t$.
    (b) With $x^{(0)} = (1, 1)^t$, we have $x^{(2)} = (0.501428, 0.869834)^t$.
    (c) With $x^{(0)} = (2, 2)^t$, we have $x^{(1)} = (1.73540, 1.80392)^t$.
    (d) With $x^{(0)} = (0, 0)^t$, we have $x^{(3)} = (-0.377890, 0.0508119)^t$.

3. (a) With $x^{(0)} = (0, 0, 0)^t$, we have $x^{(18)} = (1.03865, 1.07904, 0.928979)^t$.
    (b) With $x^{(0)} = (0, 0, 0)^t$, we have $x^{(2)} = (-0.0200570, 0.0901966, 0.994681)^t$.
    (c) With $x^{(0)} = (0, 0, 0)^t$, we have $x^{(17)} = (-1.60120, -1.20804, 0.752611)^t$.
    (d) With $x^{(0)} = (1, 1, 1)^t$, we have $x^{(3)} = (-0.002049049, 0.0075156, 0.15814)^t$.
    (e) With $x^{(0)} = (-0.5, -1, 2)^t$, we have $x^{(23)} = (-1.03044, -1.17457, 1.68028)^t$.
    (f) With $x^{(0)} = (1, 1, 1, 1)^t$, we have $x^{(8)} = (0.00953301, 0.700762, 0.706167, 1.09239)^t$.
        With $x^{(0)} = (1, -1, 1, 5)^t$, we have $x^{(4)} = (0.587303, -0.587303, 0.587303, 6.01399)^t$.
        With $x^{(0)} = (1, -1, 1, 2)^t$, we have $x^{(4)} = (0, 0, 0, 2.49517)^t$.

5. (a) With $x^{(0)} = (0.1, 0.1)^t$, we have $x^{(8)} = (5.343082, -0.6262875)^t$ and $g(x^{(8)}) = 0.006995494$.
    (b) With $x^{(0)} = (0, 0)^t$, we have $x^{(13)} = (0.6157412, 0.3768953)^t$ and $g(x^{(13)}) = 0.1481574$.
    (c) With $x^{(0)} = (0, 0, 0)^t$, we have $x^{(5)} = (-0.6633785, 0.3145720, 0.5000740)^t$ and $g(x^{(5)}) = 0.6921548$.
    (d) With $x^{(0)} = (1, 1, 1)^t$, we have $x^{(4)} = (0.04022273, 0.01592477, 0.01594401)^t$ and $g(x^{(4)}) = 1.010003$.

Exercise Set 11.2 (page 368)

1. The Linear Shooting method gives the results in the following tables:

(a)

i	x_i	w_{1i}
1	0.333333	0.5311664
2	0.666667	1.153515

(b)

i	x_i	w_{1i}
1	0.25	0.3937095
2	0.50	0.8240948
3	0.75	1.337160

3. The Linear Shooting method gives the results in the following tables:

(a)

i	x_i	w_{1i}
3	0.3	0.7833204
6	0.6	0.6023521
9	0.9	0.8568906

(b)

i	x_i	w_{1i}
5	1.0	0.00865076
10	2.0	0.00007484
15	3.0	0.00000065
20	4.0	0.00000001

(c)

i	x_i	w_{1i}
5	1.25	0.1676179
10	1.50	0.4581901
15	1.75	0.6077718
20	2.00	0.6931460

(d)

i	x_i	w_{1i}
3	0.3	-0.5185754
6	0.6	-0.2195271
9	0.9	-0.0406577

(e)

i	x_i	w_{1i}
3	1.3	0.0655336
6	1.6	0.0774590
9	1.9	0.0305619

(f)

i	x_i	w_{1i}
3	1.515485	10.094751
6	2.030969	17.547048
9	2.546454	17.380532

5. The Linear Shooting method with $h = 0.1$ gives the following results:

i	x_i	w_{1i}
3	0.3	0.05273437
5	0.5	0.00741571
8	0.8	0.00038976

The Linear Shooting method with $h = 0.05$ gives the following results:

i	x_i	w_{1i}
6	0.3	0.04990547
10	0.5	0.00676467
16	0.8	0.00033755

7. (a)

x	$w(x)$
24	0.0071265
48	0.011427
60	0.011999
72	0.011427
96	0.0071265
120	0.0000000

(b) Yes   **(c)** The actual solution satisfies, but the approximation does not.

Exercise Set 11.3 (page 373)

1. The Linear Finite-Difference method gives the results in the following tables:

(a)

i	x_i	w_i
1	0.333333	0.5343259
2	0.666667	1.1579818

(b)

i	x	w_i
1	0.25	0.3951247
2	0.50	0.8265306
3	0.75	1.3395692

3. The Linear Finite-Difference method gives the results in the following tables:

(a)

i	x_i	w_i
2	0.2	1.018096
5	0.5	0.5942743
7	0.7	0.6514520

(b)

i	x_i	w_i
5	1.0	6.332971×10^{-3}
10	2.0	4.010654×10^{-5}
15	3.0	2.539917×10^{-7}
20	4.0	1.604072×10^{-9}

(c)

i	x_i	w_i
5	1.25	0.16797186
10	1.50	0.45842388
15	1.75	0.60787335

(d) i	x_i	w_i	(e) i	x_i	w_i	(f) i	x_i	w_i
3	0.3	−0.5183084	3	1.3	0.0654387	3	1.515485	1.904530
6	0.6	−0.2192657	6	1.6	0.0773936	6	2.030969	5.415273
9	0.9	−0.0405748	9	1.9	0.0305465	9	2.546454	11.935402

5. The Linear Finite-Difference method gives the results in the following tables:

i	x_i	$w_i(h = 0.1)$	i	x_i	$w_i(h = 0.05)$
3	0.3	0.05572807	6	0.3	0.05132396
6	0.6	0.00310518	12	0.6	0.00263406
9	0.9	0.00016516	18	0.9	0.00013340

7.

i	x_i	w_i
10	10.0	0.1098549
20	20.0	0.1761424
25	25.0	0.1849608
30	30.0	0.1761424
40	40.0	0.1098549

Exercise Set 11.4 (page 379)

1. The Nonlinear Shooting method gives $w_1 = 0.405991 \approx \ln 1.5 = 0.405465$

3. The Nonlinear Shooting method gives the results in the following tables:

(a) i	x_i	w_{1i}	(b) i	x_i	w_{1i}	(c) i	x_i	w_{1i}
3	1.3	0.434783	3	1.3	2.069249	3	1.3	1.031597
6	1.6	0.384615	6	1.6	2.225015	6	1.6	1.095005
9	1.9	0.344828	9	1.9	2.426322	9	1.9	1.168170

(d) To apply the algorithm we need to redefine the initial value of *TK* to be 1.5.

i	x_i	w_{1i}	(e) i	x_i	w_{1i}
5	0.3926991	0.6600925	5	1.25	−0.7272908
10	0.7853982	1.1319596	10	1.50	−0.8000371
15	1.1780973	1.4454742	15	1.75	−0.8889473
20	1.5707963	1.5707414			

(f) To apply the algorithm we need to redefine the initial value of TK to be 2.

i	x_i	w_{1i}
5	1.25	0.4358290
10	1.50	1.3684496
15	1.75	2.9992010
20	2.00	5.5451958

Exercise Set 11.5 (page 382)

1. The Nonlinear Finite-Difference method gives the following results:

i	x_i	w_i	$\ln x_i$
1	1.5	0.406760	0.405465

3. The Nonlinear Finite-Difference method gives the results in the following tables:

(a)

i	x_i	w_i
3	1.3	0.434796
6	1.6	0.384627
9	1.9	0.344831

(b)

i	x_i	w_i
3	1.3	2.0694081
6	1.6	2.2250937
9	1.9	2.4263387

(c)

i	x_i	w_i
3	1.3	1.031970
6	1.6	1.095321
9	1.9	1.168271

(d)

i	x_i	w_i
3	0.471239	0.766923
6	0.942478	1.275944
9	1.413717	1.548057

(e)

i	x_i	w_i
5	1.25	-0.727281
10	1.50	-0.800013
15	1.75	-0.888900
20	2.00	-1.000000

(f)

i	x_i	w_i
5	1.25	0.434598
10	1.50	1.366212
15	1.75	2.996934
20	2.00	5.545177

Exercise Set 11.6 (page 392)

1. The Piecewise Linear method gives $\phi(x) = -0.07713274\phi_1(x) - 0.07442678\phi_2(x)$.

3. The Piecewise Linear method gives the results in the following tables:

(a)

i	$\phi(x_i)$
3	4.0908778×10^{-2}
6	5.8304182×10^{-2}
9	2.6538926×10^{-2}

(b)

i	$\phi(x_i)$
3	-0.212333
6	-0.241333
9	-0.090333

(c)

i	$\phi(x_i)$
3	0.1815153
6	0.1805512
9	0.05936480

(d) i	$\phi(x_i)$	(e) i	$\phi(x_i)$	(f) i	$\phi(x_i)$
5	-0.3586155	5	-0.1846134	3	-5.738551×10^{-2}
10	-0.5348645	10	-0.2737099	6	-4.974304×10^{-2}
15	-0.4510389	15	-0.2285169	9	-1.478349×10^{-2}
20	0.0000000	20	0.0000000		

5. The Cubic Spline method gives the results in the following tables:

(a) i	$\phi(x_i)$	(b) i	$\phi(x_i)$	(c) i	$\phi(x_i)$
3	4.0878126×10^{-2}	3	-0.2100000	3	0.1814269
6	5.8259848×10^{-2}	6	-0.2400000	6	0.1804753
9	2.6518200×10^{-2}	9	-0.0900000	9	0.05934321

(d) i	$\phi(x_i)$	(e) i	$\phi(x_i)$	(f) i	$\phi(x_i)$
5	-0.3585641	5	-0.1845203	3	-5.754895×10^{-2}
10	-0.5347803	10	-0.2735857	6	-4.985645×10^{-2}
15	-0.4509614	15	-0.2284204	9	-1.481176×10^{-2}
20	0.0000000	20	0.0000000		

7. $c_1 = -0.03197519$, $c_2 = -0.05488716$, $c_3 = -0.06959856$, $c_4 = -0.07688995$, $c_5 = -0.07746753$, $c_6 = -0.07197026$, $c_7 = -0.06097626$, $c_8 = -0.04500857$, $c_9 = -0.02454043$

9. A change in variable $t = (x - a)/(b - a)$ gives the boundary value problem

$$-\frac{d}{dt}(p((b-a)t+a)y') + (b-a)^2 q((b-a)t+a)y = (b-a)^2 f((b-a)t+a),$$

$$0 < t < 1, y(0) = \alpha, y(1) = \beta.$$

Then Exercise 6 can be used.

Exercise Set 12.2 (page 404)

1. The Poisson Equation Finite-Difference method gives the following results:

i	j	x_i	y_j	w_{ij}
1	1	0.5	0.5	0.5
1	2	0.5	1.0	0.25
1	3	0.5	1.5	1.0

3. The Poisson Equation Finite-Difference method gives the results in the following tables:

(a)

i	j	x_i	y_j	w_{ij}
2	2	0.4	0.4	0.159999
2	4	0.4	0.8	0.319999
4	2	0.8	0.4	0.320000
4	4	0.8	0.8	0.640000

(b)

i	j	x_i	y_j	w_{ij}
2	2	0.4	0.4	1.8579248
2	4	0.4	0.8	1.2399321
4	2	0.8	0.4	6.1038748
4	4	0.8	0.8	3.8335165

(c)

i	j	x_i	y_j	w_{ij}
4	3	0.8	0.3	1.27136
4	7	0.8	0.7	1.75084
8	3	1.6	0.3	1.61675
8	7	1.6	0.7	3.06587

(d)

i	j	x_i	y_j	w_{ij}
2	1	1.256637	0.3141593	0.2951912
2	3	1.256637	0.9424778	0.1830968
4	1	2.513274	0.3141593	-0.7721915
4	3	2.513274	0.9424778	-0.4785097

(e)

i	j	x_i	y_j	w_{ij}
2	2	1.2	1.2	0.5251533
4	4	1.4	1.4	1.3190830
6	6	1.6	1.6	2.4065150
8	8	1.8	1.8	3.8088995

(f)

i	j	x_i	y_j	w_{ij}
2	2	0.2	0.2	-0.9171063
4	4	0.4	0.4	-0.9558396
6	6	0.6	0.6	-0.9672948
8	8	0.8	0.8	-0.8841996

5.

i	j	x_i	y_j	w_{ij}
5	9	2.0	3.0	5.957716
8	3	3.2	1.0	7.915441
10	9	4.0	3.0	4.678240
12	12	4.8	4.0	2.059610

Exercise Set 12.3 (page 414)

1. The Heat Equation Backward-Difference method gives the following results:

(a)

i	j	x_i	t_j	w_{ij}
1	1	0.5	0.05	0.632952
2	1	1.0	0.05	0.895129
3	1	1.5	0.05	0.632952
1	2	0.5	0.1	0.566574
2	2	1.0	0.1	0.801256
3	2	1.5	0.1	0.566574

(b)

i	j	x_i	t_j	w_{ij}
1	1	$\frac{1}{3}$	0.05	1.59728
2	1	$\frac{2}{3}$	0.05	-1.59728
1	2	$\frac{1}{3}$	0.1	1.47300
2	2	$\frac{2}{3}$	0.1	-1.47300

3. The Forward-Difference method gives the following results:
  (a) For $h = 0.1$ and $k = 0.01$:

i	j	x_i	t_j	w_{ij}
4	50	0.4	0.5	-9.3352×10^8
10	50	1.0	0.5	-9.1860×10^8
17	50	1.7	0.5	2.6047×10^8

For $h = 0.1$ and $k = 0.005$:

i	j	x_i	t_j	w_{ij}
4	100	0.4	0.5	$3.6726805 \times 10^{-10}$
10	100	1.0	0.5	$1.0503891 \times 10^{-17}$
17	100	1.7	0.5	$-5.942522 \times 10^{-10}$

  (b) For $h = \frac{\pi}{10}$ and $k = 0.05$:

i	j	x_i	t_j	w_{ij}
3	10	0.9424778	0.5	0.4921015
6	10	1.8849556	0.5	0.5785001
9	10	2.8274334	0.5	0.1879661

  (c)

i	j	x_i	t_j	w_{ij}
4	10	0.8	0.4	1.166142
8	10	1.6	0.4	1.252404
12	10	2.4	0.4	0.4681804
16	10	3.2	0.4	-0.1027628

  (d)

i	j	x_i	t_j	w_{ij}
2	10	0.2	0.4	0.3921147
4	10	0.4	0.4	0.6344550
6	10	0.6	0.4	0.6344550
8	10	0.8	0.4	0.3921148

5. The Crank-Nicholson method gives the following results:
  (a) For $h = 0.1$ and $k = 0.01$:

i	j	x_i	t_j	w_{ij}
4	50	0.4	0.5	2.3541×10^{-9}
10	50	1.0	0.5	1.7610×10^{-17}
17	50	1.7	0.5	-3.8090×10^{-9}

For $h = 0.1$ and $k = 0.01$:

i	j	x_i	y_j	w_{ij}
4	100	0.4	0.5	2.8156746×10^{-9}
10	100	1.0	0.5	$2.4953437 \times 10^{-17}$
17	100	1.7	0.5	-4.555857×10^{-9}

  (b) For $h = \frac{\pi}{10}$ and $k = 0.05$:

i	j	x_i	t_j	w_{ij}
2	10	0.628319	0.5	0.357938
5	10	1.570796	0.5	0.608960
8	10	2.513274	0.5	0.357938

  (c)

i	j	x_i	t_j	w_{ij}
5	10	1	0.4	1.312434
10	10	2	0.4	0.9050248
15	10	3	0.4	-0.03253811

  (d)

i	j	x_i	t_j	w_{ij}
3	10	0.3	0.4	0.5440574
5	10	0.5	0.4	0.6724913
7	10	0.7	0.4	0.5440568

7.

i	j	x_i	t_j	w_{ij}
3	25	0.3	0.25	0.2883455
5	25	0.5	0.25	0.3468410
8	25	0.8	0.25	0.2169213

9.

i	j	x_i	t_j	w_{ij}
3	10	0.3	0.225	1.207730
6	10	0.75	0.225	1.836564
10	10	1.35	0.225	0.6928342

11.

i	j	x_i	t_j	w_{ij}
2	10	200	5	1.478828×10^7
5	10	500	5	4.334451×10^6
8	10	800	5	1.478828×10^7

Exercise Set 12.4 (page 421)

1. The Wave Equation Finite-Difference method gives the following results:

i	j	x_i	t_j	w_{ij}
2	1	0.25	1.0	−0.7071068
3	1	0.50	1.0	−1.0000000
4	1	0.75	1.0	−0.7071068

3. The Wave Equation Finite-Difference method with $h = \frac{\pi}{10}$ and $k = 0.05$ gives the following results:

i	j	x_i	t_j	w_{ij}
2	10	$\frac{\pi}{5}$	0.5	0.516393
5	10	$\frac{\pi}{2}$	0.5	0.878541
8	10	$\frac{4\pi}{5}$	0.5	0.516393

The Wave Equation Finite-Difference method with $h = \frac{\pi}{20}$ and $k = 0.1$ gives the following results:

i	j	x_i	t_j	w_{ij}
4	5	$\frac{\pi}{5}$	0.5	0.515916
10	5	$\frac{\pi}{2}$	0.5	0.877729
16	5	$\frac{4\pi}{5}$	0.5	0.515916

The Wave Equation Finite-Difference method with $h = \frac{\pi}{20}$ and $k = 0.05$ gives the following results:

i	j	x_i	t_j	w_{ij}
4	10	$\frac{\pi}{5}$	0.5	0.515960
10	10	$\frac{\pi}{2}$	0.5	0.877804
16	10	$\frac{4\pi}{5}$	0.5	0.515960

5.

i	j	x_i	t_j	$w_{i,j}$
2	5	0.2	0.5	-1
5	5	0.5	0.5	0
7	5	0.7	0.5	1

7.

i	j	x_i	t_j	Voltage	Current
5	2	50	0.2	77.782	3.88909
12	2	120	0.2	104.62	-1.69959
18	2	180	0.2	33.992	-5.23081
5	5	50	0.5	77.782	3.88909
12	5	120	0.5	104.62	-1.69959
18	5	180	0.5	33.992	-5.23081

Exercise Set 12.5 (page 432)

1. With $E_1 = (0.25, 0.75)$, $E_2 = (0, 1)$, $E_3 = (0.5, 0.5)$, and $E_4 = (0, 0.5)$, basis functions are

$$\phi_1(x, y) = \begin{cases} 4x & \text{on } T_1 \\ -2 + 4y & \text{on } T_2 \end{cases} \qquad \phi_2(x, y) = \begin{cases} -1 - 2x + 2y & \text{on } T_1 \\ 0 & \text{on } T_2 \end{cases}$$

$$\phi_3(x, y) = \begin{cases} 0 & \text{on } T_1 \\ 1 + 2x - 2y & \text{on } T_2 \end{cases} \qquad \phi_4(x, y) = \begin{cases} 2 - 2x - 2y & \text{on } T_1 \\ 2 - 2x - 2y & \text{on } T_2 \end{cases}$$

and $\gamma_1 = 0.323825$, $\gamma_2 = 0$, $\gamma_3 = 1.0000$, and $\gamma_4 = 0$.

3. The Finite-Element method with $K = 8$, $N = 8$, $M = 32$, $n = 9$, $m = 25$, and $NL = 0$ gives the following results:

$$\gamma_1 = 0.511023$$
$$\gamma_2 = 0.720476$$
$$\gamma_3 = 0.507898$$
$$\gamma_4 = 0.720475$$
$$\gamma_5 = 1.01885$$
$$\gamma_6 = 0.720476$$
$$\gamma_7 = 0.507897$$
$$\gamma_8 = 0.720476$$
$$\gamma_9 = 0.511023$$
$$\gamma_i = 0 \quad 10 \le i \le 25$$
$$u(0.125, 0.125) \approx 0.614187$$
$$u(0.125, 0.25) \approx 0.690343$$
$$u(0.25, 0.125) \approx 0.690343$$
$$u(0.25, 0.25) \approx 0.720475$$

(See the diagram at the top of page 495.)

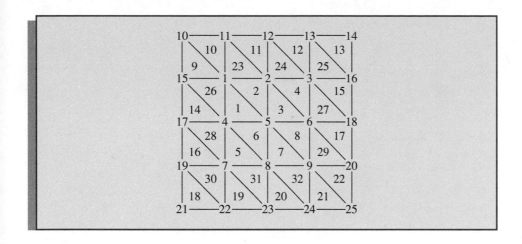

5. The Finite-Element method with $K = 0$, $N = 12$, $M = 32$, $n = 20$, $m = 27$, and $NL = 14$ gives the following results:

$$\gamma_1 = 21.40335 \quad \gamma_8 = 24.19855 \quad \gamma_{15} = 20.23334 \quad \gamma_{22} = 15$$
$$\gamma_2 = 19.87372 \quad \gamma_9 = 24.16799 \quad \gamma_{16} = 20.50056 \quad \gamma_{23} = 15$$
$$\gamma_3 = 19.10019 \quad \gamma_{10} = 27.55237 \quad \gamma_{17} = 21.35070 \quad \gamma_{24} = 15$$
$$\gamma_4 = 18.85895 \quad \gamma_{11} = 25.11508 \quad \gamma_{18} = 22.84663 \quad \gamma_{25} = 15$$
$$\gamma_5 = 19.08533 \quad \gamma_{12} = 22.92824 \quad \gamma_{19} = 24.98178 \quad \gamma_{26} = 15$$
$$\gamma_6 = 19.84115 \quad \gamma_{13} = 21.39741 \quad \gamma_{20} = 27.41907 \quad \gamma_{27} = 15$$
$$\gamma_7 = 21.34694 \quad \gamma_{14} = 20.52179 \quad \gamma_{21} = 15$$

$$u(1, 0) \approx 22.92824$$
$$u(4, 0) \approx 22.84663$$
$$u\left(\tfrac{5}{2}, \tfrac{\sqrt{3}}{2}\right) \approx 18.85895$$

(See the diagram.)

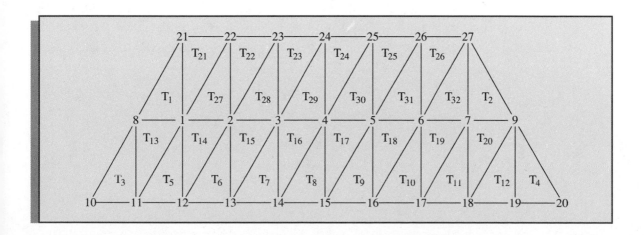

Index